The Hydraulics of Open Channel Flow: An Introduction

Basic principles, sediment motion, hydraulic modelling, design of hydraulic structures

Second Edition

D1341423

The Hydraulics of Open Channel Flow: An Introduction

Basic principles, sediment motion, hydraulic modelling, design of hydraulic structures

Second Edition

Hubert Chanson
Department of Civil Engineering
The University of Queensland, Australia

ELSEVIER
BUTTERWORTH
HEINEMANN

AMSTERDAM BOSTON HEIDELBERG LONDON NEW YORK OXFORD
PARIS SAN DIEGO SAN FRANCISCO SINGAPORE SYDNEY TOKYO

Butterworth-Heinemann is an imprint of Elsevier
The Boulevard, Langford Lane, Kidlington, Oxford, OX5 1GB
30 Corporate Drive, Suite 400, Burlington, MA 01803, USA

First edition 1999
Reprinted 2001, 2002
Second edition 2004
Reprinted 2007, 2008, 2009

British Library Cataloguing in Publication Data
A catalogue record for this book is available from the British Library

Library of Congress Cataloging-in-Publication Data
A catalog record for this book is available from the Library of Congress

ISBN: 978-0-7506-5978-9

For information on all Butterworth-Heinemann publications
visit our website at www.elsevierdirect.com

Transferred to Digital Printing in 2010.

Working together to grow
libraries in developing countries

www.elsevier.com | www.bookaid.org | www.sabre.org

ELSEVIER   BOOK AID International   Sabre Foundation

Contents

Preface to the first edition

The book is an introduction to the hydraulics of open channel flows. The material is designed for undergraduate students in civil, environmental and hydraulic engineering. It will be assumed that the students have had an introductory course in fluid mechanics and that they are familiar with the basic principles of fluid mechanics: continuity, momentum, energy and Bernoulli principles.

The book will first develop the basic principles of fluid mechanics with applications to open channels. Open channel flow calculations are more complicated than pipe flow calculations because the location of the free-surface is often unknown *a priori* (i.e. beforehand). Later the students are introduced to the basic concepts of sediment transport and hydraulic modelling (physical and numerical models). At the end of the course, the design of hydraulic structures is introduced. The book is designed to bring a basic understanding of the hydraulics of rivers, waterways and man-made canals to the reader (e.g. Plates 1–32).

The lecture material is divided into four parts of increasing complexity:

Part 1. Introduction to the basic principles. Application of the fundamental fluid mechanics principles to open channels. Emphasis on the application of the Bernoulli principle and momentum equation to open channel flows.

Part 2. Introduction to sediment transport in open channels. Basic definitions followed by simple applications. Occurrence of sediment motion in open channels. Calculations of sediment transport rate. Interactions between the sediment motion and the fluid motion.

Part 3. Modelling open channel flows. The two types of modellings are physical modelling and numerical modelling of open channel flows:

- Physical modelling: application of the basic principles of similitude and dimensional analysis to open channels.
- Numerical modelling: numerical integration of the energy equation; one-dimensional flow modelling.

Part 4. Introduction to the design of hydraulic structures for the storage and conveyance of water. Hydraulic design of dams, weirs and spillways. Design of drops and cascades. Hydraulic design of culverts: standard box culverts and minimum energy loss culvert.

Basic introduction to professional design of hydraulic structures. Application of the basic principles to real design situations. Analysis of complete systems.

Applications, tutorials and exercises are grouped into four categories: applications within the main text to illustrate the basic lecture material, exercises for each chapter within each section, revision exercises using knowledge gained in several chapters within one section and major assignments (i.e. problems) involving expertise gained in several sections: e.g. typically Part I and one or two other sections. In the lecture material, complete and detailed solutions of the applications are given. Numerical solutions of some exercises, revision exercises and problems are available on the Internet (Publisher's site: http://www.bh.com/companions/0340740671/).

A suggestion/correction form is placed at the end of the book. Comments, suggestions and critic are welcome and they will be helpful to improve the quality of the book. Readers who find an error or mistake are welcome to record the error on the page and to send a copy to the author. 'Errare Humanum Est'.[1]

---

[1] To err is human.

Preface to the second edition

Hydraulics is the branch of civil engineering related to the science of water in motion, and the interactions between the fluid and the surrounding environment. The beginnings of civil engineering as a separate discipline are often linked to the establishment of the 'École Nationale des Ponts-et-Chaussées' (France) in 1747, and it is worth noting that its directors included the famous hydraulicians A. Chézy (1717–1798) and G. de Prony (1755–1839). Other famous professors included B.F. de Bélidor (1693–1761), J.B.C. Bélanger (1789–1874), J.A.C. Bresse (1822–1883), G.G. Coriolis (1792–1843) and L.M.H. Navier (1785–1835). In a broader sense, hydraulic engineers were at the forefront of science for centuries. Although the origins of seepage water was long the subject of speculations (e.g. '*Meteorologica*' by Aristotle), the arts of tapping groundwater developed early in the antiquity: i.e. the qanats developed in Armenia and Persia. Roman aqueducts were magnificent waterworks and demonstrated the 'know-how' of Roman engineers (see Problem 4 and Plates 1 and 2). The 132 km long Carthage aqueduct was considered one of the marvels of the world by the Muslim poet El Kairouani. In China, a major navigation canal system was the Grand canal fed by the Tianping diversion weir. Completed in BC 219, the 3.9 m high, 470 m long weir diverted the Xiang River into the south and north canals allowing navigation between Guangzhou (formerly Canton), Shanghai and Beijing (Schnitter, 1994).

Hydraulic engineers have had an important role to contribute although the technical challenges are gigantic, often involving multiphase flows and interactions between fluids and biological life. The extreme complexity of hydraulic engineering is closely linked with the geometric scale of water systems, the broad range of relevant time scales, the variability of river flows from zero during droughts to gigantic floods, the complexity of basic fluid mechanics with governing equations characterized by non-linearity, natural fluid instabilities, interactions between water, solid, air and biological life, and Man's total dependence on water.

This textbook is focused on the hydraulics of open channel flows. One of the greatest challenges, in teaching open channel hydraulics, is for the students to understand that the position of the free-surface (i.e. water depth) is often unknown beforehand (Plates 1, 2, 5, 6, 8, 10, 11, 12, 16, 17, 18, 23, 24, 25, 28 and 30). In contrast, pipe flow calculations are performed with known pipe diameters and hence cross-sectional areas. Most open channel flow calculations are not straightforward. They may require solving a cubic equation (e.g. Bernoulli equation) and perform iterations (e.g. normal flow depth). One basic illustration is the sluice gate problem (Fig. I.1). Considering a sluice gate in a smooth horizontal channel, the input conditions may be the upstream depth, channel width and flow rate. What are the downstream flow depth and the force acting onto the gate? What is the correct free-surface profile upstream and downstream of the gate? Explanations are available in the text. A more detailed problem is given in the Revision exercises (Part 1).

Hydraulic engineering is an exciting discipline linking students to the basic needs of our planet (Plates 14 and 15). This second edition was prepared especially to stimulate university students and young professionals. Most material is taught at undergraduate levels at the University of Queensland. The structure of the new edition is largely based upon the successful first edition (see Comments). Updates and corrections were included. The lecture material in

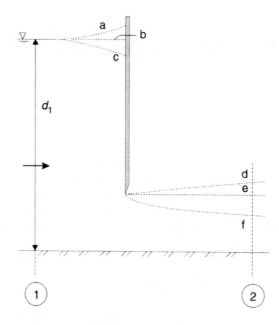

Fig. I.1. Flow beneath a sluice gate: which is the correct free-surface profile?

Part 3, modelling open channel flows was enlarged and includes two new chapters on the modelling of unsteady open channel flows. Applications, exercises and revision problems have been expanded. This edition contains further a larger number of illustrations than in the first edition.

The book is complemented by Internet references (see Additional bibliography) and a new web site {http://www.bh.com/companions/0340740671/}, while corrections and updates are posted at {http://www. uq.edu.au/~e2hchans/reprints/book3_2.htm}. The latter is updated and maintained by the writer.

Comments

First published in August 1999, the textbook *The Hydraulics of Open Channel Flow: An Introduction* has been designed for undergraduate students in civil, environmental and hydraulic engineering. The textbook is used in all six continents of our planet. It has been translated into Chinese (Hydrology Bureau of Yellow River Conservancy Committee) and Spanish (McGraw Hill Interamericana). The first edition was reviewed in several well-known international journals:

Professor S.N. Lane, University of Leeds, in *Environmental Conservation*, **27**(3), 2000, 314–315.
Without a doubt, this is the best introduction to the hydraulics of open channel flow that I have yet to read. The text deserves special credit for the explicit identification of the assumptions that exist behind relationships, something that can be (and is) easily over-looked by students whilst using other texts. As an introduction to the hydraulics of open channel flow, *I would find it difficult to recommend anything that could improve upon the approach adopted.* My overwhelming conclusion is that as an introduction to the hydraulics of open channel flow, it would be impossible to produce a better result. This will appear on both my undergraduate and postgraduate reading

lists as the core text. It is rare for me to be so readily persuaded, and Dr Chanson deserves full credit for an outstanding teaching resource.

Professor P. Bates, University of Bristol, 'Changing a winning formula: the hydraulics of open channel flow: an introduction' in *Hydrological Processes*, **15**(1), Special Issue, 2001, 159 pp.
The ultimate test of such a textbook is whether it can be useful for a range of problems and be accessible to a wide readership. To do this the hydraulics group at Bristol has been "road-testing" this volume for the past three months. [...] In that time, a diverse range of queries has been initially researched in Hubert Chanson's volume, and *it has passed each test with flying colours*. All graduate and postdoctoral researchers who have used the volume have commented favourably on its clarity and completeness, and *I can think of no better recommendation* than this. This is an excellent book for undergraduate and graduate students in civil engineering interested in open channel flow, and a very useful resource text for those interested in hydraulics outside engineering field.

Professor W.H. Hager, ETH-Zürich (Switz.), *Wasser, Energie & Luft*, Switzerland, 2000, No. 1/2, p. 55.
Another excellent piece of work. All in all, this is a well-written and carefully illustrated book which is useful for all civil and environmental engineers. It easily meets highest expectations.

Professor Ö. Starosolszky, Director, Vituki, in *Journal of Hydraulics Research*, IAHR, **39**(1), 2001, 331–332.
An original approach, which makes the book equally valuable both for students, and also practitioners in hydraulic engineering.

Professor M. Jovanovic, Belgrade, *Urban Water*, **1**(3), 270.
This book stands apart from similar previously published textbooks in two ways. Firstly, its scope has significantly been extended towards applications. Secondly, by including many exercises, notes, discussions, relevant photographs and appendices with additional information, it has an original, handbook-like presentation, very convenient for quick referencing, and use in engineering practice. Being more than a simple introductory textbook in open channel hydraulics, this book can be strongly recommended to students and engineers.

Professor D.A. Ervine, University of Glasgow, in *Chemical Engineering Research and Design*, Trans IChemE, Part A, **78**(A7), Oct. 2000, 1055.
Hubert Chanson's latest book is really designed for a civil engineering readership with its emphasis on sediment movement in rivers and also hydraulic structures for rivers and dams. All in all, a well-constructed book with many helpful examples and explanations for the student.

Acknowledgements

The author wants to thank especially Professor Colin J. Apelt, University of Queensland, for his help, support and assistance all along the academic career of the writer. He thanks Dr Sergio Montes, University of Tasmania, and Professor Nallamuthu Rajaratnam, University of Alberta for their advice and support. He acknowledges past and present students for their contributions.

He expresses his gratitude to the following people who provided photographs and illustrations of interest:

The American Society of Mechanical Engineers
Mr and Mrs Michael and Linda Burridge, Brisbane QLD, Australia
Mr and Mrs Chanson, Paris, France
Ms Chou Ya-Hui, Brisbane QLD, Australia
Mr L. Stuart Davies, Welsh Water, UK
Dr Dong, Zhiyong, Zhejiang University of Technology, China
Ms Chantal Donnelly, Brisbane QLD, Australia
Dr Chris Fielding, Brisbane QLD, Australia
Gold Coast City Council, Australia
Mrs Jenny L.F. Hacker, Brisbane QLD, Australia
The Hydro-Electric Commission (HEC) of Tasmania, Australia
Mr Patrick James, Sydney NSW, Australia
Mr D. Jeffery, Goulburn-Murray Water, Australia
Dr John Macintosh, Water Solutions, Australia
Dr Lou Maher, University of Wisconsin, USA
Mr J. Mitchell, Brisbane QLD, Australia
Mr Chris Proctor, Brisbane, Australia
Mr P.U. Hua-Chih, Taipei, Taiwan ROC
Dr R. Rankin, Rankin Publishers, Brisbane, Australia
Mr Paul Royet, Cemagref, Le Tholonet, France
Mr Marq Redeker, Ruhrverband, Germany
Sinorama-Magazine, Taipei, Taiwan, ROC.
Larry Smith, USA
Tonkin and Taylor, New Zealand
USDA, Natural Resources Conservation Service, National Design, Construction and Soil Mechanics Center, Fort Worth, Texas
US Geological Survey, USA
Mr Peter Ward
Mr Dale Young, Australia.

The author also thanks the following people who provided him with information and assistance:

Dr Shin-ichi Aoki, Toyohashi University of Technology, Japan
Professor C.J. Apelt, Brisbane QLD, Australia

Mr Noel Bedford, Tamworth NSW, Australia
Mr and Mrs Chanson, Paris, France
Ms Chou Ya-Hui, Brisbane QLD, Australia
Dr Stephen Coleman, University of Auckland, New Zealand
Concrete Pipe Association of Australasia, Sydney NSW, Australia
Mr Doug Davidson, Murwillumbah NSW, Australia
Mr L. Stuart Davies, Welsh Water, UK
Dr Michael R. Gourlay, Brisbane QLD, Australia
Mr John Grimston, Tonkin and Taylor, New Zealand
Dr D. Holloway, University of Tasmania, Australia
Mr Ralf Hornung, Germany
Mr Graham Illidge, University of Queensland, Australia
Mr Patrick James, Sydney NSW, Australia
Mr D. Jeffery, Goulburn-Murray Water, Australia
Dr Eric Jones, Proudman Oceanographic Laboratory, UK
Mr Daniel M. Leblanc, Petitcodiac RiverKeeper, Canada
Mr Steven Li, Hydro-Electric Commission of Tasmania, Australia
Ms Carolyn Litchfield, Brisbane QLD, Australia
Ms Joe McGregor, US Geological Survey (Photo Library), USA
Dr J.S. Montes, University of Tasmania, Australia
Officine Maccaferri, Italy
Professor N. Rajaratnam, University of Alberta, Canada
Mr Paul Royet, Cemagref, France
Mr Craig Savela, USDA
Mr Des Williamson, HydroTools, Canada.

The author acknowledges further numerous feedback and advice on the first edition. Thanks to all these readers.

Last, but not the least, the author thanks all the people (including colleagues, former students, students, professionals) who gave him information, feedback and comments on his lecture material. In particular, he acknowledges the support of Professor C.J. Apelt in the preparation of the book. Some material on hydraulic structure design (spillway design and culvert design more specifically) is presented and derived from Professor Apelt's personal lecture notes, while some material on the Saint-Venant equations derived from Dr Jean Cunge's lecture notes.

About the author

Hubert Chanson received a degree of 'Ingénieur Hydraulicien' from the Ecole Nationale Supérieure d'Hydraulique et de Mécanique de Grenoble (France) in 1983 and a degree of 'Ingénieur Génie Atomique' from the 'Institut National des Sciences et Techniques Nucléaires' in 1984. He worked for the industry in France as an R&D engineer at the Atomic Energy Commission from 1984 to 1986, and as a computer professional in fluid mechanics for Thomson-CSF between 1989 and 1990. From 1986 to 1988, he studied at the University of Canterbury (New Zealand) as part of a PhD project. He was awarded a Doctor of Engineering from the University of Queensland in 1999 for outstanding research achievements in gas–liquid bubbly flows. In 2003, the International Association for Hydraulic engineering and Research (IAHR) presented him the *14th Arthur Ippen Award* for outstanding achievements in hydraulic engineering.

The author is a Reader in environmental fluid mechanics and water engineering at the University of Queensland since 1990. His research interests include design of hydraulic structures, experimental investigations of two-phase flows, coastal hydrodynamics, water quality modelling, environmental management and natural resources. He is the author of four books: *Hydraulic Design of Stepped Cascades, Channels, Weirs and Spillways* (Pergamon, 1995), *Air Bubble Entrainment in Free-Surface Turbulent Shear Flows* (Academic Press, 1997), *The Hydraulics of Open Channel Flows: An Introduction* (Butterworth-Heinemann, 1999) and *The Hydraulics of Stepped Chutes and Spillways* (Balkema, 2001). He co-authored a fifth book *Fluid Mechanics for Ecologists* (IPC Press, 2002), while his textbook *The Hydraulics of Open Channel Flows: An Introduction* has already been translated into Chinese (Hydrology Bureau of Yellow River Conservancy Committee) and Spanish (McGraw Hill Interamericana). His publication record includes over 200 international refereed papers and his work was cited over 800 times since 1990. Hubert Chanson has been active also as consultant for both governmental agencies and private organizations.

He has been awarded six fellowships from the Australian Academy of Science. In 1995 he was a Visiting Associate Professor at National Cheng Kung University (Taiwan, ROC) and he was Visiting Research Fellow at Toyohashi University of Technology (Japan) in 1999 and 2001.

Hubert Chanson was invited to deliver keynote lectures at the 1998 *ASME Fluids Engineering Symposium on Flow Aeration* (Washington DC), at the *Workshop on Flow Characteristics around Hydraulic Structures* (Nihon University, Japan 1998), at the *First International Conference of the International Federation for Environmental Management System IFEMS'01* (Tsurugi, Japan 2001) and at the *6th International Conference on Civil Engineering* (Isfahan, Iran 2003). He gave invited lectures at the *International Workshop on Hydraulics of Stepped Spillways* (ETH-Zürich, 2000) and at the *30th IAHR Biennial Congress* (Thessaloniki, Greece, 2003). He lectured several short courses in Australia and overseas (e.g. Taiwan, Japan).

His Internet home page is {http://www.uq.edu.au/~e2hchans}. He also developed technical Internet resources {http://www.uq.edu.au/~e2hchans/url_menu.html} and a gallery of photographs web site {http://www.uq.edu.au/~e2hchans/photo.html} that received more than 2000 hits per month since inception.

To Ya Hui

Pour Bernard et Nicole

Glossary

Abutment   Part of the valley side against which the dam is constructed. Artificial abutments are sometimes constructed to take the thrust of an arch where there is no suitable natural abutment.

Académie des Sciences de Paris   The Académie des Sciences, Paris, is a scientific society, part of the Institut de France formed in 1795 during the French Revolution. The academy of sciences succeeded the Académie Royale des Sciences, founded in 1666 by Jean-Baptiste Colbert.

Accretion   Increase of channel bed elevation resulting from the accumulation of sediment deposits.

Adiabatic   Thermodynamic transformation occurring without loss nor gain of heat.

Advection   Movement of a mass of fluid which causes change in temperature or in other physical or chemical properties of fluid.

Aeration device (or aerator)   Device used to introduce artificially air within a liquid. Spillway aeration devices are designed to introduce air into high-velocity flows. Such aerators include basically a deflector and air is supplied beneath the deflected waters. Downstream of the aerator, the entrained air can reduce or prevent cavitation erosion.

Afflux   Rise of water level above normal level (i.e. natural flood level) on the upstream side of a culvert or of an obstruction in a channel. In the US, it is commonly referred to as maximum backwater.

Aggradation   Rise in channel bed elevation caused by deposition of sediment material. Another term is accretion.

Air   Mixture of gases comprising the atmosphere of the Earth. The principal constituents are nitrogen (78.08%) and oxygen (20.95%). The remaining gases in the atmosphere include argon, carbon dioxide, water vapour, hydrogen, ozone, methane, carbon monoxide, helium, krypton …

Air concentration   Concentration of undissolved air defined as the volume of air per unit volume of air and water. It is also called the void fraction.

Alembert (d')   Jean le Rond d'Alembert (1717–1783) was a French mathematician and philosopher. He was a friend of Leonhard Euler and Daniel Bernoulli. In 1752 he published his famous d'Alembert's paradox for an ideal-fluid flow past a cylinder (Alembert, 1752).

Alternate depth   In open channel flow, for a given flow rate and channel geometry, the relationship between the specific energy and flow depth indicates that, for a given specific energy, there is no real solution (i.e. no possible flow), one solution (i.e. critical flow) or two solutions for the flow depth. In the latter case, the two flow depths are called alternate depths. One corresponds to a subcritical flow and the second to a supercritical flow.

Analytical model   System of mathematical equations which are the algebraic solutions of the fundamental equations.

Apelt   C.J. Apelt is an Emeritus Professor in Civil Engineering at the University of Queensland (Australia).

Apron   The area at the downstream end of a weir to protect against erosion and scouring by water.

Aqueduct   A conduit for conveying a large quantity of flowing waters. The conduit may include canals, siphons, pipelines (Plates 1 and 2).

Arch dam   Dam in plan dependent on arch action for its strength.

Arched dam   Gravity dam which is curved in plan. Alternatives include 'curved-gravity dam' and 'arch-gravity dam'.

Archimedes   Greek mathematician and physicist. He lived between BC 290–280 and BC 212 (or 211). He spent most of his life in Syracuse (Sicily, Italy) where he played a major role in the defence of the city against the Romans. His treaty 'On Floating Bodies' is the first-known work on hydrostatics, in which he outlined the concept of buoyancy.

Aristotle Greek philosopher and scientist (BC 384–322), student of Plato. His work *Meteorologica* is considered as the first comprehensive treatise on atmospheric and hydrological processes.

Armouring Progressive coarsening of the bed material resulting from the erosion of fine particles. The remaining coarse material layer forms an armour, protecting further bed erosion.

Assyria Land to the north of Babylon comprising, in its greatest extent, a territory between the Euphrates and the mountain slopes east of the Tigris. The Assyrian Kingdom lasted from about BC 2300 to 606.

Avogadro number Number of elementary entities (i.e. molecules) in one mole of a substance: $6.0221367 \times 10^{23}\,\text{mol}^{-1}$. Named after the Italian physicist Amedeo Avogadro.

Backwater In a tranquil flow motion (i.e. subcritical flow) the longitudinal flow profile is controlled by the downstream flow conditions: e.g. an obstacle, a structure, a change of cross-section. Any downstream control structure (e.g. bridge piers, weir) induces a backwater effect. More generally the term backwater calculations or backwater profile refer to the calculation of the longitudinal flow profile. The term is commonly used for both supercritical and subcritical flow motion.

Backwater calculation Calculation of the free-surface profile in open channels. The first successful calculations were developed by the Frenchman J.B. Bélanger who used a finite difference step method for integrating the equations (Bélanger, 1828).

Bagnold Ralph Alger Bagnold (1896–1990) was a British geologist and a leading expert on the physics of sediment transport by wind and water. During World War II, he founded the Long Range Desert Group and organized long-distance raids behind enemy lines across the Libyan desert.

Bakhmeteff Boris Alexandrovitch Bakhmeteff (1880–1951) was a Russian hydraulician. In 1912, he developed the concept of specific energy and energy diagram for open channel flows.

Barrage French word for dam or weir, commonly used to described large dam structure in English.

Barré de Saint-Venant Adhémar Jean Claude Barré de Saint-Venant (1797–1886), French engineer of the 'Corps des Ponts-et-Chaussées', developed the equation of motion of a fluid particle in terms of the shear and normal forces exerted on it (Barré de Saint-Venant, 1871a,b).

Barrel For a culvert, central section where the cross-section is minimum. Another term is the throat.

Bazin Henri Emile Bazin was a French hydraulician (1829–1917) and engineer, member of the French 'Corps des Ponts-et-Chaussées' and later of the Académie des Sciences de Paris. He worked as an assistant of Henri P.G. Darcy at the beginning of his career.

Bed-form Channel bed irregularity that is related to the flow conditions. Characteristic bed-forms include ripples, dunes and antidunes (e.g. Plates 3, 4 and 22).

Bed-load Sediment material transported by rolling, sliding and saltation motion along the bed.

Bélanger Jean-Baptiste Ch. Bélanger (1789–1874) was a French hydraulician and Professor at the Ecole Nationale Supérieure des Ponts-et-Chaussées (Paris). He suggested first the application of the momentum principle to hydraulic jump flow (Bélanger, 1828). In the same book, he presented the first 'backwater' calculation for open channel flow.

Bélanger equation Momentum equation applied across a hydraulic jump in a horizontal channel (named after J.B.C. Bélanger).

Bélidor Bertrand Forêt de Bélidor (1693–1761) was a teacher at the Ecole Nationale des Ponts-et-Chaussées. His treatise *Architecture Hydraulique* (Bélidor, 1737–1753) was a well-known hydraulic textbook in Europe during the 18th and 19th centuries.

Bernoulli Daniel Bernoulli (1700–1782) was a Swiss mathematician, physicist and botanist who developed the Bernoulli equation in his *Hydrodynamica, de viribus et motibus fluidorum* textbook (first draft in 1733, first publication in 1738, Strasbourg).

Bessel Friedrich Wilhelm Bessel (1784–1846) was a German astronomer and mathematician. In 1810 he computed the orbit of Halley's comet. As a mathematician he introduced the Bessel functions (or circular functions) which have found wide use in physics, engineering and mathematical astronomy.

Bidone Giorgio Bidone (1781–1839) was an Italian hydraulician. His experimental investigations on the hydraulic jump were published between 1820 and 1826.

Biesel   Francis Biesel (1920–1993) was a French hydraulic engineer and a pioneer of computational hydraulics.

Blasius   H. Blasius (1883–1970) was German scientist, student and collaborator of L. Prandtl.

Boltzmann   Ludwig Eduard Boltzmann (1844–1906) was an Austrian physicist.

Boltzmann constant   Ratio of the universal gas constant ($8.3143 \, kJ^{-1}mol^{-1}$) to the Avogadro number ($6.0221367 \times 10^{23} \, mol^{-1}$). It equals: $1.380662 \times 10^{-23} \, J/K$.

Borda   Jean-Charles de Borda (1733–1799) was a French mathematician and military engineer. He achieved the rank of Capitaine de Vaisseau and participated to the US War of Independence with the French Navy. He investigated the flow through orifices and developed the Borda mouthpiece.

Borda mouthpiece   A horizontal re-entrant tube in the side of a tank with a length such that the issuing jet is not affected by the presence of the walls.

Bore   A surge of tidal origin is usually termed a bore (e.g. the Mascaret in the Seine river, France).

Bossut   Abbé Charles Bossut (1730–1804) was a French ecclesiastic and experimental hydraulician, author of a hydrodynamic treaty (Bossut, 1772).

Bottom outlet   Opening near the bottom of a dam for draining the reservoir and eventually flushing out reservoir sediments.

Boundary layer   Flow region next to a solid boundary where the flow field is affected by the presence of the boundary and where friction plays an essential part. A boundary layer flow is characterized by a range of velocities across the boundary layer region from zero at the boundary to the free-stream velocity at the outer edge of the boundary layer.

Boussinesq   Joseph Valentin Boussinesq (1842–1929) was a French hydrodynamicist and Professor at the Sorbonne University (Paris). His treatise *Essai sur la théorie des eaux courantes* (1877) remains an outstanding contribution in hydraulics literature.

Boussinesq coefficient   Momentum correction coefficient named after J.V. Boussinesq who first proposed it (Boussinesq, 1877).

Boussinesq–Favre wave   An undular surge (see Undular surge).

Boys   P.F.D. du Boys (1847–1924) was a French hydraulic engineer. He made a major contribution to the understanding of sediment transport and bed-load transport (Boys, 1879).

Braccio   Ancient measure of length (from the Italian 'braccia'). One braccio equals 0.6096 m (or 2 ft).

Braised channel   Stream characterized by random interconnected channels separated by islands or bars. By comparison with islands, bars are often submerged at large flows.

Bresse   Jacques Antoine Charles Bresse (1822–1883) was a French applied mathematician and hydraulician. He was Professor at the Ecole Nationale Supérieure des Ponts-et-Chaussées, Paris as successor of J.B.C. Bélanger. His contribution to gradually varied flows in open channel hydraulics is considerable (Bresse, 1860).

Broad-crested weir   A weir with a flat long crest is called a broad-crested weir when the crest length over the upstream head is greater than 1.5 to 3. If the crest is long enough, the pressure distribution along the crest is hydrostatic, the flow depth equals the critical flow depth $d_c = (q^2/g)^{1/3}$ and the weir can be used as a critical depth meter.

Buat   Comte Pierre Louis George du Buat (1734–1809) was a French military engineer and hydraulician. He was a friend of Abbé C. Bossut. Du Buat is considered as the pioneer of experimental hydraulics. His textbook (Buat, 1779) was a major contribution to flow resistance in pipes, open channel hydraulics and sediment transport.

Buoyancy   Tendency of a body to float, to rise or to drop when submerged in a fluid at rest. The physical law of buoyancy (or Archimedes' principle) was discovered by the Greek mathematician Archimedes. It states that any body submerged in a fluid at rest is subjected to a vertical (or buoyant) force. The magnitude of the buoyant force is equal to the weight of the fluid displaced by the body.

Buttress dam   A special type of dam in which the water face consists of a series of slabs or arches supported on their air faces by a series of buttresses.

Byewash   Ancient name for a spillway: i.e. channel to carry waste waters.

Candela   SI unit for luminous intensity, defined as the intensity in a given direction of a source emitting a monochromatic radiation of frequency $540 \times 10^{12} \, Hz$ and which has a radiant intensity in that direction of 1/683 W per unit solid angle.

Carnot Lazare N.M. Carnot (1753–1823) was a French military engineer, mathematician, general and statesman who played a key role during the French Revolution.

Carnot Sadi Carnot (1796–1832), eldest son of Lazare Carnot, was a French scientist who worked on steam engines and described the Carnot cycle relating to the theory of heat engines.

Cartesian co-ordinate One of three co-ordinates that locate a point in space and measure its distance from one of three intersecting co-ordinate planes measured parallel to that one of three straight-line axes that is the intersection of the other two planes. It is named after the French mathematician René Descartes.

Cascade (1) A steep stream intermediate between a rapid and a waterfall. The slope is steep enough to allow a succession of small drops but not sufficient to cause the water to drop vertically (i.e. waterfall). (2) A man-made channel consisting of a series of steps: e.g. a stepped fountain, a staircase chute, a stepped sewer (e.g. Plates 5, 6, 9, 14, 27, 29 and 30).

Cataract A series of rapids or waterfalls. It is usually termed for large rivers: e.g. the six cataracts of the Nile River between Karthum and Aswan.

Catena d'Acqua Italian term for 'chain of water'. Variation of the cascade developed during the Italian Renaissance. Water is channelled down the centre of an architectural ramp contained on both sides by stone carved into a scroll pattern to give a chain-like appearance. Water flows as a supercritical regime with regularly spaced increase and decrease of channel width, giving a sense of continuous motion highlighted by shock-wave patterns at the free-surface. One of the best examples is at Villa Lante, Italy. The stonework was carved into crayfish, the emblem of the owner, Cardinal Gambara.

Cauchy Augustin Louis de Cauchy (1789–1857) was a French engineer from the 'Corps des Ponts-et-Chaussées'. He devoted himself later to mathematics and he taught at Ecole Polytechnique, Paris, and at the Collège de France. He worked with Pierre-Simon Laplace and J. Louis Lagrange. In fluid mechanics, he contributed greatly to the analysis of wave motion.

Cavitation Formation of vapour bubbles and vapour pockets within a homogeneous liquid caused by excessive stress (Franc *et al.*, 1995). Cavitation may occur in low-pressure regions where the liquid has been accelerated (e.g. turbines, marine propellers, baffle blocks of dissipation basin). Cavitation modifies the hydraulic characteristics of a system, and it is characterized by damaging erosion, additional noise, vibrations and energy dissipation.

Celsius Anders Celsius (1701–1744) was a Swedish astronomer who invented the Celsius thermometer scale (or centigrade scale) in which the interval between the freezing and boiling points of water is divided into 100 degrees.

Celsius degree (or degree centigrade) Temperature scale based on the freezing and boiling points of water: 0 and 100° Celsius, respectively.

Chadar Type of narrow sloping chute peculiar to Islamic gardens and perfected by the Mughal gardens in northern India (e.g. at Nishat Bagh). These stone channels were used to carry water from one terrace garden down to another. A steep slope ($\theta \sim 20$–$35°$) enables sunlight to be reflected to the maximum degree. The chute bottom is very rough to enhance turbulence and free-surface aeration. The discharge per unit width is usually small, resulting in thin sheets of aerated waters.

Chézy Antoine Chézy (1717–1798) (or Antoine de Chézy) was a French engineer and member of the French 'Corps des Ponts-et-Chaussées'. He designed canals for the water supply of the city of Paris. In 1768 he proposed a resistance formula for open channel flows called the Chézy equation. In 1798, he became Director of the Ecole Nationale Supérieure des Ponts-et-Chaussées after teaching there for many years.

Chézy coefficient Resistance coefficient for open channel flows first introduced by the Frenchman A. Chézy. Although it was thought to be a constant, the coefficient is a function of the relative roughness and Reynolds number.

Chimu Indian of a Yuncan tribe dwelling near Trujillo on the north-west coast of Peru. The Chimu Empire lasted from AD 1250 to 1466. It was overrun by the Incas in 1466.

Choke In open channel flow, a channel contraction might obstruct the flow and induce the appearance of critical flow conditions (i.e. control section). Such a constriction is sometimes called a 'choke'.

Choking flow Critical flow in a channel contraction. The term is used for both open channel flow and compressible flow.

Chord length (1) The chord or chord length of an airfoil is the straight-line distance joining the leading and trailing edges of the foil. (2) The chord length of a bubble (or bubble chord length) is the length of the straight line connecting the two intersections of the air-bubble free-surface with the leading tip of the measurement probe (e.g. conductivity probe, conical hot-film probe) as the bubble is transfixed by the probe tip.

Clausius   Rudolf Julius Emanuel Clausius (1822–1888) was a German physicist and thermodynamicist. In 1850, he formulated the Second Law of Thermodynamics.

Clay   Earthy material that is plastic when moist and that becomes hard when baked or fired. It is composed mainly of fine particles of a group of hydrous alumino-silicate minerals (particle sizes less than 0.05 mm usually).

Clepsydra   Greek name for Water clock.

Cofferdam   Temporary structure enclosing all or part of the construction area so that construction can proceed in dry conditions. A diversion cofferdam diverts a stream into a pipe or channel.

Cohesive sediment   Sediment material of very small sizes (i.e. less than $50 \, \mu m$) for which cohesive bonds between particles (e.g. intermolecular forces) are significant and affect the material properties.

Colbert   Jean-Baptiste Colbert (1619–1683) was a French statesman. Under King Louis XIV, he was the Minister of Finances, the Minister of 'Bâtiments et Manufactures' (Buildings and Industries) and the Minister of the Marine.

Conjugate depth   In open channel flow, another name for sequent depth.

Control   Considering an open channel, subcritical flows are controlled by the downstream conditions. This is called a 'downstream flow control'. Conversely, supercritical flows are controlled only by the upstream flow conditions (i.e. 'upstream flow control').

Control section   In an open channel, cross-section where critical flow conditions take place. The concept of 'control' and 'control section' are used with the same meaning.

Control surface   Is the boundary of a control volume.

Control volume   Refers to a region in space and is used in the analysis of situations where flow occurs into and out of the space.

Coriolis   Gustave Gaspard Coriolis (1792–1843) was a French mathematician and engineer of the 'Corps des Ponts-et-Chaussées' who first described the Coriolis force (i.e. effect of motion on a rotating body).

Coriolis coefficient   Kinetic energy correction coefficient named after G.G. Coriolis who introduced first the correction coefficient (Coriolis, 1836).

Couette   M. Couette was a French scientist who measured experimentally the viscosity of fluids with a rotating viscosimeter (Couette, 1890).

Couette flow   Flow between parallel boundaries moving at different velocities, named after the Frenchman M. Couette. The most common Couette flows are the cylindrical Couette flow used to measure dynamic viscosity and the two-dimensional Couette flow between parallel plates.

Courant   Richard Courant (1888–1972), American mathematician born in Germany who made significant advances in the calculus of variations.

Courant number   Dimensionless number characterizing the stability of explicit finite difference schemes.

Craya   Antoine Craya was a French hydraulician and Professor at the University of Grenoble.

Creager profile   Spillway shape developed from a mathematical extension of the original data of Bazin in 1886–1888 (Creager, 1917).

Crest of spillway   Upper part of a spillway. The term 'crest of dam' refers to the upper part of an uncontrolled overflow.

Crib   (1) Framework of bars or spars for strengthening; (2) frame of logs or beams to be filled with stones, rubble or filling material and sunk as a foundation or retaining wall.

Crib dam   Gravity dam built up of boxes, cribs, crossed timbers or gabions, and filled with earth or rock.

Critical depth   It is the flow depth for which the mean specific energy is minimum.

Critical flow conditions   In open channel flows, the flow conditions such as the specific energy (of the mean flow) is minimum are called the critical flow conditions. With commonly used Froude number definitions, the critical flow conditions occur for $Fr = 1$. If the flow is critical, small changes in specific energy cause large changes in flow depth. In practice, critical flow over a long reach of channel is unstable.

Culvert   Covered channel of relatively short length installed to drain water through an embankment (e.g. highway, railroad, dam).

Cyclopean dam   Gravity masonry dam made of very large stones embedded in concrete.

Danel   Pierre Danel (1902–1966) was a French hydraulician and engineer. One of the pioneers of modern hydrodynamics, he worked from 1928 to his death for Neyrpic known prior to 1948 as 'Ateliers Neyret-Beylier-Piccard et Pictet'.

Darcy   Henri Philibert Gaspard Darcy (1805–1858) was a French civil engineer. He studied at Ecole Polytechnique between 1821 and 1823, and later at the Ecole Nationale Supérieure des Ponts-et-Chaussées (Brown, 2002). He performed numerous experiments of flow resistance in pipes (Darcy, 1858) and in open channels (Darcy and Bazin 1865), and of seepage flow in porous media (Darcy, 1856). He gave his name to the Darcy–Weisbach friction factor and to the Darcy law in porous media.

Darcy law   Law of groundwater flow motion which states that the seepage flow rate is proportional to the ratio of the head loss over the length of the flow path. It was discovered by H.P.G. Darcy (1856) who showed that, for a flow of liquid through a porous medium, the flow rate is directly proportional to the pressure difference.

Darcy–Weisbach friction factor   Dimensionless parameter characterizing the friction loss in a flow. It is named after the Frenchman H.P.G. Darcy and the German J. Weisbach.

Debris   Debris comprise mainly large boulders, rock fragments, gravel-sized to clay-sized material, tree and wood material that accumulate in creeks.

Degradation   Lowering in channel bed elevation resulting from the erosion of sediments.

Descartes   René Descartes (1596–1650) was a French mathematician, scientist, and philosopher. He is recognized as the father of modern philosophy. He stated: 'cogito ergo sum' (I think therefore I am).

Diffusion   The process whereby particles of liquids, gases or solids intermingle as the result of their spontaneous movement caused by thermal agitation and in dissolved substances move from a region of higher concentration to one of lower concentration. The term turbulent diffusion is used to describe the spreading of particles caused by turbulent agitation.

Diffusion coefficient   Quantity of a substance that in diffusing from one region to another passes through each unit of cross-section per unit of time when the volume concentration is unity. The units of the diffusion coefficient are m^2/s.

Diffusivity   Another name for the diffusion coefficient.

Dimensional analysis   Organization technique used to reduce the complexity of a study, by expressing the relevant parameters in terms of numerical magnitude and associated units, and grouping them into dimensionless numbers. The use of dimensionless numbers increases the generality of the results.

Diversion channel   Waterway used to divert water from its natural course.

Diversion dam   Dam or weir built across a river to divert water into a canal. It raises the upstream water level of the river but does not provide any significant storage volume.

Drag reduction   Reduction of the skin friction resistance in fluids in motion.

Drainage layer   Layer of pervious material to relieve pore pressures and/or to facilitate drainage: e.g. drainage layer in an earthfill dam.

Drop   (1) Volume of liquid surrounded by gas in a free-fall motion (i.e. dropping); (2) by extension, small volume of liquid in motion in a gas; (3) a rapid change of bed elevation also called step.

Droplet   Small drop of liquid.

Drop structure   Single step structure characterized by a sudden decrease in bed elevation.

Du Boys (or Duboys)   See P.F.D. du Boys.

Du Buat (or Dubuat)   See P.L.G. du Buat.

Dupuit   Arsène Jules Etienne Juvénal Dupuit (1804–1866) was a French engineer and economist. His expertise included road construction, economics, statics and hydraulics.

Earth dam   Massive earthen embankment with sloping faces and made watertight.

Ecole Nationale Supérieure des Ponts-et-Chaussées, Paris   French civil engineering school founded in 1747. The direct translation is: 'National School of Bridge and Road Engineering'. Among the directors there were the famous hydraulicians A. Chézy and G. de Prony. Other famous professors included B.F. de Bélidor, J.B.C. Bélanger, J.A.C. Bresse, G.G. Coriolis and L.M.H. Navier.

Ecole Polytechnique, Paris   Leading French engineering school founded in 1794 during the French Révolution under the leadership of Lazare Carnot and Gaspard Monge. It absorbed the state artillery school in 1802 and was transformed into a military school by Napoléon Bonaparte in 1804. Famous professors included Augustin Louis Cauchy, Jean-Baptiste Joseph Fourier, Siméon-Denis Poisson, Jacques Charles François Sturm, among others.

Eddy viscosity   Another name for the momentum exchange coefficient. It is also called 'eddy coefficient' by Schlichting (1979) (see Momentum exchange coefficient).

Embankment   Fill material (e.g. earth, rock) placed with sloping sides and with a length greater than its height.

Escalier d'Eau   See Water staircase.

Euler   Leonhard Euler (1707–1783) was a Swiss mathematician and physicist, and a close friend of Daniel Bernoulli.

Explicit method   Calculation containing only independent variables; numerical method in which the flow properties at one point are computed as functions of known flow conditions only.

Extrados   Upper side of a wing or exterior curve of a foil. The pressure distribution on the extrados must be smaller than that on the intrados to provide a positive lift force.

Face   External surface which limits a structure: e.g. air face of a dam (i.e. downstream face), water face (i.e. upstream face) of a weir.

Favre   H. Favre (1901–1966) was a Swiss Professor at ETH-Zürich. He investigated both experimentally and analytically positive and negative surges. Undular surges (e.g. Plate 16) are sometimes called Boussinesq–Favre waves. Later he worked on the theory of elasticity.

Fawer jump   Undular hydraulic jump.

Fick   Adolf Eugen Fick was a 19th century German physiologist who developed the diffusion equation for neutral particle (Fick, 1855).

Finite differences   Approximate solutions of partial differential equations which consists essentially in replacing each partial derivative by a ratio of differences between two immediate values: e.g. $\partial V/\partial t \approx \delta V/\delta t$. The method was first introduced by Runge (1908).

Fixed-bed channel   The bed and sidewalls are non-erodible. Neither erosion nor accretion occurs.

Flashboard   A board or a series of boards placed on or at the side of a dam to increase the depth of water. Flashboards are usually lengths of timber, concrete or steel placed on the crest of a spillway to raise the upstream water level.

Flash flood   Flood of short duration with a relatively high peak flow rate.

Flashy   Term applied to rivers and streams whose discharge can rise and fall suddenly, and is often unpredictable.

Flettner   Anton Flettner (1885–1961) was a German engineer and inventor. In 1924, he designed a rotor ship based on the Magnus effect. Large vertical circular cylinders were mounted on the ship. They were mechanically rotated to provide circulation and to propel the ship. More recently a similar system was developed for the ship 'Alcyone' of Jacques-Yves Cousteau.

Flip bucket   A flip bucket or ski-jump is a concave curve at the downstream end of a spillway, to deflect the flow into an upward direction. Its purpose is to throw the water clear of the hydraulic structure and to induce the disintegration of the jet in air.

Fog   Small water droplets near ground level forming a cloud sufficiently dense to reduce drastically visibility. The term fog refers also to clouds of smoke particles or ice particles.

Forchheimer   Philipp Forchheimer (1852–1933) was an Austrian hydraulician who contributed significantly to the study of groundwater hydrology.

Fortier   André Fortier was a French scientist and engineer. He became later Professor at the Sorbonne, Paris.

Fourier   Jean-Baptiste Joseph Fourier (1768–1830) was a French mathematician and physicist known for his development of the Fourier series. In 1794 he was offered a professorship of mathematics at the Ecole Normale in Paris and was later appointed at the Ecole Polytechnique. In 1798 he joined the expedition to Egypt lead by (then) General Napoléon Bonaparte. His research in mathematical physics culminated with the classical study 'Théorie Analytique de la Chaleur' (Fourier, 1822) in which he enunciated his theory of heat conduction.

Free-surface   Interface between a liquid and a gas. More generally a free-surface is the interface between the fluid (at rest or in motion) and the atmosphere. In two-phase gas–liquid flow, the term 'free-surface' includes also the air–water interface of gas bubbles and liquid drops.

Free-surface aeration   Natural aeration occurring at the free surface of high-velocity flows is referred to as free-surface aeration or self-aeration.

French Revolution (Révolution Française)   Revolutionary period that shook France between 1787 and 1799. It reached a turning point in 1789 and led to the destitution of the monarchy in 1791. The constitution of the First Republic was drafted in 1790 and adopted in 1791.

Frontinus   Sextus Julius Frontinus (AD 35–103 or 104) was a Roman engineer and soldier. After AD 97, he was 'Curator Aquarum' in charge of the water supply system of Rome. He dealt with discharge measurements in pipes and canals. In his analysis he correctly related the proportionality between discharge and cross-sectional area. His book *De Aquaeductu Urbis Romae* (*Concerning the Aqueducts of the City of Rome*) described the operation and maintenance of Rome water supply system.

Froude   William Froude (1810–1879) was a English naval architect and hydrodynamicist who invented the dynamometer and used it for the testing of model ships in towing tanks. He was assisted by his son Robert Edmund Froude who, after the death of his father, continued some of his work. In 1868, he used Reech's Law of Similarity to study the resistance of model ships.

Froude number   The Froude number is proportional to the square root of the ratio of the inertial forces over the weight of fluid. The Froude number is used generally for scaling free-surface flows, open channels and hydraulic structures. Although the dimensionless number was named after William Froude, several French researchers used it before. Dupuit (1848) and Bresse (1860) highlighted the significance of the number to differentiate the open channel flow regimes. Bazin (1865a) confirmed experimentally the findings. Ferdinand Reech introduced the dimensionless number for testing ships and propellers in 1852. The number is called the Reech–Froude number in France.

GK formula   Empirical resistance formula developed by the Swiss engineers E. Ganguillet and W.R. Kutter in 1869.

Gabion   A gabion consists of rockfill material enlaced by a basket or a mesh. The word 'gabion' originates from the Italian 'gabbia' cage.

Gabion dam   Crib dam built up of gabions.

Gas transfer   Process by which gas is transferred into or out of solution: i.e. dissolution or desorption, respectively.

Gate   Valve or system for controlling the passage of a fluid. In open channels the two most common types of gates are the underflow gate and the overflow gate.

Gauckler   Philippe Gaspard Gauckler (1826–1905) was a French engineer and member of the French 'Corps des Ponts-et-Chaussées'. He re-analysed the experimental data of Darcy and Bazin (1865), and in 1867 he presented a flow resistance formula for open channel flows (Gauckler–Manning formula) sometimes called improperly the Manning equation (Gauckler, 1867). His son was Directeur des Antiquités et des Beaux-Arts (Director of Antiquities and Fine Arts) for the French Republic in Tunisia and he directed an extensive survey of Roman hydraulic works in Tunisia.

Gay-lussac   Joseph-Louis Gay-lussac (1778–1850) was a French chemist and physicist.

Ghaznavid (or Ghaznevid)   One of the Moslem dynasties (10th to 12th centuries) ruling south-western Asia. Its capital city was at Ghazni (Afghanistan).

Gradually varied flow   It is characterized by relatively small changes in velocity and pressure distributions over a short distance (e.g. long waterway).

Gravity dam   Dam which relies on its weight for stability. Normally the term 'gravity dam' refers to masonry or concrete dam.

Grille d'eau   French term for 'water screen'. A series of water jets or fountains aligned to form a screen. An impressive example is 'les Grilles d'Eau' designed by A. Le Nôtre at Vaux-le Vicomte, France.

Hartree   Douglas R. Hartree (1897–1958) was an English physicist. He was a Professor of Mathematical Physics at Cambridge. His approximation to the Schrödinger equation is the basis for the modern physical understanding of the wave mechanics of atoms. The scheme is sometimes called the Hartree–Fock method after the Russian physicist V. Fock who generalized Hartree's scheme.

Hasmonean   Designing the family or dynasty of the Maccabees, in Israel. The Hasmonean Kingdom was created following the uprising of the Jews in BC 166.

Helmholtz   Hermann Ludwig Ferdinand von Helmholtz (1821–1894) was a German scientist who made basic contributions to physiology, optics, electrodynamics and meteorology.

Hennin   Georg Wilhelm Hennin (1680–1750) was a young Dutchman hired by the tsar Peter the Great to design and build several dams in Russia (DanilVeskii, 1940; Schnitter, 1994). He went to Russia in 1698 and stayed until his death in April 1750.

Hero of Alexandria   Greek mathematician (1st century AD) working in Alexandria, Egypt. He wrote at least 13 books on mathematics, mechanics and physics. He designed and experimented the first steam engine. His treatise *Pneumatica* described Hero's fountain, siphons, steam-powered engines, a water organ, and hydraulic and mechanical water devices. It influenced directly the waterworks design during the Italian Renaissance. In his book *Dioptra*, Hero stated rightly the concept of continuity for incompressible flow : the discharge being equal to the area of the cross-section of the flow times the speed of the flow.

Himyarite   Important Arab tribe of antiquity dwelling in southern Arabia (BC 700 to AD 550).

Hohokams   Native Americans in south-west America (Arizona), they built several canal systems in the Salt river valley during the period BC 350 to AD 650. They migrated to northern Mexico around AD 900 where they built other irrigation systems.

Hokusai Katsushita   Japanese painter and wood engraver (1760–1849). His *Thirty-Six Views of Mount Fuji* (1826–1833) are world known.

Huang Chun-Pi   One of the greatest masters of Chinese painting in modern China (1898–1991). Several of his paintings included mountain rivers and waterfalls: e.g. 'Red trees and waterfalls', 'The house by the waterfalls', 'Listening to the sound of flowing waters', 'Waterfalls'.

Hydraulic diameter   It is defined as the equivalent pipe diameter: i.e. four times the cross-sectional area divided by the wetted perimeter. The concept was first expressed by the Frenchman P.L.G. du Buat (1779).

Hydraulic fill dam   Embankment dam constructed of materials which are conveyed and placed by suspension in flowing water.

Hydraulic jump   Transition from a rapid (supercritical flow) to a slow flow motion (subcritical flow) (e.g. Plates 8 and 25). Although the hydraulic jump was described by Leonardo da Vinci, the first experimental investigations were published by Giorgio Bidone in 1820. The present theory of the jump was developed by Bélanger (1828) and it has been verified experimentally numerous researchers (e.g. Bakhmeteff and Matzke, 1936).

Hyperconcentrated flow   Sediment-laden flow with large suspended sediment concentrations (i.e. typically more than 1% in volume). Spectacular hyperconcentrated flows are observed in the Yellow River basin (China) with volumetric concentrations larger than 8%.

Ideal fluid   Frictionless and incompressible fluid. An ideal fluid has zero viscosity: i.e. it cannot sustain shear stress at any point.

Idle discharge   Old expression for spill or waste water flow.

Implicit method   Calculation in which the dependent variable and the one or more independent variables are not separated on opposite sides of the equation; numerical method in which the flow properties at one point are computed as functions of both independent and dependent flow conditions.

Inca   South-American Indian of the Quechuan tribes of the highlands of Peru. The Inca civilization dominated Peru between AD 1200 and 1532. The domination of the Incas was terminated by the Spanish conquest.

Inflow   (1) Upstream flow; (2) incoming flow.

Inlet   (1) Upstream opening of a culvert, pipe or channel; (2) a tidal inlet is a narrow water passage between peninsulas or islands.

Intake   Any structure in a reservoir through which water can be drawn into a waterway or pipe. By extension, upstream end of a channel.

Interface   Surface forming a common boundary of two phases (e.g. gas–liquid interface) or two fluids.

International system of units   See Système international d'unités.

Intrados   Lower side of a wing or interior curve of a foil.

Invert   (1) Lowest portion of the internal cross-section of a conduit; (2) channel bed of a spillway; (3) bottom of a culvert barrel.

Inviscid flow   It is a non-viscous flow.

Ippen   Arthur Thomas Ippen (1907–1974) was Professor in hydrodynamics and hydraulic engineering at MIT (USA). Born in London of German parents, educated in Germany (Technische Hochschule in Aachen), he moved to USA in 1932, where he obtained the MS and PhD degrees at the California

Institute of Technology. There he worked on high-speed free-surface flows with Theodore von Karman. In 1945, he was appointed at MIT until his retirement in 1973.

Irrotational flow   It is defined as a zero vorticity flow. Fluid particles within a region have no rotation. If a frictionless fluid has no rotation at rest, any later motion of the fluid will be irrotational. In irrotational flow each element of the moving fluid undergoes no net rotation, with respect to chosen co-ordinate axes, from one instant to another.

JHRC   Jump Height Rating Curve.

JHRL   Jump Height Rating Level.

Jet d'eau   French expression for water jet. The term is commonly used in architecture and landscape.

Jevons   W.S. Jevons (1835–1882) was an English chemist and economist. His work on salt finger intrusions (Jevons, 1858) was a significant contribution to the understanding of double-diffusive convection. He performed his experiments in Sydney, Australia, 23 years prior to Rayleigh's experiments (Rayleigh, 1883).

Karman   Theodore von Karman (or von Kármán) (1881–1963) was a Hungarian fluid dynamicist and aerodynamicist who worked in Germany (1906 to 1929) and later in USA. He was a student of Ludwig Prandtl in Germany. He gave his name to the vortex shedding behind a cylinder (Karman vortex street).

Karman constant (or von Karman constant)   'Universal' constant of proportionality between the Prandtl mixing length and the distance from the boundary. Experimental results indicate that K = 0.40.

Kelvin (Lord)   William Thomson (1824–1907), Baron Kelvin of Largs, was a British physicist. He contributed to the development of the Second Law of Thermodynamics, the absolute temperature scale (measured in Kelvin), the dynamical theory of heat, fundamental work in hydrodynamics.

Kelvin–Helmholtz instability   Instability at the interface of two ideal-fluids in relative motion. The instability can be caused by a destabilizing pressure gradient of the fluid (e.g. clean-air turbulence) or free-surface shear (e.g. fluttering fountain). It is named after H.L.F. Helmoltz who solved first the problem (Helmholtz, 1868) and Lord Kelvin (Kelvin, 1871).

Kennedy   Professor John Fisher Kennedy (1933–1991) was a Professor in Hydraulics at the University of Iowa. He succeeded Hunter Rouse as Head of the Iowa Institute of Hydraulic Research.

Keulegan   Garbis Hovannes Keulegan (1890–1989) was an Armenian mathematician who worked as hydraulician for the US Bureau of Standards since its creation in 1932.

Lagrange   Joseph-Louis Lagrange (1736–1813) was a French mathematician and astronomer. During the 1789 revolution, he worked on the committee to reform the metric system. He was Professor of Mathematics at the École Polytechnique from the start in 1795.

Laminar flow   This flow is characterized by fluid particles moving along smooth paths in laminas or layers, with one layer gliding smoothly over an adjacent layer. Laminar flows are governed by Newton's Law of Viscosity which relates the shear stress to the rate of angular deformation: $\tau = \mu \partial V/\partial y$.

Laplace   Pierre-Simon Laplace (1749–1827) was a French mathematician, astronomer and physicist. He is best known for his investigations into the stability of the solar system.

LDA velocimeter   Laser Doppler Anemometer system.

Left abutment   Abutment on the left-hand side of an observer when looking downstream.

Left bank (left wall)   Looking downstream, the left bank or the left channel wall is on the left.

Leonardo da Vinci   Italian artist (painter and sculptor) who extended his interest to medicine, science, engineering and architecture (AD 1452–1519).

Lining   Coating on a channel bed to provide water tightness to prevent erosion or to reduce friction.

Lumber   Timber sawed or split into boards, planks or staves.

McKay   Professor Gordon M. McKay (1913–1989) was Professor in Civil Engineering at the University of Queensland.

Mach   Ernst Mach (1838–1916) was an Austrian physicist and philosopher. He established important principles of optics, mechanics and wave dynamics.

Mach number   See Sarrau–Mach number.

Magnus   H.G. Magnus (1802–1870) was a German physicist who investigated the so-called Magnus effect in 1852.

Magnus effect   A rotating cylinder, placed in a flow, is subjected to a force acting in the direction normal to the flow direction: i.e. a lift force which is proportional to the flow velocity times the rotation speed of the cylinder. This effect, called the Magnus effect, has a wide range of applications (Swanson, 1961).

Manning   Robert Manning (1816–1897) was the Chief Engineer of the Office of Public Works, Ireland. In 1889, he presented two formulae (Manning, 1890). One was to become the so-called 'Gauckler–Manning formula' but Robert Manning did prefer to use the second formula that he gave in his paper. It must be noted that the Gauckler–Manning formula was proposed first by the Frenchman P.G. Gauckler (Gauckler, 1867).

Mariotte   Abbé Edme Mariotte (1620–1684) was a French physicist and plant physiologist. He was member of the Académie des Sciences de Paris and wrote a fluid mechanics treaty published after his death (Mariotte, 1686).

Masonry dam   Dam constructed mainly of stone, brick or concrete blocks jointed with mortar.

Meandering channel   Alluvial stream characterized by a series of alternating bends (i.e. meanders) as a result of alluvial processes.

MEL culvert   See Minimum energy loss culvert.

Metric system   See Système métrique.

Minimum energy loss culvert   Culvert designed with very smooth shapes to minimize energy losses. The design of a minimum energy loss culvert is associated with the concept of constant total head. The inlet and outlet must be streamlined in such a way that significant form losses are avoided (Apelt, 1983).

Mixing length   The mixing length theory is a turbulence theory developed by L. Prandtl, first formulated in 1925. Prandtl assumed that the mixing length is the characteristic distance travelled by a particle of fluid before its momentum is changed by the new environment.

Mochica   (1) South American civilization (AD 200–1000) living in the Moche River valley, Peru along the Pacific coastline; (2) Language of the Yuncas.

Mole   Mass numerically equal in grams to the relative mass of a substance (i.e. 12 g for carbon-12). The number of molecules in 1 mol of gas is 6.0221367×10^{23} (i.e. Avogadro number).

Momentum exchange coefficient   In turbulent flows the apparent kinematic viscosity (or kinematic eddy viscosity) is analogous to the kinematic viscosity in laminar flows. It is called the momentum exchange coefficient, the eddy viscosity or the eddy coefficient. The momentum exchange coefficient is proportional to the shear stress divided by the strain rate. It was first introduced by the Frenchman J.V. Boussinesq (1877, 1896).

Monge   Gaspard Monge (1746–1818), Comte de Péluse, was a French mathematician who invented descriptive geometry and pioneered the development of analytical geometry. He was a prominent figure during the French Revolution, helping to establish the Système métrique and the École Polytechnique, and being Minister for the Navy and Colonies between 1792 and 1793.

Moor   (1) Native of Mauritania, a region corresponding to parts of Morocco and Algeria; (2) Moslem of native North African races.

Morning-glory spillway   Vertical discharge shaft, more particularly the circular hole form of a drop inlet spillway. The shape of the intake is similar to a morning-glory flower (American native plant (*Ipomocea*)). It is sometimes called a tulip intake.

Mud   Slimy and sticky mixture of solid material and water.

Mughal (or Mughul or Mogul or Moghul)   Name or adjective referring to the Mongol conquerors of India and to their descendants. The Mongols occupied India from 1526 up to the 18th century although the authority of the Mughal emperor became purely nominal after 1707. The fourth emperor, Jahangir (1569–1627), married a Persian princess Mehr-on Nesa who became known as Nur Jahan. His son Shah Jahan (1592–1666) built the famous Taj Mahal between 1631 and 1654 in memory of his favourite wife Arjumand Banu better known by her title: Mumtaz Mahal or Taj Mahal.

Nabataean   Habitant from an ancient kingdom to the east and south-east of Palestine that included the Neguev desert. The Nabataean Kingdom lasted from around BC 312 to AD 106. The Nabataeans built a large number of soil-and-retention dams. Some are still in use today as shown by Schnitter (1994).

Nappe flow   Flow regime on a stepped chute where the water bounces from one step to the next one as a succession of free-fall jets.

Navier Louis Marie Henri Navier (1785–1835) was a French engineer who primarily designed bridges but also extended Euler's equations of motion (Navier, 1823).

Navier–Stokes equation Momentum equation applied to a small control volume of incompressible fluid. It is usually written in vector notation. The equation was first derived by L. Navier in 1822 and S.D. Poisson in 1829 by a different method. It was derived later in a more modern manner by A.J.C. Barré de Saint-Venant in 1843 and G.G. Stokes in 1845.

Negative surge A negative surge results from a sudden change in flow that decreases the flow depth. It is a retreating wave front moving upstream or downstream.

Newton Sir Isaac Newton (1642–1727) was an English mathematician and physicist. His contributions in optics, mechanics and mathematics were fundamental.

Nikuradse J. Nikuradse was a German engineer who investigated experimentally the flow field in smooth and rough pipes (Nikuradse, 1932, 1933).

Non-uniform equilibrium flow The velocity vector varies from place to place at any instant: steady non-uniform flow (e.g. flow through an expanding tube at a constant rate) and unsteady non-uniform flow (e.g. flow through an expanding tube at an increasing flow rate).

Normal depth Uniform equilibrium open channel flow depth.

Obvert Roof of the barrel of a culvert. Another name is soffit.

One-dimensional flow Neglects the variations and changes in velocity and pressure transverse to the main flow direction. An example of one-dimensional flow can be the flow through a pipe.

One-dimensional model Model defined with one spatial co-ordinate, the variables being averaged in the other two directions.

Outflow Downstream flow.

Outlet (1) Downstream opening of a pipe, culvert or canal; (2) artificial or natural escape channel.

Pascal Blaise Pascal (1623–1662) was a French mathematician, physicist and philosopher. He developed the modern theory of probability. Between 1646 and 1648, he formulated the concept of pressure and showed that the pressure in a fluid is transmitted through the fluid in all directions. He measured also the air pressure both in Paris and on the top of a mountain over-looking Clermont-Ferrand (France).

Pascal Unit of pressure named after the Frenchman B. Pascal: one Pascal equals a Newton per square-metre.

Pelton turbine (or wheel) Impulse turbine with one to six circular nozzles that deliver high-speed water jets into air which then strike the rotor blades shaped like scoop and known as bucket. A simple bucket wheel was designed by Sturm in the 17th century. The American Lester Allen Pelton patented the actual double-scoop (or double-bucket) design in 1880.

Pervious zone Part of the cross-section of an embankment comprising material of high permeability.

Pitot Henri Pitot (1695–1771) was a French mathematician, astronomer and hydraulician. He was a member of the French Académie des Sciences from 1724. He invented the Pitot tube to measure flow velocity in the Seine River (first presentation in 1732 at the Académie des Sciences de Paris).

Pitot tube Device to measure flow velocity. The original Pitot tube consisted of two tubes, one with an opening facing the flow. L. Prandtl developed an improved design (e.g. Howe, 1949) which provides the total head, piezometric head and velocity measurements. It is called a Prandtl-Pitot tube and more commonly a Pitot tube.

Pitting Formation of small pits and holes on surfaces due to erosive or corrosive action (e.g. cavitation pitting).

Plato Greek philosopher (about BC 428–347) who influenced greatly Western philosophy.

Plunging jet Liquid jet impacting (or impinging) into a receiving pool of liquid.

Poiseuille Jean-Louis Marie Poiseuille (1799–1869) was a French physician and physiologist who investigated the characteristics of blood flow. He carried out experiments and formulated first the expression of flow rates and friction losses in laminar fluid flow in circular pipes (Poiseuille, 1839).

Poiseuille flow Steady laminar flow in a circular tube of constant diameter.

Poisson Siméon Denis Poisson (1781–1840) was a French mathematician and scientist. He developed the theory of elasticity, a theory of electricity and a theory of magnetism.

Positive surge   A positive surge results from a sudden change in flow that increases the depth. It is an abrupt wave front. The unsteady flow conditions may be solved as a quasi-steady flow situation.

Potential flow   Ideal-fluid flow with irrotational motion.

Prandtl   Ludwig Prandtl (1875–1953) was a German physicist and aerodynamicist who introduced the concept of boundary layer (Prandtl, 1904) and developed the turbulent 'mixing length' theory. He was Professor at the University of Göttingen.

Preissmann   Alexandre Preissmann (1916–1990) was born and educated in Switzerland. From 1958, he worked on the development of hydraulic mathematical models at Sogreah in Grenoble.

Prismatic   A prismatic channel has an unique cross-sectional shape independent of the longitudinal distance along the flow direction. For example, a rectangular channel of constant width is prismatic.

Prony   Gaspard Clair François Marie Riche de Prony (1755–1839) was a French mathematician and engineer. He succeeded A. Chézy as Director General of the Ecole Nationale Supérieure des Ponts-et-Chaussées, Paris during the French Revolution.

Radial gate   Underflow gate for which the wetted surface has a cylindrical shape.

Rankine   William J.M. Rankine (1820–1872) was a Scottish engineer and physicist. His contribution to thermodynamics and steam engine was important. In fluid mechanics, he developed the theory of sources and sinks, and used it to improve ship hull contours. One ideal-fluid flow pattern, the combination of uniform flow, source and sink, is named after him: i.e. flow past a Rankine body.

Rapidly varied flow   It is characterized by large changes over a short distance (e.g. sharp-crested weir, sluice gate, hydraulic jump).

Rayleigh   John William Strutt, Baron Rayleigh, (1842–1919) was an English scientist who made fundamental findings in acoustics and optics. His works are the basics of wave propagation theory in fluids. He received the Nobel Prize for Physics in 1904 for his work on the inert gas argon.

Reech   Ferdinand Reech (1805–1880) was a French naval instructor who proposed first the Reech–Froude number in 1852 for the testing of model ships and propellers.

Rehbock   Theodor Rehbock (1864–1950) was a German hydraulician and Professor at the Technical University of Karlsruhe. His contribution to the design of hydraulic structures and physical modelling is important.

Renaissance   Period of great revival of art, literature and learning in Europe in the 14th, 15th and 16th centuries.

Reynolds   Osborne Reynolds (1842–1912) was a British physicist and mathematician who expressed first the Reynolds number (Reynolds, 1883) and later the Reynolds stress (i.e. turbulent shear stress).

Reynolds number   Dimensionless number proportional to the ratio of the inertial force over the viscous force.

Rheology   Science describing the deformation of fluid and matter.

Riblet   Series of longitudinal grooves. Riblets are used to reduce skin drag (e.g. on aircraft, ship hull). The presence of longitudinal grooves along a solid boundary modifies the bottom shear stress and the turbulent bursting process. Optimum groove width and depth are about 20–40 times the laminar sublayer thickness (i.e. about 10–20 μm in air, 1–2 mm in water).

Richelieu   Armand Jean du Plessis (1585–1642), Duc de Richelieu and French Cardinal, was the Prime Minister of King Louis XIII of France from 1624 to his death.

Riemann   Bernhard Georg Friedrich Riemann (1826–1866) was a German mathematician.

Right abutment   Abutment on the right-hand side of an observer when looking downstream.

Right bank (right wall)   Looking downstream, the right bank or the right channel wall is on the right.

Riquet   Pierre Paul Riquet (1604–1680) was the designer and Chief Engineer of the Canal du Midi built between 1666 and 1681. This canal provides an inland route between the Atlantic and the Mediterranean across southern France.

Rockfill   Material composed of large rocks or stones loosely placed.

Rockfill dam   Embankment dam in which more than 50% of the total volume comprise compacted or dumped pervious natural stones.

Roller   In hydraulics, large-scale turbulent eddy: e.g. the roller of a hydraulic jump.

Roller compacted concrete (RCC)   It is defined as a no-slump consistency concrete that is placed in horizontal lifts and compacted by vibratory rollers. RCC has been commonly used as construction material of gravity dams since the 1970s.

Roll wave   On steep slopes free-surface flows become unstable. The phenomenon is usually clearly visible at low flow rates. The waters flow down the chute in a series of wave fronts called roll waves.

Rouse   Hunter Rouse (1906–1996) was an eminent hydraulician who was Professor and Director of the Iowa Institute of Hydraulic Research at the University of Iowa (USA).

SAF   St Anthony's Falls Hydraulic Laboratory at the University of Minnesota (USA).

Sabaen   Ancient name of the people of Yemen in southern Arabia. Renowned for the visit of the Queen of Sabah (or Sheba) to the King of Israel around BC 950 and for the construction of the Marib dam (BC 115 to AD 575). The fame of the Marib dam was such that its final destruction in AD 575 was recorded in the Koran.

Saltation   (1) Action of leaping or jumping; (2) in sediment transport, particle motion by jumping and bouncing along the bed.

Saint-Venant   See Barré de Saint-Venant.

Sarrau   French Professor at Ecole Polytechnique, Paris, who first introduced the Sarrau–Mach number (Sarrau, 1884).

Sarrau–Mach number   Dimensionless number proportional to the ratio of inertial forces over elastic forces. Although the number is commonly named after E. Mach who introduced it in 1887, it is often called the Sarrau number after Professor Sarrau who first highlighted the significance of the number (Sarrau, 1884). The Sarrau–Mach number was once called the Cauchy number as a tribute to Cauchy's contribution to wave motion analysis.

Scalar   A quantity that has a magnitude described by a real number and no direction. A scalar means a real number rather than a vector.

Scale effect   Discrepancy betwen model and prototype resulting when one or more dimensionless parameters have different values in the model and prototype.

Scour   Bed material removal caused by the eroding power of the flow.

Sediment   Any material carried in suspension by the flow or as bed-load which would settle to the bottom in absence of fluid motion.

Sediment load   Material transported by a fluid in motion.

Sediment transport   Transport of material by a fluid in motion.

Sediment transport capacity   Ability of a stream to carry a given volume of sediment material per unit time for given flow conditions. It is the sediment transport potential of the river.

Sediment yield   Total sediment outflow rate from a catchment, including bed-load and suspension.

Seepage   Interstitial movement of water that may take place through a dam, its foundation or abutments.

Sennacherib (or Akkadian Sin-Akhkheeriba)   King of Assyria (BC 705–681), son of Sargon II (who ruled during BC 722–705). He built a huge water supply for his capital city Nineveh (near the actual Mossul, Iraq) in several stages. The latest stage comprised several dams and over 75 km of canals and paved channels.

Separation   In a boundary layer, a deceleration of fluid particles leading to a reversed flow within the boundary layer is called a separation. The decelerated fluid particles are forced outwards and the boundary layer is separated from the wall. At the point of separation, the velocity gradient normal to the wall is zero.

$$\left(\frac{\partial V_x}{\partial y}\right)_{y=0} = 0$$

Separation point   In a boundary layer, intersection of the solid boundary with the streamline dividing the separation zone and the deflected outer flow. The separation point is a stagnation point.

Sequent depth   In open channel flow, the solution of the momentum equation at a transition between supercritical and subcritical flow gives two flow depths (upstream and downstream flow depths). They are called sequent depths.

Sewage   Refused liquid or waste matter carried off by sewers. It may be a combination of water-carried wastes from residences and industries together with ground water, surface water and storm water.

Sewer   An artificial subterranean conduit to carry off water and waste matter.

Shock waves   With supercritical flows, a flow disturbance (e.g. change of direction, contraction) induces the development of shock waves propagating at the free-surface across the channel (e.g. Ippen and Harleman, 1956; Hager, 1992a). Shock waves are also called lateral shock waves, oblique hydraulic jumps, Mach waves, cross-waves, diagonal jumps.

Side-channel spillway   A side-channel spillway consists of an open spillway (along the side of a channel) discharging into a channel running along the foot of the spillway and carrying the flow away in a direction parallel to the spillway crest (e.g. Arizona-side spillway of the Hoover dam, USA).

Similitude   Correspondence between the behaviour of a model and that of its prototype, with or without geometric similarity. The correspondence is usually limited by scale effects.

Siphon   Pipe system discharging waters between two reservoirs or above a dam in which the water pressure becomes sub-atmospheric. The shape of a simple siphon is close to an omega (i.e. Ω-shape). Inverted-siphons carry waters between two reservoirs with pressure larger than atmospheric. Their design follows approximately a U-shape. Inverted-siphons were commonly used by the Romans along their aqueducts to cross valleys.

Siphon-spillway   Device for discharging excess water in a pipe over the dam crest.

Skimming flow   Flow regime above a stepped chute for which the water flows as a coherent stream in a direction parallel to the pseudo-bottom formed by the edges of the steps (Plate 9). The same term is used to characterize the flow regime of large discharges above rockfill and closely spaced large roughness elements.

Slope   (1) Side of a hill; (2) inclined face of a canal (e.g. trapezoidal channel); (3) inclination of the channel bottom from the horizontal.

Sluice gate   Underflow gate with a vertical sharp edge for stopping or regulating flow.

Soffit   Roof of the barrel of a culvert. Another name is obvert.

Specific energy   Quantity proportional to the energy per unit mass, measured with the channel bottom as the elevation datum, and expressed in metres of water. The concept of specific energy, first developed by B.A. Bakhmeteff in 1912, is commonly used in open channel flows.

Spillway   Opening built into a dam or the side of a reservoir to release (to spill) excess flood waters.

Splitter   Obstacle (e.g. concrete block, fin) installed on a chute to split the flow and to increase the energy dissipation.

Spray   Water droplets flying or falling through air: e.g. spray thrown up by a waterfall.

Stage–discharge curve   Relationship between discharge and free-surface elevation at a given location along a stream.

Stagnation point   It is defined as the point where the velocity is zero. When a streamline intersects itself, the intersection is a stagnation point. For irrotational flow a streamline intersects itself at right angle at a stagnation point.

Staircase   Another adjective for 'stepped': e.g. a staircase cascade is a stepped cascade.

Stall   Aerodynamic phenomenon causing a disruption (i.e. separation) of the flow past a wing associated with a loss of lift.

Steady flow   Occurs when conditions at any point of the fluid do not change with the time:

$$\frac{\partial V}{\partial t} = 0 \quad \frac{\partial \rho}{\partial t} = 0 \quad \frac{\partial P}{\partial t} = 0 \quad \frac{\partial T}{\partial t} = 0$$

Stilling basin   Structure for dissipating the energy of the flow downstream of a spillway, outlet work, chute or canal structure. In many cases, a hydraulic jump is used as the energy dissipator within the stilling basin.

Stokes   George Gabriel Stokes (1819–1903), British mathematician and physicist, is known for his research in hydrodynamics and a study of elasticity.

Stop-logs   Form of sluice gate comprising a series of wooden planks, one above the other, and held at each end.

Storm water   Excess water running off the surface of a drainage area during and immediately following a period of rain. In urban areas, waters drained off a catchment area during or after a heavy rainfall are usually conveyed in man-made storm waterways.

Storm waterway   Channel built for carrying storm waters.

Straub   L.G. Straub (1901–1963) was Professor and Director of the St Anthony Falls Hydraulics Laboratory at the University of Minnesota (USA).

Stream function   Vector function of space and time which is related to the velocity field as: $\vec{V} = -\overrightarrow{\text{curl}\,\varphi}$. The stream function exists for steady and unsteady flow of incompressible fluid as it does satisfy the continuity equation. The stream function was introduced by the French mathematician, Lagrange.

Streamline   It is the line drawn so that the velocity vector is always tangential to it (i.e. no flow across a streamline). When the streamlines converge the velocity increases. The concept of streamline was first introduced by the Frenchman J.C. de Borda.

Streamline maps   These maps should be drawn so that the flow between any two adjacent streamlines is the same.

Stream tube   It is a filament of fluid bounded by streamlines.

Subcritical flow   In open channel the flow is defined as subcritical if the flow depth is larger than the critical flow depth. In practice, subcritical flows are controlled by the downstream flow conditions.

Subsonic flow   Compressible flow with a Sarrau–Mach number less than unity: i.e. the flow velocity is less than the sound celerity.

Supercritical flow   In open channel, when the flow depth is less than the critical flow depth, the flow is supercritical and the Froude number is larger than one. Supercritical flows are controlled from upstream.

Supersonic flow   Compressible flow with a Sarrau–Mach number larger than unity: i.e. the flow velocity is larger than the sound celerity.

Surface tension   Property of a liquid surface displayed by its acting as if it were a stretched elastic membrane. Surface tension depends primarily on the attraction forces between the particles within the given liquid and also on the gas, solid or liquid in contact with it. The action of surface tension is to increase the pressure within a water droplet or within an air bubble. For a spherical bubble of diameter d_{ab}, the increase of internal pressure necessary to balance the tensile force caused by surface tension equals: $\Delta P = 4\sigma/d_{ab}$ where σ is the surface tension.

Surfactant (or *surface active agent*)   Substance that, when added to a liquid, reduces its surface tension thereby increasing its wetting property (e.g. detergent).

Surge   A surge in an open channel is a sudden change of flow depth (i.e. abrupt increase or decrease in depth). An abrupt increase in flow depth is called a positive surge while a sudden decrease in depth is termed a negative surge. A positive surge is also called (improperly) a 'moving hydraulic jump' or a 'hydraulic bore'.

Surge wave   Results from a sudden change in flow that increases (or decreases) the depth.

Suspended load   Transported sediment material maintained into suspension.

Swash   In coastal engineering, the swash is the rush of water up a beach from the breaking waves.

Swash line   The upper limit of the active beach reached by highest sea level during big storms.

Système international d'unités   International system of units adopted in 1960 based on the metre–kilogram–second (MKS) system. It is commonly called SI unit system. The basic seven units are: for length, the metre; for mass, the kilogram; for time, the second; for electric current, the ampere; for luminous intensity, the candela; for amount of substance, the mole; for thermodynamic temperature, the Kelvin. Conversion tables are given in Appendix A1.2.

Système métrique   International decimal system of weights and measures which was adopted in 1795 during the French Révolution. Between 1791 and 1795, the Académie des Sciences de Paris prepared a logical system of units based on the metre for length and the kilogram for mass. The standard metre was defined as 1×10^{-7} times a meridional quadrant of Earth. The gram was equal to the mass of $1\,cm^3$ of pure water at the temperature of its maximum density (i.e. 4°C) and 1 kg equalled 1000 g. The litre was defined as the volume occupied by a cube of $1 \times 10^3\,cm^3$.

TWRC   Tailwater rating curve.

TWRL   Tailwater rating level.

Tainter gate   It is a radial gate.

Tailwater depth   Downstream flow depth.

Tailwater level   Downstream free-surface elevation.

Total head   The total head is proportional to the total energy per unit mass and per gravity unit. It is expressed in metres of water.

Training wall   Sidewall of chute spillway.

Trashrack   Screen comprising metal or reinforced concrete bars located at the intake of a waterway to prevent the progress of floating or submerged debris.

Turbulence   Flow motion characterized by its unpredictable behaviour, strong mixing properties and a broad spectrum of length scales (Lesieur, 1994).

Turbulent flow   In turbulent flows the fluid particles move in very irregular paths, causing an exchange of momentum from one portion of the fluid to another. Turbulent flows have great mixing potential and involve a wide range of eddy length scales.

Turriano   Juanelo Turriano (1511–1585) was an Italian clockmaker, mathematician and engineer who worked for the Spanish Kings Charles V and later Philip II. It is reported that he checked the design of the Alicante dam for King Philip II.

Two-dimensional flow   All particles are assumed to flow in parallel planes along identical paths in each of these planes. There are no changes in flow normal to these planes. An example of two-dimensional flow can be an open channel flow in a wide rectangular channel.

USACE   United States Army Corps of Engineers.

USBR   United States Bureau of Reclamation.

Ukiyo-e (or Ukiyoe)   It is a type of Japanese painting and colour woodblock prints during the period 1803–1867.

Undular hydraulic jump   Hydraulic jump characterized by steady stationary free-surface undulations downstream of the jump and by the absence of a formed roller. The undulations can extend far downstream of the jump with decaying wave lengths, and the undular jump occupies a significant length of the channel. It is usually observed for $1 < Fr_1 < 1.5–3$ (Chanson, 1995a). The first significant study of undular jump flow can be attributed to Fawer (1937) and undular jump flows should be called Fawer jump in homage to Fawer's work.

Undular surge   Positive surge characterized by a train of secondary waves (or undulations) following the surge front. Undular surges are sometimes called Boussinesq–Favre waves in homage to the contributions of J.B. Boussinesq and H. Favre.

Uniform equilibrium flow   It occurs when the velocity is identically the same at every point, in magnitude and direction, for a given instant:

$$\frac{\partial V}{\partial s} = 0$$

in which time is held constant and ∂s is a displacement in any direction. That is, steady uniform flow (e.g. liquid flow through a long pipe at a constant rate) and unsteady uniform flow (e.g. liquid flow through a long pipe at a decreasing rate).

Universal gas constant (also called *molar gas constant* or *perfect gas constant*)   Fundamental constant equal to the pressure times the volume of gas divided by the absolute temperature for one mole of perfect gas. The value of the universal gas constant is $8.31441 \, \text{J} \, \text{K}^{-1} \text{mol}^{-1}$.

Unsteady flow   The flow properties change with the time.

Uplift   Upward pressure in the pores of a material (interstitial pressure) or on the base of a structure. Uplift pressures led to the destruction of stilling basins and even to the failures of concrete dams (e.g. Malpasset dam break in 1959).

Upstream flow conditions   Flow conditions measured immediately upstream of the investigated control volume.

VNIIG   Institute of Hydrotechnics Vedeneev in St Petersburg (Russia).

VOC   Volatile organic compound.

Validation   Comparison between model results and prototype data, to validate the model. The validation process must be conducted with prototype data that are different from that used to calibrate and to verify the model.

Vauban Sébastien Vauban (1633–1707) was Maréchal de France. He participated to the construction of several water supply systems in France, including the extension of the feeder system of the Canal du Midi between 1686 and 1687, and parts of the water supply system of the gardens of Versailles.

Velocity potential It is defined as a scalar function of space and time such that its negative derivative with respect to any direction is the fluid velocity in that direction: $V = -\text{grad}\,\Phi$. The existence of a velocity potential implies irrotational flow of ideal-fluid. The velocity potential was introduced by the French mathematician J. Louis Lagrange (1781).

Vena contracta Minimum cross-sectional area of the flow (e.g. jet or nappe) discharging through an orifice, sluice gate or weir.

Venturi meter In closed pipes, smooth constriction followed by a smooth expansion. The pressure difference between the upstream location and the throat is proportional to the velocity-square. It is named after the Italian physicist Giovanni Battista Venturi (1746–1822).

Villareal de Berriz Don Pedro Bernardo Villareal de Berriz (1670–1740) was a Basque nobleman. He designed several buttress dams, some of these being still in use (Smith, 1971).

Viscosity Fluid property which characterizes the fluid resistance to shear: i.e. resistance to a change in shape or movement of the surroundings.

Vitruvius Roman architect and engineer (BC 94–??). He built several aqueducts to supply the Roman capital with water. (*Note*: there are some incertitude on his full name: 'Marcus Vitruvius Pollio' or 'Lucius Vitruvius Mamurra', Garbrecht, 1987a.)

Von Karman constant See Karman constant.

WES Waterways Experiment Station of the US Army Corps of Engineers.

Wadi Arabic word for a valley which becomes a watercourse in rainy seasons.

Wake region The separation region downstream of the streamline that separates from a boundary is called a wake or wake region.

Warrie Australian aboriginal name for 'rushing water'.

Waste waterway Old name for a spillway, particularly used in irrigation with reference to the waste of waters resulting from a spill.

Wasteweir A spillway. The name refers to the waste of hydro-electric power or irrigation water resulting from the spill. A 'staircase' wasteweir is a stepped spillway.

Water Common name applied to the liquid state of the hydrogen–oxygen combination H_2O. Although the molecular structure of water is simple, the physical and chemical properties of H_2O are unusually complicated. Water is a colourless, tasteless and odourless liquid at room temperature. One most important property of water is its ability to dissolve many other substances: H_2O is frequently called the universal solvent. Under standard atmospheric pressure, the freezing point of water is 0°C (273.16 K) and its boiling point is 100°C (373.16 K).

Water clock Ancient device for measuring time by the gradual flow of water through a small orifice into a floating vessel. The Greek name is *Clepsydra*.

Waterfall Abrupt drop of water over a precipice characterized by a free-falling nappe of water. The highest waterfalls are the Angel fall (979 m) in Venezuela (Churún Merú), Tugel fall (948 m) in South Africa, Mtarazi (762 m) in Zimbabwe.

Water-mill Mill (or wheel) powered by water.

Water staircase (or '*Escalier d'Eau*') It is the common architectural name given to a stepped cascade with flat steps.

Weak jump A weak hydraulic jump is characterized by a marked roller, no free-surface undulation and low energy loss. It is usually observed after the disappearance of undular hydraulic jump with increasing upstream Froude numbers.

Weber Moritz Weber (1871–1951) was a German Professor at the Polytechnic Institute of Berlin. The Weber number characterizing the ratio of inertial force over surface tension force was named after him.

Weber number Dimensionless number characterizing the ratio of inertial forces over surface tension forces. It is relevant in problems with gas–liquid or liquid–liquid interfaces.

Weir Low river dam used to raise the upstream water level (e.g. Plate 8). Measuring weirs are built across a stream for the purpose of measuring the flow.

Weisbach Julius Weisbach (1806–1871) was a German applied mathematician and hydraulician.

Wen Cheng-Ming Chinese landscape painter (1470–1559). One of his famous works is the painting of *Old trees by a cold waterfall*.

WES standard spillway shape Spillway shape developed by the US Army Corps of Engineers at the Waterways Experiment Station.

Wetted perimeter Considering a cross-section (usually selected normal to the flow direction), the wetted perimeter is the length of wetted contact between the flowing stream and the solid boundaries. For example, in a circular pipe flowing full, the wetted perimeter equals the circle perimeter.

Wetted surface In open channel, the term 'wetted surface' refers to the surface area in contact with the flowing liquid.

White waters Non-technical term used to design free-surface aerated flows. The refraction of light by the entrained air bubbles gives the 'whitish' appearance to the free-surface of the flow (e.g. Plates 5, 6, 7, 25 and 30).

White water sports These include canoe, kayak and rafting racing down swift-flowing turbulent waters.

Wing wall Sidewall of an inlet or outlet.

Wood I.R. Wood is an Emeritus Professor in Civil Engineering at the University of Canterbury (New Zealand).

Yen Professor Ben Chie Yen (1935–2001) was a Professor in Hydraulics at the University of Illinois at Urbana-Champaign, although born and educated in Taiwan.

Yunca Indian of a group of South American tribes of which the Chimus and the Chinchas are the most important. The Yunca civilization developed a pre-Inca culture on the coast of Peru.

List of symbols

A	flow cross-sectional area (m^2)
A_s	particle cross-sectional area (m^2)
B	open channel free-surface width (m)
B_{max}	inlet lip width (m) of MEL culvert (Chapter 21)
B_{min}	(1) minimum channel width (m) for onset of choking flow
	(2) barrel width (m) of a culvert (Chapter 21)
C	(1) celerity (m/s): e.g. celerity of sound in a medium, celerity of a small disturbance at a free-surface
	(2) dimensional discharge coefficient (Chapters 19 and 21)
$\mathcal{C}$	discharge parameter of box culvert with submerged entrance
Ca	Cauchy number
$C_{Chézy}$	Chézy coefficient (m$^{1/2}$/s)
C_D	dimensionless discharge coefficient (SI units) (Chapter 19)
C_L	lift coefficient
C_d	(1) skin friction coefficient (also called drag coefficient)
	(2) drag coefficient
C_{des}	design discharge coefficient (SI units) (Chapter 19)
C_o	initial celerity (m/s) of a small disturbance (Chapter 17)
C_p	specific heat at constant pressure (J kg^{-1}K^{-1}): $C_p = \left(\dfrac{\partial h}{\partial T}\right)_p$
C_s	mean volumetric sediment concentration
$(C_s)_{mean}$	mean sediment suspension concentration
C_{sound}	sound celerity (m/s)
C_v	specific heat at constant volume (J kg^{-1}K^{-1})
c_s	sediment concentration
D	(1) circular pipe diameter (m)
	(2) culvert barrel height (m) (Chapter 21)
D_H	hydraulic diameter (m), or equivalent pipe diameter, defined as:

$$D_H = 4 \times \frac{\text{cross-sectional area}}{\text{wetted perimeter}} = \frac{4 \times A}{P_w}$$

D_s	sediment diffusivity (m^2/s)
D_t	diffusion coefficient (m^2/s)
D_1, D_2	characteristics of velocity distribution in turbulent boundary layer
d	flow depth (m) measured perpendicular to the channel bed
d_{ab}	air-bubble diameter (m)
d_b	brink depth (m) (Chapter 20)

d_c	critical flow depth (m)
d_{charac}	characteristic geometric length (m);
d_{conj}	conjugate flow depth (m)
d_o	(1) uniform equilibrium flow depth (m): i.e. normal depth
	(2) initial flow depth (m) (Chapter 17)
d_p	pool depth (m) (Chapter 20)
d_s	(1) sediment size (m)
	(2) dam break wave front thickness (m) (Chapter 17)
d_{tw}	tailwater flow depth (m)
d_{50}	median grain size (m) defined as the size for which 50% by weight of the material is finer
d_{84}	sediment grain size (m) defined as the size for which 84% by weight of the material is finer
d_i	characteristic grain size (m), where $i = 10, 16, 50, 75, 84, 90$
d_*	dimensionless particle parameter: $d_* = d_s \sqrt[3]{(\rho_s/\rho - 1)g/\nu^2}$
e	internal energy per unit mass (J/kg)
E	mean specific energy (m) defined as: $E = H - z_0$
E	local specific energy (m) defined as: $E = \dfrac{P}{\rho g} + (z - z_0) + \dfrac{v^2}{2g}$
Eu	Euler number
E_b	bulk modulus of elasticity (Pa): $E_b = \rho \dfrac{\partial P}{\partial \rho}$
E_{co}	compressibility (1/Pa): $E_{co} = \dfrac{1}{\rho} \dfrac{\partial \rho}{\partial P}$
E_{min}	minimum specific energy (m)
E	total energy (J) of system
f	Darcy friction factor (also called head loss coefficient)
F	force (N)
$\vec{F}$	force vector
F_b	buoyant force (N)
F_d	drag force (N)
F_{fric}	friction force (N)
F_p	pressure force (N)
F_p'	pressure force (N) acting on the flow cross-sectional area
F_p''	pressure force (N) acting on the channel side boundaries
F_{visc}	viscous force (N/m³)
F_{vol}	volume force per unit volume (N/m³)
Fr	Froude number
F_r	ratio of prototype to model forces: $F_r = F_p/F_m$ (Chapter 14)
g	gravity constant (m/s²) (see Appendix A1.1); in Brisbane, Australia: $g = 9.80 \, \text{m/s}^2$
$g_{centrif}$	centrifugal acceleration (m/s²)
h	specific enthalpy (i.e. enthalpy per unit mass) (J/kg): $h = e + \dfrac{P}{\rho}$
h	(1) dune bed-form height (m)
	(2) step height (m)
H	(1) mean total head (m): $H = d \cos\theta + z_0 + \alpha \dfrac{V_{mean}^2}{2g}$ assuming a hydrostatic pressure distribution
	(2) depth-averaged total head (m) defined as: $H = \dfrac{1}{d} \displaystyle\int_0^d H \, dy$

H	local total head (m) defined as: $H = \dfrac{P}{pg} + z + \dfrac{v^2}{2g}$
H_{dam}	reservoir height (m) at dam site (Chapter 17)
H_{des}	design upstream head (m) (Chapter 19 and 20)
H_{res}	residual head (m)
H_1	upstream total head (m)
H_2	downstream total head (m)
i	integer subscript
i	imaginary number: $i = \sqrt{-1}$
JHRL	jump height rating level (m RL)
k	permeability (m²) of a soil
k_{Bazin}	Bazin resistance coefficient
$k_{Strickler}$	Strickler resistance coefficient (m$^{1/3}$/s)
k_s	equivalent sand roughness height (m)
K	hydraulic conductivity (m/s) of a soil
K	von Karman constant (i.e. K = 0.4)
K_M	empirical coefficient (s) in the Muskingum method
K'	head loss coefficient: $K' = \Delta H/(0.5V^2/g)$
l	(1) dune bed-form length (m)
	(2) step length (m)
L	length (m)
L_{crest}	crest length (m)
L_{culv}	culvert length (m) measured in the flow direction (Chapter 21)
L_d	drop length (m)
L_{inlet}	inlet length (m) measured in the flow direction (Chapter 21)
L_r	(1) length of roller of hydraulic jump (m) (Chapters 4 and 19)
	(2) ratio of prototype to model lengths: $L_r = L_p/L_m$ (Chapter 14)
M	momentum function (m²)
Ma	Sarrau–Mach number
Mo	Morton number
M_r	ratio of prototype to model masses (Chapter 14)
M	total mass (kg) of system
$\dot{m}$	mass flow rate per unit width (kg s^{-1}m^{-1})
$\dot{m}_s$	sediment mass flow rate per unit width (kg s^{-1}m^{-1})
N	inverse of velocity distribution exponent
No	Avogadro constant: No = 6.0221367×10^{23} mole^{-1}
Nu	Nusselt number
N_{bl}	inverse of velocity distribution exponent in turbulent boundary layer (Appendix A4.1)
$n_{Manning}$	Gauckler–Manning coefficient (s/m$^{1/3}$)
P	absolute pressure (Pa)
P_{atm}	atmospheric pressure (Pa)
$P_{centrif}$	centrifugal pressure (Pa)
P_r	ratio of prototype to model pressures (Chapter 14)
P_{std}	standard atmosphere (Pa) or normal pressure at sea level
P_v	vapour pressure (Pa)
P_w	wetted perimeter (m)
P_o	porosity factor
q	discharge per metre width (m²/s)

q_{des}	design discharge (m^2/s) per unit width (Chapters 19–21)
q_{max}	maximum flow rate per unit width (m^2/s) in open channel for a constant specific energy
q_s	sediment flow rate per unit width (m^2/s)
q_h	heat added to a system per unit mass (J/kg)
Q	total volume discharge (m^3/s)
Q_{des}	design discharge (m^3/s) (Chapters 19–21)
Q_{max}	maximum flow rate (m^3/s) in open channel for a constant specific energy
Q_r	ratio of prototype to model discharges (Chapter 14)
Q_h	heat added to a system (J)
R	invert curvature radius (m)
R	fluid thermodynamic constant (J/kg/K) also called gas constant: $P = \rho R T$ perfect gas law (i.e. Mariotte law)
Re	Reynolds number
Re_*	shear Reynolds number
R_H	hydraulic radius (m) defined as: $R_H = \dfrac{\text{cross-sectional area}}{\text{wetted perimeter}} = \dfrac{A}{P_w}$
Ro	universal gas constant: Ro = $8.3143\,\mathrm{J\,K^{-1}\,mol^{-1}}$
r	radius of curvature (m)
s	curvilinear co-ordinate (m) (i.e. distance measured along a streamline and positive in the flow direction)
s	relative density of sediment: s = ρ_s/ρ (Chapters 7–12)
S	sorting coefficient of a sediment mixture: $S = \sqrt{d_{90}/d_{10}}$
S	specific entropy (i.e. entropy per unit mass) (J/K/kg): $dS = \left(\dfrac{dq_h}{T}\right)_{rev}$
S_c	critical slope
S_f	friction slope defined as: $S_f = -\dfrac{\partial H}{\partial s}$
S_o	bed slope defined as: $S_o = -\dfrac{\partial z_o}{\partial s} = \sin\theta$
S_t	transition slope for a multi-cell minimum energy loss culvert (Chapter 21)
t	time (s)
t_r	ratio of prototype to model times: $t_r = t_p/t_m$ (Chapter 14)
t_s	sedimentation time scale (s)
T	thermodynamic (or absolute) temperature (K)
T_o	reference temperature (K)
TWRL	tailwater rating level (m RL)
U	(1) volume force potential (m^2/s^2) (Chapter 2)
	(2) wave celerity (m/s) for an observed standing on the bank (Chapter 17)
V	flow velocity (m/s)
V	(local) velocity (m/s)
$\vec{V}$	velocity vector; in Cartesian co-ordinates the velocity vector equals: $\vec{V} = (V_x, V_y, V_z)$
V_c	critical flow velocity (m/s)
V_H	characteristic velocity (m/s) defined in terms of the dam height (Chapter 17)
V_{mean}	mean flow velocity (m/s): $V_{mean} = Q/A$
V_{max}	maximum velocity (m/s) in a cross-section; in fully developed open channel flow, the velocity is maximum near the free-surface
V_r	ratio of prototype to model velocities: $V_r = V_p/V_m$ (Chapter 14)
V_s	average speed (m/s) of sediment motion

V_{srg}	surge velocity (m/s) as seen by an observer immobile on the channel bank
V_o	(1) uniform equilibrium flow velocity (m/s)
	(2) initial flow velocity (m/s) (Chapter 17)
V'	depth-averaged velocity (m/s): $V' = \dfrac{1}{d}\displaystyle\int_0^d V\,dy$
V_*	shear velocity (m/s) defined as: $V_* = \sqrt{\dfrac{\tau_o}{\rho}}$
Vol	volume (m³)
v_s	particle volume (m³)
w_o	(1) particle settling velocity (m/s)
	(2) fall velocity (m/s) of a single particle in a fluid at rest
w_s	settling velocity (m/s) of a suspension
W_s	work done by shear stress per unit mass (J/kg)
W	(1) channel bottom width (m)
	(2) channel width (m) at a distance y from the invert
We	Weber number
W_p	work (J) done by the pressure force
W_s	work (J) done by the system by shear stress (i.e. torque exerted on a rotating shaft)
W_t	total work (J) done by the system
Wa	Coles wake function
X	horizontal co-ordinate (m) measured from spillway crest (Chapter 19)
X_M	empirical coefficient in the Muskingum method
X_r	ratio of prototype to model horizontal distances (Chapter 14)
x	Cartesian co-ordinate (m)
$\vec{x}$	Cartesian co-ordinate vector: $\vec{x} = (x,y,z)$
x_s	dam break wave front location (m) (Chapter 17)
Y	(1) vertical co-ordinate (m) measured from spillway crest (Chapter 19)
	(2) free-surface elevation (m): $Y = z_o + d$ (Chapters 16 & 17)
y	(1) distance (m) measured normal to the flow direction
	(2) distance (m) measured normal to the channel bottom
	(3) Cartesian co-ordinate (m)
$y_{channel}$	channel height (m)
y_s	characteristic distance (m) from channel bed
Z_r	ratio of prototype to model vertical distances (Chapter 14)
z	(1) altitude or elevation (m) measured positive upwards
	(2) Cartesian co-ordinate (m)
z_{apron}	apron invert elevation (m)
z_{crest}	spillway crest elevation (m)
z_o	(1) reference elevation (m)
	(2) bed elevation (m).

Greek symbols

α	Coriolis coefficient or kinetic energy correction coefficient
β	momentum correction coefficient (i.e. Boussinesq coefficient)
δ	(1) sidewall slope
	(2) boundary layer thickness (m)
δ_{ij}	identity matrix element

δ_s	bed-load layer thickness (m)
δt	small time increment (s)
δx	small distance increment (m) along the x-direction
Δ	angle between the characteristics and the x-axis (Chapters 16 and 17)
Δd	change in flow depth (m)
ΔE	change in specific energy (m)
ΔH	head loss (m): i.e. change in total head
ΔP	pressure difference (Pa)
Δq_s	change in sediment transport rate (m^2/s)
Δs	small distance (m) along the flow direction
Δt	small time change (s)
ΔV	change in flow velocity (m/s)
Δx	small distance (m) along the x-direction
Δz_o	change in bed elevation (m)
Δz_o	weir height (m) above natural bed level
ε_{ij}	velocity gradient element (m/s^2)
ϕ	(1) sedimentological size parameter
	(2) velocity potential (m^2/s)
ϕ_s	angle of repose
φ	stream function (m^2/s)
γ	specific heat ratio: $\gamma = C_p/C_v$
μ	dynamic viscosity (Pa s)
ν	kinematic viscosity (m^2/s): $\nu = \mu/\rho$
Π	wake parameter
π	$\pi = 3.141\,592\,653\,589\,793\,238\,462\,643$
θ	channel slope
ρ	density (kg/m^3)
ρ_r	density (kg/m^3)
ρ_s	ratio of prototype to model densities (Chapter 14)
ρ_{sed}	sediment mixture density (kg/m^3)
σ	surface tension (N/m)
σ_e	effective stress (Pa)
σ_{ij}	stress tensor element (Pa)
σ_g	geometric standard deviation of sediment size distribution: $\sigma_g = \sqrt{d_{84}/d_{16}}$
τ	shear stress (Pa)
τ_{ij}	shear stress component (Pa) of the i-momentum transport in the j-direction
τ_o	average boundary shear stress (Pa)
$(\tau_o)_c$	critical shear stress (Pa) for onset of sediment motion
τ'_o	skin friction shear stress (Pa)
τ''_o	bed-form shear stress (Pa)
τ_1	yield stress (Pa)
τ_*	Shields parameter: $\tau_* = \dfrac{\tau_o}{\rho g(\rho_s/\rho - 1)d_s}$
$(\tau_*)_c$	critical Shields parameter for onset of sediment motion

Subscript

air	air
bl	bed-load

c	critical flow conditions
conj	conjugate flow property
des	design flow conditions
dry	dry conditions
exit	exit flow condition
i	characteristics of section $\{i\}$ (in the numerical integration process) (Chapter 15)
inlet	inlet flow condition
m	model
mean	mean flow property over the cross-sectional area
mixt	sediment–water mixture
model	model conditions
o	(1) uniform equilibrium flow conditions
	(2) initial flow conditions (Chapter 17)
outlet	outlet flow condition
p	prototype
prototype	prototype conditions
r	ratio of prototype to model characteristics
s	(1) component in the s-direction
	(2) sediment motion
	(3) sediment particle property
sl	suspended load
t	flow condition at time t
tw	tailwater flow condition
w	water
wet	wet conditions
x	x-component
y	y-component
z	z-component
1	upstream flow conditions
2	downstream flow conditions

Abbreviations

CS	control surface
CV	control volume
D/S	downstream
GVF	gradually varied flow
Hg	mercury
RVF	rapidly varied flow
SI	Système international d'unités (International System of Units)
THL	total head line
U/S	upstream

Reminder

1. At 20°C, the density and dynamic viscosity of water (at atmospheric pressure) are: $\rho_w = 998.2\,\text{kg/m}^3$ and $\mu_w = 1.005 \times 10^{-3}\,\text{Pa s}$.
2. Water at atmospheric pressure and 20.2°C has a kinematic viscosity of exactly $10^{-6}\,\text{m}^2/\text{s}$.

3. Water in contact with air has a surface tension of about 0.0733 N/m at 20°C.
4. At 20°C and atmospheric pressure, the density and dynamic viscosity of air are about 1.2 kg/m³ and 1.8×10^{-5} Pa s, respectively.

Dimensionless numbers

Ca   Cauchy number: $Ca = \dfrac{\rho V^2}{E_b}$

        Note: the Sarrau–Mach number equals: $Ma \sim \sqrt{Ca}$

C_d (1) drag coefficient for bottom friction (i.e. friction drag):

$$C_d = \frac{\tau_o}{\frac{1}{2}\rho V^2} = \frac{\text{shear stress}}{\text{dynamic pressure}}$$

        Note: another notation is C_f (e.g. Comolet, 1976)
        (2) drag coefficient for a structural shape (i.e. form drag):

$$C_d = \frac{F_d}{\frac{1}{2}\rho V^2 A} = \frac{\text{drag force per unit cross-sectional area}}{\text{dynamic pressure}}$$

        where A is the projection of the structural shape (i.e. body) in the plane normal to the flow direction.

Eu   Euler number defined as: $Eu = \dfrac{V}{\sqrt{\Delta P / \rho}}$

Fr   Froude number defined as: $Fr = \dfrac{V}{\sqrt{g d_{charac}}} \propto \dfrac{\sqrt{\text{inertial force}}}{\text{weight}}$

        Note: some authors use the notation:

$$F_r = \frac{V^2}{g d_{charac}} = \frac{\rho V^2 A}{\rho g A d_{charac}} \propto \frac{\text{inertial force}}{\text{weight}}$$

Ma   Sarrau–Mach number: $Ma = \dfrac{V}{C}$

Mo   Morton number defined as: $Mo = \dfrac{g \mu_w^4}{\rho_w \sigma^3}$

        The Morton number is a function only of fluid properties and gravity constant. If the same fluids (air and water) are used in both model and prototype, Mo may replace the Weber, Reynolds or Froude number as: $Mo = \dfrac{We^3}{Fr^2 Re^4}$

Nu   Nusselt number: $Nu = \dfrac{H d_{charac}}{\lambda} \propto \dfrac{\text{heat transfer by convection}}{\text{heat transfer by conduction}}$

        where H is the heat transfer coefficient $(\text{W m}^{-2}\text{K}^{-1})$ and λ is the thermal conductivity $(\text{W m}^{-1}\text{K}^{-1})$

Re   Reynolds number: $Re = \dfrac{V d_{charac}}{\nu} \propto \dfrac{\text{inertial forces}}{\text{viscous forces}}$

$Re*$ shear Reynolds number: $Re_* = \dfrac{V_* k_s}{\nu}$

We Weber number: $We = \dfrac{V}{\sqrt{\sigma / \rho d_{charac}}} \quad \propto \quad \sqrt{\dfrac{\text{inertial force}}{\text{surface tension force}}}$

     Note: some authors use the notation: $We = \dfrac{V^2}{\sigma / \rho d_{charac}} \quad \propto \quad \dfrac{\text{inertial force}}{\text{surface tension force}}$

τ_* Shields parameter characterizing the onset of sediment motion:

$$\tau_* = \frac{\tau_o}{\rho g \left(\rho_s / \rho - 1 \right) d_s} \quad \propto \quad \frac{\text{destabilizing force moment}}{\text{stabilizing moment of weight force}}$$

Notes

The variable d_{charac} is the characteristic geometric length of the flow field: e.g. pipe diameter, flow depth, sphere diameter. Some examples are listed below:

Flow	d_{charac}	Comments
Circular pipe flow	D	Pipe diameter
Flow in pipe of irregular cross-section	D_H	Hydraulic diameter
Flow resistance in open channel flow	D_H	Hydraulic diameter
Wave celerity in open channel flow	d	Flow depth
Flow past a cylinder	D	Cylinder diameter

Part 1 Basic Principles of Open Channel Flows

Introduction

1

Summary

The introduction chapter reviews briefly the basic fluid properties and some important results for fluids at rest. Then the concept of open channel flow is defined and some applications are described.

1.1 PRESENTATION

The term 'hydraulics' is related to the application of the Fluid Mechanics principles to water engineering structures, civil and environmental engineering facilities, especially hydraulic structures (e.g. canal, river, dam, reservoir and water treatment plant).

In the book, we consider open channels in which liquid (i.e. water) flows with a free surface. Examples of open channels are natural streams and rivers. Man-made channels include irrigation and navigation canals, drainage ditches, sewer and culvert pipes running partially full, and spillways.

The primary factor in open channel flow analysis is the location of the free surface, which is unknown beforehand (i.e. *a priori*). The free surface rises and falls in response to perturbations to the flow (e.g. changes in channel slope or width). The main parameters of a hydraulic study are the geometry of the channel (e.g. width, slope and roughness), the properties of the flowing fluid (e.g. density and viscosity) and the flow parameters (e.g. velocity and flow depth).

1.2 FLUID PROPERTIES

The *density* ρ of a fluid is defined as its mass per unit volume. All real fluids resist any force tending to cause one layer to move over another, but this resistance is offered only while the movement is taking place. The resistance to the movement of one layer of fluid over an adjoining one is referred to as the *viscosity* of the fluid. Newton's law of viscosity postulates that, for the straight parallel motion of a given fluid, the tangential stress between two adjacent layers is proportional to the velocity gradient in a direction perpendicular to the layers:

$$\tau = \mu \frac{dv}{dy} \qquad (1.1)$$

where τ is the shear stress between adjacent fluid layers, μ is the dynamic viscosity of the fluid, v is the velocity and y is the direction perpendicular to the fluid motion. Fluids that obey Newton's law of viscosity are called *Newtonian fluids*.

At the interface between a liquid and a gas, a liquid and a solid, or two immiscible liquids, a tensile force is exerted at the surface of the liquid and tends to reduce the area of this surface to the greatest possible extent. The *surface tension* is the stretching force required to form the film: i.e. the tensile force per unit length of the film in equilibrium.

The basic properties of air and water are detailed in Appendix A1.1.

Notes

1. Isaac Newton (1642–1727) was an English mathematician.
2. The kinematic viscosity is the ratio of viscosity to mass density:

$$v = \frac{\mu}{\rho}$$

3. A Newtonian fluid is one in which the shear stress, in one-directional flow, is proportional to the rate of deformation as measured by the velocity gradient across the flow (i.e. equation (1.1)). The common fluids such as air, water and light petroleum oils, are Newtonian fluids. Non-Newtonian fluids will not be considered any further in Part I. They will be briefly mentioned in Part II (i.e. hyperconcentrated flows).

Application

At atmospheric pressure and 20°C the density and dynamic viscosity of water are:

$$\rho_w = 998.2 \, kg/m^3$$

$$\mu_w = 1.005 \times 10^{-3} \, Pa\,s$$

and the density of air is around:

$$\rho_{air} = 1.2 \, kg/m^3$$

Water in contact with air has a surface tension of about 0.0733 N/m at 20°C.

Considering a spherical gas bubble (diameter d_{ab}) in a liquid, the increase of gas pressure required to balance the tensile force caused by surface tension equals: $\Delta P = 4\sigma/d_{ab}$.

1.3 STATIC FLUIDS

Considering a fluid at rest (Fig. 1.1), the pressure at any point within the fluid follows Pascal's law. For any small control volume, there is no shear stress acting on the control surface. The only forces acting on the control volume of fluid are the gravity and the pressure[1] forces.

In a static fluid, the pressure at one point in the fluid has an unique value, independent of the direction. This is called Pascal's law. The pressure variation in a static fluid follows:

$$\frac{dP}{dz} = -\rho g \qquad (1.2)$$

where P is the pressure, z is the vertical elevation positive upwards, ρ is the fluid density and g is the gravity constant (see Appendix A1.1).

---

[1] By definition, the pressure acts always normal to a surface. That is, the pressure force has no component tangential to the surface.

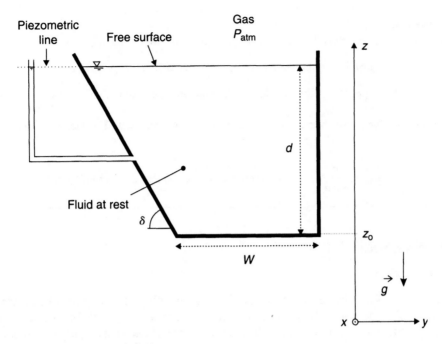

Fig. 1.1 Pressure variation in a static fluid.

For a body of fluid at rest with a free surface (e.g. a lake) and with a constant density, the pressure variation equals:

$$P(x, y, z) = P_{atm} - \rho g(z - (z_o + d)) \tag{1.3}$$

where P_{atm} is the atmospheric pressure (i.e. air pressure above the free surface), z_o is the elevation of the reservoir bottom and d is the reservoir depth (Fig. 1.1). $(d + z_o)$ is the free-surface elevation. Equation (1.3) implies that the pressure is independent of the horizontal co-ordinates (x, y). The term $\{-\rho g(z - (z_o + d))\}$ is positive within the liquid at rest. It is called the hydrostatic pressure.

The pressure force acting on a surface of finite area which is in contact with the fluid is distributed over the surface. The resultant force is obtained by integration:

$$F_p = \int P \, dA \tag{1.4}$$

where A is the surface area.

Note

Blaise Pascal (1623–1662) was a French mathematician, physicist and philosopher. He developed the modern theory of probability. He also formulated the concept of pressure (between 1646 and 1648) and showed that the pressure in a fluid is transmitted through the fluid in all directions (i.e. Pascal's law).

Application

In Fig. 1.1, the pressure force (per unit width) applied on the sides of the tank are:

$$F_p = \rho g d W \qquad \text{Pressure force acting on the bottom per unit width}$$

$$F_p = \frac{1}{2}\rho g d^2 \qquad \text{Pressure force acting on the right-wall per unit width}$$

For the left-wall, the pressure force acts in the direction normal to the wall. The integration of equation (1.4) yields:

$$F_p = \frac{1}{2}\rho g \frac{d^2}{\sin\delta} \qquad \text{Pressure force acting on the left-wall per unit width}$$

1.4 OPEN CHANNEL FLOW

1.4.1 Definition

An open channel is a waterway, canal or conduit in which a liquid flows with a free surface. An open channel flow describes the fluid motion in open channel (Fig. 1.2). In most applications, the liquid is water and the air above the flow is usually at rest and at standard atmospheric pressure (see Appendix A1.1).

Notes

1. In some practical cases (e.g. a closed conduit flowing partly full), the pressure of the air above the flow might become sub-atmospheric.
2. Next to the free surface of an open channel flow, some air is entrained by friction at the free surface. That is, the no-slip condition at the air–water interface induces the air motion. The term 'air boundary layer' is sometimes used to describe the atmospheric region where air is entrained through momentum transfer at the free surface.
3. In a clear-water open channel flow, the free surface is clearly defined: it is the interface between the water and the air. For an air–water mixture flow (called 'white waters'), the definition of the free surface (i.e. the interface between the flowing mixture and the surrounding atmosphere) becomes somewhat complicated (e.g. Wood, 1991; Chanson, 1997).

1.4.2 Applications

Open channel flows are found in Nature as well as in man-made structures (Fig. 1.3, Plates 1 to 32). In Nature, tranquil flows are observed in large rivers near their estuaries: e.g. the Nile River between Alexandria and Cairo, the Brisbane River in Brisbane. Rushing waters are encountered in mountain rivers, river rapids and torrents. Classical examples include the cataracts of the Nile River, the Zambesi rapids in Africa and the Rhine waterfalls.

Man-made open channels can be water-supply channels for irrigation, power supply and drinking waters, conveyor channel in water treatment plants, storm waterways, some public fountains, culverts below roads and railways lines.

Open channel flows are observed in small-scale as well as large-scale situations. For example, the flow depth can be between few centimetres in water treatment plants and over 10 m in large

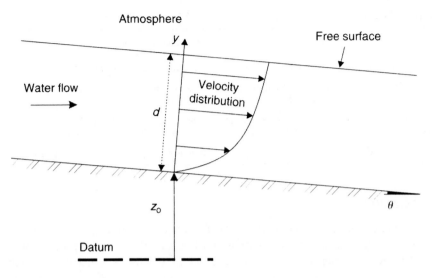

Fig. 1.2 Sketch of open channel flow.

Fig. 1.3 Wisconsin River and sand bars in Aug, 1966 (Courtesy of Dr Lou Maher) – looking upstream.

rivers. The mean flow velocity may range from less than 0.01 m/s in tranquil waters to above 50 m/s in high-head spillway. The range of total discharges[2] may extend from $Q \sim 0.001$ l/s in chemical plants to $Q > 10\,000\,\mathrm{m}^3/\mathrm{s}$ in large rivers or spillways. In each flow situation, however, the location of the free surface is unknown beforehand and it is determined by applying the continuity and momentum principles.

---

[2] In hydraulics of open channels, the water flow is assumed incompressible and the volume discharge is commonly used.

Table 1.1 Basic differences between pipe flow and open channel flow of an incompressible fluid

	Pipe flow	Open channel flow
Flow driven by	Pressure work	Gravity (i.e. potential energy)
Flow cross-section	Known (fixed by pipe geometry)	Unknown in advance because the flow depth is unknown beforehand
Characteristic flow parameters	Velocity deduced from continuity equation	Flow depth and velocity deduced by solving simultaneously the continuity and momentum equations
Specific boundary conditions		Atmospheric pressure at the flow free surface

1.4.3 Discussion

There are characteristic differences between open channel flow and pipe flow (Table 1.1). In an open channel, the flow is driven by gravity in most cases rather than by pressure work as with pipe flow. Another dominant feature of open channel flow is the presence of a free surface:

- the position of the free surface is unknown 'in advance',
- its location must be deduced by solving simultaneously the continuity and momentum equations, and
- the pressure at the free surface is atmospheric.

1.5 EXERCISES

Give the values (and units) of the specified fluid and physical properties:

(a) Density of water at atmospheric pressure and 20°C.
(b) Density of air at atmospheric pressure and 20°C.
(c) Dynamic viscosity of water at atmospheric pressure and 20°C.
(d) Kinematic viscosity of water at atmospheric pressure and 20°C.
(e) Kinematic viscosity of air at atmospheric pressure and 20°C.
(f) Surface tension of air and water at atmospheric pressure and 20°C.
(g) Acceleration of gravity in Brisbane.

What is the Newton's law of viscosity?

In a static fluid, express the pressure variation with depth.

Considering a spherical air bubble (diameter d_{ab}) submerged in water with hydrostatic pressure distribution:

(a) Will the bubble rise or drop?
(b) Is the pressure inside the bubble greater or smaller than the surrounding atmospheric pressure?
(c) What is the magnitude of the pressure difference (between inside and outside the bubble)?

Note: the last question requires some basic calculation.

Fundamental equations

Summary

In this chapter, the fundamental equations are developed and applied to open channel flows. It is shown that the momentum equation leads to the Bernoulli equation.

2.1 INTRODUCTION

In open channel flow the free surface is always at a constant absolute pressure (usually atmospheric) and the driving force of the fluid motion is gravity. In most practical situations, open channels contain waters. The general principles of open channel flow calculations developed in this chapter are, however, applicable to other liquids. Specific results (e.g. flow resistance) are based primarily upon experimental data obtained mostly with water.

2.2 THE FUNDAMENTAL EQUATIONS

2.2.1 Introduction

The law of conservation of mass states that the mass within a closed system remains constant with time (disregarding relativity effects):

$$\frac{DM}{Dt} = 0 \qquad (2.1)$$

where M is the total mass and D/Dt is the absolute differential (see Appendix A1.3, section on Differential and differentiation). Equation (2.1) leads to the continuity equation.

The expression of Newton's second law of motion for a system is:

$$\sum \vec{F} = \frac{D}{Dt}\left(M \times \vec{V}\right) \qquad (2.2)$$

where $\sum \vec{F}$ refers to the resultant of all external forces acting on the system including body forces such as gravity, and $\vec{V}$ is the velocity of the centre of mass of the system. The application of equation (2.2) is called the motion equation.

Equations (2.1) and (2.2) must be applied to a control volume. A *control volume* is a specific region of space selected for analysis. It refers to a region in space where fluid enters and leaves (e.g. Fig. 2.1). The boundary of a control volume is its *control surface*. The concept of control volume used in conjunction with the differential form of the continuity, momentum and energy equations, is an open system.

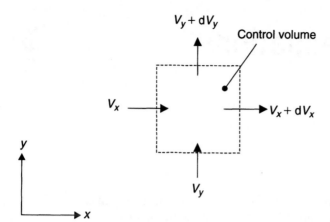

Fig. 2.1 Sketch of control volume for a two-dimensional flow.

All flow situations including open channel flows are subject to the following relationships (e.g. Streeter and Wylie, 1981: pp. 88–89):

1. the first and second laws of thermodynamics,
2. the law of conservation of mass (i.e. the continuity relationship),
3. Newton's law of motion,
4. the boundary conditions.

Other relations (such as a state equation, Newton's law of viscosity) may apply.

Note

A control volume may be either infinitesimally small or finite. It may move or remain fixed in space. It is an *imaginary* volume and *does not interfere* with the flow.

2.2.2 The continuity equation

The principle of conservation of mass states that the mass within a closed system remains constant with time:

$$\frac{DM}{Dt} = \frac{D}{Dt} \int_x \int_y \int_z \rho \, dx \, dy \, dz = 0 \tag{2.3}$$

where ρ is the fluid density and (x, y, z) are three components of a Cartesian system of co-ordinates.

For an infinitesimal small control volume the continuity equation is:

$$\frac{\partial \rho}{\partial t} + \mathrm{div}(\rho \vec{V}) = 0 \tag{2.4a}$$

and in Cartesian co-ordinates:

$$\frac{\partial \rho}{\partial t} + \frac{\partial(\rho V_x)}{\partial x} + \frac{\partial(\rho V_y)}{\partial y} + \frac{\partial(\rho V_z)}{\partial z} = 0 \tag{2.4b}$$

where V_x, V_y and V_z are the velocity components in the x-, y- and z-directions, respectively.

For an incompressible flow (i.e. ρ = constant) the *continuity equation* becomes:

$$\mathrm{div}\ \vec{V} = 0 \tag{2.5a}$$

and in Cartesian co-ordinates:

$$\frac{\partial V_x}{\partial x} + \frac{\partial V_y}{\partial y} + \frac{\partial V_z}{\partial z} = 0 \tag{2.5b}$$

For an incompressible fluid (e.g. open channel flow), the inflow (i.e. amount of fluid entering into a control volume) equals the outflow.

Application

Considering an open channel flow with no flow across the side and bottom walls, equation (2.5) can be integrated between two cross-sections of areas A_1 and A_2. It yields:

$$Q = \int_{A_1} V\ \mathrm{d}A = \int_{A_2} V\ \mathrm{d}A$$

where Q is the total discharge (i.e. volume discharge). Integration of the equation leads to:

$$Q = V_1 A_1 = V_2 A_2$$

where V_1 and V_2 are the mean velocities across the cross-sections A_1 and A_2 respectively.

2.2.3 The momentum equation

The Navier–Stokes equation

Newton's law of motion is used as a basis for developing the momentum equation for a control volume:

$$\sum \vec{F} = \frac{\mathrm{D}}{\mathrm{D}t}\left(M \times \vec{V}\right) = \frac{\partial}{\partial t}\left(\int_{\mathrm{CV}} \rho \vec{V}\ \mathrm{d}Volume\right) + \int_{\mathrm{CS}} \rho \vec{V}\vec{V}\ \mathrm{d}Area \tag{2.6}$$

where CV and CS refer to the control volume and control surface, respectively. Basically, it states that the change of momentum equals the sum of all forces applied to the control volume.

The forces acting on the control volume are (1) the surface forces (i.e. pressure and shear forces) acting on the control surface and (2) the volume force (i.e. gravity) applied at the centre of mass of the control volume. For an infinitesimal small volume the momentum equation applied to the *i*-component of the vector equation is:

$$\frac{\mathrm{D}(\rho V_i)}{\mathrm{D}t} = \left(\frac{\partial(\rho V_i)}{\partial t} + \sum_j V_j \frac{\partial(\rho V_i)}{\partial x_j}\right) = \rho F_{\mathrm{vol}\,i} + \sum_j \frac{\partial \sigma_{ij}}{\partial x_j} \tag{2.7}$$

where $\mathrm{D}/\mathrm{D}t$ is the absolute differential (or absolute derivative, see Appendix A1.3), V_i is the velocity component in the *i*-direction, F_{vol} is the resultant of the volume forces (per unit volume) and σ_{ij} is the stress tensor (see notes below). The subscripts *i* and *j* refer to the Cartesian co-ordinate components (e.g. *x*, *y*).

If the volume forces $\vec{F}_{vol}$ derives from a potential U, they can be rewritten as: $\vec{F}_{vol} = -\overrightarrow{grad}\,U$ (e.g. gravity force $\vec{F}_{vol} = -\overrightarrow{grad}\,(gz)$). Further, for a Newtonian fluid the stress forces are (1) the pressure forces and (2) the resultant of the viscous forces on the control volume. Hence for a Newtonian fluid and for volume force deriving from a potential, the momentum equation becomes:

$$\frac{D\left(\rho \times \vec{V}\right)}{Dt} = \rho\vec{F}_{vol} - \overrightarrow{grad}\,P + \vec{F}_{visc} \tag{2.8}$$

where P is the pressure and $\vec{F}_{visc}$ is the resultant of the viscous forces (per unit volume) on the control volume.

In Cartesian co-ordinates (x, y, z):

$$\left(\frac{\partial(\rho V_x)}{\partial t} + \sum_j V_j \frac{\partial(\rho V_x)}{\partial x_j}\right) = \rho F_{vol_x} - \frac{\partial P}{\partial x} + F_{visc_x} \tag{2.9a}$$

$$\left(\frac{\partial(\rho V_y)}{\partial t} + \sum_j V_j \frac{\partial(\rho V_y)}{\partial x_j}\right) = \rho F_{vol_y} - \frac{\partial P}{\partial y} + F_{visc_y} \tag{2.9b}$$

$$\left(\frac{\partial(\rho V_z)}{\partial t} + \sum_j V_j \frac{\partial(\rho V_z)}{\partial x_j}\right) = \rho F_{vol_z} - \frac{\partial P}{\partial z} + F_{visc_z} \tag{2.9c}$$

where the subscript j refers to the Cartesian co-ordinate components (i.e. $j = x, y, z$). In equation (2.9), the term on the left side is the sum of the momentum accumulation $\partial(\rho V)/\partial t$ plus the momentum flux $V\partial(\rho V)/\partial x$. The left term is the sum of the forces acting on the control volume: body force (or volume force) acting on the mass as a whole and surface forces acting at the control surface.

For an incompressible flow (i.e. $\rho = $ constant), for a Newtonian fluid and assuming that the viscosity is constant over the control volume, the motion equation becomes:

$$\rho\left(\frac{\partial V_x}{\partial t} + \sum_j V_j \frac{\partial V_x}{\partial x_j}\right) = \rho F_{vol_x} - \frac{\partial P}{\partial x} + F_{visc_x} \tag{2.10a}$$

$$\rho\left(\frac{\partial V_y}{\partial t} + \sum_j V_j \frac{\partial V_y}{\partial x_j}\right) = \rho F_{vol_y} - \frac{\partial P}{\partial y} + F_{visc_y} \tag{2.10b}$$

$$\rho\left(\frac{\partial V_z}{\partial t} + \sum_j V_j \frac{\partial V_z}{\partial x_j}\right) = \rho F_{vol_z} - \frac{\partial P}{\partial z} + F_{visc_z} \tag{2.10c}$$

where ρ, the fluid density, is assumed constant in time and space. Equation (2.10) is often called the *Navier–Stokes equation*.

Considering a *two-dimensional flow* in the (x, y) plane and for *gravity forces*, the Navier–Stokes equation becomes:

$$\rho\left(\frac{\partial V_x}{\partial t} + V_x\frac{\partial V_x}{\partial x} + V_y\frac{\partial V_x}{\partial y}\right) = -\rho g\frac{\partial z}{\partial x} - \frac{\partial P}{\partial x} + F_{visc_x} \tag{2.11a}$$

$$\rho\left(\frac{\partial V_y}{\partial t} + V_x\frac{\partial V_y}{\partial x} + V_y\frac{\partial V_y}{\partial y}\right) = -\rho g\frac{\partial z}{\partial y} - \frac{\partial P}{\partial y} + F_{\text{visc}_y} \qquad (2.11b)$$

where z is aligned along the vertical direction and positive upward. Note that the x- and y-directions are perpendicular to each other and they are independent of (and not necessarily orthogonal to) the vertical direction.

Notes

1. For gravity force the volume force potential U is:

$$U = \vec{g} \times \vec{x}$$

where $\vec{g}$ is the gravity acceleration vector and $\vec{x} = (x, y, z)$, z being the vertical direction positive upward. It yields that the gravity force vector equals:

$$\vec{F}_{\text{vol}} = -\overrightarrow{\text{grad}}(gz)$$

2. A Newtonian fluid is characterized by a linear relation between the magnitude of shear stress τ and the rate of deformation $\partial V/\partial y$ (equation (1.1)), and the stress tensor is:

$$\sigma_{ij} = -P\delta_{ij} + \tau_{ij}$$

$$\tau_{ij} = -\frac{2\mu}{3}\varepsilon\delta_{ij} + 2\mu\varepsilon_{ij}$$

where P is the static pressure, τ_{ij} is the shear stress component of the i-momentum transported in the j-direction, δ_{ij} is the identity matrix element: $\delta_{ii} = 1$ and $\delta_{ij} = 0$ (for i different from j),

$$\varepsilon_{ij} = \frac{1}{2}\left(\frac{\partial V_i}{\partial x_j} + \frac{\partial V_j}{\partial x_i}\right) \quad \text{and} \quad \varepsilon = \text{div}\,\vec{V} = \sum_i \frac{\partial V_i}{\partial x_i}$$

3. The vector of the viscous forces is:

$$F_{\text{visc}_i} = \text{div}\,\tau_i = \sum_j \frac{\partial \tau_{ij}}{\partial x_j}$$

For an incompressible flow the continuity equation gives: $\varepsilon = \text{div}\,\vec{V} = 0$. And the viscous force per unit volume becomes:

$$F_{\text{visc}_i} = \sum_j \mu\frac{\partial^2 V_i}{\partial x_j\,\partial x_j}$$

where μ is the dynamic viscosity of the fluid. Substituting this into equation (2.11), yields:

$$\rho\left(\frac{\partial V_x}{\partial t} + V_x\frac{\partial V_x}{\partial x} + V_y\frac{\partial V_x}{\partial y}\right) = -\rho g\frac{\partial z}{\partial x} - \frac{\partial P}{\partial x} + \sum_j \mu\frac{\partial^2 V_x}{\partial x_j\,\partial x_j} \qquad (2.12a)$$

$$\rho\left(\frac{\partial V_y}{\partial t} + V_x\frac{\partial V_y}{\partial x} + V_y\frac{\partial V_y}{\partial y}\right) = -\rho g\frac{\partial z}{\partial y} - \frac{\partial P}{\partial y} + \sum_j \mu\frac{\partial^2 V_y}{\partial x_j\,\partial x_j} \qquad (2.12b)$$

Equation (2.12) is the *original* Navier–Stokes equation.

4. Equation (2.12) was first derived by Navier in 1822 and Poisson in 1829 by an entirely different method. Equations (2.10)–(2.12) were derived later in a manner similar as above by Barré de Saint-Venant in 1843 and Stokes in 1845.

5. Louis Navier (1785–1835) was a French engineer who not only primarily designed bridge but also extended Euler's equations of motion. Siméon Denis Poisson (1781–1840) was a French mathematician and scientist. He developed the theory of elasticity, a theory of electricity and a theory of magnetism. Adhémar Jean Claude Barré de Saint-Venant (1797–1886), French engineer, developed the equations of motion of a fluid particle in terms of the shear and normal forces exerted on it. George Gabriel Stokes (1819–1903), British mathematician and physicist, is known for his research in hydrodynamics and a study of elasticity.

Application

Considering an open channel flow in a rectangular channel (Fig. 2.2), we assume a one-dimensional flow, with uniform velocity distribution, a constant channel slope θ and a constant channel width B. The Navier–Stokes equation in the s-direction is:

$$\rho\left(\frac{\partial V}{\partial t} + V\frac{\partial V}{\partial s}\right) = -\rho g\frac{\partial z}{\partial s} - \frac{\partial P}{\partial s} + F_{\text{visc}}$$

where V is the velocity along a streamline. Integrating the Navier–Stokes equation over the control volume (Fig. 2.2), the forces acting on the control volume shown in Fig. 2.2 in the s-direction are:

$$\int_{\text{CV}} -\rho g\frac{\mathrm{d}z}{\mathrm{d}s} = +\rho g A\,\Delta s\,\sin\theta \qquad \text{Volume force (i.e. weight)}$$

$$\int_{\text{CV}} -\frac{\mathrm{d}P}{\mathrm{d}s} = -\rho g d\,\Delta dB\cos\theta \qquad \text{Pressure force (assuming hydrostatic pressure distribution)}$$

$$\int_{\text{CV}} F_{\text{vis}} = -\tau_0 P_{\text{w}}\,\Delta s \qquad \text{Friction force (i.e. boundary shear)}$$

where A is the cross-sectional area (i.e. $A = Bd$ for a rectangular channel), d is the flow depth, Δs is the length of the control volume, τ_0 is the average bottom shear stress and P_{w} is the wetted perimeter.

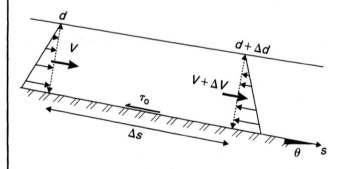

Channel cross-sections

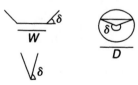

Fig. 2.2 Control volume for an open channel flow.

Assuming a *steady* flow, the change in momentum equals:

$$\int_{CV} \rho v \frac{dv}{ds} = \rho A \, \Delta s V \frac{\Delta V}{\Delta s}$$

where V is the mean flow velocity. The term (ρV) is the momentum per unit volume.

The integration of the Navier–Stokes equation for a one-dimensional steady open channel flow yields:

$$\rho A V \, \Delta V = +\rho g A \, \Delta s \sin \theta - \rho g d \, \Delta dB \cos \theta - \tau_o P_w \, \Delta s$$

The term on the left is the gradient of momentum flux: $\Delta[(1/2)(\rho V)(VA)]$ (i.e. the rate of change of momentum). On the right side of the equation, $(+\rho g A \sin \theta \Delta s)$ is the gravity force (potential energy), $(-\rho g d \, \Delta dB \cos \theta)$ is the pressure term (work of the flow) and $(-\tau_o P_w \Delta s)$ is the friction force (losses).

Application of the momentum equation

Considering an arbitrary control volume, it is advantageous to select a volume with a control surface perpendicular to the flow direction denoted s. For a steady and incompressible flow the forces acting on the control volume in the s-direction are equal to the rate of change in the flow momentum (i.e. no momentum accumulation). The momentum equation gives:

$$\sum F_s = \rho_2 A_2 V_2 V_{s2} - \rho_1 A_1 V_1 V_{s1} \qquad (2.13a)$$

where $\sum F_s$ is the resultant of all the forces in the s-direction, the subscripts 1 and 2 refer to the upstream and downstream cross-sections, respectively, and $(V_{si})_{i=1,2}$ is the velocity component in the s-direction. Combining with the continuity equation for a steady and incompressible flow, it yields:

$$\sum F_s = \rho Q(V_{s2} - V_{s1}) \qquad (2.13b)$$

In simple terms, the momentum equation states that the change in momentum flux is equal to the sum of all forces (volume and surface forces) acting on the control volume.

Note
For a steady incompressible flow, the momentum flux equals $(\rho Q V)$.

Application: hydraulic jump

In open channels, the transition from a rapid flow to a slow flow is called a hydraulic jump (Fig. 2.3, Plate 25). The transition occurs suddenly and is characterized by a sudden rise of liquid surface.

The forces acting on the control volume are the hydrostatic pressure forces at each end of the control volume, the gravity force (i.e. the weight of water) and the bottom friction. Considering a horizontal rectangular open channel of constant channel width B, and neglecting the shear stress at the channel bottom, the resultant of the forces acting in the s-direction are the result of hydrostatic pressure at the ends of the control volume. The continuity equation and momentum

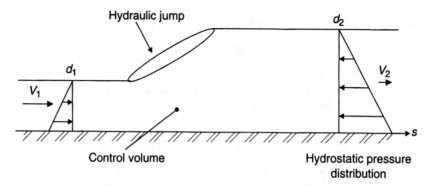

Fig. 2.3 Hydraulic jump in a rectangular channel.

equations are:

$$V_1 d_1 B = V_2 d_2 B \qquad (2.14)$$

$$\rho Q(\beta_2 V_2 - \beta_1 V_1) = \left(\frac{1}{2}\rho g d_1^2 B_1\right) - \left(\frac{1}{2}\rho g d_2^2 B_2\right) \qquad (2.15)$$

where B is the channel width (assumed constant) and Q is the total discharge (i.e. $Q = VdB$).

Notes

If the velocity distribution is not uniform over the cross-section, the average velocity can be used by introducing the momentum correction factor β defined as:

$$\beta = \frac{\int_A \rho v^2 \, dA}{\rho V^2 A}$$

where A is the cross-sectional area (normal to the flow direction) and V is the mean flow velocity ($V = Q/A$).

Let us consider a practical case. For an steady flow in a horizontal open channel of rectangular cross-section, and assuming a hydrostatic pressure distribution, the momentum equation (2.15) becomes:

$$\rho Q(\beta_1 V_2 - \beta_2 V_1) = \left(\frac{1}{2}\rho g d_1^2 B_1\right) - \left(\frac{1}{2}\rho g d_2^2 B_2\right)$$

where Q is the discharge (m³/s), d is the flow depth, B is the channel width and β the momentum correction coefficient. If the shape of the velocity distribution does not change substantially between the cross-sections 1 and 2, the variations of momentum correction coefficient are negligible. The momentum equation becomes:

$$\rho \beta Q(V_2 - V_1) = \left(\frac{1}{2}\rho g d_1^2 B_1\right) - \left(\frac{1}{2}\rho g d_2^2 B_2\right)$$

The Bernoulli equation

The 'local form' of the Bernoulli equation can be deduced from the Navier–Stokes equation.

Considering {H1} the flow is along a *streamline*, assuming that {H2} the fluid is *frictionless* (i.e. $F_{\text{visc}} = 0$), {H3} the volume force *potential* (i.e. *gravity*) *is independent of the time* (i.e. $\partial U/\partial t = 0$); for {H4} a *steady* flow (i.e. $\partial V/\partial t = 0$) and {H5} an *incompressible* flow (i.e. ρ = constant), the Navier–Stokes equation (2.10) along the streamline becomes:

$$\rho V \frac{dV}{ds} = -\rho g \frac{dz}{ds} - \frac{dP}{ds} \tag{2.16}$$

where V is the velocity along the streamline, s is the direction along the streamline. A streamline is defined as an imaginary line that is everywhere tangent to the fluid velocity vector. There is no flow across a streamline and the velocity is aligned in the s-direction. The above equation can be re-arranged as:

$$\rho V \, dV = -\rho g \, dz - dP$$

and it can be rewritten as:

$$\frac{dP}{\rho} + g \, dz + d\left(\frac{V^2}{2}\right) = 0 \tag{2.17}$$

The integration of equation (2.17) along a streamline yields:

$$\frac{P}{\rho} + gz + \frac{V^2}{2} = \text{constant} \tag{2.18}$$

Equation (2.18) is the *Bernoulli equation*. Equation (2.17) is called the differential form of the Bernoulli equation. Each term of the Bernoulli equation may be interpreted by analogy as a form of energy:

1. P/ρ is analogous to the flow work per unit of mass of flowing fluid (net work done by the fluid element on its surroundings while it is flowing),
2. $U = gz$ is similar to the potential energy per unit mass,
3. $V^2/2$ is related to the kinetic energy per unit mass.

If there are no friction losses, the sum of the fluid's potential energy, kinetic energy and pressure work is a constant. Along a streamline the flow 'energy' may be re-arranged between kinetic energy (i.e. velocity), potential energy (i.e. altitude) and pressure work (i.e. flow depth) but the sum of all the terms must remain constant.

Notes
1. The Bernoulli equation is named after the Swiss mathematician Daniel Bernoulli (1700–1782) who developed the equation in his 'Hydrodynamica, de viribus et motibus fluidorum' textbook (first draft in 1733, first publication in 1738, Strasbourg) (Carvill, 1981; Garbrecht, 1987a: pp. 245–258).
2. Under particular conditions each of the assumptions underlying the Bernoulli equation may be abandoned:
   (a) For a gas flow such that the change of pressure is only a small fraction of the absolute pressure (i.e. less than 5%), the gas may be considered incompressible. Equation (2.18) may be applied with an average density ρ.
   (b) For unsteady flow with gradually changing conditions (i.e. slow emptying a reservoir), the Bernoulli equation may be applied without noticeable error.

(c) For real fluid, the Bernoulli equation is used by first neglecting viscous shear to obtain theoretical results. The resulting equation may then be affected by an experimental coefficient to
   correct the analytical solution so that it conforms to the actual physical case.

3. Liggett (1993) developed a superb discussion of the complete integration process of the
   Navier–Stokes equation leading to the Bernoulli equation.

2.2.4 The energy equation

The first law of thermodynamics for a system states that the net energy (e.g. heat and potential
energy) supplied to the system equals the increase in energy of the system plus the energy that
leaves the system as work is done:

$$\frac{DE}{Dt} = \frac{\Delta Q_h}{\Delta t} - \frac{\Delta W_t}{\Delta t} \tag{2.19}$$

where E is the total energy of the system, ΔQ_h is the heat transferred *to* the system and ΔW_t is
the work done *by* the system. The energy of the system is the sum of (1) the potential energy
term gz, (2) the kinetic energy term $v^2/2$ and (3) the internal energy e.

The work done by the system on its surroundings includes the work done by the pressure forces:

$$\Delta W_p = \Delta t \int_{CS} Pv\, dA$$

and the work done by shear forces (i.e. on a rotating shaft) is ΔW_s.

For a steady and one-dimensional flow through a control volume the first law of thermodynamics becomes:

$$\frac{\Delta Q_h}{\Delta t} + \left(\frac{P_1}{\rho_1} + gz_1 + \frac{V_1^2}{2} \right) \rho_1 V_1 A_1 = \frac{\Delta W_s}{\Delta t} + \left(\frac{P_2}{\rho_2} + gz_2 + \frac{V_2^2}{2} \right) \rho_2 V_2 A_2 \tag{2.20}$$

where the subscripts 1 and 2 refer to the upstream and downstream flow conditions, respectively.

Since the flow is steady the conservation of mass implies:

$$\rho_1 V_1 A_1 = \rho_2 V_2 A_2 \tag{2.21}$$

and dividing the first law of thermodynamics by (ρVA) the energy equation in differential form
becomes:

$$dq_h - dw_s = d\left(\frac{P}{\rho} \right) + (g\, dz + v\, dv + de) \tag{2.22}$$

where e is the internal energy per unit mass, q_h is the heat added to the system per unit mass
and w_s is the work done (by shear forces) by the system per unit mass. For a frictionless fluid
(reversible transformation) the first law of thermodynamics may be written in terms of the
entropy S as:

$$de = T\, dS - P\, d\left(\frac{1}{\rho} \right) \tag{2.23}$$

The Clausius inequality states that:

$$dS > \frac{dq_h}{T}$$

Replacing the internal energy by the above equation and calling 'losses' the term $(T\,dS - dq_h)$, the energy equation becomes:

$$\left(\frac{dP}{\rho} + g\,dz + v\,dv \right) + dw_s + d(\text{losses}) = 0 \qquad (2.24)$$

In absence of work of shear forces (i.e. $w_s = 0$) this equation differs from the differentiation of the Bernoulli equation (2.17) by the loss term only.

Notes

1. For non-steady flows the energy equation (for two particular cases) becomes:

   (a) for perfect and frictionless gas without heat added to the system, the energy equation is:

   $$\rho C_p \frac{dT}{dt} = \frac{dP}{dt}$$

   (b) for a incompressible and undilatable fluid, the energy equation is:

   $$\rho C_p \frac{dT}{dt} = \kappa\,\Delta T + \Psi + \Phi$$

   where C_p is the specific heat at constant pressure, κ is the thermal diffusivity, Ψ is the volume density of heat added to the system and Φ is the dissipation rate. The knowledge of the density ρ (from the continuity equation) and the pressure distribution (from the Navier–Stokes equation) enables the calculation of the temperature distribution.

2. If work is done on the fluid in the control volume (i.e. pump) the work done by shear forces w_s is negative.

2.3 EXERCISES

Considering a circular pipe (diameter 2.2 m), the total flow rate is 1600 kg/s. The fluid is slurry (density 1360 kg/m³). What is the mean flow velocity?

Water flows in a trapezoidal open channel (1V:3H sideslopes, 1 m bottom width) with a 1.2 m/s mean velocity and a 0.9 m water depth (on the channel centreline). (a) Calculate the mass flow rate. (b) Compute the volume discharge.

During a cyclonic event, a debris flow (density 1780 kg/m³) discharges down a trapezoidal open channel (1V:1H sideslopes, 2 m bottom width). The estimated mass flow rate is 4700 kg/s and the mean velocity is about 1.7 m/s. (a) Calculate the volume flow rate. (b) Compute the water depth on the channel centreline.

Note: For more information on debris flow, see Chapters 9 and 11.

For a two-dimensional steady incompressible flow, write the Navier–Stokes equation in Cartesian co-ordinates.

Considering a sluice gate in a horizontal smooth rectangular channel, write the momentum and Bernoulli equations as functions of the flow rate, channel width, upstream and downstream depths and the force of the gate onto the fluid only.

Considering a spherical air bubble (diameter d_{ab}) submerged in water with hydrostatic pressure distribution and rising freely at a constant speed v (a) develop the relationship between buoyancy and forces. (b) If the drag force is expressed as:

$$\text{Drag} = C_d \frac{\rho}{2} v^2$$

where ρ is the water density and C_d is the drag coefficient, express the relationship between the bubble rise velocity, the fluid properties, the bubble diameter and the drag coefficient. (c) For a bubble diameter of 0.5 mm, calculate the bubble rise velocity in still water. Deduce the drag coefficient from Fig. 7.4, where the drag coefficient is plotted as a function of the particle Reynolds number: $v d_{ab}/v$.

Applications of the Bernoulli equation to open channel flows

3

Summary

Applications of the Bernoulli equation to open channel flow situations (neglecting the flow resistance) are developed. In this chapter it is assumed that the fluid is frictionless. The concept of specific energy is introduced for free-surface flows and critical flow conditions are defined. The Froude number is introduced.

3.1 INTRODUCTION

The Bernoulli equation is derived from the Navier–Stokes equation, considering the flow along a streamline, assuming that the volume force potential is independent of the time, for a frictionless and incompressible fluid, and for a steady flow. For gravity forces the differential form of the Bernoulli equation is:

$$\left(\frac{\mathrm{d}P}{\rho} + g\,\mathrm{d}z + V\,\mathrm{d}V \right) = 0 \tag{2.17}$$

where V is the velocity, g is the gravity constant, z is the altitude, P is the pressure and ρ is the fluid density.

This chapter describes hydraulic applications in which the Bernoulli equation is valid: e.g. smooth channel, short transitions and sluice gate.

3.2 APPLICATION OF THE BERNOULLI EQUATION – SPECIFIC ENERGY

3.2.1 Bernoulli equation

Summary

The Bernoulli equation (2.18) was obtained for a frictionless steady flow of incompressible fluid. It states that, along a streamline,

$$\frac{V^2}{2} + gz + \frac{P}{\rho} = \text{constant} \tag{2.18}$$

Dividing by the gravity constant, the Bernoulli equation becomes:

$$\frac{V^2}{2g} + z + \frac{P}{\rho g} = \text{constant} \tag{3.1}$$

where the constant is expressed in metres.

Along any streamline, the *total head* is defined as:

$$H = \frac{V^2}{2g} + z + \frac{P}{\rho g} \qquad \text{Total head along a streamline} \qquad (3.2a)$$

where V, z and P are the local fluid velocity, altitude and pressure. The term $(P/(\rho g) + z)$ is often called the *piezometric head*.

Note

If the flow is accelerated the following equation must be used:

$$\frac{D}{Dt}\left(\frac{V^2}{2}\right) + \frac{DU}{Dt} + \frac{1}{\rho}\frac{DP}{Dt} = \frac{1}{\rho}\frac{\partial P}{\partial t}$$

where U is the volume force potential (i.e. gravity force $U = gz$).

If the velocity is not uniform, the total head H is a function of the distance y from the channel:

$$H(y) = \frac{V(y)^2}{2g} + z(y) + \frac{P(y)}{\rho g} \qquad (3.2b)$$

where y is measured perpendicular to the channel bottom. Assuming a hydrostatic pressure gradient (see next section), the local elevation $z(y)$ and pressure $P(y)$ can be transformed:

$$z(y) = z_0 + y\cos\theta$$

$$P(y) = P_{atm} + \rho g(d - y)\cos\theta$$

where z_0 is the bottom elevation, θ is the channel slope and d is the flow depth (Fig. 3.1).

With uniform or non-uniform velocity distributions, the piezometric head is a constant *assuming a hydrostatic pressure distribution*:

$$\left(\frac{P}{\rho g} + z\right) = \frac{P_{atm}}{\rho g} + z_0 + d\cos\theta \qquad \text{Piezometric head} \qquad (3.3)$$

In summary: at a given cross-section, although the total head varies with respect to the distance y from the channel bottom, the piezometric head in open channel is constant if the pressure is hydrostatic.

Application of the Bernoulli equation

Hydrostatic pressure distribution in open channel flow

If the flow is not accelerated (i.e. $dV = 0$) the differential form of the Bernoulli equation (2.17) gives the pressure distribution:

$$\frac{dP}{\rho} + g\,dz = 0$$

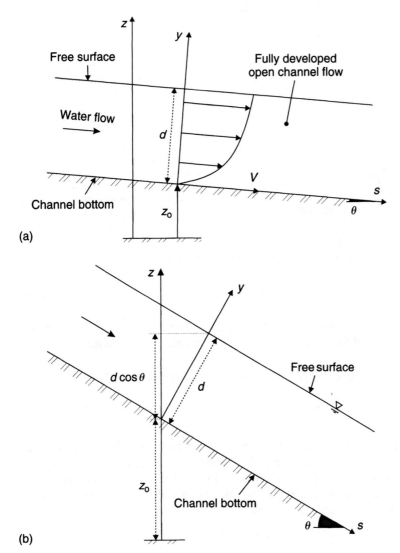

Fig. 3.1 Sketch of open channel flow: (a) general case and (b) steep channel flow.

For a horizontal channel it leads to:

$$P = P_{atm} - \rho g(z - (z_o + d)) \tag{3.4}$$

where z is along the vertical direction and positive upward, z_o is the bottom elevation and d is the flow depth. Equation (3.4) is simply a rewriting of the hydrostatic pressure distribution in static fluid (i.e. equation (1.3)).

Pressure distribution in open channel flow

In open channel flow, the assumption of hydrostatic pressure distribution is valid in gradually varied flow along flat slope. Equation (3.4) is not valid if the flow acceleration is important, if there is a marked acceleration perpendicular to the flow, if the slope is steep or if the streamline curvature is pronounced.

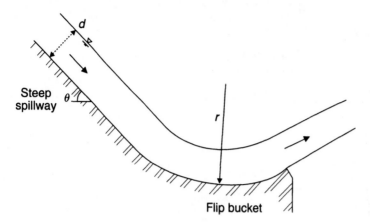

Steep
spillway

Flip bucket

Fig. 3.2 Open channel flow in a curved channel.

For the flow along a curved streamline, the centrifugal acceleration equals:

$$g_{centrif} = \pm \frac{V^2}{r}$$ (3.5)

where $g_{centrif}$ is the centrifugal acceleration due to the streamline curvature, r is the radius of curvature of the streamline, (+) is used for concave streamline curvature, (−) is used for convex curvature.

Considering a curved channel (Fig. 3.2), the centrifugal pressure equals approximately:

$$P_{centrif} = \pm \frac{\rho V^2 d}{r}$$ (3.6)

where $P_{centrif}$ is the pressure rise due to the channel bottom curvature (e.g. flip bucket at spillway toe, Henderson, 1966: p. 189), V is the mean flow velocity, r is the radius of curvature of the invert, (+) is used for concave boundary curvature, (−) is used for convex curvature.

Considering now an open channel flow down a steep slope θ (e.g. Fig. 3.1(b)), the pressure at the channel bottom is (in absence of channel curvature):

$$P = P_{atm} + \rho g d \cos\theta$$ (3.7)

where d is the flow depth measured perpendicular to the channel bottom and θ is the channel slope.

In the general case, the pressure at the channel bottom equals:

$$P = P_{atm} + \rho d \left(g \cos\theta \pm \frac{V^2}{r} \right)$$ (3.8)

where θ is the bed slope and r is the radius of curvature of the invert.

Mean total head

In an open channel flow, it is often convenient to consider the depth-averaged total head H defined as:

$$H = \frac{1}{d}\int_0^d H(y)\,\mathrm{d}y = \frac{1}{d}\int_0^d \left(\frac{v(y)^2}{2g} + z(y) + \frac{P(y)}{\rho g} \right)\mathrm{d}y$$

Replacing the pressure P and elevation z by their respective expression, the mean total head becomes:

$$H = \frac{V^2}{2g} + z_o + \frac{d}{g}\left(g\cos\theta \pm \frac{V^2}{r}\right) + \frac{P_{atm}}{\rho g} \tag{3.9a}$$

where z_o is the bottom elevation and a uniform velocity distribution (i.e. $v(y) = V$) is assumed.

In open channel hydraulics the local pressure is often taken as the pressure relative to the atmospheric pressure P_{atm} and the mean total head is rewritten as:

$$H = \frac{V^2}{2g} + z_o + \frac{d}{g}\left(g\cos\theta \pm \frac{V^2}{r}\right) \tag{3.9b}$$

where V is the mean flow velocity, z_o is the bed elevation, d is the flow depth and θ is the bed slope.

Notes
1. An important and practical result to remember is that *the pressure gradient is not hydrostatic if the curvature of the streamlines is important.* In absence of additional information, the free-surface curvature and/or the bottom curvature can be used to give some indication of the streamline curvature. The free surface and the channel bottom are both streamlines (i.e. no flow across the free surface nor the bottom).

   Any rapid change of flow stage (i.e. a short transition) is always accompanied with a change of streamline curvature under which circumstances it is seldom possible to assume hydrostatic pressure distribution. Rouse (1938, pp. 319–321) discussed in detail one particular example: the broad-crested weir.
2. A flip bucket (e.g. Fig. 3.2) or ski-jump is a deflector located at the end of a spillway and designed to deflect the flow away from the spillway foundation. The waters take off at the end of the deflector as a free-fall flow and impact further downstream in a dissipation pool.

Pitot tube

A direct application of the Bernoulli equation is a device for measuring fluid velocity called the Pitot tube. Aligned along a streamline, the Pitot tube enables to measure the total head H and the pressure head (i.e. $z + P/(\rho g)$) denoted H_1 and H_2 respectively in Fig. 3.3. The fluid velocity is then deduced from the total head difference.

Notes
1. The Pitot tube is named after the Frenchman H. Pitot. The first presentation of the concept of the Pitot tube was made in 1732 at the French Academy of Sciences by Henri Pitot.
2. The original Pitot tube included basically a total head reading. L. Prandtl improved the device by introducing a pressure (or piezometric head) reading. The modified Pitot tube (illustrated in Fig. 3.3) is sometimes called a Pitot–Prandtl tube.
3. The Pitot tube is also called the Pitot static tube.
4. For many years, aeroplanes used Pitot tubes to estimate their relative velocity.

Application: smooth transition

Considering a smooth flow transition (e.g. a step and a weir), for a steady, incompressible, frictionless flow, between an upstream section (1) and a downstream section (2), the continuity and Bernoulli equations applied to the open channel flow become:

$$V_1 A_1 = V_2 A_2 = Q \qquad (3.10a)$$

$$z_{o1} + d_1 \cos \theta_1 + \frac{V_1^2}{2g} = z_{o2} + d_2 \cos \theta_2 + \frac{V_2^2}{2g} \qquad (3.11)$$

assuming a hydrostatic pressure distribution and where z_{o1} and z_{o2} are the bottom elevations, d_1 and d_2 are the flow depths measured normal to the channel bottom (Fig. 3.4). For a rectangular channel, the continuity equations becomes:

$$V_1 d_1 B_1 = V_2 d_2 B_2 = Q \qquad (3.10b)$$

If the discharge Q, the bed elevations (z_{o1}, z_{o2}), channel widths (B_1, B_2) and channel slopes (θ_1, θ_2) are known, the downstream flow conditions (d_2, V_2) can be deduced from the upstream flow conditions (d_1, V_1) using equations (3.10) and (3.11).

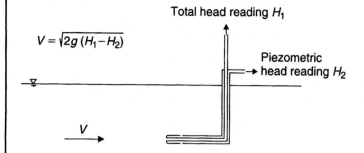

Total head reading H_1

$$V = \sqrt{2g\,(H_1 - H_2)}$$

Piezometric head reading H_2

V

Fig. 3.3 Sketch of a Pitot tube.

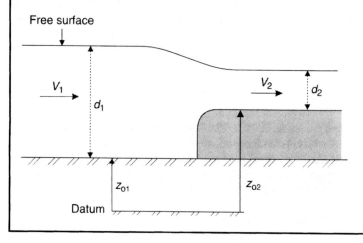

Free surface

V_1

d_1

V_2

d_2

z_{o1}

z_{o2}

Datum

Fig. 3.4 Sketch of a smooth transition.

3.2.2 Influence of the velocity distribution

Introduction
The above developments were obtained assuming (or implying) uniform velocity distributions: i.e. the velocity was assumed constant over the entire cross-section. In practice, the velocity distribution is not uniform because of the bottom and sidewall friction.

It will be shown that most results are still valid using the mean flow velocity (i.e. $V = Q/A$) instead of the uniform velocity by introducing velocity correction coefficients: the Boussinesq or momentum correction coefficient, and the Coriolis or kinetic energy correction coefficient.

Velocity distribution
The velocity distribution in fully developed turbulent open channel flows is given approximately by Prandtl's power law:

$$\frac{v}{V_{max}} = \left(\frac{y}{d}\right)^{1/N} \tag{3.12}$$

where the exponent $1/N$ varies from 1/4 down to 1/12 depending upon the boundary friction and cross-sectional shape. The most commonly used power law formulae are the one-sixth power (1/6) and the one-seventh power (1/7) formula.

Notes
1. Chen (1990) developed a complete analysis of the velocity distribution in open channel and pipe flow with reference to flow resistance. Moreover, he showed that the Gauckler–Manning formula (see Chapter 4) implies a 1/6 power formula while the 1/7 power formula derives from Blasius resistance formula (i.e. for smooth turbulent flow). In uniform equilibrium flows, the velocity distribution exponent is related to the flow resistance:

$$N = K\sqrt{\frac{8}{f}}$$

   where f is the Darcy friction factor and K is the von Karman constant (K = 0.4).
2. In practical engineering applications, N can range from 4 (for shallow waters in wide rough channels) up to 12 (smooth narrow channel). A value $N = 6$ is reasonably representative of open channel flows in smooth-concrete channels. However, it must be remembered that N is a function of the flow resistance.
3. For a wide rectangular channel, the relationship between the mean flow velocity V and the free surface velocity V_{max} derives from the continuity equation:

$$q = Vd = \int_0^d v \, dy = \frac{N}{N+1} V_{max} d$$

4. Further discussions on velocity profiles in turbulent boundary layers and open channel flows are developed in Section 3.2.2.

Velocity coefficients

Momentum correction coefficient
If the velocity distribution is not uniform over the cross-section, a correction coefficient must be introduced in the momentum equation if the average velocity is used. The momentum correction

factor β is defined as (e.g. Streeter and Wylie, 1981: p. 116):

$$\beta = \frac{\int_A \rho v^2 \mathrm{d}A}{\rho V^2 A} \qquad (3.13)$$

where V is the mean velocity over a cross-section (i.e. $V = Q/A$). The momentum correction coefficient is sometimes called the Boussinesq coefficient.

Considering a steady flow in a horizontal channel and assuming a hydrostatic pressure distribution (e.g. Fig. 2.3), the momentum equation is often rewritten as:

$$\rho Q(\beta_2 V_2 - \beta_1 V_1) = \left(\frac{1}{2}\rho g d_1^2 B_1\right) - \left(\frac{1}{2}\rho g d_2^2 B_2\right)$$

where Q is the discharge (m³/s), d is the flow depth, B is the channel width and β is the momentum correction coefficient.

Notes
1. The Boussinesq coefficient is named after J. Boussinesq, French mathematician (1842–1929) who proposed it first (Boussinesq, 1877).
2. The momentum correction coefficient is always larger than unity. ($\beta = 1$) implies an uniform velocity distribution.
3. For a $1/N$ power law velocity distribution (equation (3.12)), the momentum correction coefficient equals:

$$\beta = \frac{(N+1)^2}{N(N+2)}$$

Kinetic energy correction coefficient
If the velocity varies across the section, the mean of the velocity head $(v^2/2g)_{\mathrm{mean}}$ is not equal to $V^2/2g$, where the subscript mean refers to the mean value over the cross-section. The ratio of these quantities is called the *Coriolis coefficient*, denoted α and defined as:

$$\alpha = \frac{\int_A \rho v^3 \mathrm{d}A}{\rho V^3 A} \qquad (3.14)$$

If the energy equation is rewritten in term of the *mean total head*, the later must be transformed as:

$$H = \alpha \frac{V^2}{2g} + z_o + d \cos \theta$$

assuming a hydrostatic pressure distribution.

Notes
1. The Coriolis coefficient is named after G.G. Coriolis, French engineer, who introduced first the correction coefficient (Coriolis, 1836).
2. The Coriolis coefficient is also called the kinetic energy correction coefficient.

---

3. α is equal or larger than 1 but rarely exceeds 1.15 (see discussion by Li and Hager, 1991). For an uniform velocity distribution $\alpha = 1$.
4. For a $1/N$ power law velocity distribution (equation (3.12)), the kinetic energy coefficient equals:

$$\alpha = \frac{(N + 1)^3}{N^2(N + 3)}$$

Correction coefficient in the Bernoulli equation

As shown in Chapter 2, the Bernoulli equation derives from the Navier–Stokes equation: i.e. from the conservation of momentum. Application of the Bernoulli equation to open channel flow (within the frame of relevant assumptions) yields:

$$\beta \frac{V'^2}{2g} + z_0 + d = \text{constant} \tag{3.15}$$

where V' is the depth-averaged velocity:

$$V' = \frac{1}{d} \int_0^d V \, dy$$

Note that V' differs from V usually.

Equation (3.15) is the depth-averaged Bernoulli equation in which the correction factor is the Boussinesq coefficient.

Notes

1. Some textbooks introduce incorrectly the kinetic energy correction coefficient in the Bernoulli equation. This is *incorrect*.
2. The reader is referred to the excellent discussion of Liggett (1993) for further details.

3.2.3 Specific energy

Definition

The *specific energy* E is defined as:

$$E(y) = \frac{P(y)}{\rho g} + \frac{V(y)^2}{2g} + (z(y) - z_0) \tag{3.16}$$

where P is the pressure, V is the velocity, z is the elevation and z_0 is the bed elevation. The specific energy is similar to the energy per unit mass, measured with the channel bottom as the datum. The specific energy changes along a channel because of changes of bottom elevation and energy losses (e.g. friction loss).

The mean specific energy is defined as:

$$E = H - z_0 \tag{3.17a}$$

where H is the mean total head. For a *flat channel*, assuming a *hydrostatic pressure distribution* (i.e. $P = \rho g d$) and an uniform velocity distribution, it yields:

$$E = d + \frac{V^2}{2g} \qquad \text{Flat channel and hydrostatic pressure distribution} \qquad (3.17b)$$

where d is the flow depth and V is the mean flow velocity.

Notes

It must be emphasized that, even if the total head H is constant, the specific energy varies with the bed elevation z_o (or bed altitude) as:

$$H = E + z_o$$

For a non-uniform velocity distribution, assuming a hydrostatic pressure distribution, the *mean specific energy* as used in the energy equation becomes:

$$E = d \cos \theta + \alpha \frac{V^2}{2g}$$

where θ is the channel slope, α is the Coriolis coefficient and V is the mean flow velocity ($V = Q/A$).

For a horizontal channel, the Bernoulli equation implies that the specific energy is constant. This statement is true only within the assumptions of the Bernoulli equation: i.e. for an incompressible, frictionless and steady flow along a streamline.

For a rectangular channel it is convenient to combine the continuity equation and the specific energy definition. Using the total discharge Q, the expression of the specific energy becomes (for a flat channel):

$$E = d + \frac{Q^2}{2gd^2 B^2} \qquad (3.17c)$$

where B is the free-surface width.

Analysis of the specific energy

Relationship flow depth versus specific energy

The specific energy is usually studied as a function of the flow depth d. It is convenient to plot the relationship $d = f(E)$ as shown in Fig. 3.5. In rectangular channels there is only one specific energy–flow depth curve for a given discharge per unit width Q/B.

For a tranquil and slow flow, the velocity is small and the flow depth is large. The kinetic energy term $V^2/2g$ is very small and the specific energy tends to the flow depth d (i.e. asymptote $E = d$).

For a rapid flow (e.g. a torrent), the velocity is large and, by continuity, the flow depth is small. The pressure term $P/\rho g$ (i.e. flow depth) is small compared to the kinetic energy term. The specific energy term tends to an infinite value when d tends to zero (i.e. asymptote $d = 0$).

At any cross-section, the specific energy has a unique value. For a given value of specific energy and a given flow rate, there may be zero, one or two possible flow depths (Fig. 3.5, Appendix A1.4).

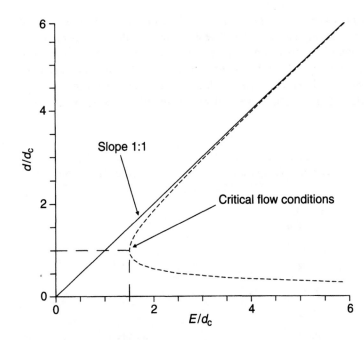

Fig. 3.5 Dimensionless specific energy curve for a flat rectangular channel (equation (3.20)).

Critical flow conditions

For a constant discharge Q and a given cross-section, the relationship $E = f(d)$ indicates the existence of a minimum specific energy (Fig. 3.5). The flow conditions (d_c, V_c), such that the mean specific energy is minimum, are called the *critical flow conditions*. They take place for:

$$\left(\frac{\partial E}{\partial d} \right)_{(Q \text{ constant})} = 0$$

The critical flow conditions may be expressed in terms of the discharge Q and geometry of the channel (cross-section A, free-surface width B) after transformation of equation (3.16). For a rectangular flat channel of constant width, the minimum specific energy E_{min} and the critical flow depth are respectively:

$$E_{min} = \frac{3}{2} d_c \qquad (3.18)$$

$$d_c = \sqrt[3]{\frac{Q^2}{gB^2}} \qquad (3.19)$$

The specific energy can be rewritten in dimensionless term (for a flat channel) as:

$$\frac{E}{d_c} = \frac{d}{d_c} + \frac{1}{2} \left(\frac{d_c}{d} \right)^2 \qquad (3.20)$$

Equation (3.20) is plotted in Fig. 3.5. Note that equation (3.20) is an unique curve: it is valid for any discharge.

The relationship specific energy versus flow depth indicates two trends (Fig. 3.5). When the flow depth is greater than the critical depth (i.e. $d/d_c > 1$), an increase in specific energy causes an increase in depth. For $d > d_c$, the flow is termed *subcritical*. When the flow depth is less than the critical depth, an increase in specific energy causes a decrease in depth. The flow is *supercritical*.

If the flow is critical (i.e. the specific energy is minimum), small changes in specific energy cause large changes in depth. In practice, critical flow over a long reach of channel is unstable.

The definition of specific energy and critical flow conditions are summarized in the table below for flat channels of irregular cross-section (area A) and channels of rectangular cross-section (width B):

Variable	Channel of irregular cross-section	Rectangular channel
Specific energy E	$d + \dfrac{Q^2}{2gA^2}$	$d + \dfrac{q^2}{2gd^2}$
$\dfrac{\partial E}{\partial d}$	$1 - \dfrac{Q^2 B}{gA^3}$	$1 - \dfrac{q^2}{gd^3}$
$\dfrac{\partial^2 E}{\partial d^2}$	$\dfrac{Q^2}{gA^3}\left(\dfrac{3B^2}{A} - \dfrac{\partial B}{\partial d}\right)$	$\dfrac{3q^2}{gd^4}$
$\lim\limits_{d \to 0^+} E$	Infinite	Infinite
$\lim\limits_{d \to +\infty} E$	d	d
Critical depth d_c		$\sqrt[3]{\dfrac{q^2}{g}}$
Critical velocity V_c	$\sqrt{g\dfrac{A_c}{B_c}}$	$\sqrt{gd_c}$
Minimum specific energy E_{min}	$d_c + \dfrac{1}{2}\dfrac{A_c}{B_c}$	$\dfrac{3}{2}d_c$
Froude number Fr	$\dfrac{V}{\sqrt{g\dfrac{A}{B}}}$	$\dfrac{V}{\sqrt{gd}}$
$\dfrac{\partial E}{\partial d}$	$1 - Fr^2$	$1 - Fr^2$

Notes: A: cross-sectional area; B: free-surface width; Q: total discharge; q: discharge per unit width.

Notes

1. In open channel flow, the Froude number is defined such as it equals 1 for critical flow conditions. The corollary is that critical flow conditions are reached if $Fr = 1$.
2. A general dimensionless expression of the specific energy is:

$$\frac{E}{d_c} = \frac{d}{d_c}\left(\cos\theta + \frac{1}{2}Fr^2\right)$$

For a flat channel it yields:

$$\frac{E}{d_c} = \frac{d}{d_c}\left(1 + \frac{1}{2}Fr^2\right)$$

3. The ratio A/B of cross-sectional area over free-surface width is called sometimes the mean depth.

Application of the specific energy

For a frictionless flow in a horizontal channel, the specific energy is constant along the channel. The specific energy concept can be applied to predict the flow under a sluice gate as a function of the gate operation (Fig. 3.6). For a given gate opening, a specific discharge ($q = Q/B$) takes place and there is only one specific energy/flow depth curve.

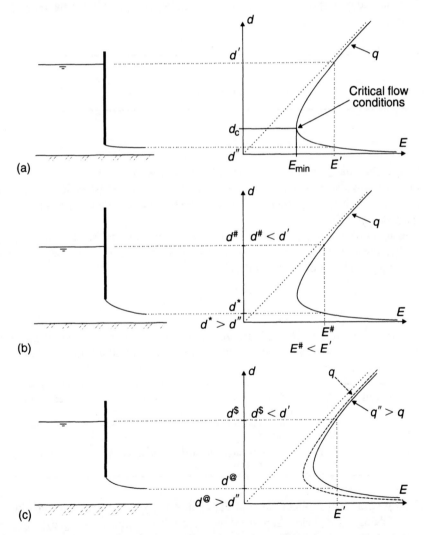

Fig. 3.6 Specific energy diagrams for flow under a sluice gate: (a) sluice gate; (b) larger gate opening for the same flow rate; (c) constant specific energy and larger flow rate ($q'' > q$).

The upstream and downstream values of the specific energy are equal (by application of the Bernoulli equation). For this value of specific energy, there are two possible values of the flow depth: one subcritical depth (i.e. upstream depth) and one supercritical depth (i.e. downstream depth). If the value of the specific energy is known, the upstream and downstream flow depths are deduced from the specific energy curve (Fig. 3.6a).

If the specific energy is not known *a priori*, the knowledge of the upstream flow depth and of the flow rate fixes the specific energy, and the downstream flow depth is deduced by graphical solution.

If the discharge remains constant but for a larger gate opening (Fig. 3.6b), the upstream depth is smaller and the downstream flow depth is larger.

In Fig. 3.6c, the specific energy is the same as in Fig. 3.6a but the flow rate is larger. The upstream and downstream flow depths are located on a curve corresponding to the larger flow rate. The graphical solution of the specific energy/flow depth relationship indicates that the upstream flow depth is smaller and the downstream depth is larger than for the case in Fig. 3.6a. This implies that the gate opening in Fig. 3.6c must be larger than in Fig. 3.6a to compensate the larger flow rate.

Note

A sluice gate is a device used for regulating flow in open channels.

Discussion

Change in specific energy associated with a fixed discharge

Considering a rectangular channel, Figs 3.5 and 3.6 show the relationship flow depth versus specific energy. For a specific energy such as $E > E_{min}$, the two possible depths are called the *alternate depths* (Appendix A1.4).

Let us consider now a change in bed elevation. For a constant total head H, the specific energy decreases when the channel bottom rises. For a tranquil flow (i.e. $Fr < 1$ or $d > d_c$), a decrease in specific energy implies a decrease in flow depth as:

$$\frac{\partial E}{\partial d} = 1 - Fr^2 > 0 \tag{3.21}$$

For rapid flow (i.e. $Fr > 1$ or $d < d_c$), the relation is inverted: a decrease of specific energy implies an increase of flow depth. The results are summarized in the following table:

Flow depth	Fr	Type of flow	Bed raised	Bed lowered
$d > d_c$	<1	slow, tranquil, fluvial *Subcritical*	d decreases	d increases
$d < d_c$	>1	fast, shooting, torrential *Supercritical*	d increases	d decreases

Figure 3.7 shows practical applications of the specific energy concept to flow transitions in open channel. For a change of bed elevation, the downstream flow conditions (E_2, d_2) can be deduced directly from the specific energy/flow depth curve. They are functions of the upstream flow conditions (E_1, d_1) and the change of bed elevation Δz_o.

Notes
1. Examples of application of the relationship E versus d are shown in Figs 3.6 and 3.7: sluice gate and transition with bed elevation.
2. Rouse (1946: p. 139) illustrated with nice photographs the effects of change in bed elevation for subcritical ($d > d_c$) and supercritical ($d < d_c$) flows.

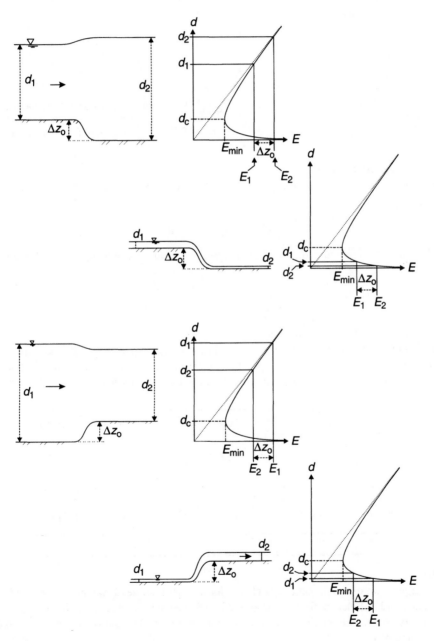

Fig. 3.7 Specific energy and transition problem.

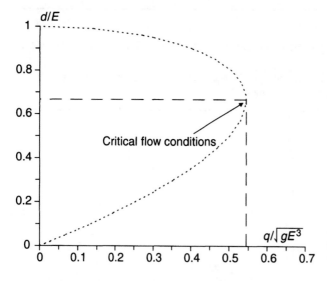

Fig. 3.8 Dimensionless discharge-depth curve for given specific energy in a horizontal rectangular channel (equation (3.22b)).

Case of fixed specific energy

For a fixed specific energy E, the relationship flow rate versus flow depth in a rectangular flat channel is:

$$Q = Bd\sqrt{2g(E - d)} \tag{3.22a}$$

where d is the flow depth and B is the channel width. In dimensionless terms, it becomes:

$$\frac{Q}{B\sqrt{gE^3}} = \frac{d}{E}\sqrt{2\left(1 - \frac{d}{E}\right)} \tag{3.22b}$$

Equation (3.22b) is plotted in Fig. 3.8.

Figure 3.8 indicates that there is a maximum discharge Q_{max} such as the problem has a real solution. It can be shown that, for a fixed specific energy E, the maximum discharge is obtained for critical flow conditions (i.e. $d = d_c$) (see notes below). Results are detailed in the following table:

E constant	General section	Rectangular channel
Critical depth d_c	$d_c = E - \dfrac{A_c}{2B_c}$	$d_c = \dfrac{2}{3}E$
Maximum discharge Q_{max}	$Q_{max} = \sqrt{g\dfrac{A_c^3}{B_c}}$	$Q_{max} = B\sqrt{\left(\dfrac{2}{3}\right)^3 gE^3}$

Notes

1. The expression of the maximum discharge for a given total head was first formulated by the Frenchman J.B. Bélanger for the flow over a broad-crested weir (Bélanger, 1849). He showed in particular that maximum flow rate is achieved for $\partial Q/\partial d = 0$ (if the streamlines are parallel to the weir crest).

2. Assuming a rectangular channel of constant width, equation (3.17) can be rewritten as:

$$Q = Bd\sqrt{2g(E - d)} \tag{3.22}$$

For a constant specific energy, the maximum discharge is obtained for $\partial Q/\partial d = 0$, i.e.:

$$\frac{2gE - 3gd}{\sqrt{2g(E - d)}} = 0$$

or

$$E = \frac{3}{2}d$$

Such a relationship between specific energy and flow depth is obtained only and only at critical flow conditions (i.e. equation (3.18), Fig. 3.5). The result implies that $E = E_{min}$ and $d = d_c$ (critical flow depth) for $Q = Q_{max}$.

3.2.4 Limitations of the Bernoulli equation

The application of the Bernoulli equation is valid only within the range of assumptions (i.e. steady frictionless flow of incompressible fluid).

For *short and smooth transitions* the energy losses are negligible and the Bernoulli equation may be applied quite successfully. Short transitions may be gates (e.g. sluice or radial gates), weirs and steps (e.g. Figs 3.6, 3.7 and 3.9).

If energy losses (e.g. friction loss) occur, they must be taken in account in the energy equation. The Bernoulli equation (i.e. equation (3.1)) is no longer valid. For example, the Bernoulli equation is not valid at a hydraulic jump, where turbulent energy losses are significant.

There is, however, another limitation of the use of the Bernoulli equation. At short transitions the velocity and flow depth may *not* be *uniform* across a section (e.g. flow through bridge piers) and the complete energy equation must be used (Henderson, 1966: pp. 49–50).

Note
At short transitions (e.g. gate and weir intake), there is a rapid change in velocity and pressure distributions. It is often observed that the assumptions of uniform velocity distributions and hydrostatic pressure distributions are not valid.

3.3 FROUDE NUMBER

3.3.1 Definition

The *Froude number* is a dimensionless number proportional to the square root of the ratio of the inertial forces over the weight of fluid:

$$Fr = \frac{V}{\sqrt{gd_{charac}}} = \sqrt{\frac{\rho V^2 A}{\rho g A L}} \propto \sqrt{\frac{\text{inertial force}}{\text{weight}}}$$

Fig. 3.9 Overflow above a broad-crested weir (weir height: 0.067 m, channel width: 0.25 m, crest length: 0.42 m): (a) flow from the left to the right – note the supercritical flow downstream of the weir crest; (b) undular flow above the weir crest, flow from the right to the left.

where d_{charac} is a characteristic dimension (see notes below). For open channel flows, the relevant characteristic length is the flow depth d. For a rectangular channel, the Froude number is defined as:

$$Fr = \frac{V}{\sqrt{gd}} \qquad \text{Rectangular channel} \qquad (3.23a)$$

Notes
1. The Froude number is used for scaling open channel flows and free-surface flows.
2. The length d_{charac} is the characteristic geometric dimension: internal diameter of pipe for pipe flows, flow depth for open channel flow in a rectangular channel, ship draught, hydrofoil thickness, etc.
3. For a channel of irregular cross-sectional shape, the Froude number is usually defined as:

$$Fr = \frac{V}{\sqrt{g(A/B)}} \qquad (3.23b)$$

where V is the mean flow velocity, A is the cross-sectional area and B is the free-surface width.

For a horizontal channel, equation (3.23b) implies $Fr = 1$ at critical flow conditions (see Section 3.2.3, in this chapter).
4. Some authors use the notation:

$$Fr = \frac{V^2}{gd_{charac}} = \frac{\rho V^2 A}{\rho g A d_{charac}} \quad \propto \quad \frac{\text{inertial force}}{\text{weight}}$$

3.3.2 Similarity and Froude number

In problems of fluid flow there are always at least four parameters involved: pressure difference ΔP, velocity V, length d and fluid density ρ. In open channel hydraulics another important parameter is the gravity acceleration g.

Considering a closed pipe circuit, and provided that the liquid remains liquid (i.e. no cavitation), the piezometric pressure $(P + \rho gz)$ is merely affected by the gravity effect as the pressure P is free to take any value at all. In an open channel, through the existence of a free surface where the pressure must take a prescribed value P_{atm}, the gravity can influence the flow pattern. For P_{atm} fixed, the gravity determines the piezometric pressure and hence the flow properties. In any flow situations where the gravity effects are significant, the appropriate dimensionless number (i.e. the Froude number) must be considered. In practice, model studies of open channel flows and hydraulic structures are performed using a Froude similitude. That is, the Froude number is the same for the model and the prototype (see Part III).

Notes
1. For open channel flow (one-phase and one-component flow) the Froude number plays a very important part in the appropriate flow equations. This has been already noted at critical flow conditions defined as $d = d_c$ for which $Fr = 1$.
2. For two-phases (e.g. ice–water flows) or two-components (e.g. air–water flows) open channel flows, the Froude number might not be a significant number.
3. If the viscosity μ plays an effective part, the Reynolds number is another relevant parameter in dimensional analysis (see Part III).

Application
A prototype channel is 1000 m long and 12 m wide. The channel is rectangular and the flow conditions are: $d = 3$ m and $Q = 15$ m³/s. What would be the size and discharge of a 1/25 scale model using a Froude similitude?

Solution

On the 1/25 scale model, all the geometric dimensions (including the flow depth) are scaled by a factor 1/25:

Length	$L_{model} = L_{prototype}/25$
Width	$B_{model} = B_{prototype}/25$
Flow depth	$d_{model} = d_{prototype}/25$

The flow velocity on the model is deduced by a Froude similitude. The Froude similitude implies that the Froude number is the same on both model and prototype:

Froude similitude	$Fr_{model} = Fr_{prototype}$

It yields:

Velocity	$V_{model} = V_{prototype}/\sqrt{25}$

For the discharge, the continuity equation $Q = VdB$ provides the scaling ratio:

Discharge	$Q_{model} = Q_{prototype}/\sqrt[5]{25}$

As a result the model characteristics are:

Length	$L_{model} = 40\,\text{m}$
Width	$B_{model} = 0.25\,\text{m}$
Flow depth	$d_{model} = 0.12\,\text{m}$
Velocity	$V_{model} = 0.083\,\text{m/s}$
Discharge	$Q_{model} = 0.0048\,\text{m}^3/\text{s}$
Froude similitude	$Fr_{model} = Fr_{prototype} = 0.077$

3.3.3 Critical conditions and wave celerity

Considering an oscillatory wave[1], the velocity C or celerity[2] of waves of length l is given by:

$$C^2 = \frac{gl}{2\pi}\tanh\left(\frac{2\pi d}{l}\right) \tag{3.24}$$

where d is the flow depth and tanh is the hyperbolic tangent function (Henderson, 1966: p. 37; Liggett, 1994: p. 394). For a long wave of small amplitude (i.e. $d/l \ll 1$) this expression becomes:

$$C^2 = gd$$

A similar result can be deduced from the continuity equation applied to a weak free-surface gravity wave of height $\Delta d \ll d$ in an open channel (Streeter and Wylie, 1981: p. 514). The wave celerity equals:

$$C = \sqrt{gd} \tag{3.25}$$

where C is the celerity of a small free-surface disturbance in open channel flow.

The Froude number may be written in a form analogous to the Sarrau–Mach number (see Section 3.3.4) in gas flow, as the ratio of the flow velocity divided by the celerity of small disturbances:

$$Fr = \frac{V}{C} \tag{3.26}$$

[1]The term progressive wave can be also used (e.g. Liggett, 1994).
[2]The speed of a disturbance is called celerity (e.g. sound celerity, wave celerity).

where $C = \sqrt{gd}$ is the celerity of small waves (or small disturbance) at the free surface in open channels.

For Froude or Sarrau–Mach numbers less than unity, disturbances at a point are propagated to all parts of the flow; however, for Froude and Sarrau–Mach numbers of greater than unity, disturbances propagate downstream[3] only. In other words, in a supercritical flow, small disturbances propagate only in the downstream flow direction. While in subcritical flows, small waves can propagate in both the upstream and downstream flow directions.

Note
Oscillatory waves (i.e. progressive waves) are characterized by no net mass transfer.

3.3.4 Analogy with compressible flow

In compressible flows, the pressure and the fluid density depend on the velocity magnitude relative to the celerity of sound in the fluid C_{sound}. The compressibility effects are often expressed in term of the Sarrau–Mach number $Ma = V/C_{sound}$. Both the Sarrau–Mach number and the Froude number are expressed as the ratio of the fluid velocity over the celerity of a disturbance (celerity of sound and celerity of small wave respectively).

Dimensional analysis shows that dynamic similarity in compressible flows is achieved with equality of both the Sarrau–Mach and Reynolds numbers, and equal value of the specific heat ratio.

The propagation of pressure waves (i.e. sound waves) in a compressible fluid is comparable to the movement of small amplitude waves on the surface of an open channel flow. It was shown (e.g. Thompson, 1972; Liggett, 1994) that the combination of motion equation for two-dimensional compressible flow with the state equation produces the same basic equation as for open channel flow (of incompressible fluid) in which the gas density is identified with the flow depth (i.e. free-surface position). Such a result is obtained however assuming: an inviscid flow, a hydrostatic pressure gradient (and zero channel slope), and the ratio of specific heat γ must equal 2.

The formal analogy and correspondence of flow parameters are summarized in the following table:

	Open channel flow	Compressible flow
Basic parameters	Flow depth d Velocity V d^{γ} $d^{(\gamma-1)}$	Gas density ρ Velocity V Absolute pressure P Absolute temperature T
Other parameters	$\sqrt{gd}$ Froude number Gravity acceleration g Channel width B	Sound celerity C Sarrau–Mach number $1/2\,(P/\rho^{\gamma})$ Flow area A
Flow analogies	Hydraulic jump Oblique shock wave[4]	Normal shock wave Oblique shock wave
Basic assumptions	Inviscid flow Hydrostatic pressure gradient	$\gamma = 2$

[3]Downstream means in the flow direction.
[4]Also called oblique jump or diagonal jump.

Application

The study of two-dimensional supercritical flow in open channel is very similar to the study of supersonic gas flow. Liggett (1994) developed the complete set of flow equations. The analogy was applied with some success during the early laboratory studies of supersonic flows.

Notes

1. At the beginning of high-speed aerodynamics (i.e. first half of the 20th century), compressible flows were investigated experimentally in open channels using water. For example, the propagation of oblique shock waves in supersonic (compressible) flows was deduced from the propagation of oblique shock waves at the free surface of supercritical open channel flows. Interestingly the celerity C in open channel flow is slow (compared to the sound celerity) and it can be easily observed.

   With the development of high-speed wind tunnels in the 1940s and 1950s, some compressible flow experimental results were later applied to open channel flow situations. Nowadays the analogy is seldom applied because of limitations.

2. The main limitations of the compressible flow/open channel flow analogy are
   (a) The ratio of specific heat must equal 2. For real gases the maximum possible value for γ is 5/3 (see Appendix A1.1). For air, $\gamma = 1.4$. The difference in specific heat ratio (between the analogy and real gases) implies that the analogy can only be approximate.
   (b) The accuracy of free-surface measurements is disturbed by surface tension effects and the presence of capillary waves at the free surface.
   (c) Other limitations of the analogy include the hydraulic jump case. The hydraulic jump is analogue to a normal shock wave. Both processes characterize a flow discontinuity with energy dissipation (i.e. irreversible energy loss). But in a hydraulic jump, the ratio of the sequent depths (i.e. upstream and downstream depth) is not identical to the density ratio across a normal shock wave (except for $Fr = 1$).

3.3.5 Critical flows and controls

Occurrence of critical flow – control section

For an open channel flow, the basic equations are the continuity equation (i.e. conservation of mass), and the motion equation or the Bernoulli equation which derives from the Navier–Stokes equation. The occurrence of critical flow conditions provide one additional equation:

$$V = \sqrt{gd} \qquad \text{Flat rectangular channel} \qquad (3.27)$$

in addition to the continuity equation and the Bernoulli equation. These three conditions fix all the flow properties at the location where critical flow occurs, called the *control section*.

For a given discharge, the flow depth and velocity are fixed at a control section, independently of the upstream and downstream flow conditions. For a rectangular channel, it yields:

$$d = d_c = \sqrt[3]{\frac{Q^2}{gB^2}} \qquad (3.28)$$

$$V = V_c = \sqrt[3]{g\frac{Q}{B}} \qquad (3.29)$$

where B is the channel width.

Corollary: at a control section the discharge can be calculated once the depth is known: e.g. the critical depth meter.

Hydraulic structures that cause critical flow (e.g. gates and weirs) are used as control sections for open channel flows. Examples of control sections include: spillway crest, weir, gate, overfall, etc. (e.g. Henderson, 1966: Chapter 6). Some hydraulic structures are built specifically to create critical flow conditions (i.e. critical depth meter). Such structures provide means to record the flow rates simply by measuring the critical flow depth: e.g. gauging stations, broad-crested weir, sharp-crested weirs (e.g. Rouse, 1938: pp. 319–326; Henderson, 1966: pp. 210–214; Bos, 1976; Bos *et al.*, 1991).

A control section 'controls' the upstream flow if it is subcritical and controls also the downstream flow if it is supercritical. A classical example is the sluice gate (Fig. 3.6).

Upstream and downstream controls

In subcritical flow a disturbance travelling at a celerity C can move upstream and downstream because the wave celerity C is larger than the flow velocity V (i.e. $V/C < 1$). A control mechanism (e.g. sluice gate) can make its influence on the flow upstream of the control. Any small change in the downstream flow conditions affects tranquil (subcritical) flows. Therefore subcritical flows are controlled by downstream conditions. This is called a *downstream flow control*.

Conversely, a disturbance cannot travel upstream in a supercritical flow because the celerity is less than the flow velocity (i.e. $V/C > 1$). Hence supercritical flows can only be controlled from upstream (i.e. *upstream flow control*). All rapid flows are controlled by the upstream flow conditions.

Discussion

All supercritical flow computations (i.e. 'backwater' calculations) must be started at the upstream end of the channel (e.g. gate and critical flow). Tranquil flow computations are started at the downstream end of the channel and are carried upstream.

Considering a channel with a steep slope upstream followed by a mild slope downstream (see definitions in Chapter 5), critical flow conditions occur at the change of slope. Computations must proceed at the same time from the upstream end (supercritical flow) and from the downstream end (subcritical flow). Near the break in grade, there is a transition from a supercritical flow to a subcritical flow. This transition is called a hydraulic jump. It is a flow discontinuity that is solved by applying the momentum equation across the jump (see Chapter 4).

Application: influence of the channel width

For an incompressible open channel flow, the differential form of the continuity equation ($Q = VA$ = constant) along a streamline in the s-direction gives:

$$V\frac{\partial A}{\partial s} + A\frac{\partial V}{\partial s} = 0 \tag{3.30a}$$

where A is the cross-sectional area. Equation (3.30a) is valid for any shape of cross-section. For a channel of rectangular cross-section, it becomes:

$$q\frac{\partial B}{\partial s} + B\frac{\partial q}{\partial s} = 0 \tag{3.30b}$$

where B is the channel surface width and q is the discharge per unit width ($q = Q/B$).

For a rectangular and horizontal channel (z_o constant and H fixed) the differentiation of the Bernoulli equation is:

$$\frac{\partial d}{\partial s} - \frac{q^2}{gd^3}\frac{\partial d}{\partial s} + \frac{q}{gd^2}\frac{\partial q}{\partial s} = 0 \tag{3.31}$$

Introducing the Froude number and using the continuity equation (3.30) it yields:

$$(1 - Fr^2)\frac{\partial d}{\partial s} = Fr^2\frac{d}{B}\frac{\partial B}{\partial s} \tag{3.32}$$

In a horizontal channel, equation (3.32) provides a mean to predict the flow depth variation associated with an increase or a decrease in channel width.

Considering an upstream subcritical flow (i.e. $Fr < 1$) the flow depth decreases if the channel width B decreases: i.e. $\partial d/\partial s < 0$ for $\partial B/\partial s < 0$. As a result the Froude number increases and when $Fr = 1$ critical flow conditions occur for a channel width B_{min}.

Considering an upstream supercritical flow (i.e. $Fr > 1$), a channel contraction (i.e. $\partial B/\partial s < 0$) induces an increase of flow depth. As a result the Froude number decreases and when $Fr = 1$ critical flow conditions occur for a particular downstream channel width B_{min}.

With both types of upstream flow (i.e. sub- and supercritical flow), a constriction which is severe enough may induce critical flow conditions at the throat. The characteristic channel width B_{min} for which critical flow conditions occur is deduced from the Bernoulli equation:

$$B_{min} = \frac{Q}{\sqrt{\dfrac{8}{27}gE_1{}^3}} \tag{3.33}$$

where E_1 is the upstream specific energy. Note that equation (3.33) is valid for horizontal channel of rectangular cross-section. B_{min} is the minimum channel width of the contracted section for the appearance of critical flow conditions. For $B > B_{min}$, the channel contraction does not induce critical flow and it is not a control section. With $B < B_{min}$ critical flow takes place, and the flow conditions at the control section may affect (i.e. modify) the upstream flow conditions.

Notes
1. At the location of critical flow conditions, the flow is sometimes referred to as 'choking'.
2. Considering a channel contraction such as the critical flow conditions are reached (i.e. $B = B_{min}$) the flow downstream of that contraction will tend to be subcritical if there is a downstream control or supercritical in absence of downstream control (Henderson, 1966: pp. 47–49).
3. The symbol used for the channel width may be:

   B channel free-surface width (m) (e.g. Henderson, 1966),
   W channel bottom width (m), also commonly used for rectangular channel width.

3.4 PROPERTIES OF COMMON OPEN-CHANNEL SHAPES

3.4.1 Properties

In practice, natural and man-made channels do not have often a rectangular cross-section. The most common shapes of artificial channels are circular (i.e. made of pipes) or trapezoidal.

The geometrical characteristics of the most common open channel shapes are summarized in the next table.

Shape	Flow depth d	Free-surface width B	Cross-sectional area A	Wetted perimeter P_w	Hydraulic diameter D_H
Trapezoidal		$W + 2d \cot \delta$	$d(W + d \cot \delta)$	$W + \dfrac{2d}{\sin \delta}$	$4 \dfrac{A}{P_w}$
Triangular		$2d \cot \delta$	$d^2 \cot \delta$	$\dfrac{2d}{\cot \delta}$	$2d \cos \delta$
Circular	$\dfrac{D}{2}\left(1 - \cos \dfrac{\delta}{2}\right)$	$D \sin \dfrac{\delta}{2}$	$\dfrac{D^2}{8}(\delta - \sin \delta)$	$\delta \dfrac{D}{2}$	$D\left(1 - \dfrac{\sin \delta}{\delta}\right)$
Rectangular		W	Wd	$W + 2d$	$4 \dfrac{Bd}{B + 2d}$

Notes: A: cross-sectional area; B: free-surface width; D: pipe diameter (circular pipe); D_H: hydraulic diameter; d: flow depth on the channel centreline; P_w: wetted perimeter.

Note

The hydraulic diameter (also called equivalent pipe diameter) D_H is defined as:

$$D_H = 4 \frac{A}{P_w}$$

where A is the cross-sectional area and P_w is the wetted perimeter.

3.4.2 Critical flow conditions

Critical flow conditions are defined as the flow conditions for which the mean specific energy is minimum. In a horizontal channel and assuming hydrostatic pressure distribution, critical flow conditions imply (Section 3.3.3):

$$\frac{A_c^3}{B_c} = \frac{Q^2}{g} \tag{3.34a}$$

where A_c and B_c are respectively the cross-sectional area and free-surface width at critical flow conditions. Typical results are summarized in the following table:

Channel cross-sectional shape	Critical flow depth d_c	Minimum specific energy E_{min}
General shape	$\dfrac{A_c^3}{B_c} = \dfrac{Q^2}{g}$	$d_c + \dfrac{1}{2} \dfrac{A_c}{B_c}$

(continued)

Channel cross-sectional shape	Critical flow depth d_c	Minimum specific energy E_{min}
Trapezoidal	$$\dfrac{(d_c(W + d_c \cot \delta))^3}{W + 2d_c \cot \delta} = \dfrac{Q^2}{g}$$	$$d_c + \dfrac{1}{2} \dfrac{d_c(W + d_c \cot \delta)}{W + 2d_c \cot \delta}$$
Triangular	$$d_c^5 = 2\dfrac{Q^2}{g(\cot \delta)^2}$$	$$\dfrac{5}{4}d_c$$
Rectangular	$$d_c^3 = \dfrac{Q^2}{gB^2}$$	$$\dfrac{3}{2}d_c$$

For a rectangular channel the critical flow depth is:

$$d_c = \sqrt[3]{\dfrac{Q^2}{gB^2}} \qquad \text{Rectangular channel} \qquad (3.34b)$$

3.5 EXERCISES

In a 3.5 m wide rectangular channel, the flow rate is $14 \, m^3/s$. Compute the flow properties in the three following cases:

	Case 1	Case 2	Case 3	Units
Flow depth	0.8	1.15	3.9	m
Cross-sectional area				
Wetted perimeter				
Mean flow velocity				
Froude number				
Specific energy				

Considering a two-dimensional flow in a rectangular channel, the velocity distribution follows a power law:

$$\frac{V}{V_{max}} = \left(\frac{y}{d}\right)^{1/N}$$

where d is the flow depth and V_{max} is a characteristic velocity.

(a) Develop the relationship between the maximum velocity V_{max} and the mean flow velocity Q/dB.
(b) Develop the expression of the Coriolis coefficient and the momentum correction coefficient as functions of the exponent N only.

Explain in words the meaning and significance of the momentum correction coefficient. Develop a simple flow situation for which you write the fundamental equation(s) using the momentum correction coefficient. Using this example, discuss in words the correction resulting from taking into account the momentum correction coefficient.

Considering water discharging in a wide inclined channel of rectangular cross-section, the bed slope is 30° and the flow is a fully developed turbulent shear flow. The velocity distribution follows a 1/6th power law:

$$\text{i.e. } V(y) \text{ is proportional to } \left(\frac{y}{d}\right)^{1/6}$$

where y is the distance from the channel bed measured perpendicular to the bottom, $V(y)$ is the velocity at a distance y normal from the channel bottom and d is the flow depth. The discharge per unit width in the channel is $2\,m^2/s$. At a gauging station (in the gradually varied flow region), the observed flow depth equals $0.9\,m$. At that location, compute and give the values (and units) of the specified quantities in the following list: (a) velocity at $y = 0.1\,m$, (b) velocity at the free surface, (c) momentum correction coefficient, (d) Coriolis coefficient, (e) bottom pressure, (f) specific energy.

Considering a smooth transition in a rectangular open channel, compute the downstream flow properties as functions of the upstream conditions in the following cases:

	Upstream conditions	D/S conditions	Units
d	5		m
V	0.5		m/s
B	20	18	m
Q			
z_0	6	6	m
E			
d	0.1		m
V			
B	300	300	m
Q	12	12	m^3/s
z_0	0	−1	m
E			
d	1		m
V	2		m/s
B	0.55	0.65	m
Q			m^3/s
z_0	3	2.2	m
E			

Considering a horizontal channel of rectangular cross-section, develop the relationship between E, Q and d. Demonstrate that, for a fixed specific energy E, the maximum discharge (in a horizontal channel) is obtained if and only if the flow conditions are critical.

Considering an un-gated spillway crest, the reservoir free-surface elevation above the crest is $0.17\,m$. The spillway crest is rectangular and $5\,m$ long. Assuming a smooth spillway crest, calculate the water discharge into the spillway.

Considering a broad-crested weir, draw a sketch of the weir. What is the main purpose of a broad-crest weir? A broad-crested weir is installed in a horizontal and smooth channel of rectangular cross-section. The channel width is 10 m. The bottom of the weir is 1.5 m above the channel bed. The water discharge is 11 m³/s. Compute the depth of flow upstream of the weir, above the sill of the weir and downstream of the weir (in absence of downstream control), assuming that critical flow conditions take place at the weir crest.

A prototype channel is 1000 m long and 12 m wide. The channel is rectangular and the flow conditions are $d = 3$ m and $Q = 15$ m³/s. Calculate the size (length, width and flow depth) and discharge of a 1/25 scale model using a Froude similitude.

Considering a rectangular and horizontal channel (z_o constant and H fixed), investigate the effects of change in channel width. Using continuity and Bernoulli equations, deduce the relationship between the longitudinal variation of flow depth ($\Delta d/\Delta s$), the variation in channel width ($\Delta B/\Delta s$) and the Froude number.
   Application: The upstream flow conditions being: $d_1 = 0.05$ m, $B_1 = 1$ m, $Q = 10$ l/s and the downstream channel width being $B_2 = 0.8$ m, compute the downstream flow depth and downstream Froude number. Is the downstream flow sub-, super- or critical?

A rectangular channel is 23 m wide and the water flows at 1.2 m/s. The channel contracts smoothly to 17.5 m width without energy loss.

• If the flow rate is 41 m³/s, what is the depth and flow velocity in the channel contraction?
• If the flow rate is 0.16 m³/s, what is the depth and flow velocity in the channel contraction?

A broad-crested weir is a flat-crested structure with a crest length large compared to the flow thickness. The ratio of crest length to upstream head over crest must be typically greater than 1.5–3 (Chapter 19). Critical flow conditions occur at the weir crest. If the crest is 'broad' enough for the flow streamlines to be parallel to the crest, the pressure distribution above the crest is hydrostatic and critical depth is recorded on the weir.
   Considering a horizontal rectangular channel ($B = 15$ m), the crest of the weir is 1.2 m above the channel bed. Investigate the two upstream flow conditions (assume supercritical d/s flow) and complete the following table:

	Case 1	Case 2
Upstream flow depth	1.45 m	2.1 m
Upstream Froude number		
Flow depth above the weir crest		
Upstream specific energy		
Velocity above the crest		
Specific energy above the crest		
Downstream flow depth		
Downstream Froude number		
Downstream specific energy		
Volume discharge	3.195 m³/s	

Notes: For a horizontal rectangular channel, the Froude number is defined as: $Fr = V/\sqrt{gd}$. The solution of Case 2 requires the solution of a cubic equation (Appendix A1.4).

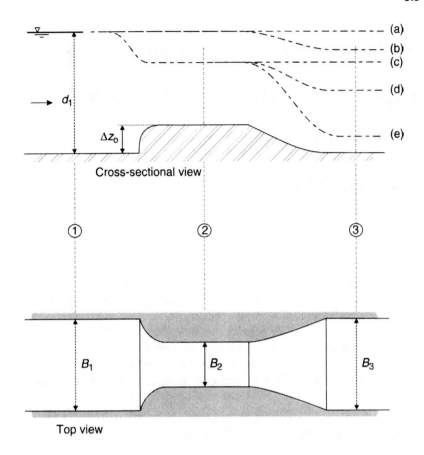

Fig. E.3.1 Sketch of the measuring flume.

Solution
Case 2: $d_1 = 2.1$ m, $\Delta z = 1.2$ m, $Q = 22.8$ m^3/s, $d_3 = 0.25$ m (*d/s* depth).

A measuring flume is a short channel contraction designed to induce critical flow conditions and to measure (at the throat) the discharge. Considering the measuring flume sketched in Fig. E.3.1, the inlet, throat and outlet sections are rectangular. The channel bed is smooth and it is horizontal at sections 1, 2 and 3. The channel width equals $B_1 = B_3 = 15$ m, $B_2 = 8.5$ m. The weir height is: $\Delta z_o = 1.1$ m. The total flow rate is 88 m^3/s. We shall assume that critical flow conditions (and hydrostatic pressure distribution) occur at the weir crest (i.e. section 2). We shall investigate the channel flow conditions (assuming a supercritical downstream flow).

(a) At section 2, compute: flow depth, specific energy.
(b) In Fig. E.3.1, five water surface profiles are labelled a, b, c, d and e. Which is the correct one?
(c) At section 1, compute the following properties: specific energy, flow depth and flow velocity.
(d) At section 3, compute the specific energy, flow depth and flow velocity.

Applications of the momentum principle: hydraulic jump, surge and flow resistance in open channels

4

Summary

In this chapter the momentum equation is applied to open channel flows. The momentum principle is always used for hydrodynamic force calculations: e.g. force acting on a gate, flow resistance in uniform equilibrium flow. Other applications include the hydraulic jump, surge and bore. Hydraulic jump and positive surge calculations are developed for frictionless flow. Later, the momentum equation is combined with flow resistance formulations. Several calculations of friction loss in open channel are discussed with analogy to pipe flow. Head losses calculations are developed with the Darcy, Chézy, Manning and Strickler coefficients.

4.1 MOMENTUM PRINCIPLE AND APPLICATION

4.1.1 Introduction

In Chapter 3, subcritical and supercritical flows, controls and control sections were introduced. The basic results may be summarized as follows:

(a) critical flow occurs for the minimum specific energy; at critical flow conditions the Froude number is unity;

(b) at a control section critical flow conditions take place and this fixes a unique relationship between depth and discharge in the vicinity (e.g. sluice gate and weir);

(c) subcritical flows are controlled from downstream (e.g. reservoir) while supercritical flows have upstream control (e.g. spillway and weir);

(d) a control influences both the flows upstream and downstream of the control section: i.e. downstream flow control and upstream flow control, respectively.

There are, however, flow situations for which the use of Bernoulli equation is not valid (e.g. significant friction loss and energy loss) or cannot predict the required parameters (e.g. force exerted by the fluid onto a structure). In such cases the momentum principle may be used in conjunction with the continuity equation.

4.1.2 Momentum principle

The momentum principle states that, for a selected control volume, the rate of change in momentum flux equals the sum of the forces acting on the control volume. For a horizontal

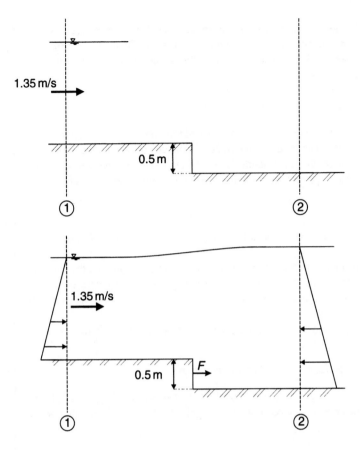

Fig. 4.1 Application of the momentum equation to a negative step.

rectangular open channel, the momentum equation applied in the flow direction leads to:

$$\rho Q V_2 - \rho Q V_1 = \frac{1}{2}\rho g d_1^2 B - \frac{1}{2}\rho g d_2^2 B - F_{\text{fric}} \qquad (4.1)$$

where V and d are, respectively, the mean velocity and flow depth, the subscripts 1 and 2 refer to the upstream and downstream cross-sections respectively (e.g. Fig. 4.1), ρ is the fluid density, B is the channel width, Q is the total flow rate and F_{fric} is the boundary friction force.

In addition, the continuity equation gives:

$$Q = V_1 d_1 B = V_2 d_2 B \qquad (4.2)$$

Discussion

The total momentum flux across a section equals: $\rho V V A = \rho Q V$ (see Section 2.2.3). The rate of change in momentum flux is then $(\rho Q V_2 - \rho Q V_1)$ in a horizontal channel. The forces acting on a control volume include the pressure forces, friction force and gravity force. For a horizontal rectangular channel with hydrostatic distribution, the pressure force at a section is $\rho g d^2 B/2$.

Further discussions on the momentum principle have been developed in Section 2.2.3. If the velocity distribution is not uniform, the momentum correction coefficient or Boussinesq coefficient must be taken into account (see Section 3.2.2). The momentum flux is then expressed as $\beta \rho V V A = \beta \rho Q V$ where β is the Boussinesq coefficient.

Application

The backward-facing step, sketched in Fig. 4.1, is in a 5 m wide channel of rectangular cross-section. The total discharge is 55 m³/s. The fluid is water at 20°C. The bed of the channel, upstream and downstream of the step, is horizontal and smooth. Compute the pressure force acting on the vertical face of the step and indicate the direction of the force (i.e. upstream or downstream).

Solution

First we sketch a free-surface profile between sections 1 and 2 (we might have to modify it after further calculations).

Secondly we must compute the upstream and downstream flow properties. Applying the continuity equation, the upstream flow depth equals $d_1 = 8.15$ m. Assuming a hydrostatic pressure distribution, the upstream specific energy equals $E_1 = 8.24$ m. Note that the upstream flow is subcritical ($Fr_1 = 0.15$).

The downstream specific energy is deduced from the Bernoulli principle assuming a smooth transition:

$$E_1 + \Delta z_o = E_2 \quad \text{Bernoulli equation}$$

where Δz_o is the drop height (0.5 m). It gives $E_2 = 8.74$ m. The downstream flow depth is deduced from the definition of the specific energy (in a rectangular channel assuming a hydrostatic pressure distribution):

$$E_2 = d_2 + \frac{Q^2}{2gB^2 d_2^2}$$

where B is the channel width. The solution of the equation is $d_2 = 8.66$ m. Note that the drop in bed elevation yields an increase of flow depth. Indeed the upstream flow being subcritical ($d_1 = 8.15$ m $> d_c = 2.31$ m), the relationship specific energy versus flow depth (Fig. 3.5) predicts such an increase in flow depth for an increasing specific energy.

Now let us consider the forces acting on the control volume contained between sections 1 and 2: the pressure forces at sections 1 and 2, the weight of the control volume, the reaction force of the bed (equal, in magnitude, to the weight in absence of vertical fluid motion) and the pressure force F exerted by the vertical face of the step on the fluid. Friction force is zero as the channel is assumed smooth (i.e. frictionless).

The momentum equation as applied to the control volume between Sections 4.1 and 4.2 is:

$$\rho Q V_2 - \rho Q V_1 = \frac{1}{2} \rho g d_1^2 B - \frac{1}{2} \rho g d_2^2 B + F \quad \text{Momentum principle}$$

in the horizontal direction.

The solution of the momentum equation is $F = +205$ kN (remember F is the force exerted by the step onto the fluid). In other words, the pressure force exerted by the fluid onto the vertical face of the step acts in the upstream direction and equals 205 kN.

4.1.3 Momentum function

The momentum function in a rectangular channel is defined as (e.g. Henderson, 1966: pp. 67–70):

$$M = \frac{\rho QV + F_p}{\rho g B} \tag{4.3a}$$

where F_p is the pressure force acting in the flow direction on the channel cross-sectional area (i.e. $F_p = 1/2\rho g d^2 B$, in a horizontal channel). For a flat rectangular channel and assuming a hydrostatic pressure distribution, the momentum function becomes:

$$M = \frac{d^2}{2} + \frac{q^2}{gd} \tag{4.3b}$$

where q is the discharge per unit width. In dimensionless terms, it yields:

$$\frac{M}{d_c^2} = \frac{1}{2}\left(\frac{d}{d_c}\right)^2 + \frac{d_c}{d} \tag{4.3c}$$

Application

The momentum equation (4.1) can be rewritten in terms of the momentum function:

$$M_2 - M_1 = -\frac{F_{fric}}{\rho g}$$

Note

The momentum function is minimum at critical flow conditions. For a momentum function such as $M > M_{min}$ there are two possible depths called the *sequent* or *conjugate depths*. For a non-rectangular channel the momentum function M becomes (Henderson, 1966: pp. 72–73):

$$M = \bar{d}A + \frac{Q^2}{gA}$$

where $\bar{d}$ is the depth from the surface to the centroid of the section.

4.2 HYDRAULIC JUMP

4.2.1 Presentation

An open channel flow can change from subcritical to supercritical in a relatively 'low-loss' manner at gates or weirs. In these cases the flow regime evolves from subcritical to supercritical with the occurrence of critical flow conditions associated with relatively small energy loss (e.g. broad-crested weir). The transition from supercritical to subcritical flow is, on the other hand, characterized by a strong dissipative mechanism. It is called a hydraulic jump (Figs 4.2 and 4.3).

Note

An example of subcritical-to-supercritical transition is the flow at the crest of a steep spillway. Upstream of the channel intake, the flow motion in the reservoir is tranquil. Along the steep chute, the flow is supercritical. Critical flow conditions are observed near the crest of the spillway. In practice, it is reasonable to assume critical flow at the spillway crest although the pressure distribution is not hydrostatic at the crest depending upon the crest shape (e.g. see discussions Rouse, 1938: p. 325; Creager *et al.*, 1945: Vol. II, p. 371).

Definition

Considering a channel with both an upstream control (e.g. sluice gate) leading to a supercritical flow (downstream of the control section) and a downstream control (e.g. reservoir) imposing a subcritical flow at the downstream end, the channel conveys an upstream supercritical flow, a downstream subcritical flow and a transition flow.

The transition from a supercritical flow to a subcritical flow is called a *hydraulic jump*. A hydraulic jump is extremely turbulent. It is characterized by the development of large-scale turbulence, surface waves and spray, energy dissipation and air entrainment (Fig. 4.3). The large-scale turbulence region is usually called the 'roller'. A hydraulic jump is a region of rapidly varied flow.

The flow within a hydraulic jump is extremely complicated (Hager, 1992b) and it is not required usually to consider its details. To evaluate the basic flow properties and energy losses in such a region, the momentum principle is used.

4.2.2 Basic equations

For a *steady flow* in a *horizontal rectangular* channel of *constant channel width*, the three fundamental equations become:

(A) *Continuity equation*

$$Q = V_1 d_1 B = V_2 d_2 B \tag{4.2}$$

where V_1 and d_1 are, respectively, the velocity and flow depth at the upstream cross-section (Fig. 4.4), V_2 and d_2 are defined at the downstream cross-section, B is the channel width and Q is the total flow rate.

Fig. 4.2 Hydraulic jump in a natural waterway: Bald Rock Creek at the Junction Qld, Australia (9 November 1997) – flow from the left to the right.

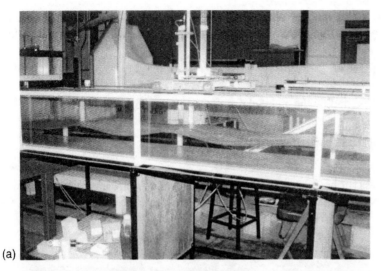

(a)

(b)

(c)

Fig. 4.3 Photographs of hydraulic jump in a rectangular channel (a) Undular hydraulic jump: $Fr_1 = 125$, $d_1 = 0.081$ m, $B = 0.5$ m, flow from the left to the right (courtesy of Ms Chantal Donnely); (b) Steady/strong jump: $Fr_1 = 9.4$, $d_1 = 0.013$ m, $B = 0.5$ m, flow from the left to the right; (c) Strong jump: $Fr_1 = 13.2$, $d_1 = 0.013$ m, $B = 0.5$ m, flow from the left to the right.

(B) *Momentum equation (Bélanger equation)*[1]

The momentum equation states that the sum of all the forces acting on the control volume equals the change in momentum flux across the control volume. For a hydraulic jump, it yields:

$$\left(\frac{1}{2}\rho g d_1^2 - \frac{1}{2}\rho g d_2^2 \right) B - F_{\text{fric}} = \rho Q (V_2 - V_1) \tag{4.3}$$

where F_{fric} is a drag force exerted by the channel roughness on the flow (Fig. 4.4).

(C) *Energy equation*

The energy equation (2.24) can be transformed as:

$$H_1 = H_2 + \Delta H \tag{4.4a}$$

where ΔH is the energy loss (or head loss) at the jump, and H_1 and H_2 are upstream and downstream total heads, respectively. Assuming a hydrostatic pressure distribution and taking the channel bed as the datum, equation (4.4a) becomes:

$$\frac{V_1^2}{2g} + d_1 = \frac{V_2^2}{2g} + d_2 + \Delta H \tag{4.4b}$$

Note that equations (4.1)–(4.4) were developed assuming hydrostatic pressure distributions at both the upstream and downstream ends of the control volume (Fig. 4.4). Furthermore, the upstream and downstream velocity distributions were assumed uniform for simplicity.

Notes

1. In the momentum equation, (ρV) is the momentum per unit volume.
2. In the simple case of uniform velocity distribution, the term (ρVV) is the momentum flux per unit area across the control surface. (ρVVA) is the total momentum flux across the control surface.

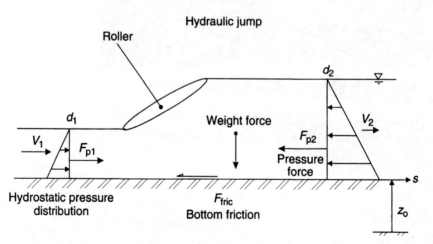

Fig. 4.4 Application of the momentum equation to a hydraulic jump.

---

[1] Jean-Baptiste Bélanger (1789–1874) was the first to suggest the application of the momentum principle to the hydraulic jump flow (Bélanger, 1828). The momentum equation applied across a hydraulic jump is often called the Bélanger equation.

Neglecting the drag force on the fluid, the continuity and momentum equations provide a relationship between the upstream and downstream flow depths as:

$$d_2 = \sqrt{\left(\frac{d_1}{2}\right)^2 + \frac{2Q^2}{gd_1B^2}} - \frac{d_1}{2}$$ (4.5a)

or in dimensionless terms:

$$\frac{d_2}{d_1} = \frac{1}{2}\left(\sqrt{1 + 8Fr_1^2} - 1\right)$$ (4.5b)

where Fr_1 is the upstream Froude number: $Fr_1 = V_1/\sqrt{gd_1}$. It must be noted that $Fr_1 > 1$. The depths d_1 and d_2 are referred to as *conjugate depths* (or sequent depths). Using equation (4.5) the momentum equation yields:

$$Fr_2 = \frac{2^{3/2} Fr_1}{\left(\sqrt{1 + 8Fr_1^2} - 1\right)^{3/2}}$$ (4.6)

where Fr_2 is the downstream Froude number. The energy equation gives the head loss:

$$\Delta H = \frac{(d_2 - d_1)^3}{4d_1d_2}$$ (4.7a)

and in dimensionless terms:

$$\frac{\Delta H}{d_1} = \frac{\left(\sqrt{1 + 8Fr_1^2} - 3\right)^3}{16\left(\sqrt{1 + 8Fr_1^2} - 1\right)}$$ (4.7b)

Equations (4.5)–(4.7) are summarized in Fig. 4.5. Figure 4.5 provides means to estimate rapidly the jump properties as functions of the upstream Froude number. For example, for $Fr_1 = 5$, we can deduce $Fr_2 \sim 0.3$, $d_2/d_1 \sim 6.5$, $\Delta H/d_1 \sim 7$.

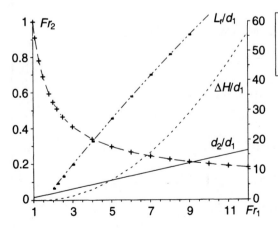

Fig. 4.5 Flow properties downstream of a hydraulic jump in a rectangular horizontal channel.

Notes

1. In a hydraulic jump, the downstream flow depth d_2 is always larger than the upstream flow depth d_1.
2. The main parameter of a hydraulic jump is its upstream Froude number Fr_1.
3. If only the downstream flow conditions are known, the solution of the continuity and momentum equations is given by:

$$\frac{d_1}{d_2} = \frac{1}{2}\left(\sqrt{1 + 8Fr_2^2} - 1\right)$$

4. A hydraulic jump is a very effective way of dissipating energy. It is also an effective mixing device. Hydraulic jumps are commonly used at the end of spillway or in dissipation basin to destroy much of the kinetic energy of the flow. The hydraulic power dissipated in a jump equals; $\rho g Q \Delta H$ where ΔH is computed using equation (4.7).
5. Hydraulic jumps are characterized by air entrainment. Air is entrapped at the impingement point of the supercritical inflow with the roller. The rate of air entrainment may be approximated as:

$$\frac{Q_{air}}{Q} \approx 0.018\,(Fr_1 - 1)^{1.245} \qquad \text{(Rajaratnam, 1967)}$$

$$\frac{Q_{air}}{Q} \approx 0.014(Fr_1 - 1)^{1.4} \qquad \text{(Wisner, 1965)}$$

Wood (1991) and Chanson (1997) presented complete reviews of these formulae.
If the jump is in a closed duct then, as the air is released, it could accumulate on the roof of the duct. This phenomena is called 'blowback' and has caused failures in some cases (Falvey, 1980).

6. Recent studies showed that the flow properties (including air entrainment) of hydraulic jumps are not only functions of the upstream Froude number but also of the upstream flow conditions: i.e. partially developed or fully developed inflow. The topic is currently under investigation.

Application

Considering a hydraulic jump in a horizontal rectangular channel, write the continuity equation and momentum principle. Neglecting the boundary shear, deduce the relationships $d_2/d_1 = f(Fr_1)$ and $Fr_2 = f(Fr_1)$.

Solution

The continuity equation and the momentum equation (in the flow direction) are respectively:

$$q = V_1 d_1 = V_2 d_2 \qquad \text{[C]}$$

$$\frac{1}{2}\rho g d_1^2 - \frac{1}{2}\rho g d_2^2 = \rho q\,(V_2 - V_1) \qquad \text{[M]}$$

where q is the discharge per unit width.

[C] implies $V_2 = V_1 d_1/d_2$. Replacing [C] into [M], it yields:

$$\frac{1}{2}\rho g d_1^2 - \frac{1}{2}\rho g d_2^2 = \rho V_1^2 d_1\left(\frac{d_1}{d_2}\right) - \rho V_1^2 d_1$$

Dividing by $(\rho g d_1^2)$ it becomes:

$$\frac{1}{2} - \frac{1}{2}\left(\frac{d_2}{d_1}\right)^2 = Fr_1^2\left(\frac{d_1}{d_2}\right) - Fr_1^2$$

After transformation we obtain a polynomial equation of degree three in terms of d_2/d_1:

$$\frac{1}{2}\left(\frac{d_2}{d_1}\right)^3 - \left(\frac{1}{2} + Fr_1^2\right)\left(\frac{d_2}{d_1}\right) + Fr_1^2 = 0$$

or

$$\frac{1}{2}\left(\frac{d_2}{d_1} - 1\right)\left[\left(\frac{d_2}{d_1}\right)^2 + \left(\frac{d_2}{d_1}\right) - 2Fr_1^2\right] = 0$$

The solutions of the momentum equation are the obvious solution $d_2 = d_1$ and the solutions of the polynomial equation of degree two:

$$\left(\frac{d_2}{d_1}\right)^2 + \left(\frac{d_2}{d_1}\right) - 2Fr_1^2 = 0$$

The (only) meaningful solution is:

$$\frac{d_2}{d_1} = \frac{1}{2}\left(\sqrt{1 + 8Fr_1^2} - 1\right)$$

Using the continuity equation in the form $V_2 = V_1 d_1/d_2$, and dividing by $\sqrt{gd_2}$, it yields:

$$Fr_2 = \frac{V_2}{\sqrt{gd_2}} = \frac{V_1}{\sqrt{gd_1}}\left(\frac{d_1}{d_2}\right)^{3/2} = Fr_1 \frac{2^{3/2}}{\left(\sqrt{1 + 8Fr_1^2} - 1\right)^{3/2}}$$

Application
A hydraulic jump takes place in a 0.4 m wide laboratory channel. The upstream flow depth is 20 mm and the total flow rate is 31 l/s. The channel is horizontal, rectangular and smooth. Calculate the downstream flow properties and the energy dissipated in the jump. If the dissipated power could be transformed into electricity, how many 100 W bulbs could be lighted with the jump?

Solution
The upstream flow velocity is deduced from the continuity equation:

$$V_1 = \frac{Q}{Bd_1} = \frac{31 \times 10^{-3}}{0.4 \times (20 \times 10^{-3})} = 3.875 \text{ m/s}$$

The upstream Froude number equals $Fr_1 = 8.75$ (i.e. supercritical upstream flow). The downstream flow properties are deduced from the above equation:

$$\frac{d_2}{d_1} = \frac{1}{2}\left(\sqrt{1 + 8Fr_1^2} - 1\right) = 11.9$$

$$Fr_2 = Fr_1 \frac{2^{3/2}}{\left(\sqrt{1 + 8Fr_1^2} - 1\right)^{3/2}} = 0.213$$

Hence $d_2 = 0.238$ m and $V_2 = 0.33$ m/s.
    The head loss in the hydraulic jump equals:

$$\Delta H = \frac{(d_2 - d_1)^3}{4d_1 d_2} = 0.544 \text{ m}$$

The hydraulic power dissipated in the jump equals:

$$\rho g Q \Delta H = 998.2 \times 9.8 \times 31 \times 10^{-3} \times 0.544 = 165 \text{ W}$$

where ρ is the fluid density (kg/m^3), Q is in m^3/s and ΔH is in m. In the laboratory flume, the dissipation power equals 165 W: one 100 W bulb and one 60 W bulb could be lighted with the jump power.

4.2.3 Discussion

Types of hydraulic jump
Experimental observations highlighted different types of hydraulic jumps, depending upon the upstream Froude number Fr_1 (e.g. Chow, 1973: p. 395). For hydraulic jumps in rectangular horizontal channels, Chow's classification is commonly used and it is summarized in the following table:

Fr_1	Definition (Chow, 1973)	Remarks
1	Critical flow	No hydraulic jump
1–1.7	Undular jump	Free-surface undulations developing downstream of jump over considerable distances. Negligible energy losses. Also called Fawer jump in homage to the work of Fawer (1937)
1.7–2.5	Weak jump	Low energy loss
2.5–4.5	Oscillating jump	Wavy free surface. Unstable oscillating jump. Production of large waves of irregular period. Each irregular oscillation produces a large wave which may travel far downstream, damaging and eroding the banks. *To be avoided* if possible
4.5–9	Steady jump	45–70% of energy dissipation. Steady jump. Insensitive to downstream conditions (i.e. tailwater depth). *Best economical design*
>9	Strong jump	Rough jump. Up to 85% of energy dissipation. *Risk of channel bed erosion*. To be avoided

Notes
1. The above classification must be considered as *rough* guidelines.
    For example, experiments performed at the University of Queensland (Australia) showed the existence of undular hydraulic jumps for upstream Froude number between 1 and 3 (Chanson,

1995a). Furthermore, the inflow conditions (uniform velocity distribution, partially developed or fully developed) modify substantially the flow properties and affect the classification of jumps. Also, the shape of the channel cross-section affects the hydraulic jump characteristics (Hager, 1992b). Note that the above table is given for a rectangular cross-section only.

2. A hydraulic jump is a very unsteady flow. Experimental measurements of bottom pressure fluctuations indicated that the mean pressure is quasi-hydrostatic below the jump *but large pressure fluctuations* are observed (see reviews in Hager, 1992b and Chanson, 1995b). The re-analysis of bottom pressure fluctuation records below hydraulic jumps over long periods of time indicates that the extreme minimum pressures might become negative (i.e. below atmospheric pressure) and could lead to uplift pressures on the channel bottom. The resulting uplift loads on the channel bed might lead to substantial damage, erosion and destruction of the channel.

Length of the roller

Chow (1973) proposed some guidelines to estimate the length of the roller of hydraulic jump as a function of the upstream flow conditions. Hager *et al.* (1990) reviewed a broader range of data and correlations. For wide channel (i.e. $d_1/B < 0.10$), they proposed the following correlation:

$$\frac{L_r}{d_1} = 160 \tanh\left(\frac{Fr_1}{20}\right) - 12 \qquad 2 < Fr_1 < 16 \tag{4.8}$$

where L_r is the length of the roller. Equation (4.8) is valid for rectangular horizontal channels with $2 < Fr_1 < 16$. Such a correlation can be used when designing energy dissipation basins (see Fig. 4.5).

Notes

1. The hyperbolic tangent tanh is defined as:

$$\tanh(x) = \frac{e^x - e^{-x}}{e^x + e^{-x}}$$

2. Equation (4.8) is an empirical correlation based upon model and prototype data. It fits reasonably well experimental data for hydraulic jumps with partially developed inflow conditions (in rectangular channels).

Application: energy dissipation basin

Hydraulic jumps are known for their energy dissipation performances. Prior to late 19th century, designers tried to avoid hydraulic jumps whenever possible to minimize the risks of channel destruction. Since the beginning of the 20th century and with the introduction of high-resistance materials (e.g. reinforced concrete), hydraulic jumps are used to dissipate flow energy downstream of supercritical flow structures (e.g. spillways and bottom outlets) (e.g. Henderson, 1966: pp. 221–225).

In practice, energy dissipation structures[2] are designed to induce a steady jump or a strong jump. The lowest design (inflow) Froude number must be above 4.5. The selection of a strong jump requires a careful analysis of the risks of bed erosion.

---

[2] Energy dissipation structures are also called stilling basin, transition structures or energy dissipators.

Application

Considering a dissipation basin at the downstream end of a spillway, the total discharge is $Q = 2000\,\text{m}^3/\text{s}$. The energy dissipation structure is located in a horizontal rectangular channel (25 m wide). The flow depth at the downstream end of the spillway is 2.3 m. Compute the energy dissipation in the basin.

Solution

The upstream flow conditions of the jump are:

$$d_1 = 2.3\,\text{m}$$

$$V_1 = 34.8\,\text{m/s}$$

$$Fr_1 = 7.3 \qquad \text{(i.e. steady jump)}$$

The downstream flow conditions are:

$$d_2 = 22.707\,\text{m}$$

$$V_2 = 3.52\,\text{m/s}$$

$$Fr_2 = 0.236$$

The head loss across the jump equals:

$$\Delta H = 40.7\,\text{m}$$

The power dissipated in the jump is:

$$\rho g Q \Delta H = 796 \times 10^6\,\text{W} \qquad \text{(i.e. nearly 800 MW!)}$$

In a dissipation basin with flat horizontal bottom, the location of the jump may change as a function of the upstream and downstream flow conditions. That is, with a change of (upstream or downstream) flow conditions, the location of the jump changes in order to satisfy the Bélanger equation (4.1). The new location might not be suitable and could require a very long and uneconomical structure.

In practice, design engineers select desirable features to make the jump stable and as short as possible. Abrupt rise and abrupt drop, channel expansion and channel contraction, chute blocks introduce additional flow resistance and tend to promote the jump formation. Hager (1992b) described a wide range of designs.

Application

Considering a hydraulic jump in a horizontal rectangular channel located immediately upstream of an abrupt rise (Fig. 4.6), estimate the downstream flow depth d_3 (see Fig. 4.6) for the design flow conditions $d_1 = 0.45\,\text{m}$, $V_1 = 10.1\,\text{m/s}$. The step height equals $\Delta z_o = 0.5\,\text{m}$.

Solution

First, we will assume that the complete jump is located upstream of the bottom rise as sketched in Fig. 4.6.

The continuity equation between sections 1 and 3 is

$$q = V_1 d_1 = V_3 d_3 \qquad \text{Continuity equation}$$

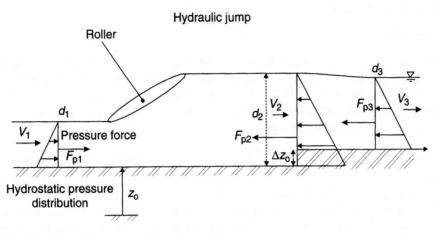

Fig. 4.6 Sketch of hydraulic jump at an abrupt bottom rise.

The momentum equation between the section 1 (upstream flow) and section 2 (i.e. immediately *upstream* of the abrupt rise) is:

$$\left(\frac{1}{2}\rho g d_1^2 - \frac{1}{2}\rho g d_2^2 \right) = \rho q (V_2 - V_1) \qquad \text{Momentum equation}$$

where q is the discharge per unit width. The momentum equation implies hydrostatic pressure distribution at section 2.

The momentum equation between the section 2 (taken immediately *downstream* of the bottom rise) and section 3 is:

$$\left(\frac{1}{2}\rho g (d_2 - \Delta z_0)^2 - \frac{1}{2}\rho g d_3^2 \right) = \rho q (V_3 - V_2) \qquad \text{Momentum equation}$$

The solution of the non-linear system of equations is:

$$q = 4.54 \, \text{m}^2/\text{s}$$

$$Fr_1 = 4.8$$

$$d_2/d_1 = 6.32 \qquad \text{(Equation 4.5)}$$

$$d_2 = 2.84 \, \text{m}$$

$$V_2 = 1.6 \, \text{m/s} \qquad \text{Continuity equation}$$

$$d_3 = 2.26 \, \text{m}$$

$$V_3 = 2.0 \, \text{m/s}$$

$$Fr_3 = 0.43$$

Comments

1. $(d_2 - \Delta z_0)$ is not equal to d_3. At section 2, the velocity V_2 is indeed slower than that at section 3.
2. The hydraulic jump remains confined upstream of the abrupt rise as long as $d_3 > d_c$ where d_c is the critical flow depth (in this case $d_c = 1.28 \, \text{m}$).
3. Experimental data showed that the pressure distribution at section 2 is not hydrostatic. In practice, the above analysis is somehow oversimplified although it provides a good order of magnitude.

4.3 SURGES AND BORES

4.3.1 Introduction

A *surge* wave results from a sudden change in flow (e.g. a partial or complete closure of a gate) that increases the depth: such a flow situation is called a positive surge (see also Section 4.3.4). In such a case the application of the momentum principle to unsteady flow is simple: the unsteady flow conditions are solved as a quasi-steady flow situation using the momentum equation developed for the hydraulic jump.

Notes
1. A positive surge is an abrupt wave front (see Section 4.3.4).
2. When the surge is of tidal origin it is usually termed a *bore*. The difference of name does not mean a difference in principle. Hydraulic bores results from the upstream propagation of tides into estuaries and river mouths (Lynch, 1982). Tricker (1965) presented numerous photographs of interest. Classical examples are described in Table 4.1 and shown in Fig. 4.7 and Plate 16.

4.3.2 Equations

Considering a positive surge in a rectangular channel (Fig. 4.8), the surge is a unsteady flow situation for an observer standing on the bank (Fig. 4.8 left). But the surge is seen by an observer travelling at the surge speed V_{srg} as a steady-flow case called a *quasi-steady hydraulic jump*.

Table 4.1 Examples of positive surges (bores) in estuaries

River	Country	Name of the bore	Reference – comments
Amazon river	Brazil	*Pororoca*	Bazin (1865b: p. 624). $\Delta d = 3.7$ to 5 m.
Qiantang river	China	Hang-chow bore	Chow (1973, p. 558), Tricker (1965). $V_{srg} = 4.1$ to 5.7 m/s, $\Delta d = 1$ to 3.7 m
Severn river	England (near Gloucester)		Tricker (1965). $V_{srg} = 3.1$ to 6 m/s, $\Delta d = 1$ to 1.8 m
Seine river	France	*Mascaret* or *La Barre*	Bazin (1865b: p. 623), Tricker (1965), Malandain (1988). $V_{srg} = 2$ to 10 m/s, $\Delta d = 2$ to 7.3 m
Dordogne river	France	*Mascaret*	At St-Pardon, Vayres, near Blaye and Macau. $\Delta d = 0.3$ to 1.2 m
Garonne river	France	*Mascaret*	At Le Tourne, Cambes, Langoiran and Cadillac
Channel Sée-Sélune	France (Bay of Mt-St-Michel)	*Mascaret*	Larsonneur (1989). Near Grouin du Sud. $\Delta d = 0.2$ to 0.7 m
Ganges river	India	Hoogly bore	Bazin (1865b: p. 623). $V_{srg} = 10$ m/s, $\Delta d = 3.7$ to 4.5 m
Petitcodiac river	Canada (Bay of Fundy)		Tricker (1965). $\Delta d = 1$ to 1.5 m. Bore affected by construction of upstream barrage
Turnagain Arm and Knik Arm	Alaska		Smaller bores after 1964 earthquake during which the bed subsided by 2.4 m (Molchan-Douthit, 1998)
Bamu river	PNG		Beaver (1920). $\Delta d = 2.7$ m

Note: V_{srg}: is the velocity of the surge and Δd is the height of the wave front.

Fig. 4.7 Photograph of tidal bore. Tidal bore at Truro, Nova Scotia, Canada (Courtesy of Larry Smith) – Cobequod Bay (Indian name), being called Salmon River upstream – looking downstream at the incoming bore.

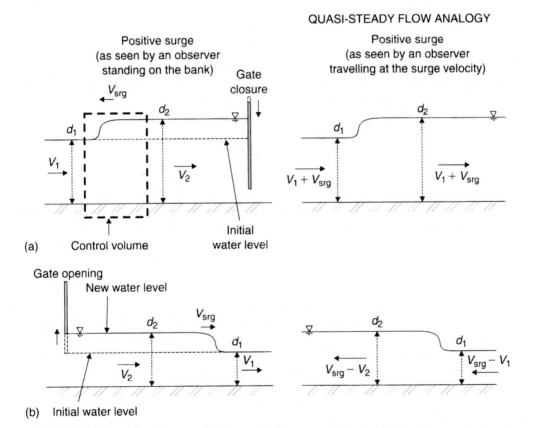

Fig. 4.8 Positive surges and wave front propagation: (a) advancing front moving upstream and (b) advancing front moving downstream.

For a rectangular horizontal channel and considering a control volume across the front of the surge travelling at a velocity V_{srg} (Fig. 4.8a), the continuity equation is:

$$(V_1 + V_{srg})\,d_1 = (V_2 + V_{srg})\,d_2 \tag{4.9a}$$

where V_{srg} is the surge velocity, as seen by an observer immobile (standing) on the channel bank, the subscript 1 refers to the initial flow conditions and the subscript 2 refers to the new flow conditions.

By analogy with the hydraulic jump (Fig. 4.4), the momentum equation for the control volume, *neglecting friction loss*, yields (see Section 4.2.2):

$$Fr_2 = \frac{2^{3/2}\,Fr_1}{\left(\sqrt{1 + 8Fr_1^2} - 1\right)^{3/2}} \tag{4.6}$$

where the Froude numbers Fr_1 and Fr_2 are defined as (Fig. 4.8a):

$$Fr_1 = \frac{V_1 + V_{srg}}{\sqrt{gd_1}}$$

$$Fr_2 = \frac{V_2 + V_{srg}}{\sqrt{gd_2}}$$

This is a system of two equations (4.9) and (4.6) with five variables (i.e. d_1, d_2, V_1, V_2, V_{srg}). Usually the upstream conditions V_1, d_1 are known and the new flow rate Q_2/Q_1 is determined by the rate of closure of the gate (e.g. complete closure $Q_2 = 0$).

Note that the continuity equation provides an estimate of the velocity of the surge:

$$V_{srg} = \frac{Q_1 - Q_2}{(d_2 - d_1)\,B} \tag{4.10}$$

Equations (4.9) and (4.6) can be solved graphically or numerically to provide the new flow depth d_2 and velocity V_2, and the surge velocity V_{srg} as functions of the initial flow conditions (i.e. d_1, V_1) and the new flow rate Q_2 (e.g. Henderson, 1966: pp. 75–77).

Notes
1. A stationary surge (i.e. $V_{srg} = 0$) is a hydraulic jump.
2. A surge can be classified as for a hydraulic jump as a function of its 'upstream' Froude number Fr_1. As an example, a surge with $Fr_1 = (V_1 + V_{srg})/\sqrt{gd_1} = 1.4$ is called an undular surge.
3. Equations (4.9) and (4.10) are valid for a positive surging moving upstream (Fig. 4.8a). For a positive surging moving downstream (Fig. 4.8b), equation (4.9) becomes:

$$(V_1 - V_{srg})\,d_1 = (V_2 - V_{srg})\,d_2 \tag{4.9b}$$

Note that V_{srg} must be larger than V_1 since the surge is moving downstream in the direction of the initial flow.

Application

Considering the flow upstream of a gate (Fig. 4.8, top), the gate suddenly closes. The initial flow conditions were $Q = 5000\,m^3/s$, $d = 5\,m$, $B = 100\,m$. The new discharge is $Q = 3000\,m^3/s$. Compute the new flow depth and flow velocity.

Solution

The surge is an advancing wave front (i.e. positive surge). Using the quasi-steady flow assumption, the flow conditions upstream of the surge front are (notation defined in Fig. 4.8) $d_1 = 5\,m$, $V_1 = 10\,m/s$, $Q_1 = 5000\,m^3/s$ (i.e. $Fr = 1.43$). The flow conditions downstream of the front surge are $Q_2 = 3000\,m^3/s$ and $B = 100\,m$.

To start the calculations, it may be assumed $V_{srg} = 0$ (i.e. stationary surge or hydraulic jump). In this particular case (i.e. $V_{srg} = 0$), the continuity equation (4.9) becomes:

$$V_1 d_1 = V_2 d_2$$

Using the definition of the Froude number, it yields:

$$Fr_1 \sqrt{gd_1^3} = Fr_2 \sqrt{gd_2^3}$$

For an initialization step where $V_{srg} = 0$, the above equation is more practical than equation (4.10).

Notation Equation	V_{srg}	Fr_1	d_2 Equation (4.10)	Fr_2 Equation (4.6)	V_2 Definition of Fr_2	$V_2 d_2 B$ Continuity equation
1st iteration	0.0	1.43	6.34[a]	0.72	10.9	2540
2nd iteration	2	1.71	15	0.62	5.5	8229
3rd iteration	3	1.86	11.7	0.58	3.2	3730
Solution	3.26	1.89	11.1	0.57	2.70	3000

Note: [a] During the initialization step, it is assumed: $d_2 = d_c$. That is, the positive surge would be a steady hydraulic jump. Note that with an upstream Froude number close to unity, the downstream flow depth is slightly greater than the critical depth (equation (4.5b)).

Comments

It must be noted that the initial flow conditions are supercritical (i.e. $V_1/\sqrt{gd_1} = 1.43$). The surge is a large disturbance travelling upstream against a supercritical flow. After the passage of the surge, the flow becomes subcritical (i.e. $V_2/\sqrt{gd_2} = 0.3$). The surge can be characterized as a weak surge ($Fr_1 = 1.89$). For a surge flow, engineer should not be confused between the surge Froude numbers $(V_1 + V_{srg})/\sqrt{gd_1}$ (and) $(V_2 + V_{srg})/\sqrt{gd_2}$, and the initial and new channel Froude numbers ($V_1/\sqrt{gd_1}$ and $V_2/\sqrt{gd_2}$ respectively). Positive surge calculations are performed with the surge Froude numbers.

4.3.3 Discussion

Considering the simple case of a positive surge travelling upstream of a sluice gate (after the gate closure), the flow sketch is sketched in Fig. 4.8 (top). Several important results derive from the basic equations and they are summarized as follows:

(a) For an observer travelling with the flow upstream of the surge front (i.e. at a velocity V_1), the celerity of the surge (relative to the upstream flow) is:

$$V_1 + V_{srg} = \sqrt{gd_1} \sqrt{\frac{1}{2}\frac{d_2}{d_1}\left(1 + \frac{d_2}{d_1}\right)} \tag{4.11}$$

Note that if $d_2 > d_1$ then $(V_1 + V_{srg}) > \sqrt{gd_1}$. For a small wave (i.e. $d_2 = d_1 + \Delta d$), the term $V_1 + V_{srg}$ tends to the celerity of a small disturbance $\sqrt{gd_1}$ (see Section 3.3.3, critical conditions and wave celerity).

(b) As $(V_1 + V_{srg}) > \sqrt{gd_1}$ a surge can move upstream even if the upstream (initial) flow is supercritical. Earlier in the monograph it was stated that a *small* disturbance celerity C cannot move upstream in a supercritical flow. However, a large disturbance can make its way against supercritical flow provided it is *large enough* and in so doing the flow becomes subcritical (see example in previous section). A positive surge is a large disturbance.

(c) Considering the flow upstream of the surge front: for an observer moving at the same speed as the upstream flow, the celerity of a small disturbance travelling upstream is $\sqrt{gd_1}$. For the same observer, the celerity of the surge (i.e. a large disturbance) is: $(V_1 + V_{srg}) > \sqrt{gd_1}$. As a result, the surge travels faster than a small disturbance. The surge overtakes and absorbs any small disturbances that may exist at the free surface of the upstream water (i.e. in front of the surge).

Relative to the downstream water the surge travels more slowly than small disturbances as $(V_2 + V_{srg}) < \sqrt{gd_2}$. Any small disturbance, downstream of the surge front and moving upstream toward the wave front, overtakes the surge and is absorbed into it. So the wave absorbs random disturbances on both sides of the surge and this makes the positive surge *stable* and *self-perpetuating*.

(d) A lower limit of the surge velocity V_{srg} is set by the fact that $(V_1 + V_{srg}) > \sqrt{gd_1}$. It may be used as the initialization step of the iterative process for solving the equations. An upper limit of the surge celerity exists but it is a function of d_2 and V_2, the unknown variables.

Note

A positive surge can travel over very long distance without losing much energy because it is self-perpetuating. In natural and artificial channels, observations have shown that the wave front may travel over dozens of kilometres.

In water supply channels, brusque operation of controls (e.g. gates) may induce large surge which might overtop the channel banks, damaging and eroding the channel. In practice, rapid operation of gates and controls must be avoided. The theoretical calculations of positive surge development are detailed in Chapter 17.

4.3.4 Positive and negative surges

Definitions
Positive surges

A positive surge is an *advancing wave front* resulting from an increase of flow depth moving upstream (i.e. closure of a downstream gate) or downstream (i.e. opening of an upstream gate and dam break) (Fig. 4.8).

Negative surges

A negative surge is a retreating wave front resulting from a decrease in flow depth moving upstream (e.g. opening of a downstream gate) or downstream (e.g. closure of an upstream gate) (Fig. 4.9) (e.g. Henderson, 1966; Montes, 1988).

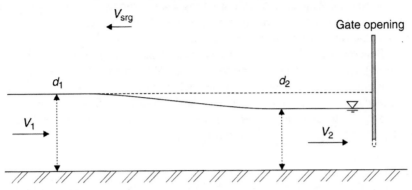

Fig. 4.9 Negative surge.

Discussion

Positive surge flows are solved using the quasi-steady flow analogy (Section 4.3.2). For negative surges, the flow is unsteady and no quasi-steady flow analogy exists. The complete unsteady analysis is necessary (e.g. Henderson, 1966; Liggett, 1994). This will be developed in Chapters 16 and 17.

For practising engineers, it is *important* to recognize between positive and negative surge cases. The table below summarizes the four possible cases:

Surge wave:	Positive surge		Negative surge	
Moving	downstream	upstream	downstream	upstream
Front of	deep water	deep water	shallow water	shallow water
Energy balance	loss of energy	loss of energy	gain of energy	gain of energy
Wave front stability	stable	stable	unstable	unstable
Analysis	quasi-steady	quasi-steady	unsteady	unsteady

Notes
1. A positive surge is characterized by a steep advancing front. It is easy to recognize. Negative surges are more difficult to notice as the free-surface curvature is very shallow.
2. *For a negative surge, the quasi-steady analysis is not valid* (Chapters 16 and 17).

4.4 FLOW RESISTANCE IN OPEN CHANNELS

4.4.1 Presentation and definitions

Introduction

In a real fluid flow situation, energy is continuously dissipated. In open channel flows, flow resistance can be neglected over a short transition[3] as a first approximation, and the continuity and

---

[3] In open channel hydraulics, the flow below a sluice gate, a hydraulic jump, the flow above a weir or at an abrupt drop or rise may be considered as short transitions.

Bernoulli equations can be applied to estimate the downstream flow properties as functions of the upstream flow conditions and boundary conditions. However, the approximation of frictionless flow is no longer valid for long channels. Considering a water supply canal extending over several kilometres, the bottom and sidewall friction retards the fluid, and, at equilibrium, the friction force counterbalances exactly the weight force component in the flow direction.

The laws of flow resistance in open channels are essentially the same as those in closed pipes (Henderson, 1966). In an open channel, the calculations of the boundary shear stress are complicated by the existence of the free surface and the wide variety of possible cross-sectional shapes. Another difference is the propulsive force acting in the direction of the flow. In closed pipes, the flow is driven by a pressure gradient along the pipe whereas, in open channel flows, the fluid is propelled by the weight of the flowing water resolved down a slope.

Head loss

For open channel flow as for pipe flow, the head loss ΔH over a distance Δs (along the flow direction) is given by the *Darcy equation*:

$$\Delta H = f \frac{\Delta s}{D_H} \frac{V^2}{2g} \tag{4.12}$$

where f is the Darcy coefficient[4], V is the mean flow velocity and D_H is the hydraulic diameter or equivalent pipe diameter. In open channels and assuming hydrostatic pressure distribution, the energy equation can be conveniently rewritten as:

$$d_1 \cos \theta_1 + z_{o1} + \alpha_1 \frac{V_1^2}{2g} = d_2 \cos \theta_2 + z_{o2} + \alpha_2 \frac{V_2^2}{2g} + \Delta H \tag{4.13}$$

where the subscripts 1 and 2 refer to the upstream and downstream cross-section of the control volume, d is the flow depth measured normal to the channel bottom, θ is the channel slope, z_o is the bed elevation, V is the mean flow velocity and α is the kinetic energy correction coefficient (i.e. Coriolis coefficient).

Notes
1. Henri P.G. Darcy (1805–1858) was a French civil engineer. He gave his name to the Darcy–Weisbach friction factor.
2. *Hydraulic diameter and hydraulic radius*
   The hydraulic diameter and hydraulic radius are defined respectively as:

$$D_H = 4 \frac{\text{cross-sectional area}}{\text{wetted perimeter}} = \frac{4A}{P_w}$$

$$R_H = \frac{\text{cross-sectional area}}{\text{wetted perimeter}} = \frac{A}{P_w} \quad \text{(e.g. Henderson, 1966: p. 91)}$$

   where the subscript H refers to the *hydraulic* diameter or radius, A is the cross-sectional area and P_w is the wetted perimeter.

---

[4] Also called the Darcy–Weisbach friction factor or head loss coefficient.

The hydraulic diameter is also called the equivalent pipe diameter. Indeed it is noticed that:

$$D_H = \text{pipe diameter } D \text{ for a circular pipe}$$

The hydraulic radius was used in the early days of hydraulics as a mean flow depth. We note that:

$$R_H = \text{flow depth } d \text{ for an open channel flow in a wide rectangular channel}$$

The author of the present textbook believes that it is preferable to use the hydraulic diameter rather than the hydraulic radius, as the friction factor calculations are done with the hydraulic diameter (and not the hydraulic radius).

Bottom shear stress and shear velocity

The average shear stress on the wetted surface or *boundary shear stress* equals:

$$\tau_o = C_d \frac{1}{2} \rho V^2 \tag{4.14a}$$

where C_d is the skin friction coefficient[5] and V is the mean flow velocity. In open channel flow, it is common practice to use the Darcy friction factor f, which is related to the skin friction coefficient by:

$$f = 4C_d$$

It yields:

$$\tau_o = \frac{f}{8} \rho V^2 \tag{4.14b}$$

The *shear velocity* V_* is defined as (e.g. Henderson, 1966: p. 95):

$$V_* = \sqrt{\frac{\tau_o}{\rho}} \tag{4.15}$$

where τ_o is the boundary shear stress and ρ is the density of the flowing fluid. The shear velocity is a measure of shear stress and velocity gradient near the boundary.

As for pipe flows, the flow regime in open channels can be either laminar or turbulent. In industrial applications, it is commonly accepted that the flow becomes turbulent for Reynolds numbers larger than 2000–3000, the *Reynolds number* being defined for pipe and open channel flows as:

$$Re = \rho \frac{V D_H}{\mu} \tag{4.16}$$

where μ is the dynamic viscosity of the fluid, D_H is the hydraulic diameter and V is the mean flow velocity.

---

[5] Also called drag coefficient or Fanning friction factor (e.g. Liggett, 1994).

Most open channel flows are turbulent. There are three types of turbulent flows: smooth, transition and fully rough. Each type of turbulent flow can be distinguished as a function of the *shear Reynolds number* defined as:

$$Re_* = \frac{V_* k_s}{\nu} \tag{4.17}$$

where k_s is the average surface roughness (e.g. Henderson, 1966: p. 95–96). For turbulent flows, the transition between smooth turbulence and fully rough turbulence is approximately defined as:

Flow situation (Reference)	Open channel flow (Henderson, 1966)	Pipe flow (Schlichting, 1979)
Smooth turbulent	$Re_* < 4$	$Re_* < 5$
Transition	$4 < Re_* < 100$	$5 < Re_* < 75$
Fully rough turbulent	$100 < Re_*$	$75 < Re_*$

Notes
1. The shear velocity being a measure of shear stress and velocity gradient near the boundary, a large shear velocity V_* implies large shear stress and large velocity gradient. The shear velocity is commonly used in sediment-laden flows to calculate the sediment transport rate.
2. The shear velocity may be rewritten as:

$$\frac{V_*}{V} = \sqrt{\frac{f}{8}}$$

where V is the mean flow velocity.

Friction factor calculation

For open channel flow the effect of turbulence becomes sensible for $Re > 2000$–3000 (e.g. Comolet, 1976). In most practical cases, open channel flows are turbulent and the friction factor (i.e. Darcy coefficient) may be estimated from the *Colebrook–White formula* (Colebrook, 1939):

$$\frac{1}{\sqrt{f}} = -2.0 \log_{10}\left(\frac{k_s}{3.71 D_H} + \frac{2.51}{Re\sqrt{f}} \right) \tag{4.18}$$

where k_s is the equivalent sand roughness height, D_H is the hydraulic diameter and Re is the Reynolds number defined as:

$$Re = \rho \frac{V D_H}{\mu} \tag{4.16}$$

Equation (4.18) is a non-linear equation in which the friction factor f is present on both the left- and right-sides. A graphical solution of the Colebrook–White formula is the Moody diagram (Moody, 1944) given in Fig. 4.10.

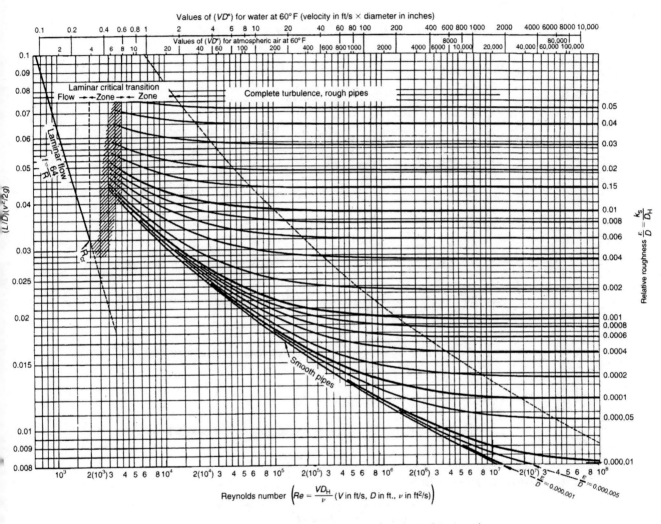

4.10 Moody diagram (after Moody, 1944, with permission of the American Society of Mechanical Engineers).

Notes

1. The Colebrook–White formula is valid *only* for turbulent flows. It can be applied to any type of turbulent flows (i.e. smooth, transition and fully rough).
2. More generally, the Darcy friction factor of open channel flows can be estimated as for pipe flows:

Laminar flow

$$f = \frac{64}{Re} \qquad Re < 2000 \qquad \text{(e.g. Streeter and Wylie, 1981: p. 238)}$$

Smooth turbulent flow

$$f = \frac{0.3164}{Re^{1/4}} \qquad Re < 1 \times 10^5 \quad \text{(Blasius' formula)}$$

$$\text{(e.g. Schlichting, 1979; Streeter and Wylie, 1981: p. 236)}$$

$$\frac{1}{\sqrt{f}} = 2.0 \log_{10}\left(Re\sqrt{f}\right) - 0.8 \qquad Re > 1 \times 10^5 \quad \text{(Karman–Nikuradse's formula)}$$

$$\text{(e.g. Henderson, 1966)}$$

Turbulent flow

$$\frac{1}{\sqrt{f}} = -2.0 \log_{10}\left(\frac{k_s}{3.71 D_H} + \frac{2.51}{Re\sqrt{f}}\right) \qquad \text{Colebrook–White's formula (Colebrook, 1939)}$$

A less-accurate formula which can be used to initialize the calculation with the Colebrook–White formula is the Altsul's formula:

$$f = 0.1\left(1.46\frac{k_s}{D_H} + \frac{100}{Re}\right)^{1/4} \qquad \text{Altsul's formula (Idelchik, 1969, 1986)}$$

Fully rough turbulent flow

$$\frac{1}{\sqrt{f}} = 2.0 \log_{10}\left(\frac{D_H}{k_s}\right) + 1.14 \qquad \text{deduced from the Colebrook's formula}$$

3. Roughness height
   Typical roughness heights are:

k_s (mm)	Material
0.01–0.02	PVC (plastic)
0.02	Painted pipe
1–10	Riveted steel
0.25	Cast iron (new)
1–1.5	Cast iron (rusted)
0.3–3	Concrete
3–10	Untreated shot-concrete
0.6–2	Planed wood
5–10	Rubble masonry
3	Straight uniform earth channel

4. The calculation of the friction factor in turbulent flow is an iterative process. In practice, the Moody diagram (Fig. 4.10) is often used to estimate the friction factor as:

$$f = f\left(\rho\frac{VD_H}{\mu}; \frac{k_s}{D_H}\right)$$

5. For calculations of head losses, Darcy coefficients and roughness heights, a recommended reference textbook for professional engineers is Idelchik (1969, 1986).

Application

Considering an open channel flow in a rectangular channel, the total discharge is 2.4 m³/s and the flow depth is 3.1 m. The channel width is 5 m and the bottom and sidewalls are made of smooth concrete. Estimate the Darcy friction factor and the average boundary shear stress. The fluid is water at 20°C.

Solution

The flow is subcritical ($Fr = 0.03$). For smooth concrete, it is reasonable to assume $k_s = 1$ mm. The flow velocity, hydraulic diameter, relative roughness and Reynolds number are respectively:

$$V = 0.155\ \text{m/s}$$
$$D_H = 5.54\ \text{m}$$
$$k_s/D_H = 1.81 \times 10^{-4}\quad \text{Relative roughness}$$
$$Re = 8.51 \times 10^5\quad \text{Reynolds number (Equation (4.16))}$$

The flow is turbulent ($Re > 2000$–3000).
Using the Colebrook–White's formula, the friction factor and bottom shear stress are:

$$f = 0.015\quad \text{(Equation (4.18))}$$
$$\tau_0 = 0.044\ \text{Pa}\quad \text{(Equation (4.14))}$$

Comments

The shear velocity and shear Reynolds number are:

$$V_* = 0.0066\ \text{m/s}\quad \text{(Equation (4.15))}$$
$$Re_* = 6.6\quad \text{(Equation (4.17))}$$

The flow is turbulent in the transition region between smooth and rough turbulent.

Historically, flow resistance in open channels was investigated earlier than in pipe flows. The first successful empirical resistance formulae were obtained in the 18th century (Chézy formula) and 19th century (Gauckler–Manning formula). Such empirical formulae are inaccurate!

In the following section the momentum equation for uniform equilibrium open channel flows is developed. The flow resistance is analysed as for pipe flows. The well-known empirical resistance formulae are presented. The correspondence between the empirical coefficients and the Darcy friction factor is also detailed.

4.4.2 Flow resistance of open channel flows

Momentum equation in steady uniform equilibrium open channel flow

The fundamental problem of steady uniform equilibrium flow is determining the relation between the flow velocity, the uniform flow depth, the channel slope and the channel geometry.

For a *steady* and *uniform equilibrium* flow the flow properties are independent of time and of position along the flow direction. That is:

$$\frac{\partial V}{\partial t} = 0$$

and

$$\frac{\partial V}{\partial s} = 0$$

where t is the time and s is the co-ordinate in the flow direction.

The momentum equation along a streamline states the exact balance between the shear forces and the gravity component. Considering a control volume as shown in Fig. 4.11, the momentum equation yields:

$$\tau_o P_w \Delta s = \rho g A \Delta s \sin \theta \qquad (4.19a)$$

where τ_o is the bottom shear stress, P_w is the wetted perimeter and Δs is the length of the control volume, A is the cross-sectional area and θ is the channel slope. Replacing the bottom shear stress by its expression (equation (4.14)), the momentum equation for uniform equilibrium flows becomes:

$$V_o = \sqrt{\frac{8g}{f}} \sqrt{\frac{(D_H)_o}{4} \sin \theta} \qquad (4.19b)$$

where V_o is the uniform (equilibrium) flow velocity and $(D_H)_o$ is the hydraulic diameter of uniform equilibrium flows.

The momentum equation for steady uniform open channel flow (equation (4.19a)) is rewritten usually as:

$$S_f = S_o \qquad (4.19c)$$

where S_f is called the friction slope and S_o is the channel slope defined as:

$$S_f = -\frac{\partial H}{\partial s} = \frac{4\tau_o}{\rho g D_H} \qquad (4.20)$$

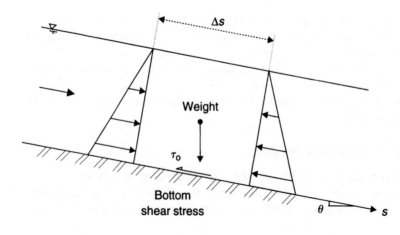

Weight

τ_o

Bottom
shear stress

θ s

Fig. 4.11 Application of the momentum equation to uniform equilibrium open channel flow.

$$S_o = -\frac{\partial z_0}{\partial s} = \sin\theta \tag{4.21}$$

where H is the mean total head and z_0 is the bed elevation. Note that the definitions of the friction and bottom slope are general and applied to both uniform equilibrium and gradually varied flows.

Notes

1. In sediment transport calculations, the momentum equation is usually expressed as equation (4.19a). Equation (4.19b) is more often used in clear-water flows.
2. For flat channel, the channel slope S_o (or bed slope) is almost equal to the slope tangent:

$$S_o = -\frac{\partial z_0}{\partial s} = \sin\theta \sim \tan\theta$$

S_o might be denoted 'i' (Comolet, 1976).

3. For uniform equilibrium and gradually varied flows the shear stress τ_0 can be expressed as:

$$\tau_0 = \rho g \frac{D_H}{4} S_f$$

4. Equations (4.19)–(4.21) are valid for any shape of channel cross-section.

Combining the definitions of bottom shear stress (equation (4.14)) and of friction slope, the momentum equation can be rewritten as:

$$V = \sqrt{\frac{8g}{f}} \sqrt{\frac{D_H}{4} S_f} \tag{4.22}$$

This relationship (i.e. equation (4.22)) is valid for both uniform equilibrium and gradually varied flows.

Chézy coefficient

The dependence of the flow velocity on channel slope and hydraulic diameter can be deduced from equations (4.14) and (4.19). Replacing τ_0 by its expression (equation (4.14)), the momentum equation (equation (4.19b)) can be transformed to give the Chézy equation:

$$V = C_{\text{Chézy}} \sqrt{\frac{D_H}{4} \sin\theta} \tag{4.23}$$

where $C_{\text{Chézy}}$ is the Chézy coefficient (units $m^{1/2}/s$), D_H is the hydraulic diameter and θ is the channel slope (e.g. Henderson, 1966: pp. 91–96; Streeter and Wylie, 1981: p. 229).

The Chézy equation (4.23) was first introduced in A.D. 1768 as an empirical correlation. Equation (4.23) is defined for uniform equilibrium and non-uniform gradually varied flows. The Chézy coefficient ranges typically from $30\,m^{1/2}/s$ (small rough channel) up to $90\,m^{1/2}/s$ (large smooth channel). Equations (4.19b) and (4.23) look similar. But it must be emphasized that equation (4.19b) was deduced from the momentum equation for uniform equilibrium flows. Equation (4.19b) is not valid for non-uniform equilibrium flows for which equation (4.22) should be used.

Notes
1. The Chézy equation was introduced in 1768 by the French engineer A. Chézy when designing canals for the water supply of the city of Paris.
2. The Chézy equation applies for *turbulent* flows. Although A. Chézy gave several values for $C_{\text{Chézy}}$, several researchers assumed later that $C_{\text{Chézy}}$ was independent of the flow conditions. Further studies showed clearly its dependence upon the Reynolds number and channel roughness.
3. The Chézy equation is valid for uniform equilibrium and non-uniform (gradually varied) turbulent flows.
4. For *uniform equilibrium flows* the boundary shear stress τ_0 can be rewritten in term of the Chézy coefficient as:

$$\tau_0 = \rho g \frac{V^2}{(C_{\text{Chézy}})^2}$$

The Chézy equation in uniform equilibrium flows can be rewritten in term of the shear velocity as:

$$\frac{V}{V_*} = \frac{C_{\text{Chézy}}}{\sqrt{g}}$$

and the Chézy coefficient and the Darcy friction factor are related by:

$$C_{\text{Chézy}} = \sqrt{\frac{8g}{f}}$$

5. At uniform equilibrium and in fully rough turbulent flows (e.g. natural streams), the Chézy coefficient becomes:

$$C_{\text{Chézy}} = 17.7 \log_{10}\left(\frac{D_{\text{H}}}{k_{\text{s}}}\right) + 10.1 \qquad \text{Uniform equilibrium fully rough turbulent flow}$$

This expression derives from Colebrook–White formula and it is very close to Keulegan formula (see below).
6. In *non-uniform gradually varied flows* the combination of equations (4.22) and (4.23) indicates that the Chézy coefficient and the Darcy friction factor are related by:

$$C_{\text{Chézy}} = \sqrt{\frac{8g}{f}} \sqrt{\frac{S_{\text{f}}}{\sin\theta}}$$

7. Empirical estimates of the Chézy coefficient include the Bazin and Keulegan formulae:

$$C_{\text{Chézy}} = \frac{87}{1 + k_{\text{Bazin}}/\sqrt{D_{\text{H}}/4}} \qquad \text{Bazin formula}$$

where $k_{\text{Bazin}} = 0.06$ (very smooth boundary: cement and smooth wood), 0.16 (smooth boundary: wood, brick, freestone), 0.46 (stonework), 0.85 (fine earth), 1.30 (raw earth), 1.75 (grassy boundary, pebble bottom, and grassed channel) (e.g. Comolet, 1976); and

$$C_{\text{Chézy}} = 18.0 \log_{10}\left(\frac{D_{\text{H}}}{k_{\text{s}}}\right) + 8.73 \qquad \text{Keulegan formula}$$

which is valid for fully rough turbulent flows and where k_{s} (in mm) = 0.14 (cement), 0.5 (planed wood), 1.2 (brick), 10–30 (gravel) (Keulegan, 1938).
Interestingly Keulegan formula was validated with Bazin's data (Bazin, 1865a).

The Gauckler–Manning coefficient

Natural channels have irregular channel bottom, and information on the channel roughness is not easy to obtain. An empirical formulation, called the Gauckler–Manning formula, was developed for turbulent flows in rough channels.

The Gauckler–Manning formula is deduced from the Chézy equation by setting:

$$C_{\text{Chézy}} = \frac{1}{n_{\text{Manning}}} \left(\frac{D_{\text{H}}}{4} \right)^{1/6} \tag{4.24}$$

and equation (4.23) becomes:

$$V = \frac{1}{n_{\text{Manning}}} \left(\frac{D_{\text{H}}}{4} \right)^{2/3} \sqrt{\sin \theta} \tag{4.25}$$

where n_{Manning} is the *Gauckler–Manning coefficient* (units $s/m^{1/3}$), D_{H} is the hydraulic diameter and θ is the channel slope.

The Gauckler–Manning coefficient is an *empirical* coefficient, found to be a characteristic of the surface roughness alone. Such an approximation might be reasonable as long as the water is not too shallow nor the channel too narrow.

Notes

1. Equation (4.25) is improperly called the 'Manning formula'. In fact it was first proposed by Gauckler (1867) based upon the re-analysis of experimental data obtained by Darcy and Bazin (1865).
2. Philippe Gaspard Gauckler (1826–1905) was a French engineer and member of the French 'Corps des Ponts-et-Chaussées'.
3. Robert Manning (1816–1897) was chief engineer at the Office of Public Works, Ireland. He presented two formulas in 1890 in his paper 'On the flow of water in open channels and pipes' (Manning, 1890). One of the formula was the 'Gauckler–Manning' formula (equation 4.25) but Robert Manning did prefer to use the second formula that he gave in the paper. Further information on the 'history' of the Gauckler–Manning formula was reported by Dooge (1991).
4. The Gauckler–Manning equation is valid for uniform equilibrium and non-uniform (gradually varied) flows.
5. Equation (4.25) is written in SI units. The units of the Gauckler–Manning coefficient n_{Manning} is $s/m^{1/3}$. A main critic of the (first) Manning formula (equation (4.25)) is its dimensional aspect: i.e. n_{Manning} is not dimensionless (Dooge, 1991).
6. The Gauckler–Manning equation applies for *fully rough turbulent* flows and *water* flows. It is an *empirical* relationship but has been found *reasonably* reliable.
7. Typical values of n_{Manning} (in SI Units) are:

n_{Manning}	
0.010	Glass and plastic
0.012	Planed wood
0.013	Unplanned wood
0.012	Finished concrete
0.014	Unfinished concrete
0.025	Earth
0.029	Gravel
0.05	Flood plain (light brush)
0.15	Flood plain (trees)

Yen (1991b) proposed an extensive list of values for a wide range of open channels.

The Strickler's coefficient

In Europe, the Strickler equation is used by defining:

$$C_{\text{Chézy}} = k_{\text{Strickler}} \left(\frac{D_H}{4} \right)^{1/6} \tag{4.26}$$

and equation (4.23) becomes the Strickler's equation:

$$V = k_{\text{Strickler}} \left(\frac{D_H}{4} \right)^{2/3} \sqrt{\sin \theta} \tag{4.27}$$

where $k_{\text{Strickler}}$ is only a function of the surfaces.

Notes

1. The Gauckler–Manning and Strickler coefficients are related as:

$$k_{\text{Strickler}} = 1/n_{\text{Manning}}$$

2. The Strickler equation is used for pipes, gallery and channel carrying water. This equation is preferred to the Gauckler–Manning equation in Europe.
3. The coefficient $k_{\text{Strickler}}$ varies from 20 (rough stone and rough surface) to $80 \, \text{m}^{1/3}/\text{s}$ (smooth concrete and cast iron).
4. The Strickler equation is valid for uniform and non-uniform (gradually varied) flows.

Particular flow resistance approximations

In Nature, rivers and streams do not exhibit regular uniform bottom roughness. The channel bed consists often of unsorted sand, gravels and rocks. Numerous researchers attempted to relate the equivalent roughness height k_s to a characteristic grain size (e.g. median grain size d_{50}). The analysis of numerous experimental data suggested that:

$$k_s \propto d_{50} \tag{4.28}$$

where the constant of proportionality k_s/d_{50} ranges from 1 to well over 6 (see Table 12.2)! Obviously it is extremely difficult to relate grain size distributions and bed forms to a single parameter (i.e. k_s).

For gravel-bed streams Henderson (1966) produced a relationship between the Gauckler–Manning coefficient and the gravel size:

$$n_{\text{Manning}} = 0.038 d_{75}^{1/6} \tag{4.29}$$

where the characteristic grain size d_{75} is in metres and n_{Manning} is in $\text{s/m}^{1/3}$. Equation (4.29) was developed for $k_s/D_H < 0.05$.

For *flood plains* the vegetation may be regarded as a kind of roughness. Chow (1973) presented several empirical formulations for grassed channels as well as numerous photographs to assist in the choice of a Gauckler–Manning coefficient.

Notes

1. Strickler (1923) proposed the following empirical correlation for the Gauckler–Manning coefficient of rivers:

$$n_{\text{Manning}} = 0.041 d_{50}^{1/6}$$

where d_{50} is the median grain size (in m).

2. In torrents and mountain streams the channel bed might consists of gravels, stones and boulders with size of the same order of magnitude as the flow depth. In such cases, the overall flow resistance results from a combination of skin friction drag, form drag and energy dissipation in hydraulic jumps behind large boulders. Neither the Darcy friction factor nor the Gauckler–Manning coefficient should be used to estimate the friction losses. Experimental investigations should be performed to estimate an overall Chézy coefficient (for each discharge).

4.5 FLOW RESISTANCE CALCULATIONS IN ENGINEERING PRACTICE

4.5.1 Introduction

The transport of real fluids is associated with energy losses and friction losses. With pipe systems, pumps or high-head intakes are needed to provide the required energy for the fluid transport (kinetic energy and potential energy) and the associated energy loss (flow resistance). Two types of flow regimes are encountered: laminar flow at low Reynolds numbers and turbulent flows. In turbulent flows, the head loss can be estimated from the Darcy equation (4.12) in which the friction factor f is a function of the Reynolds number Re and relative roughness k_s/D_H. An extremely large number of experiments were performed in pipe flow systems to correlate f with Re and k_s/D_H. Usually the friction factor in turbulent flows is calculated with the Colebrook–White formula (4.18) or from the Moody diagram (Fig. 4.10).

Flow resistance calculations in open channels

In open channels, the Darcy equation is valid using the hydraulic diameter as equivalent pipe diameter. It is the only sound method to estimate the energy loss.

For various reasons (mainly historical reasons), empirical resistance coefficients (e.g. Chézy coefficient) were and are still used. The Chézy coefficient was introduced in 1768 while the Gauckler–Manning coefficient was first presented in 1865: i.e. well before the classical pipe flow resistance experiments in the 1930s. Historically both the Chézy and the Gauckler–Manning coefficients were expected to be constant and functions of the roughness only. But it is now well recognized that these coefficients are only constant for a range of flow rates (e.g. Chow, 1973; Chen 1990; Yen, 1991a). Most friction coefficients (except perhaps the Darcy friction factor) are estimated 100%-empirically and they apply only to fully rough turbulent water flows.

In practice, the Chézy equation is often used to compute open channel flow properties. The Chézy equation can be related to the Darcy equation using equations (4.22) and (4.23). As a lot of experimental data are available to estimate the friction factor f (or Darcy coefficient), accurate estimate of the Darcy friction factor and Chézy coefficient is possible for standard geometry and material. But the data do not apply to natural rivers with vegetation, trees, large stones, boulders and complex roughness patterns (e.g. Fig. 4.12, Plates 10, 11 and 12) and with movable boundaries (see also Chapter 12).

(a)

(b)

Fig. 4.12 Examples of natural rivers and flood plains. (a) Lance Creek and it's flood plain, looking upstream (by H.E. Malde, with permission of US Geological Survey, Ref. 3994ct). Note the driftwood on the flood terrace that is flooded about every 10 years. (b) Burdekin River Qld, Australia (10 July 1983) (Courtesy of Mrs J. Hacker). From Herveys Development Road Bridge, looking South.

4.5.2 Selection of a flow resistance formula

Flow resistance calculations in open channels must be performed in term of the *Darcy friction factor*. First the type of flow regime (laminar or turbulent) must be determined. Then the friction factor is estimated using the classical results (Section 4.3.1).

In turbulent flows, the choice of the boundary equivalent roughness height is important. Hydraulic handbooks (e.g. Idelchik, 1969, 1986) provide a selection of appropriate roughness heights for standard materials.

Fig. 4.12 (c) Flood plain with trees of the South Alligator River, Kakadu National Park, Northern Territory, Australia in March 1998 (Courtesy of Dr R. Rankin). (d) Flood of the Mississippi River at Prairie du Chien, WI on 23 April 1969 (Courtesy of Dr Lou Maher) – The high bridge (on left) carries Highway 18 to Iowa.

The main limitations of the Darcy equation for turbulent flows are:

- the friction factor can be estimated for relative roughness k_s/D_H less than 0.05,
- classical correlations for f were validated for uniform-size roughness and regular roughness patterns.

In simple words, the Darcy equation cannot be applied to complex roughness patterns: e.g. vegetation and trees (in flood plains), shallow waters over rough channels.

For *complex channel bed roughness*, practising engineers might estimate the flow resistance by combining the Chézy equation (4.23) with an 'appropriate' Gauckler–Manning or Chézy coefficient. Such an approximation is valid only for fully rough turbulent flows.

Discussion

Great care must be taken when using the Gauckler–Manning equation (or Strickler equation) as the values of the coefficient are empirical. It is well known to river engineers that the estimate of the Gauckler–Manning (or Strickler) coefficient is a most difficult choice. Furthermore, it must be emphasized that the Gauckler–Manning formula is valid *only* for fully rough turbulent flows of water. It should not be applied to smooth (or transition) turbulent flows. It is not valid for fluids other than water. The Gauckler–Manning equation was developed and 'validated' for clear-water flows only.

In practice, it is recommended to calculate the flow resistance using the Darcy friction factor. Empirical correlations such as the Bazin, Gauckler–Manning and/or Strickler formula could be used to check the result. If there is substantial discrepancy between the sets of results, experimental investigations must be considered.

Note

Several computer models of river flows using the unsteady flow equations are based on the Gauckler–Manning equation. Professionals (engineers, designers and managers) must not put too much confidence in the results of these models as long as the resistance coefficients have not been checked and verified with experimental measurements.

Applications

1. In a rectangular open channel (boundary roughness: PVC), the uniform equilibrium flow depth equals 0.5 m. The channel width is 10 m and the channel slope is 0.002°. Compute the discharge. The fluid is water at 20°C.

Solution

The problem must be solved by iterations. First we will assume that the flow is turbulent (we will need to check this assumption later) and we assume $k_s = 0.01$ mm (PVC).

An initial velocity (e.g. 0.1 m/s) is assumed to estimate the Reynolds number and hence the Darcy friction factor at the first iteration. The mean flow velocity is calculated using the momentum equation (or the Chézy equation). The full set of calculations are summarized in following table. The total flow rate equals 1.54 m³/s. The results indicate that the flow is subcritical ($Fr = 0.139$) and turbulent ($Re = 5.6 \times 10^6$). The shear Reynolds number Re_* equals 0.12: i.e. the flow is smooth turbulent.

Note: As the flow is not fully rough turbulent, the Gauckler–Manning equation must not be used.

Iteration	V (initialization) (m/s)	f (equation (4.18))	$C_{Chézy}$ (m$^{1/2}$/s) (equation (4.23))	V (m/s) (equation (4.23))
1	0.1	0.0160	69.8	0.28
2	0.28	0.0134	76.6	0.305
3	0.305	0.0131	77.2	0.31
4	0.31	0.0131	77.3	0.31

2. Considering an uniform equilibrium flow in a rectangular concrete channel ($B = 2$ m), the total discharge is 10 m³/s. The channel slope is 0.02°. Compute the flow depth. The fluid is water at 20°C.

Solution

The problem must be solved again by iterations. We will assume a turbulent flow and $k_s = 1$ mm (concrete).

At the first iteration, we need to assume a flow depth d (e.g. $d = 0.5$ m) to estimate the relative roughness k_s/D_H, the mean velocity (by continuity) and the Reynolds number. We deduce then the friction factor, the Chézy coefficient and the mean flow velocity (Chézy equation). The new flow depth is deduced from the continuity equation.

The iterative process is repeated until convergence.

Iteration	d (initialization) (m)	f (equation (4.18))	$C_{Chézy}$ (m$^{1/2}$/s) (equation (4.23))	V (m/s) (equation (4.23))	d (m)
1	0.5	0.018	65.4	0.705	7.09
2	7.1	0.0151	72.1	1.26	4.0
3	4.0	0.0153	71.7	1.20	4.18
4	4.2	0.0152	71.7	1.20	4.15
5	4.15	0.0152	71.7	1.20	4.16
6	4.16	0.0152	71.7	1.20	4.16

The calculations indicate that the uniform equilibrium flow depth equals 4.16 m. The flow is subcritical ($Fr = 0.19$) and turbulent ($Re = 3.8 \times 10^6$). The shear Reynolds number equals 52.1: i.e. the flow is at transition between smooth turbulent and fully-rough-turbulent.

Note: as the flow is not fully rough turbulent, the Gauckler–Manning equation must not be used.

3. Considering a rectangular concrete channel ($B = 12$ m), the flow rate is 23 m^3/s. The channel slope is 1°. Estimate the uniform equilibrium flow depth using both the Darcy friction factor and the Gauckler–Manning coefficient (if the flow is fully rough turbulent). Compare the results. Investigate the sensitivity of the results upon the choice of the roughness height and Gauckler–Manning coefficient.

Solution

The calculations are performed in a similar manner as the previous example. We will detail the effects of roughness height and Gauckler–Manning coefficient upon the flow depth calculation.

For concrete, the equivalent roughness height varies from 0.3 to 3 mm for finish concrete and from 3 to 10 mm for rough concrete. For damaged concrete the equivalent roughness height might be greater than 10 mm. The Gauckler–Manning coefficient for concrete can be between 0.012 (finished concrete) and 0.014 s/m$^{1/3}$ (unfinished concrete). The results of the calculation are summarized in the following table.

Calculation	Surface	k_s (mm)	f	$n_{Manning}$ (s/m$^{1/3}$)	$C_{Chézy}$ (m$^{1/2}$/s)	d (m)
Darcy friction factor	Finished	0.3	0.0143		74.1	0.343
		1	0.0182		65.7	0.373
	Unfinished	3	0.0223		58.0	0.406
		10	0.032		49.5	0.452
Gauckler–Manning formula	Finished			0.012	69.6	0.359
	Unfinished			0.014	74.8	0.341

First, the Reynolds number is typically within the range $7 \times 10^6 - 7.2 \times 10^6$ (i.e. turbulent flow). The shear Reynolds number is between 70 and 2700. That is, the flow is fully rough turbulent and within the validity range of the Gauckler–Manning formula. Secondly, the flow depth increases with increasing roughness height. The increase in flow depth results from a decrease of flow velocity with increasing flow resistance. Thirdly, let us observe the discrepancy of results for unfinished concrete between Darcy friction factor calculations and the Gauckler–Manning formula. Furthermore, note that the calculations (using the Darcy friction factor) are sensitive upon the choice of roughness height.

4.5.3 Flow resistance in a flood plain

A practical problem is the flow resistance calculations of a river channel and the adjacent flood plain (Figs 4.12 and 4.13, Plates 17 and 24). Usually the flood plain is much rougher than the river channel and the flow depth is much smaller. The energy slope of both portions must be the same. For uniform equilibrium flow, this implies that the bed slope S_o is the same in the channel and plain.

Two practical applications of such calculations are:

1. assuming that the total discharge is known, estimate the flow depths in the river d_1 and in the flood plain d_2, or
2. assuming known flow depths (d_1 and d_2) find the total discharge in the flood plain and channel.

In practice, the complete hydraulic calculations are an iterative process. The main equations are summarized in the following table:

	River channel	Flood plain
Flow depth	d_1	d_2
Flow velocity	V_1	V_2
Width	B_1	B_2
Bed altitude	z_{o1}	$z_{o2} = z_{o1} + y_{channel}$
Wetted perimeter	$P_{w1} = B_1 + d_1 + y_{channel}$	$P_{w2} = B_2 + d_2$
Cross-sectional area	$A_1 = d_1 B_1$	$A_2 = d_2 B_2$
Hydraulic diameter	$D_{H1} = 4\dfrac{A_1}{P_{w1}}$	$D_{H2} = 4\dfrac{A_2}{P_{w2}}$
Equivalent roughness height	k_{s1}	k_{s2}
Darcy friction factor	$f_1 = f\left(\dfrac{k_{s1}}{D_{H1}}; \dfrac{V_1 D_{H1}}{\nu}\right)$	$f_2 = f\left(\dfrac{k_{s2}}{D_{H2}}; \dfrac{V_2 D_{H2}}{\nu}\right)$
Momentum equation	$V_1 = \sqrt{\dfrac{8g}{f_1}}\sqrt{\dfrac{D_{H1}}{4}}\sqrt{\sin\theta}$	$V_2 = \sqrt{\dfrac{8g}{f_2}}\sqrt{\dfrac{D_{H2}}{4}}\sqrt{\sin\theta}$
Continuity equation	$Q_1 = V_1 A_1$	$Q_2 = V_2 A_2$

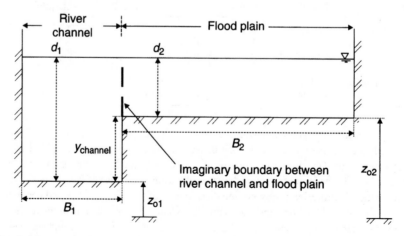

Fig. 4.13 Flood plain cross-section.

For a given flow depth the total discharge equals:

$$Q = Q_1 + Q_2 \qquad\qquad [C]$$

The flow depths are deduced by iterative calculations.

Notes

1. The longitudinal bed slope ($\sin\theta$) is the same for each portion (i.e. main channel and flood plain).
2. The flow depths d_1 and d_2 are related as: $d_1 = d_2 + y_{channel}$.
3. Note that friction (and energy loss) is assumed zero at the interface between the river channel flow and the flood plain flow. In practice, turbulent energy losses and secondary currents are observed at the transition between the channel flow and the much slower flood plain flow.

4.6 EXERCISES

Momentum equation

The backward-facing step, sketched in Fig. E.4.1, is in a 5 m wide channel of rectangular cross-section. The total discharge is 55 m³/s. The fluid is water at 20°C. The bed of the channel, upstream and downstream of the step, is horizontal and smooth. Compute the pressure force acting on the vertical face of the step and indicate the direction of the force (i.e. upstream or downstream).

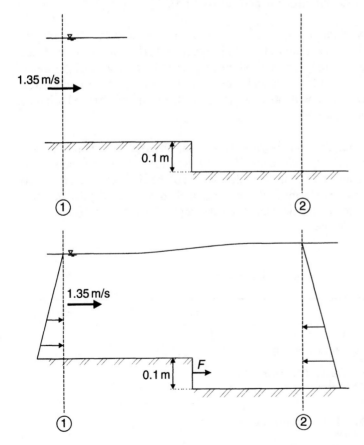

Fig. E.4.1 Sketch of a negative step.

Solution: Applying the continuity equation, the upstream flow depth equals $d_1 = 8.15$ m. Assuming a hydrostatic pressure distribution, the upstream specific energy equals $E_1 = 8.24$ m. Note that the upstream flow is subcritical ($Fr_1 = 0.15$).

The downstream specific energy is deduced from the Bernoulli principle assuming a smooth transition:

$$E_1 + \Delta z_o = E_2 \qquad \text{Bernoulli equation}$$

where Δz_o is the drop height (0.1 m). It gives $E_2 = 8.34$ m. The downstream flow depth is deduced from the definition of the specific energy:

$$E_2 = d_2 + \frac{Q^2}{2gB^2 d_2^2}$$

where B is the channel width. The solution is: $d_2 = 8.25$ m.

Considering the forces acting on the control volume contained between sections 1 and 2, the momentum equation as applied to the control volume between sections 1 and 2 is:

$$\rho Q V_2 - \rho Q V_1 = \frac{1}{2}\rho g d_1^2 B - \frac{1}{2}\rho g d_2^2 B + F \qquad \text{Momentum principle}$$

in the horizontal direction. Note that the friction is zero as the bed is assumed smooth.

The solution of the momentum equation is $F = +40.3$ kN (*Remember*: F is the force exerted by the step onto the fluid). In other words, the pressure force exerted by the fluid onto the vertical face of the step acts in the upstream direction and equals 40.3 kN.

Considering a broad-crested weir laboratory model (0.25 m wide rectangular channel), the upstream flow conditions (measured upstream of the weir) are $d_1 = 0.18$ m and $V_1 = 0.36$ m/s. The bed of the channel is horizontal and smooth (both upstream and downstream of the weir).

Assuming critical flow conditions at the crest and supercritical flow downstream of the weir, compute the following quantities: (a) flow depth downstream of the weir and (b) the horizontal force acting on the weir (i.e. sliding force).

Considering a sluice gate in a rectangular channel ($B = 0.80$ m), the observed upstream and downstream flow depth are respectively: $d_1 = 0.450$ m and $d_2 = 0.118$ m. (a) Derive the expression of the flow rate in terms of the upstream and downstream depths, and of the channel width; (b) derive the expression of the force on the gate in terms of the upstream and downstream depths, the channel width and the flow rate; (c) compute the flow rate and (d) compute the force on the gate.

Assume of smooth horizontal channel.

Hydraulic jump

Considering a hydraulic jump, answer the following questions:

Question	Yes	No	Correct answer
The upstream Froude number is less than 1			
The downstream velocity is less than the upstream velocity			
The energy loss term is always positive			
The upstream depth is larger than the downstream flow depth			
Hydraulic jump can occur in closed-conduit flow			
Air entrainment occurs in a hydraulic jump			
Hydraulic jump is dangerous			
If the upstream flow is such as $Fr_1 = 1$ can hydraulic jump occur?			
What is then the downstream Froude number?	–	–	

What is the definition of alternate depths, conjugate depths and sequent depths? In each case, explain your answer in words and use appropriate example(s) if necessary.

Considering a hydraulic jump in a rectangular horizontal channel, sketch the flow and write the following basic equations across the jump: (a) continuity equation, (b) momentum equation and (c) energy equation. Deduce the expression of the downstream flow depth as a function of the upstream flow depth and upstream Froude number only.

Application: The upstream flow depth is 1.95 m. The channel width is 2 m. The flow rate is 70 m³/s. Compute the downstream flow depth and the head loss across the jump.

Considering a hydraulic jump in a horizontal channel of trapezoidal cross-section. The channel cross-section is symmetrical around the channel centreline, the angle between the sidewall and the horizontal being δ. Write the continuity and momentum equations for the hydraulic jump. Use the subscript 1 to refer to the upstream flow conditions and the subscript 2 for the downstream flow conditions. Neglect the bottom friction.

A hydraulic jump flow takes place in a horizontal rectangular channel. The upstream flow conditions are $d = 1$ m and $q = 11.2$ m²/s. Calculate the downstream flow depth, downstream Froude number, head loss in the jump and roller length.

Considering the upstream flow $Q = 160$ m³/s and $B = 40$ m, design a hydraulic jump dissipation structure. Select an appropriate upstream flow depth for an optimum energy dissipation. What would be the upstream and downstream Froude numbers?

Considering a hydraulic jump in a horizontal rectangular channel of constant width and neglecting the friction force, apply the continuity, momentum and energy equations to the following case: $Q = 1500$ m³/s, $B = 50$ m. The energy dissipation in the hydraulic jump is $\Delta H = 5$ m. Calculate the upstream and downstream flow conditions (d_1, d_2, Fr_1, Fr_2). Discuss the type of hydraulic jump.

Considering a rectangular horizontal channel downstream of a radial gate, a stilling basin (i.e. energy dissipation basin) is to be designed immediately downstream of the gate. Assume no flow contraction at the gate. The design conditions are $Q = 65$ m³/s and $B = 5$ m. Select an appropriate gate opening for an optimum design of a dissipation basin. Explain in words what gate opening would you choose while designing this hydraulic structure to dissipate the energy in a hydraulic jump. Calculate the downstream Froude number. What is the minimum length of the dissipation basin? What hydraulic power would be dissipated in the jump at design flow conditions? How many 100 W bulbs could be powered (ideally) with the hydraulic power dissipated in the jump?

Surges and bores
What is the difference between a surge, a bore and a wave? Give examples of bore.

Considering a flow downstream of a gate in an horizontal and rectangular channel, the initial flow conditions are $Q = 500$ m³/s, $B = 15$ m and $d = 7.5$ m. The gate suddenly opens and provides the new flow conditions $Q = 700$ m³/s. (a) Can the flow situation be described using the quasi-steady flow analogy? (Sketch the flow and justify in words your answer.) (b) Assuming a frictionless flow, compute the surge velocity and the new flow depth. (c) Are the new flow conditions supercritical? (d) Is the wave stable? (If a small disturbance start from the gate (i.e. gate vibrations) after the surge wave what will happen? Would the wave become unstable?)

Considering a flow upstream of a gate in an horizontal and rectangular channel, the initial flow conditions are $Q = 100\,m^3/s$, $B = 9\,m$ and $d = 7.5\,m$. The gate suddenly opens and provides the new flow conditions $Q = 270\,m^3/s$. Characterize the surge: i.e. positive or negative. Is the wave stable?

Considering the flow upstream of a gate, the gate suddenly closes. The initial flow conditions were $Q = 5000\,m^3/s$, $d = 5\,m$ and $B = 100\,m$. The new discharge is $Q = 3000\,m^3/s$. Compute the new flow depth and flow velocity.

Solution: The surge is an advancing wave front (i.e. positive surge). Using the quasi-steady flow assumption, the initial flow conditions are the flow conditions upstream of the surge front $d_1 = 5\,m$, $V_1 = 10\,m/s$ and $Fr = 1.43$. To start the calculations, it may be assumed $V_s = 0$ (i.e. stationary surge also called hydraulic jump). In this particular case (i.e. $V_{srg} = 0$), the continuity equation becomes $V_1 d_1 = V_2 d_2$.

Notation equation	V_{srg}	Fr_1	d_2 [C]	Fr_2 [M]	V_2 Def. Fr_2	$V_2 d_2 B$ Q_2
1st iteration	0.0	1.43	23.4	0.72	10.9	2540
2nd iteration	2	1.71	15	0.62	5.5	8229
Solution	3.26	1.89	11.1	0.57	2.70	3000

Notes
1. For an initialization step (where $V_{srg} = 0$), the continuity equation can be rewritten as $Fr_1\sqrt{gd_1^3} = Fr_2\sqrt{gd_2^3}$ using the definition of the Froude number. The above equation is more practical than the general continuity equation for surge.
2. It must be noted that the initial flow conditions are supercritical. The surge is a large disturbance travelling upstream against a supercritical flow.
3. The surge can be classified as a hydraulic jump using the (surge) upstream Froude number Fr_1. For this example, the surge is a weak surge.

Flow resistance
Considering an uniform equilibrium flow down a rectangular channel, develop the momentum equation. For a wide rectangular channel, deduce the expression of the normal depth as a function of the Darcy friction factor, discharge per unit width and bed slope.

For an uniform equilibrium flow down an open channel: (a) Write the Chézy equation. Define clearly all your symbols. (b) What are the SI units of the Chézy coefficient? (c) Give the expression of the Chézy coefficient as a function of the Darcy–Weisbach friction factor. (d) Write the Gauckler–Manning equation. Define clearly your symbols. (e) What is the SI units of the Gauckler–Manning coefficient?

Considering a gradually varied flow in a rectangular open channel, the total discharge is $2.4\,m^3/s$ and the flow depth is $3.1\,m$. The channel width is $5\,m$, and the bottom and sidewalls are made of smooth concrete. Estimate: (a) the Darcy friction factor and (b) the average boundary shear stress. The fluid is water at $20°C$.

A rectangular ($5.5\,m$ width) concrete channel carries a discharge of $6\,m^3/s$. The longitudinal bed slope is $1.2\,m/km$. (a) What is the normal depth at uniform equilibrium? (b) At uniform equilibrium what is the average boundary shear stress? (c) At normal flow conditions, is the flow subcritical, supercritical or critical? Would you characterize the channel as mild, critical or steep? *For man-made channels, perform flow resistance calculations based upon the Darcy–Weisbach friction factor.*

Solution: (a) $d = 0.64$ m, (b) $\tau_o = 6.1$ Pa and (c) $Fr = 0.68$: near-critical flow, although subcritical (hence mild slope) (see discussion on near critical flow in Section 5.1.2).

In a rectangular open channel (boundary roughness: PVC), the uniform equilibrium flow depth equals 0.9 m. The channel is 10 m wide and the bed slope is 0.0015°. The fluid is water at 20°C. Calculate (a) the flow rate, (b) Froude number, (c) Reynolds number, (d) relative roughness, (e) Darcy friction factor and (f) mean boundary shear stress.

Considering an uniform equilibrium flow in a trapezoidal grass waterway (bottom width: 15 m, sidewall slope: 1V:5H), the flow depth is 5 m and the longitudinal bed slope is 3 m/km. Assume a Gauckler–Manning coefficient of 0.05 s/m$^{1/3}$ (flood plain and light brush). The fluid is water at 20°C. Calculate: (a) discharge, (b) critical depth, (c) Froude number, (d) Reynolds number and (e) Chézy coefficient.

Norman Creek, in Southern Brisbane, has the following channel characteristics during a flood event: water depth: 1.16 m, width: 55 m, bed slope: 0.002 and short grass: $k_s = 3$ mm. *Assume uniform equilibrium flow conditions in a rectangular channel.* Calculate the hydraulic characteristics of the stream in flood.

Solution: $Q = 200$ m^3/s and $V = 3.1$ m/s.

During a flood, measurement in Oxley Creek, in Brisbane, gave: water depth: 1.16 m, width: 55 m, bed slope: 0.0002 and short grass: $k_s = 0.003$ m. Assuming uniform equilibrium flow conditions in a quasi-rectangular channel, compute the flow rate.

Solution: $V = 0.99$ m/s and $V_* = 0.047$ m/s.

Considering an uniform equilibrium flow in a trapezoidal concrete channel (bottom width: 2 m and sidewall slope: 30°), the total discharge is 10 m^3/s. The channel slope is 0.02°. The fluid is water at 20°C. Estimate: (a) critical depth, (b) normal depth, (c) Froude number, (d) Reynolds number and (e) Darcy friction factor.

Considering the flood plain, sketched in Fig. E.4.2, the mean channel slope is 0.05°. The river channel is lined with concrete and the flood plain is riprap material (equivalent roughness height: 8 cm). The fluid is water with a heavy load of suspended sediment (fluid density: 1080 kg/m^3). The flow is assumed to be uniform equilibrium. Compute and give the values (and units) of the following quantities: (a) Volume discharge in the river channel. (b) Volume discharge in the flood plain. (c) Total volume discharge (river channel + flood plain). (d) Total mass flow rate (river channel + flood plain). (e) Is the flow subcritical or supercritical? Justify your answer clearly. (Assume no friction (and energy loss) at the interface between the river channel flow and the flood plain flow.)

Considering a river channel with a flood plain in each side (Fig. E.4.3), the river channel is lined with finished concrete. The lowest flood plain is liable to flooding (it is land used as flood water retention system) and its bed consists of gravel ($k_s = 20$ mm). The right bank plain is a grassed area (centipede grass, $n_{Manning} \sim 0.06$ SI units). The longitudinal bed slope of the river is 2.5 km.

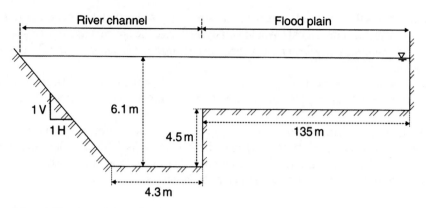

Fig. E.4.2 Sketch of a flood plain.

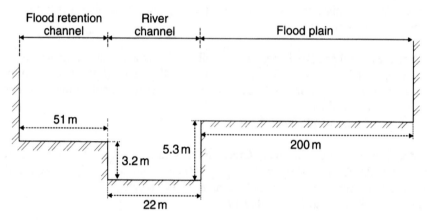

Fig. E.4.3 Sketch of a flood plain.

For the 1-in-50-years flood ($Q = 500\,\text{m}^3/\text{s}$), compute (a) the flow depth in the main channel, (b) the flow depth in the right flood plain, (c) the flow rate in the main channel, (d) the flow rate in the flood water retention system and (e) the flow rate in the right flood plain. (Assume no friction (and energy loss) at the interface between the river channel flow and the flood plain flows.)

Considering the river channel and flood plain sketched below (Fig. E.4.4), the main dimensions of the channel and flood plain are: $W_1 = 2.5\,\text{m}$, $W_2 = 25\,\text{m}$ and $\Delta z_{12} = 1.2\,\text{m}$. The main channel is lined with finished concrete. The flood plain is liable to flooding. (It is used as a flood water retention system.) The flood retention plain consists of light bush. (Field observations suggested $n_{\text{Manning}} = 0.04\,\text{s/m}^{1/3}$.) The longitudinal bed slope of the river is 1.5 m/km.

During a storm event, the observed water depth in the main (deeper) channel is 1.9 m. Assuming uniform equilibrium flow conditions, compute (a) the flow rate in the main channel, (b) the flow rate in the flood water retention system, (c) the total flow rate and (d) What the flow Froude number in the deep channel and in the retention system? (Assume energy loss at the interface between the river channel flow and the flood plain flows.)

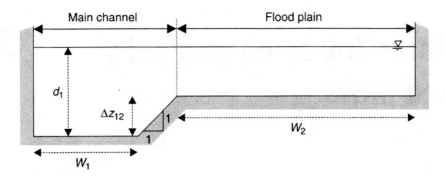

Fig. E.4.4 Sketch of flood plain.

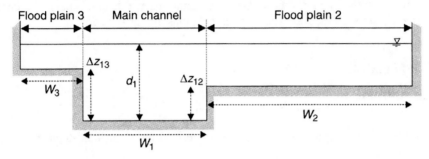

Fig. E.4.5 Sketch of flood plain.

For man-made channels, DO perform flow resistance calculations based upon the Darcy–Weisbach friction factor.

Solution: (a) 18.4 m³/s, (b) 13.1 m³/s, (c) 31.5 m³/s and (d) $Fr = 0.7$ and 0.3, respectively (see Section 3.4).

Considering a river channel with a flood plain in each side (Fig. E.4.5), the cross-sectional characteristics are $W_1 = 5$ m, $W_2 = 14$ m, $\Delta z_{12} = 0.95$ m, $W_3 = 58$ m, $\Delta z_{13} = 1.35$ m. The river channel is lined with finished concrete. The lowest flood plain (Flood plain No. 2) is liable to flooding and its bed consists of gravel ($k_s = 20$ mm). The left bank plain (Flood plain No. 2) is a grassed area (centipede grass, $n_{Manning} \sim 0.06$ SI units). The longitudinal bed slope of the river is 3.2 m/km. For the 1-in-50-years flood ($Q = 150$ m³/s), compute (a) the flow depth in the main channel, (b) the flow depth in the right flood plain, (c) the flow rate in the main channel, (d) the flow rate in the flood water retention system and (e) the flow rate in the right flood plain.

Assume no friction (and no energy loss) at the interface between the river channel flow and the flood plain flows. For man-made channels, DO perform flow resistance calculations based upon the Darcy–Weisbach friction factor.

Solution: (a) $d_1 = 2.17$ m, (d) $Q_2 = 53.2$ m³/s and (e) $Q_3 = 39.2$ m³/s.

Uniform flows and gradually varied flows

5

Summary

First, the properties of uniform equilibrium open channel flows are described (for steady flows). Then the gradually varied flow (GVF) assumptions are detailed. Later the energy equation is applied to GVFs, and backwater calculations are introduced.

5.1 UNIFORM FLOWS

5.1.1 Presentation

Definition

Uniform equilibrium open channel flows are characterized by a constant depth and constant mean flow velocity:

$$\frac{\partial d}{\partial s} = 0 \quad \text{and} \quad \frac{\partial V}{\partial s} = 0$$

where s is the co-ordinate in the flow direction. Uniform equilibrium open channel flows are commonly called 'uniform flows' or 'normal flows'. The expression should not be confused with 'uniform velocity distribution flows'. In open channels, the uniform equilibrium flow regime is 'equivalent' to fully developed pipe flow regime.

Uniform equilibrium flow can occur only in a straight channel with a constant channel slope and cross-sectional shape, and a constant discharge. The depth corresponding to uniform flow in a particular channel is called the *normal depth* or uniform flow depth.

Basic equations

For an uniform open channel flow, the shear forces (i.e. flow resistance) exactly balance the gravity force component. Considering a control volume as shown in Fig. 4.9, the momentum equation becomes (see Section 4.4):

$$\tau_o P_w \Delta s = \rho g A \Delta s \sin \theta \tag{4.19a}$$

where τ_o is the average shear stress on the wetted surface, P_w is the wetted perimeter, θ is the channel slope, A is the cross-sectional area and Δs is the length of the control volume.

For steady uniform open channel flow, this equation is equivalent to the energy equation:

$$\frac{\partial H}{\partial s} = \frac{\partial z_o}{\partial s} \tag{5.1}$$

where H is the mean total head and z_o is the bed elevation. This is usually rewritten as:

$$S_f = S_o$$

where S_o is the *bed slope* and S_f is the *friction slope*.

Notes

1. For flat channel the bed slope is:

$$S_o = \sin\theta = -\frac{\partial z_o}{\partial s} \sim \tan\theta$$

2. For uniform or non-uniform flows the *friction slope* S_f is defined as:

$$S_f = -\frac{\partial H}{\partial s} = \frac{4\tau_o}{\rho g D_H}$$

3. At any cross-section the relationship between the Darcy coefficient and the friction slope leads to:

$$S_f = f\frac{1}{D_H}\frac{V^2}{2g} = \frac{Q^2 P_w f}{8gA^3}$$

4. The uniform flow depth or *normal depth* is denoted d_o.

5.1.2 Discussion

Mild and steep slopes

A channel slope is usually 'classified' by comparing the uniform flow depth d_o with the critical flow depth d_c. When the uniform flow depth is larger than the critical flow depth, the uniform equilibrium flow is tranquil and subcritical. The slope is called a *mild slope*. For $d_o < d_c$, the uniform flow is supercritical and the slope is *steep*.

$d_o > d_c$	Mild slope	Uniform flow: $Fr_o < 1$ (subcritical flow)
$d_o = d_c$	Critical slope	Uniform flow: $Fr_o = 1$ (critical flow)
$d_o < d_c$	Steep slope	Uniform flow: $Fr_o > 1$ (supercritical flow)

Note

For a wide rectangular channel, the ratio d_o/d_c can be rewritten as:

$$\frac{d_o}{d_c} = \sqrt[3]{\frac{f}{8\sin\theta}}$$

where f is the friction factor and θ is the channel slope. The above result shows that the notion of steep and mild slope is not only a function of the bed slope but is also a function of the flow resistance: i.e. of the flow rate and roughness height.

Critical slope

A particular case is the situation where $d_o = d_c$: i.e. the uniform equilibrium flow is critical. The channel slope, for which the uniform flow is critical, is called the *critical slope* and is denoted S_c.

Critical slopes are seldom found in nature because critical flows and near-critical flows[1] are unstable. They are characterized by flow instabilities (e.g. free-surface undulations and waves) and the flow becomes rapidly unsteady and non-uniform.

---

[1]Near-critical flows are characterized by a specific energy only slightly greater than the minimum specific energy and by a Froude number close to unity (i.e. $0.7 < Fr < 1.5$ typically). Such flows are unstable, as any small change in specific energy (e.g. bed elevation and roughness) induces a large variation of flow depth.

Note

In the general case, the critical slope satisfies:

$$S_c = \sin\theta_c = S_f = f\frac{V_o^2}{2g(D_H)_o}$$

where V_o and $(D_H)_o$ are the uniform flow velocity and hydraulic diameter respectively which must satisfy also: $V_o = V_c$ and $(D_H)_o = (D_H)_c$.

For a wide rectangular channel, the critical slope satisfies:

$$S_c = \sin\theta_c = \frac{f}{8}$$

Application: most efficient cross-sectional shape

Uniform flows seldom occur in Nature. Even in artificial channels of uniform section (e.g. Figs 5.1 and 5.2), the occurrence of uniform flows is not frequent because of the existence of controls (e.g. weirs and sluice gates) which rule the relationship between depth and discharge. However, uniform equilibrium flow is of importance as a reference. Most channels are analysed and designed for uniform equilibrium flow conditions.

During the design stages of an open channel, the channel cross-section, roughness and bottom slope are given. The objective is to determine the flow velocity, depth and flow rate, given any one of them. The design of channels involves selecting the channel shape and bed slope to convey a given flow rate with a given flow depth. For a given discharge, slope and roughness, the designer aims to minimize the cross-sectional area A in order to reduce construction costs (e.g. Henderson, 1966: p. 101).

The most 'efficient' cross-sectional shape is determined for uniform flow conditions. Considering a given discharge Q, the velocity V is maximum for the minimum cross-section A. According to the Chézy equation (4.23) the hydraulic diameter is then maximum. It can be shown that:

1. the wetted perimeter is also minimum (e.g. Streeter and Wylie, 1981: pp. 450–452),
2. the semi-circle section (semi-circle having its centre in the surface) is the best hydraulic section (e.g. Henderson, 1966: pp. 101–102).

For several types of channel cross-sections the 'best' design is shown in Fig. 5.1 and summarized in the following table:

Cross-section	Optimum width B	Optimum cross-sectional area A	Optimum wetted perimeter P_w	Optimum hydraulic diameter D_H
Rectangular	$2d$	$2d^2$	$4d$	$2d$
Trapezoidal	$\frac{2}{\sqrt{3}}d$	$\sqrt{3}d^2$	$2\sqrt{3}d$	$2d$
Semi-circle	$2d$	$\frac{\pi}{2}d^2$	πd	$2d$

5.1.3   Uniform flow depth in non-rectangular channels

For any channel shape, the uniform flow conditions are deduced from the momentum equation (4.19a). The uniform flow depth must satisfy:

$$\frac{A_o^3}{(P_w)_o} = \frac{fQ^2}{8g\sin\theta} \tag{5.2a}$$

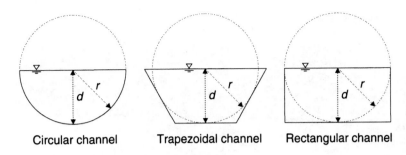

Circular channel   Trapezoidal channel   Rectangular channel

Fig. 5.1 Cross-sections of maximum flow rate: i.e. 'optimum design'.

(a)

(b)

Fig. 5.2 Examples of waterways. (a) Storm waterway with small baffle chute in background, Oxenford QLD, Gold Coast (April 2003) – looking upstream – concrete-lined low flow channel and grass-lined flood plains. (b) The McPartlan canal connecting Lake Pedler to Lake Gordon, Tasmania (20 December 1992). Trapezoidal concrete-lined channel – flow from the left to the top background.

(c)

(d)

Fig. 5.2 (c) Norman Creek in Brisbane on 13 May 2002 – looking downstream at the trapezoidal low flow channel and grass-lined flood plains on each side – note the bridge in background (Fig. 5.2d). (d) Norman Creek in Brisbane on 31 December 2001 during a small flood ($Q \sim$ 60 m³/s) – looking upstream from the bridge seen in Fig. 5.2c – note the hydraulic jump, in the foreground, induced by the bridge constriction.

where the cross-sectional area A_o and wetted perimeter $(P_w)_o$ for uniform flow conditions are functions of the normal flow depth d_o. Typical examples of relationships between A, P_w and d have been detailed in Section 3.4.

For a wide rectangular channel equation (5.2a) yields:

$$d_o = \sqrt[3]{\frac{q^2 f}{8g \sin \theta}} \qquad \text{Wide rectangular channel} \qquad (5.2b)$$

where q is the discharge per unit width. Equation (5.2b) applies only to rectangular channels such as $B > 10d_o$.

Notes
1. The calculation of the normal depth (equation (5.2)) is an iterative process (see next application). In practice, the iterative method converges rapidly. The Chézy equation (4.23) may also be used (see Sections 4.4.2 and 4.5.2).
2. Calculations of the normal flow depth always converge to an unique solution for either laminar or turbulent flows. But near the transition between laminar and turbulent flow regimes, the calculations might not converge. Indeed flow conditions near the transition laminar–turbulent flow are naturally unstable.

Applications
1. Considering a rectangular channel, the channel slope is 0.05°. The bed is made of bricks ($k_s = 10$ mm) and the width of the channel is 3 m. The channel carries water (at 20°C). For a 2 m³/s flow rate at uniform equilibrium flow conditions, is the flow super- or subcritical?

Solution
We must compute both the normal and critical depths to answer the question.

For a rectangular channel, the critical depth equals (Chapter 3):

$$d_c = \sqrt[3]{\frac{Q^2}{gB^2}}$$

It yields $d_c = 0.357$ m.

The uniform equilibrium flow depth (i.e. normal depth) is calculated by an iterative method. At the first iteration, we assume a flow depth. For that flow depth, we deduce the velocity (by continuity), the Reynolds number, relative roughness and Darcy friction factor. The flow velocity is computed from the momentum equation:

$$V_o = \sqrt{\frac{8g}{f}} \sqrt{\frac{(D_H)_o}{4}} \sin\theta \qquad (4.19b)$$

Note that the Chézy equation (4.23) may also be used. The flow depth d_o is deduced by continuity:

$$A_o = \frac{Q}{V_o}$$

where A_o is the cross-sectional area for normal flow conditions (i.e. $A_o = Bd_o$ for a rectangular cross-sectional channel).

The process is repeated until convergence:

Iteration	d initialization (m)	Re	k_s/D_H	f	V_o (m/s)	d_o (m)
1	0.357	2.14×10^6	0.0087	0.0336	0.738	0.903
2	0.903	1.65×10^6	0.0044	0.0284	1.146	0.581
Solution		1.84×10^6	0.0054	0.0299	1.005	0.663

The normal flow is a fully rough turbulent flow ($Re = 1.8 \times 10^6$, $Re_* = 622$). This observation justifies the calculation of the Darcy friction factor using the Colebrook–White correlation. The normal depth is 0.663 m.

As the normal depth is greater than the critical depth, the uniform equilibrium flow is subcritical.
2. For the same channel and the same flow rate as above, compute the critical slope S_c.

Solution

The question can be re-worded as: for $Q = 2\,\text{m}^3/\text{s}$ in a rectangular ($B = 3\,\text{m}$) channel made of bricks ($k_s = 10\,\text{mm}$), what is the channel slope θ for which the uniform equilibrium flow is critical?

The critical depth equals $0.357\,\text{m}$ (see previous example). For that particular depth, we want to determine the channel slope such as $d_o = d_c = 0.357\,\text{m}$. Equation (4.19) can be transformed as:

$$\sin \theta_c = \frac{V_o^2}{(8g/f)((D_H)_o/4)}$$

in which $d_o = d_c = 0.357\,\text{m}$ and $V_o = V_c = 1.87\,\text{m/s}$.

It yields $S_c = \sin \theta_c = 0.00559$. The critical slope is $0.32°$.

Comments

For wide rectangular channels, the critical slope can be approximated as: $S_c = f/8$. In our application, calculations indicate that, for the critical slope, $f \sim 0.036$ leading to: $S_c \sim 0.0045$ and $\theta_c \sim 0.259°$. That is, the approximation of a wide channel induces an error of about 25% on the critical slope S_c.

3. For a triangular channel ($26.6°$ wall slope, i.e. 1V:2H) made of concrete, compute the (centreline) flow depth for uniform flow conditions. The flow rate is $2\,\text{m}^3/\text{s}$ and the channel slope is $0.05°$. Is channel slope mild or steep?

Solution

The uniform flow depth is calculated by successive iterations. We assume $k_s \sim 1\,\text{mm}$ in absence of further information on the quality of the concrete:

Iteration	d initialization (m)	Re	k_s/D_H	f	V_o (m/s)	d_o (m)
1	0.7	2.5×10^6	0.0008	0.0187	1.07	0.97
2	0.97	9.2×10^5	0.0006	0.0177	1.29	0.88
Solution		9.9×10^5	0.00062	0.0180	1.62	0.899

The normal (centreline) depth equals: $d_o = 0.899\,\text{m}$. The uniform flow is turbulent ($Re = 9.9 \times 10^5$, $Re_* = 58.2$). To characterize the channel slope as mild or steep, we must first calculate the critical depth. For a non-rectangular channel, the critical depth must satisfy a minimum specific energy. It yields (Chapter 3):

$$\frac{A^3}{B} = \frac{Q^2}{g} \tag{3.34a}$$

where A is the cross-sectional area and B is the free-surface width. For a triangular channel, $A = 0.5Bd$ and $A = 2d \cot \delta$ where δ is the wall slope ($\delta = 26.6°$). Calculations indicate that the critical depth equals: $d_c = 0.728\,\text{m}$. As the normal depth is larger than the critical depth, the channel slope is *mild*.

5.2 NON-UNIFORM FLOWS

5.2.1 Introduction

In most practical cases, the cross-section, depth and velocity in a channel vary along the channel and the uniform flow conditions are not often reached. In this section, we analyse steady

GVFs in open channels. Flow resistance, and changes in bottom slope, channel shape, discharge and surface conditions are taken into account. Examples are shown in Fig. 5.3.

Figure 5.3 illustrates typical longitudinal free-surface profiles. Upstream and downstream control can induce various flow patterns. In some cases, a hydraulic jump might take place (e.g. Fig. 5.2d). A jump is a rapidly varied flow (RVF) and calculations were developed in Chapter 4 (Section 4.3.2). However, it is also a control section and it affects the free-surface profile upstream and downstream.

It is usual to denote d the actual depth (i.e. the non-uniform flow depth), d_o the normal depth (i.e. uniform flow depth) and d_c the critical depth.

Note

A non-uniform flow is such as along a streamline in the s-direction:

$$\frac{\partial V}{\partial s} \neq 0 \quad \text{or} \quad \frac{\partial d}{\partial s} \neq 0$$

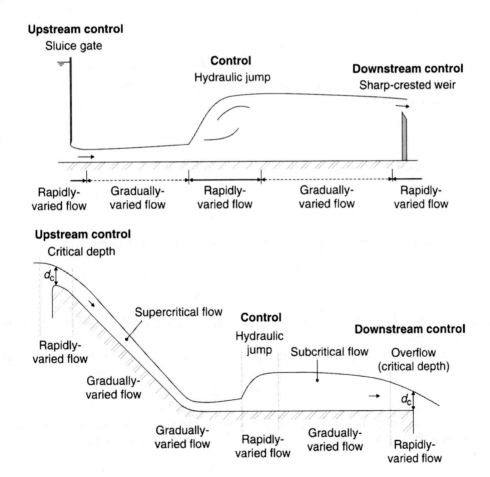

Fig. 5.3 Examples of non-uniform flows.

5.2.2 Equations for GVF: backwater calculation

For subcritical flows the flow situation is governed (controlled) by the downstream flow conditions. A downstream hydraulic structure (e.g. bridge piers and gate) will increase the upstream depth and create a 'backwater' effect. The term 'backwater calculations' refers more generally to the calculations of the longitudinal free-surface profile for both sub- and supercritical flows.

The backwater calculations are developed assuming:

[H1] a *non-uniform* flow,
[H2] a *steady* flow,
[H3] that the flow is *gradually varied*,
[H4] that, at a given section, the *flow resistance is the same as for an uniform flow* for the same depth and discharge, regardless of trends of the depth.

The GVF calculations do not apply to uniform equilibrium flows, nor to unsteady flows, nor to RVFs. The last assumption [H4] implies that the Darcy, Chézy or Gauckler–Manning equation may be used to estimate the flow resistance, although these equations were originally developed for uniform equilibrium flows only.

Along a streamline in the s-direction, the energy equation, written in terms of mean total head becomes:

$$\frac{\partial H}{\partial s} = -f \frac{1}{D_H} \frac{V^2}{2g} \tag{5.3a}$$

where V is the fluid velocity, f is the Darcy friction factor, D_H is the hydraulic diameter. Using the definition of the friction slope, it yields:

$$\frac{\partial H}{\partial s} = -S_f \tag{5.3b}$$

where S_f is the friction slope defined as:

$$S_f = f \frac{1}{D_H} \frac{V^2}{2g} \tag{4.20}$$

Assuming a hydrostatic pressure gradient, equation (5.3a) can be transformed as:

$$\frac{\partial H}{\partial s} = \frac{\partial}{\partial s}\left(z_o + d\cos\theta + \frac{V^2}{2g} \right) = -S_f \tag{5.3c}$$

where z_o is the bed elevation, d is the flow depth measured normal to the channel bottom and θ is the channel slope.

Introducing the mean specific energy (i.e. $H = E + z_o$), equation (5.3) can be rewritten as:

$$\frac{\partial E}{\partial s} = S_o - S_f \tag{5.4}$$

where S_o is the bed slope ($S_o = \sin\theta$). For any cross-sectional shape it was shown that the differentiation of the mean specific energy E with respect to d equals (see Section 3.2.3):

$$\frac{\partial E}{\partial d} = 1 - Fr^2 \tag{5.5}$$

where Fr is the Froude number (defined such as $Fr = 1$ for critical flow conditions). Hence the energy equation leads to:

$$\frac{\partial d}{\partial s}(1 - Fr^2) = S_o - S_f \tag{5.6}$$

Notes

1. The first backwater calculations were developed by the Frenchman J.B. Bélanger who used a finite difference step method of integration (Bélanger, 1828).

2. The assumption: *'the flow resistance is the same as for an uniform flow for the same depth and discharge'* implies that the Darcy equation, the Chézy equation or the Gauckler–Manning equation may be used to estimate the flow resistance, although these equations were initially developed for normal flows only. For RVFs these equations are no longer valid.

   In practice, the friction factor and friction slope are calculated as in uniform equilibrium flow but using the non-uniform flow depth d.

   It is well recognized that accelerating flows are characterized by lesser flow resistance than normal flows for the same flow properties (V, d). This phenomenon is sometimes called drag reduction in accelerating flow. For decelerating flows, the flow resistance is larger than that for uniform equilibrium flow with identical flow properties.

3. For non-uniform flows we have:

$$
\begin{array}{lll}
d > d_o & => & S_f < S_o \\
d = d_o & => & S_f = S_o \\
d < d_o & => & S_f > S_o
\end{array}
$$

and at the same time:

$$
\begin{array}{lll}
d > d_c & \text{Subcritical flow} & Fr < 1 \\
d = d_c & \text{Critical flow} & Fr = 1 \\
d < d_c & \text{Supercritical flow} & Fr > 1
\end{array}
$$

4. For uniform or non-uniform flows the bed slope S_o and the friction slope S_f are defined respectively as:

$$S_f = f\frac{1}{D_H}\frac{V^2}{2g} \tag{4.20}$$

$$S_o = \sin\theta = -\frac{\partial z_o}{\partial s} \tag{4.21}$$

5.2.3 Discussion

Singularity of the energy equation

The energy equations (5.3), (5.4) or (5.6) can be applied to non-uniform flow (GVF) situations as long as the friction slope differs from the bed slope. Indeed equation (5.6) may be transformed as:

$$\frac{\partial d}{\partial s} = \frac{S_o - S_f}{1 - Fr^2} \tag{5.7}$$

where S_f and Fr are functions of the flow depth d for a given value of Q. Equation (5.7) emphasizes three possible singularities of the energy equation. Their physical meaning is summarized as follows:

Case	Singularity condition	Physical meaning
[1]	$\dfrac{\partial d}{\partial s} = 0$	Uniform equilibrium flow: i.e. $S_f = S_o$
[2]	$Fr = 1$	Critical flow conditions: i.e. $S_f = S_c$
[3]	$\dfrac{\partial d}{\partial s} = 0$ and $Fr = 1$	Uniform equilibrium flow conditions and critical slope: i.e. $S_f = S_o = S_c$. This case is seldom encountered because critical (and near-critical) flows are unstable

Free-surface profiles

With aid of equations (5.6) it is possible to establish the behaviour of $(\partial d/\partial s)$ as a function of the magnitudes of the flow depth d, the critical depth d_c and the normal depth d_o. The sign of $(\partial d/\partial s)$ and the direction of depth change can be determined by the actual depth relative to both the normal and critical flow depths. Altogether, there are 12 different types of free-surface profiles (of non-uniform GVFs), excluding uniform flow. They are listed in the following table:

S_o (1)	d/d_c (2)	d_o/d_c (3)	d/d_o (4)	Fr (5)	S_f/S_o (6)	d, d_o, d_c (7)	Name[a] (8)
>0	<1	<1	<1	>1	>1	$d < d_o < d_c$	S3
			>1	>1	<1	$d_o < d < d_c$	S2
		>1	<1	>1	>1	$d < d_c < d_o$	M3
			>1	>1	<1	Not possible	
	>1	<1	<1	<1	>1	Not possible	
			>1	<1	<1	$d_o < d_c < d$	S1
		>1	<1	<1	>1	$d_c < d < d_o$	M2
			>1	<1	<1	$d_c < d_o < d$	M1
= 0	<1	N/A	N/A	>1	N/A	$d < d_c$	H3
	>1	N/A	N/A	<1	N/A	$d > d_c$	H2
<0	<1	N/A	N/A	>1	N/A	$d < d_c$	A3
	>1	N/A	N/A	<1	N/A	$d > d_c$	A2

Remarks: [a] Name of the free-surface profile where the letter is descriptive for the slope: H for horizontal, M for mild, S for steep and A for adverse (negative slope) (e.g. Chow, 1973).

For a given free-surface profile, the relationships between the flow depth d, the normal depth d_o and the critical depth d_c give the shape of the longitudinal profile. The relationship between d and d_c enables to predict the Froude number and the relationship between d and d_o gives the sign of $(S_o - S_f)$. Combining with the differential form of the mean specific energy (i.e. equation (5.6)), we can determine how the behaviour of $(\Delta d/\Delta s)$ (i.e. longitudinal variation of flow depth) is affected by the relative magnitude of d, d_o and d_c (see table below):

Name[a] (1)	Case (2)	$S_o - S_f$ (3)	Fr (4)	$\Delta d/\Delta s$ (5)	Remarks (6)
M1	$d > d_o > d_c$	$S_o > S_f$	$Fr < 1$	Positive	The water surface is asymptotic to a horizontal line (backwater curve behind a dam).
M2	$d_o > d > d_c$	$S_o < S_f$	$Fr < 1$	Negative	S_f tends to S_o and we reach a transition from subcritical to critical flow (i.e. overfall).
M3	$d_o > d_c > d$	$S_o < S_f$	$Fr > 1$	Positive	Supercritical flow that tends to a hydraulic jump.
S1	$d > d_c > d_o$	$S_o > S_f$	$Fr < 1$	Positive	
S2	$d_c > d > d_o$	$S_o > S_f$	$Fr > 1$	Negative	
S3	$d_c > d_o > d$	$S_o < S_f$	$Fr > 1$	Positive	
H2	$d > d_c$	$S_o = 0 < S_f$	$Fr < 1$	Negative	
H3	$d_c > d$	$S_o = 0 < S_f$	$Fr > 1$	Positive	
A2	$d > d_c$	$S_o < 0 < S_f$	$Fr < 1$	Negative	
A3	$d_c > d$	$S_o < 0 < S_f$	$Fr > 1$	Positive	

Remarks: [a]Name of the free-surface profile where the letter is descriptive for the slope: H for horizontal, M for mild, S for steep and A for adverse. $\Delta d/\Delta s$ is negative for gradually accelerating flow. For $\Delta d/\Delta s > 0$ the flow is decelerating by application of the continuity principle.

Notes

1. The classification of backwater curves was developed first by the French professor J.A.C. Bresse (Bresse, 1860). Bresse originally considered only wide rectangular channels.
2. The friction slope may be re-formulated as:

	Expression	Comments
Friction slope	$S_f = \dfrac{Q^2 P_w f}{8 g A^3}$	f is the Darcy friction factor
	$S_f = \dfrac{Q^2 P_w}{(C_{\text{Chézy}})^2 A^3}$	$C_{\text{Chézy}}$ is the Chézy coefficient

3. The Froude number can be re-formulated as:

	Expression	Comments
Froude number	$Fr = \dfrac{Q}{\sqrt{g(A^3/B)}}$	General case (i.e. any cross-sectional shape)
	$Fr = \dfrac{Q}{\sqrt{g d^3 B^2}}$	Rectangular channel (channel width B)

Application

Considering a mild slope channel, discuss all possible cases of free-surface profiles (non-uniform flows). Give some practical examples.

Solution

Considering a GVF down a mild slope channel, the normal depth must be large than the critical depth (i.e. definition of mild slope). There are three cases of non-uniform flows:

Case	d_o/d_c	d/d_c	Fr	S_f	$\Delta d/\Delta s$	Remarks
$d > d_o > d_c$	>1	>1	<1	$<S_o$	>0	The water surface is asymptotic to a horizontal line (e.g. backwater curve behind a dam)
$d_o > d > d_c$	>1	>1	<1	$>S_o$	<0	S_f tends to S_o and we reach a transition from subcritical to critical flow (e.g. overfall)
$d_o > d_c > d$	>1	<1	>1	$>S_o$	>0	Supercritical flow that tends to a hydraulic jump (e.g. flow downstream of sluice gate with a downstream control)

Discussion

The *M1-profile* is a subcritical flow. It occurs when the downstream of a long mild channel is submerged in a reservoir of greater depth than the normal depth d_o of the flow in the channel.

The *M2-profile* is also a subcritical flow ($d > d_c$). It can occur when the bottom of the channel at the downstream end is submerged in a reservoir to a depth less than the normal depth d_o.

The *M3-profile* is a supercritical flow. This profile occurs usually when a supercritical flow enters a mild channel (e.g. downstream of a sluice gate and downstream of a steep channel).

The *S1-profile* is a subcritical flow in a steep channel.

The *S2-profile* is a supercritical flow. Examples are the profiles formed on a channel as the slope changes from steep to steeper and downstream of an enlargement of a step channel.

The *S3-profile* is a supercritical flow. An example is the flow downstream of a sluice gate in a steep slope when the gate opening is less than the normal depth.

5.2.4 Backwater computations

The backwater profiles are obtained by combining the differential equation (5.6), the flow resistance calculations (i.e. friction slope) and the boundary conditions. The solution is then obtained by numerical computations or graphical analysis.

Discussion

To solve equation (5.6) it is *essential* to determine correctly the *boundary conditions*. By boundary conditions we mean the flow conditions imposed upstream, downstream and along the channel. These boundary conditions are in practice:

(a) *control devices*: sluice gate, weir, reservoir, that impose flow conditions for the depth, the discharge, a relationship between the discharge and the depth, etc., and
(b) *geometric characteristics*: bed altitude, channel width, local friction factor, etc. (e.g. Fig. 5.3).

The backwater profile is calculated using equation (5.6):

$$\frac{\partial d}{\partial s}(1 - Fr^2) = S_o - S_f \tag{5.6}$$

It must be emphasized that the results depend critically on the *assumed friction factor f*. As discussed earlier a great uncertainty applies to the assumed friction factor.

> **Notes**
> 1. Some textbooks suggest to use equation (5.7) to compute the backwater profile. But equation (5.7) presents a numerical singularity for critical flow conditions (i.e. $Fr = 1$). In practice, equation (5.7) *should not* be used, unless it is known *a priori* (i.e. beforehand) that critical flow condition does not take place anywhere along the channel.
>   It is suggested *very strongly* to compute the backwater curves using equation (5.6).
> 2. Note that equation (5.6) has a solution for: $S_f = S_o$ (i.e. uniform flow).
> 3. The backwater equation is not applicable across a hydraulic jump. Backwater computations may be conducted upstream and downstream of the jump, but the momentum principle must be applied at the hydraulic jump (see Section 4.3.2).

Several integration methods exists: analytical solutions are rare but numerous numerical integration methods exist. For simplicity, we propose to develop the standard step method (distance computed from depth). This method is simple, extremely reliable and very stable. It is *strongly* recommended to practising engineers who are not necessarily hydraulic experts.

Standard step method (distance calculated from depth)

Equation (5.4) or (5.6) can be rewritten in a finite difference form as:

$$\Delta s = \frac{\Delta E}{(S_o - S_f)_{\text{mean}}} \tag{5.8}$$

where the subscript 'mean' indicates the *mean value over an interval* and ΔE is the change in mean specific energy along the distance Δs for a given change of flow depth Δd.

In practice, calculations are started at a location of known flow depth d (e.g. a control section). The computations can be performed either in the upstream or the downstream flow direction. A new flow depth $d + \Delta d$ is selected. The flow properties for the new flow depth are computed. We can deduce the change in specific energy ΔE and the average difference $(S_o - S_f)$ over the control volume (i.e. interval). The longitudinal position of the new flow depth is deduced from equation (5.8).

A practical computation example is detailed below to explain the step method. The flow conditions are flow discharge $Q = 100\,\text{m}^3/\text{s}$, channel slope $\theta = 5°$ (i.e. $S_o = 0.0875$), roughness $k_s = 5\,\text{mm}$, rectangular channel, and known flow depth $d = 4\,\text{m}$ at $s = 0$.

d (m)	B (m)	A (m²)	P_w (m)	D_H (m)	f	V (m/s)	Fr	S_f	S_o	$S_o - S_f$	ΔE (m)	Δs (m)	s (m)
4.0	10.0	40.0	18.0	8.89	0.015	2.50	0.399	5.38×10^{-4}	8.75×10^{-2}				0.0
										0.08689	0.0829	0.9544	
4.1	9.0	36.9	17.2	8.58	0.015	2.71	0.427	6.54×10^{-4}	8.75×10^{-2}				0.954
										0.08685	0.0824	0.9485	
4.2	9.0	37.8	17.4	8.69	0.015	2.65	0.412	6.16×10^{-4}	8.75×10^{-2}				1.903

where:
A is the channel cross-sectional area ($A = dB$ for a rectangular channel), B is the channel free-surface width, d is the flow depth, D_H is the hydraulic diameter ($D_H = 4A/P_w$), Fr is the Froude number defined in term of the flow depth d ($Fr = V/\sqrt{gd}$ for a rectangular channel), f is the Darcy coefficient, computed as a function of the Reynolds number Re and the relative roughness k_s/D_H, P_w is the wetted perimeter ($P_w = B + 2d$ for a rectangular channel), Re is the Reynolds number defined in terms of the average flow velocity and hydraulic diameter D_H ($Re = VD_H/\nu$), V is the mean flow velocity ($V = Q/A$), V_* is the shear velocity ($V_* = \sqrt{f\,V^2/8}$), Re_* is the shear Reynolds number ($Re_* = (V_* k_s/\nu)$, S_f is the friction slope ($S_f = (Q^2 P_w f)/(8gA^3)$), S_o is the bed slope ($S_o = \sin\theta$), $\Delta E = \Delta d(1 - Fr^2)$ (mean value over the interval), and s is the curvilinear co-ordinate.

Warning

This step method applies only to the energy equation and does not necessary converge (e.g. uniform flow conditions). Do not use this numerical method for any other equations.

Notes

1. The above method is called the 'step method/distance calculated from depth' (e.g. Henderson, 1966: pp. 126–130). The method, based on equation (5.8), is numerically valid for subcritical, critical and supercritical flows. It is not valid for RVFs (e.g. hydraulic jumps).
2. Another method is based on equation (5.7) as:

$$\Delta d = \frac{S_o - S_f}{1 - Fr^2} \Delta s$$

It is called the 'step method/depth calculated from distance' (e.g. Henderson, 1966: pp. 136–140). This method has the great disadvantage to be singular (i.e. unsolvable) at critical flow conditions.

5.3 EXERCISES

What are the basic assumptions for GVF calculations?

What is the definition of the normal depth?

What is the definition of a steep slope? Does the notion of steep slope depend only on the bed slope? Discuss your answer. A discharge of $100 \, \text{m}^3/\text{s}$ flows in a 12 m wide concrete channel. The channel slope is 5 m/km. Calculate: (a) normal depth, (b) critical depth and (c) indicate the slope type: steep or mild.

A smooth-concrete channel carries a water flow. The channel cross-section is rectangular with a 2 m bottom width. The longitudinal bed slope is 1.3 m/km (i.e. $S_o = 0.0013$). The uniform equilibrium flow depth equals 3.1 m.

(a) Compute the following flow properties: flow velocity, flow rate, average bed shear stress and critical flow depth.
(b) For the uniform equilibrium flow, where would be the optimum location of a control?

Assume $k_s = 1 \, \text{mm}$ for the finished concrete lining.

An artificial canal carries a discharge of $25 \, \text{m}^3/\text{s}$. The channel cross-section is trapezoidal and symmetrical, with a 0.5 m bottom width and 1V:2.5H sidewall slopes. The longitudinal bed slope is 8.5 m/km. The channel bottom and sidewall consist of a mixture of fine sands ($k_s = 0.3 \, \text{mm}$). (a) What is the normal depth at uniform equilibrium? (b) At uniform equilibrium what is the average boundary shear stress? (c) At uniform equilibrium what is the shear velocity?

A gauging station is set at a bridge crossing the waterway. The observed flow depth, at the gauging station, is 2.2 m. (c) Compute the flow velocity at the gauging station. (d) Calculate the Darcy friction factor (at the gauging station). (e) What is the boundary shear stress (at the gauging station)? (f) How would you describe the flow at the gauging station? (g) At the gauging station, from where is the flow controlled? Why?

Solution: (a) $d_o = 1.23$ m, (b) $\tau_o = 51$ Pa, (c) $V_* = 0.226$ m/s, (d) $f = 0.0115$, (e) $\tau_o = 5.2$ Pa, (f) turbulent, transition between smooth-turbulent to rough-turbulent (Section 4.4.1) and (g) downstream because $Q/\sqrt{gA^3/B} = 0.56$ (subcritical flow) (see Section 3.3.4).

The Cotter dam (Canberra ACT) is a 27 m high un-gated weir. The spillway is 60 m wide and a 50° slope. At the toe of the spillway, the slope changes from 50° to 0.05°. The weir is followed by a 20 km long channel with a 60 m width. The channel slope is 0.05°. Assuming that the spillway and the channel are made of smooth concrete, compute the free-surface profiles above the spillway and in the channel for a discharge of 500 m³/s.

Solution: In the reservoir upstream of the weir, the flow motion is tranquil. The flow above the spillway is expected to be rapid (to be checked). Hence a transition from slow to rapid flow motion occurs at the spillway crest: i.e. critical flow conditions. The channel downstream of the weir has a long section of constant slope, constant roughness, constant cross-sectional shape and constant shape. Uniform flow conditions will be obtained at the end of the channel. *Note*: if the uniform flow conditions are subcritical, a transition from supercritical to subcritical flow is expected between the end of the spillway and end of the channel.

Critical and uniform flow calculations
There is only one discharge and channel shape, and it yields to only one set of critical flow conditions $d_c = 1.92$ m, and $V_c = 4.34$ m/s.

Assuming a roughness height $k_s = 1$ mm, the uniform flow conditions for the spillway and the downstream channel are:

	d_o (m)	V_o (m/s)	Fr_o
Spillway	0.28	29.7	17.9
Channel	2.27	3.67	0.78

Hydraulic controls
The preliminary calculations indicate that (1) the spillway slope is *steep* and (2) the channel slope is *mild*. Hence critical flow conditions occur at two locations (A) at the spillway crest, transition from subcritical to supercritical flow and (B) in the channel, transition from supercritical to subcritical flow (i.e. hydraulic jump).

Backwater calculations
The easiest location to start the backwater calculations are the spillway crest. The flow is critical and hence we know the flow depth and flow velocity. By applying the energy equation, we can deduce the free-surface location at any position along the spillway.

In the channel downstream of the spillway, the first calculations predict the occurrence of a hydraulic jump. In such a case, we can assume that the flow depth downstream of the hydraulic jump equals the uniform flow depth. We can deduce then the flow depth immediately upstream of the jump ($d = 1.61$ m, $Fr = 1.30$) by applying the momentum and continuity equations across the hydraulic jump.

The location of the hydraulic jump is deduced from the backwater calculations. Starting at the bottom of the spillway, the flow is decelerated along the channel. When the flow depth equal the sequent depth (i.e. 1.61 m), the hydraulic jump takes place.

Considering a 20 m high un-gated weir with a stepped chute spillway, the spillway is 20 m wide and the slope is 45°. At the toe of the spillway, the slope changes from 45° to 0.1°. The spillway is followed by a dissipation basin (10 m long and 20 m wide) consisting of concrete baffle blocks and then by a 800 m long concrete channel with a 25 m width. The channel slope is 0.1°.

Assume that the friction factor of the stepped spillway is 1.0, and the friction factor of the baffle block lining is $f = 2.5$. Compute the free-surface profiles above the spillway and in the channel for a discharge of 160 m³/s. Is the dissipation basin operating properly?

Notes
1. A dissipation basin or stilling basin is a short length of paved channel placed at the foot of a spillway or any other source of supercritical flow. The aim of the designer is to make a hydraulic jump within the basin so that the flow is converted to subcritical before it reaches the exposed riverbed or channel downstream (Henderson, 1966).
2. Baffle blocks or chute blocks consist of reinforced concrete blocks installed to increase the resistance to the flow and to dissipate part of the kinetic energy of the flow.

Part 1 Revision exercises

REVISION EXERCISE NO. 1

The undershot sluice gate sketched in Fig. R.1 is in a channel 5 m wide and the discharge, Q, is $4\,\mathrm{m^3/s}$. The upstream flow depth is 1.2 m. The bed of the channel is horizontal and smooth.

- Sketch on Fig. R.1 the variation of the pressure with depth at sections 1 and 2, and on the upstream face of the sluice gate. Sections 1 and 2 are located far enough from the sluice gate for the velocity to be essentially horizontal and uniform.
- Show on Fig. R.1 the forces acting on the control volume contained between sections 1 and 2. Show also your choice for the positive direction of distance and of force.
- Write the momentum equation as applied to the control volume between sections 1 and 2, using the sign convention you have chosen. Show on Fig. R.1 the forces and velocities used in the momentum equation.
- Three water surface profiles on the upstream side of the sluice gate are labelled a, b and c. Which is the correct one? Three water surface profiles downstream of the sluice gate are labelled d, e and f. Which is the correct one?

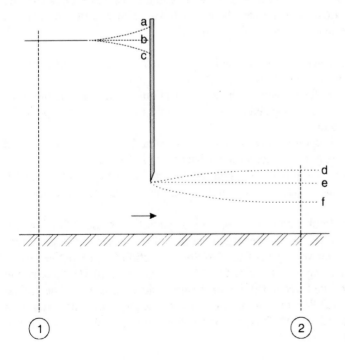

Fig. R.1 Sketch of a sluice gate.

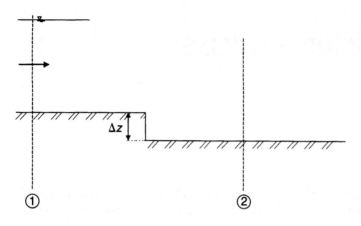

Fig. R.2 Sketch of a stepped transition.

- Compute and give the values (and units) of the specified quantities in the following list: (a) Velocity of flow at section 1. (b) Specific energy at section 1. (c) Specific energy at section 2. (d) Assumption used in answer (c). (e) Depth of flow at section 2. (f) Velocity of flow at section 2. (g) The force acting on the sluice gate. (h) The direction of the force in (g): i.e. upstream or downstream. (i) The critical depth for the flow in Fig. R.1. (j) What is maximum possible discharge per unit width for a flow with specific energy entered at (b). (This question is a general question, not related to the sluice gate.)

REVISION EXERCISE NO. 2

A channel step, sketched in Fig. R.2, is in a 12 m wide channel and the discharge, Q, is 46 m³/s. The flow depth at section 1 is 1.6 m and the step height is 0.4 m. The bed of the channel, upstream and downstream of the step, is horizontal and smooth.

- Sketch on Fig. R.2 the free-surface profile between sections 1 and 2. (You might have to modify it after your calculations in the following questions.)
- Sketch on Fig. R.2 the variation of the pressure with depth at sections 1 and 2, and on the vertical face of the step. Sections 1 and 2 are located far enough from the step for the velocity to be essentially horizontal and uniform.
- Show on Fig. R.2 the forces acting on the control volume contained between sections 1 and 2. Show also your choice for the positive direction of distance and of force.
- Write the momentum equation as applied to the control volume between sections 1 and 2, using the sign convention you have chosen. Show on Fig. R.2 the forces and velocities used in the momentum equation.
- Compute and give the values (and units) of the specified quantities in the following list: (a) Velocity of flow at section 1. (b) Specific energy at section 1. (c) Specific energy at section 2. (d) Assumption used in answer (c). (e) Depth of flow at section 2. (f) Plot the correct free-surface profile on Fig. R.2 using (e). (g) Velocity of flow at section 2. (h) The force acting on the step. (i) The direction of the force in (h): i.e. upstream or downstream. (j) The critical depth for the flow in Fig. R.2. (k) What is maximum possible discharge per unit width for a flow with specific energy entered at (b). (This question is a general question, not related to the stepped channel.)

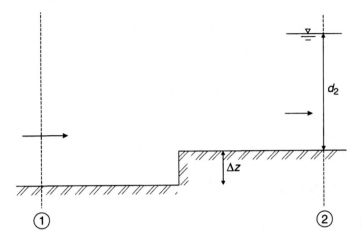

Fig. R.3 Sketch of an upward-step.

REVISION EXERCISE NO. 3

A channel step, sketched in Fig. R.3, is in a 5 m wide channel of rectangular cross-section. The total discharge is 64 000 kg/s. The fluid is slurry (density 1200 kg/m^3). The downstream flow depth (i.e. at section 2) is 1.5 m and the step height is 0.8 m. The bed of the channel, upstream and downstream of the step, is horizontal and smooth. The flow direction is from section 1 to section 2.

- Compute and give the values (and units) of the following quantities: (a) Velocity of flow at section 2. (b) Froude number at section 2. (c) Specific energy at section 2.
- For the flow at section 2, where would you look for the hydraulic control?
- Sketch on Fig. R.3 the free-surface profile between sections 1 and 2. You might have to modify it after your calculations in the following questions.
- Show on Fig. R.3 *all* the forces acting on the control volume contained between sections 1 and 2. Show also your choice for the positive direction of distance and of force.
- Compute and give the values (and units) of the specified quantities in the following list: (a) Specific energy at section 1. (b) Depth of flow at section 1. (c) Velocity of flow at section 1. (d) Froude number at section 1. (e) Plot the correct free-surface profile on Fig. R.3 using (b). (f) The force acting on the step. (g) The direction of the force in (f): i.e. upstream or downstream. (h) The critical depth for the flow in Fig. R.3.

REVISION EXERCISE NO. 4

The horizontal rectangular channel sketched in Fig. R.4 is used as a stilling basin in which energy dissipation takes place in a hydraulic jump. The channel is equipped with four baffle blocks. The width of the channel is 20 m. The bed of the channel is horizontal and smooth. The inflow conditions are $d_1 = 4.1$ m, $V_1 = 22$ m/s, and the *observed* downstream flow depth is $d_2 = 19.7$ m.

- Sketch on Fig. R.4 an appropriate control volume between the upstream and downstream flow locations (on the cross-sectional view).
- Sketch on Fig. R.4 (cross-sectional view) the variation of the pressure with depth at sections 1 and 2. Sections 1 and 2 are located far enough from the hydraulic jump for the velocity to be essentially horizontal and uniform.

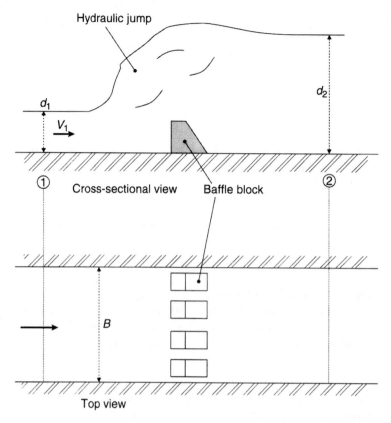

Fig. R.4 Sketch of a stilling basin.

- Show on Fig. R.4 *all* the forces acting on the control volume contained between sections 1 and 2. Show also your choice for the positive direction of distance and of force.
- Write the momentum equation as applied to the control volume between sections 1 and 2, using the sign convention that you have chosen.
- Compute and give the values (and units) of the specified quantities in the following list: (a) Total flow rate at section 1. (b) Specific energy at section 1. (c) Froude number at section 1. (d) Total force acting on the baffle blocks. (e) Velocity of flow at section 2. (f) Specific energy at section 2. (g) Energy loss between section 1 and 2. (h) Force acting on a single block. (i) The direction of the force in (h) and (g): i.e. upstream or downstream.

REVISION EXERCISE NO. 5

The 'Venturi' flume sketched on Fig. R.5 is in a rectangular channel and the discharge is $2 \, \text{m}^3/\text{s}$. The channel bed upstream and downstream of the throat is horizontal and smooth. The channel width is $B_1 = B_3 = 5 \, \text{m}$. The throat characteristics are $\Delta z_o = 0.5 \, \text{m}$ and $B_2 = 2.5 \, \text{m}$. The upstream flow depth is $d_1 = 1.4 \, \text{m}$.

- Sketch on Fig. R.5 a free-surface profile in the Venturi flume between sections 1 and 3. (You might have to modify it after your calculations in the following questions.)
- Sketch on Fig. R.5 the variation of the pressure with depth at sections 1 and 3. Sections 1 and 2 are located far enough from the throat for the velocity to be essentially horizontal and uniform.

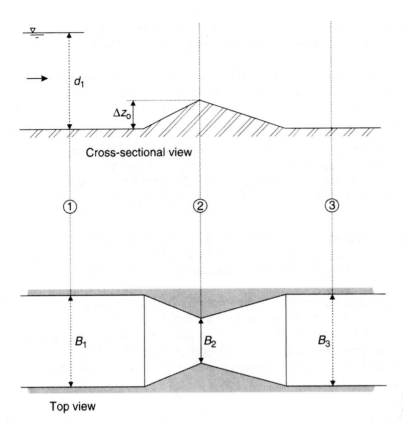

Fig. R.5 Sketch of a Venturi flume.

- Compute and give the values (and units) of the specified quantities in the following list: (a) Velocity of flow at section 1. (b) Specific energy at section 1. (c) Specific energy at section 2 (i.e. at throat). (d) Assumption used in answer (c). (e) Depth of flow at section 2 (i.e. at throat). (f) Velocity of flow at section 2 (i.e. at throat). (g) Depth of flow at section 3. (h) Velocity of flow at section 3. (i) Specific energy at section 3. (j) The critical depth for the flow at section 2 (i.e. at throat). (k) What is maximum possible discharge per unit width for a flow with specific energy entered at (c) (i.e. at throat)? (This question is a general question, not related to the sluice gate.)

REVISION EXERCISE NO. 6

An (undershot) *sluice gate* is in a rectangular channel 0.8 m wide. The flow depth upstream of the gate is 0.45 m and, downstream of the sluice gate, the free-surface elevation is 0.118 m above the channel bed. The bed of the channel is horizontal and smooth.

- Sketch the sluice gate and the free-surface profiles upstream and downstream of the sluice gate.
- Sketch on your figure the variation of the pressure with depth at a section 1 (upstream of the gate) and a section 2 (downstream of the gate), and on the upstream face of the sluice gate. Sections 1 and 2 are located far enough from the sluice gate for the velocity to be essentially horizontal and uniform.
- Compute and give the values (and units) of the following quantities:
  (a) Flow rate. (b) Assumption used in answer (a). (c) Velocity of flow at section 1. (d) Specific energy at section 1. (e) Specific energy at section 2. (f) Velocity of flow at section 2.

- Write the momentum equation as applied to the control volume between sections 1 and 2, using a consistent sign convention. Show on your figure the forces and velocities used in the momentum equation.
- Compute and give the values (and units) of the specified quantities in the following list (a) The force acting on the sluice gate. (b) The direction of the force in (a): i.e. upstream or downstream. (c) The critical depth for the flow.

A *broad-crested weir* is to be constructed downstream of the sluice gate. The purpose of the gate is to force the dissipation of the kinetic energy of the flow, downstream of the sluice gate, between the gate and the weir: i.e. to induce a (fully developed) hydraulic jump between the gate and the weir. The channel between the gate and the weir is horizontal and smooth. The weir crest will be set at a vertical elevation Δz above the channel bed, to be determined for the flow conditions in the above questions.

- Sketch the channel, the sluice gate, the hydraulic jump and the broad-crested weir.
- Write the momentum equation between section 3 (located immediately upstream of the jump and downstream of the gate) and section 4 (located immediately downstream of the jump and upstream of the weir).
- For the above flow conditions, compute and give the values (and units) of the specified quantities in the following list: (a) Flow depth upstream of the hydraulic jump (section 3). (b) Specific energy upstream of the jump (section 3). (c) Flow depth downstream of the jump (section 4). (d) Specific energy downstream of the jump (section 4). (e) Flow depth above the crest. (f) Specific energy of the flow at the weir crest. (g) Weir crest elevation Δz. (h) Assumptions used in answer (g).

REVISION EXERCISE NO. 7

A settling basin is a deep and wide channel in which heavy sediment particles may settle, to prevent siltation of the downstream canal. Settling basins are commonly located at the upstream end of an irrigation canal. In this section we shall investigate the flow characteristics of the outlet of the settling basin (Fig. R.6).

The outlet section is a rectangular channel sketched in Fig. R.6. The channel bed is smooth and, at sections 1, 2 and 3, it is horizontal. The channel width equals $B_1 = 12\,\text{m}$, $B_2 = B_3 = 4.2\,\text{m}$. The change in bed elevation is $\Delta z_0 = 0.4\,\text{m}$. The total flow rate is 48 200 kg/s. The fluid is slurry (density 1405 kg/m³). The inflow conditions are $d_1 = 4\,\text{m}$.

(a) Compute and give the value of velocity of flow at section 1. (b) Compute the specific energy at section 1. (c) Calculate the specific energy at section 2. (d) In answer (c), what basic principle did you *not* use? (e) Compute the depth of flow at section 2. (f) Calculate the velocity of flow at section 2. (g) Compute the depth of flow at section 3. (h) Compute the velocity of flow at section 3. (i) Calculate the specific energy at section 3. (j) Compute the critical depth for the flow at section 2. (k) What is maximum possible discharge per unit width for a flow with specific energy entered at question (i). (This question is a general question, not related to the settling basin.)

REVISION EXERCISE NO. 8

A broad-crested weir is to be designed to operate with critical flow above its crest and no downstream control. The maximum discharge capacity will be 135 m³/s. The weir crest will be 2.5 m above the natural channel bed and its width will be 15 m. (Assume the channel to be prismatic, smooth and rectangular.)

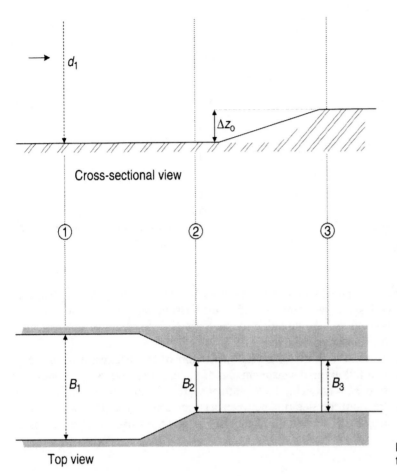

Fig. R.6 Sketch of the settling basin outlet.

Sketch the free-surface profile between sections 1 and 3 located respectively upstream and downstream of the weir. (You might have to modify it after your calculations in the following questions.) Sketch also the variation of the pressure with depth at sections 1 and 3. Sections 1 and 3 are located far enough from the throat for the velocity to be essentially horizontal and uniform.

Calculate (a) the upstream flow depth at maximum flow rate, (b) the upstream specific energy at maximum flow rate, (c) the specific energy at the crest at maximum flow rate, (d) the downstream specific energy at maximum flow rate, (e) the downstream flow depth at maximum flow rate. (f) At maximum flow rate, what will be the longitudinal free-surface profile? (g) Calculate the horizontal component of the sliding force acting on the broad-crested weir at maximum flow rate. (h) What is maximum possible discharge per unit width for a flow with specific energy entered at (b) (i.e. upstream). (This question is a general question, not related to the broad-crested weir.)

REVISION EXERCISE NO. 9

The backward-facing step sketched in Fig. R.7 is in a 1.3 m wide channel and the discharge is 190 l/s. The bed of the channel is horizontal and smooth (both upstream and downstream of the step). The flow direction is from section 1 to section 2.

(a) Show in Fig. R.7 *all* the forces acting on the control volume contained between sections 1 and 2. Show also your choice for the positive direction of distance, of velocity and of force.

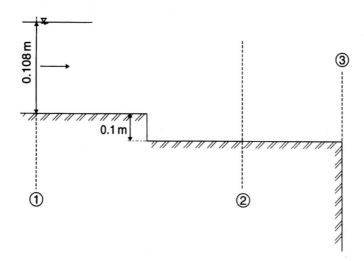

Fig. R.7 Sketch of stepped channel and overfall.

(b) Compute and give the values (and units) of the specified quantities in the following list: specific energy at sections 1 and 2, assumption used in the answer, depth of flow at section 2, the force acting on the step and the direction of the force (i.e. upstream or downstream). Then plot the correct free-surface profile in Fig. R.7.

(c) A free overfall is located at a short distance downstream of the backward-facing step. Considering now the overfall at the downstream end of the channel (Fig. R.7). What is the specific energy at section 3? What is the flow depth at section 3? Plot in Fig. R.7 the free-surface profile downstream of the step (i.e. between sections 2 and 3) AND downstream of the overfall (i.e. downstream of section 3). Sketch in Fig. R.7 the variation of the pressure with depth at section 3.

Appendices to Part 1

A1.1 CONSTANTS AND FLUID PROPERTIES

A1.1.1 Acceleration of gravity

Standard acceleration of gravity

The standard acceleration of gravity equals:

$$g = 9.80665 \text{ m/s}^2 \tag{A1.1}$$

This value is roughly equal to that at sea level and at 45° latitude. The gravitational acceleration varies with latitude and elevation owing to the form and rotation of the earth and may be estimated as:

$$g = 9.806056 - 0.025027 \cos(2 \text{ latitude}) - 3 \times 10^{-6} z \tag{A1.2}$$

where z is the altitude with the sea level as the origin, and the latitude is in degrees (Rouse, 1938).

Altitude z (m) (1)	g (m/s^2) (2)
-1000	9.810
0 (sea level)	9.807
1000	9.804
2000	9.801
3000	9.797
4000	9.794
5000	9.791
6000	9.788
7000	9.785
10 000	9.776

Absolute gravity values

Gravity also varies with the local geology and topography. Measured values of g are reported in the following table:

Location (1)	g (m/s^2) (2)	Location (1)	g (m/s^2) (2)	Location (1)	g (m/s^2) (2)
Addis Ababa, Ethiopia	9.7743	Brisbane, Australia	9.794	Kuala Lumpur, Malaysia	9.78034
Algiers, Algeria	9.79896	Buenos Aires, Argentina	9.7949	La Paz, Bolivia	9.7745
Anchorage, USA	9.81925	Christchurch, NZ	9.8050	Lisbon, Portugal	9.8007
Ankara, Turkey	9.79925	Denver, USA	9.79598	Manila, Philippines	9.78382
Aswan, Egypt	9.78854	Edmonton, Canada	9.81145	Mexico City, Mexico	9.77927
Bangkok, Thailand	9.7830	Guatemala, Guatemala	9.77967	Nairobi, Kenya	9.77526
Bogota, Colombia	9.7739	Helsinki, Finland	9.81090	New Delhi, India	9.79122

(*continued*)

Location (1)	g (m/s^2) (2)	Location (1)	g (m/s^2) (2)	Location (1)	g (m/s^2) (2)
Paris, France	9.80926	Quito, Ecuador	9.7726	Thule, Greenland	9.82914
Perth, Australia	9.794	Sapporo, Japan	9.80476	Tokyo, Japan	9.79787
Port-Moresby, PNG	9.782	Reykjavik, Iceland	9.82265	Vancouver, Canada	9.80921
Pretoria, South Africa	9.78615	Taipei, Taiwan	9.7895	Ushuaia, Argentina	9.81465
Québec, Canada	9.80726	Teheran, Iran	9.7939		

Reference: Morelli (1971).

A1.1.2 Properties of water

Temperature (°C) (1)	Density ρ_w (kg/m^3) (2)	Dynamic viscosity μ_w (Pa s) (3)	Surface tension σ (N/m) (4)	Vapour pressure P_v (Pa) (5)	Bulk modulus of elasticity (E_b) (Pa) (6)
0	999.9	1.792×10^{-3}	0.0762	0.6×10^3	2.04×10^9
5	1000.0	1.519×10^{-3}	0.0754	0.9×10^3	2.06×10^9
10	999.7	1.308×10^{-3}	0.0748	1.2×10^3	2.11×10^9
15	999.1	1.140×10^{-3}	0.0741	1.7×10^3	2.14×10^9
20	998.2	1.005×10^{-3}	0.0736	2.5×10^3	2.20×10^9
25	997.1	0.894×10^{-3}	0.0726	3.2×10^3	2.22×10^9
30	995.7	0.801×10^{-3}	0.0718	4.3×10^3	2.23×10^9
35	994.1	0.723×10^{-3}	0.0710	5.7×10^3	2.24×10^9
40	992.2	0.656×10^{-3}	0.0701	7.5×10^3	2.27×10^9

Reference: Streeter and Wylie (1981).

A1.1.3 Gas properties

Basic equations

The *state equation* of perfect gas is:

$$P = \rho R T \tag{A1.3}$$

where P is the absolute pressure (in Pascal), ρ is the gas density (in kg/m^3), T is the absolute temperature (in Kelvin) and R is the gas constant (in J/kg K) (see table below).

For a perfect gas, the *specific heat* at constant pressure C_p and the specific heat at constant volume C_v are related to the gas constant as:

$$C_p = \frac{\gamma}{\gamma - 1} R \tag{A1.4a}$$

$$C_p = C_v R \tag{A1.4b}$$

where γ is the specific heat ratio (i.e. $\gamma = C_p / C_v$).

During an *isentropic transformation* of perfect gas, the following relationships hold:

$$\frac{P}{\rho^\gamma} = \text{constant} \tag{A1.5a}$$

$$T P^{(1-\gamma)/\gamma} = \text{constant} \tag{A1.5b}$$

Physical properties

Gas	Formula	Gas constant R (J/kg K)	Specific heat		Specific heat ratio γ
			C_p (J/kg K)	C_v (J/kg K)	
(1)	(2)	(3)	(4)	(5)	(6)
Perfect gas					
Mono-atomic gas	(e.g. He)		(5/2)R	(3/2)R	5/3
Di-atomic gas	(e.g. O_2)		(7/2)R	(5/2)R	7/5
Poly-atomic gas	(e.g. CH_4)		4R	3R	4/3
Real gas[a]					
Air		287	1.004	0.716	1.40
Helium	He	2077.4	5.233	3.153	1.67
Nitrogen	N_2	297	1.038	0.741	1.40
Oxygen	O_2	260	0.917	0.657	1.40
Water vapour	H_2O	462	1.863	1.403	1.33

Note: [a] at low pressures and at 299.83 K (Streeter and Wylie, 1981).

Compressibility and bulk modulus of elasticity

The compressibility of a fluid is a measure of change in volume and density when the fluid is subjected to a change of pressure. It is defined as:

$$E_{co} = \frac{1}{\rho}\frac{\partial \rho}{\partial P} \tag{A1.6}$$

The reciprocal function of the compressibility is called the bulk modulus of elasticity:

$$E_b = \rho\frac{\partial P}{\partial \rho} \tag{A1.7a}$$

For a perfect gas, the bulk modulus of elasticity equals:

$$E_b = \gamma P \quad \text{Adiabatic transformation for a perfect gas} \tag{A1.7b}$$

$$E_b = P \quad \text{Isothermal transformation for a perfect gas} \tag{A1.7c}$$

Celerity of sound

Introduction
The celerity of sound in a medium is:

$$C_{sound} = \sqrt{\frac{\partial P}{\partial \rho}} \tag{A1.8}$$

where P is the pressure and ρ is the density. It may be rewritten in terms of the bulk modulus of elasticity E_b:

$$C_{sound} = \sqrt{\frac{E_b}{\rho}} \tag{A1.9}$$

Equation (A1.7) applies to both liquids and gases.

Sound celerity in gas

For an isentropic process and a perfect gas, equation (A1.9) yields:

$$C_{sound} = \sqrt{\gamma R T} \tag{A1.10}$$

where γ and R are the specific heat ratio and gas constant respectively (see above).

The dimensionless velocity of compressible fluid is called the Sarrau–Mach number:

$$Ma = \frac{V}{C_{sound}} \tag{A1.11}$$

Classical values

Celerity of sound in water at 20°C: 1485 m/s.
Celerity of sound in dry air and sea level at 20°C: 343 m/s.

A1.1.4 Atmospheric parameters

Air pressure

The standard atmosphere or normal pressure at sea level equals:

$$P_{std} = 1\,atm = 360\,mm\,of\,Hg = 101\,325\,Pa \tag{A1.12}$$

where Hg is the chemical symbol of mercury. Unit conversion tables are provided in Appendix A1.2. The atmospheric pressure varies with the elevation above the sea level (i.e. altitude). For dry air, the atmospheric pressure at the altitude z equals:

$$P_{atm} = P_{std}\,\exp\left(-\int_0^z \frac{0.0034841g}{T}\,dz\right) \tag{A1.13}$$

where T is the absolute temperature in Kelvin and equation (A1.13) is expressed in SI units.
Reference: Miller (1971).

Air temperature

In the troposphere (i.e. $z < 10\,000\,m$), the air temperature decreases with altitude, on the average, at a rate of $6.5 \times 10^3\,K/m$ (i.e. 6.5 K/km).

Table A1.1 presents the distributions of average air temperatures (Miller, 1971) and corresponding atmospheric pressures with the altitude (equation (A1.13)).

Viscosity of air

Viscosity and density of air at 1.0 atm:

Temperature (K) (1)	μ_{air} (Pa s) (2)	ρ_{air} (kg/m^3) (3)
300	18.4×10^{-6}	1.177
400	22.7×10^{-6}	0.883
500	26.7×10^{-6}	0.705
600	29.9×10^{-6}	0.588

Table A1.1 Distributions of air temperature and air pressure as functions of the altitude (for dry air and standard acceleration of gravity)

Altitude z (m) (1)	Mean air temperature (K) (2)	Atmospheric pressure (equation (A1.13)) (Pa) (3)	Atmospheric pressure (equation (A1.13)) (atm) (4)
0	288.2	1.013×10^5	1.000
500	285.0	9.546×10^4	0.942
1000	281.7	8.987×10^4	0.887
1500	278.4	8.456×10^4	0.834
2000	275.2	7.949×10^4	0.785
2500	272.0	7.468×10^4	0.737
3000	268.7	7.011×10^4	0.692
3500	265.5	6.576×10^4	0.649
4000	262.2	6.164×10^4	0.608
4500	259.0	5.773×10^4	0.570
5000	255.7	5.402×10^4	0.533
5500	252.5	5.051×10^4	0.498
6000	249.2	4.718×10^4	0.466
6500	246.0	4.404×10^4	0.435
7000	242.8	4.106×10^4	0.405
7500	239.5	3.825×10^4	0.378
8000	236.3	3.560×10^4	0.351
8500	233.0	3.310×10^4	0.327
9000	229.8	3.075×10^4	0.303
9500	226.5	2.853×10^4	0.282
10 000	223.3	2.644×10^4	0.261

The viscosity of air at standard atmosphere is commonly fitted by the Sutherland formula (Sutherland, 1883):

$$\mu_{air} = 17.16 \times 10^{-6} \left(\frac{T}{273.1}\right)^{3/2} \frac{383.7}{T + 110.6} \tag{A1.14}$$

A simpler correlation is:

$$\frac{\mu_{air}(T)}{\mu_{air}(T_o)} = \left(\frac{T}{T_o}\right)^{0.76} \tag{A1.15}$$

where μ_{air} is in Pa s, and the temperature T and reference temperature T_o are expressed in Kelvin.

*   *   *

A1.2 UNIT CONVERSIONS

A1.2.1 Introduction

The systems of units derived from the metric system have gradually been replaced by a single system, called the Système International d'Unités (SI unit system, or International System of Units). The present lecture notes are presented in SI units.

Some countries continue to use British and American units, and conversion tables are provided in this appendix. Basic references in unit conversions include Degremont (1979) and ISO (1979).

Principles and rules

Unit symbols are written in small letters (i.e. m for metre, kg for kilogramme) but a capital is used for the first letter when the name of the unit derives from a surname (e.g. Pa after Blaise Pascal, N after Isaac Newton).

Multiples and submultiples of SI units are formed by adding one prefix to the name of the unit: e.g. km for kilometre, cm for centimetre, dam for decametre, μm for micrometre (or micron).

Multiple/submultiple factor	Prefix	Symbol
1×10^9	giga	G
1×10^6	mega	M
1×10^3	kilo	k
1×10^2	hecto	d
1×10^1	deca	da
1×10^{-1}	deci	d
1×10^{-2}	centi	c
1×10^{-3}	milli	m
1×10^{-6}	micro	μ
1×10^{-9}	nano	n

The basic SI units are the metre, kilogramme, second, Ampere, Kelvin, mole and candela. Supplementary units are the radian and the steradian. All other SI units derive from the basic units.

A1.2.2 Units and conversion factors

Quantity (1)	Unit (symbol) (2)	Conversion (3)	Comments (4)
Length	1 inch (in)	$= 25.4 \times 10^{-3}$ m	Exactly
	1 foot (ft)	$= 0.3048$ m	Exactly
	1 yard (yd)	$= 0.9144$ m	Exactly
	1 mil	$= 25.4 \times 10^{-6}$ m	1/1000 inch
	1 mile	$= 1.609344$ m	Exactly
Area	1 square inch (in²)	$= 6.4516 \times 10^{-4}$ m²	Exactly
	1 square foot (ft²)	$= 0.09290306$ m²	Exactly
Volume	1 litre (l)	$= 1.0 \times 10^{-3}$ m³	Exactly. Previous symbol: L
	1 cubic inch (in³)	$= 16.387064 \times 10^{-6}$ m³	Exactly
	1 cubic foot (ft³)	$= 28.3168 \times 10^{-3}$ m³	Exactly
	1 gallon UK (gal UK)	$= 4.54609 \times 10^{-3}$ m³	
	1 gallon US (gal US)	$= 3.78541 \times 10^{-3}$ m³	
	1 barrel US	$= 158.987 \times 10^{-3}$ m³	For petroleum, etc.
Velocity	1 foot per second (ft/s)	$= 0.3048$ m/s	Exactly
	1 mile per hour (mph)	$= 0.44704$ m/s	Exactly
Acceleration	1 foot per second squared (ft/s²)	$= 0.3048$ m/s²	Exactly
Mass	1 pound (lb or lbm)	$= 0.45359237$ kg	Exactly
	1 ton UK	$= 1016.05$ kg	
	1 ton US	$= 907.185$ kg	
Density	1 pound per cubic foot (lb/ft³)	$= 16.0185$ kg/m³	
Force	1 kilogram-force (kgf)	$= 9.80665$ N (exactly)	Exactly
	1 pound force (lbf)	$= 4.4482216152605$ N	
Moment of force	1 foot pound force (ft lbf)	$= 1.35582$ N m	
Pressure	1 Pascal (Pa)	$= 1$ N/m²	
	1 standard atmosphere (atm)	$= 101\,325$ Pa	Exactly
		$= 760$ mm of Mercury at normal pressure (i.e. mm of Hg)	
	1 bar	$= 10^5$ Pa	Exactly
	1 torr	$= 133.322$ Pa	
	1 conventional metre of water (m of H₂O)	$= 9.80665 \times 10^3$ Pa	Exactly
	1 conventional metre of Mercury (m of Hg)	$= 1.333224 \times 10^5$ Pa	
	1 pound per square inch (PSI)	$= 6.8947572 \times 10^3$ Pa	

(continued)

Quantity (1)	Unit (symbol) (2)	Conversion (3)	Comments (4)
Temperature	T (°C)	$= T$ (K) $- 273.16$	0°C is 0.01 K below the temperature of the triple point of water
	T (Fahrenheit)	$= T$ (°C)(9/5) $+ 32$	
	T (Rankine)	$= (9/5)T$ (K)	
Dynamic viscosity	1 Pa s	$= 0.006\,720$ lbm/ft/s	
	1 Pa s	$= 10$ Poises	Exactly
	1 N s/m^2	$= 1$ Pa s	Exactly
	1 Poise (P)	$= 0.1$ Pa s	Exactly
	1 milliPoise (mP)	$= 1.0 \times 10^4$ Pa s	Exactly
Kinematic viscosity	1 square foot per second (ft^2/s)	$= 0.092\,903\,0$ m^2/s	
	1 m^2/s	$= 10.7639$ ft^2/s	
	1 m^2/s	$= 10^4$ Stokes	
Surface tension	1 dyne/cm	$= 0.99987 \times 10^{-3}$ N/m	
	1 dyne/cm	$= 5.709 \times 10^{-6}$ pound/inch	
Work energy	1 Joule (J)	$= 1$ N m	
	1 Joule (J)	$= 1$ W s	
	1 Watt hour (Wh)	$= 3.600 \times 10^3$ J	Exactly
	1 electronvolt (eV)	$= 1.60219 \times 10^{-19}$ J	
	1 Erg	$= 10^{-7}$ J	Exactly
	1 foot pound force (ft lbf)	$= 1.355\,82$ J	
Power	1 Watt (W)	$= 1$ J/s	
	1 foot pound force per second (ft lbf/s)	$= 1.355\,82$ W	
	1 horsepower (hp)	$= 745.700$ W	

*   *   *

A1.3 MATHEMATICS
Summary

1. Introduction
2. Vector operations
3. Differential and differentiation
4. Trigonometric functions
5. Hyperbolic functions
6. Complex numbers
7. Polynomial equations

A1.3.1 Introduction

References

Beyer, W.H. (1982). "*CRC Standard Mathematical Tables.*" CRC Press Inc., Boca Raton, Florida, USA.

Korn, G.A. and Korn, T.M. (1961). "*Mathematical Handbook for Scientist and Engineers.*" McGraw-Hill, New York, USA.

Spiegel, M.R. (1968). "*Mathematical Handbook of Formulas and Tables.*" McGraw-Hill Inc., New York, USA.

Notation

x, y, z	Cartesian coordinates
r, θ, z	polar coordinates
$\dfrac{\partial}{\partial x}$	partial differentiation with respect to the x-coordinate
$\dfrac{\partial}{\partial y}, \dfrac{\partial}{\partial z}$	partial differential (Cartesian coordinate)

$\dfrac{\partial}{\partial r}, \dfrac{\partial}{\partial \theta}$ partial differential (polar coordinate)

$\dfrac{\partial}{\partial t}$ partial differential with respect of the time t

$\dfrac{D}{Dt}$ absolute derivative

δ_{ij} identity matrix element: $\delta_{ii} = 1$ and $\delta_{ij} = 0$ (for i different of j)

$N!$ N-factorial: $N! = 1 \times 2 \times 3 \times 4 \times \cdots \times (N-1) \times N$

Constants

e   constant such as $\ln(e) = 1$: e = 2.718 281 828 459 045 235 360 287

π π = 3.141 592 653 589 793 238 462 643

$\sqrt{2}$ $\sqrt{2}$ = 1.414 213 562 373 095 048 8

$\sqrt{3}$ $\sqrt{3}$ = 1.732 050 807 568 877 293 5

A1.3.2 Vector operations

Definitions

Considering a three-dimensional space, the coordinates of a point M or of a vector $\vec{A}$ can be expressed in a Cartesian system of coordinates or a cylindrical (or polar) system of coordinates as:

	Cartesian system of coordinates	Cylindrical system of coordinates
Point M	(x, y, z)	(r, θ, z)
Vector $\vec{A}$	(A_x, A_y, A_z)	(A_r, A_θ, A_z)

The relationship between the cartesian coordinates and the polar coordinates of any point M are:

$$r^2 = x^2 + y^2$$

$$\tan \theta = \frac{y}{x}$$

Vector operations

Scalar product of two vectors

$$\vec{A} \times \vec{B} = |A| \times |B| \times \cos(\vec{A}, \vec{B})$$

where $|A| = \sqrt{A_x^2 + A_y^2 + A_z^2}$. Two non-zero vectors are perpendicular to each other if and only if their scalar product is null.

Vector product

$$\vec{A} \wedge \vec{B} = \vec{i}\,(A_y \times B_z - A_z \times B_y) + \vec{j}(A_z \times B_x - A_x \times B_z) + \vec{k}(A_x \times B_y - A_y \times B_x)$$

where $\vec{i}, \vec{j}$ and $\vec{k}$ are the unity vectors in the x-, y- and z-directions, respectively.

A1.3.3 Differential and differentiation

Absolute differential

The absolute differential D/Dt of a scalar $\Phi(r)$ along the curve: $r = r(t)$ is, at each point (r) of the curve, the rate of change of $\Phi(r)$ with respect to the parameter t as r varies as a function of t:

$$\frac{D\Phi}{Dt} = \left(\frac{Dr}{Dt} \times \nabla\right)\Phi = \frac{\partial x}{\partial t} \times \frac{\partial \Phi}{\partial x} + \frac{\partial y}{\partial t} \times \frac{\partial \Phi}{\partial y} + \frac{\partial z}{\partial t} \times \frac{\partial \Phi}{\partial z}$$

If Φ depends explicitly on t [$\Phi = \Phi(r, t)$] then:

$$\frac{D\Phi}{Dt} = \frac{\partial \Phi}{\partial t} + \left(\frac{Dr}{Dt} \times \nabla\right)\Phi = \frac{\partial \Phi}{\partial t} + \frac{\partial x}{\partial t} \times \frac{\partial \Phi}{\partial x} + \frac{\partial y}{\partial t} \times \frac{\partial \Phi}{\partial y} + \frac{\partial z}{\partial t} \times \frac{\partial \Phi}{\partial z}$$

and this may be rewritten as:

$$\frac{D\Phi}{Dt} = \frac{\partial \Phi}{\partial t} + \vec{V} \times \nabla\Phi = \frac{\partial \Phi}{\partial t} + V_x \times \frac{\partial \Phi}{\partial x} + V_y \times \frac{\partial \Phi}{\partial y} + V_z \times \frac{\partial \Phi}{\partial z}$$

In the above equations ∇ is the nabla differential operator. It is considered as a vector:

$$\nabla = \vec{i}\left(\frac{\partial}{\partial x}\right) + \vec{j}\left(\frac{\partial}{\partial y}\right) + \vec{k}\left(\frac{\partial}{\partial z}\right)$$

Differential operators

Gradient

$$\overrightarrow{\text{grad }} \Phi(x, y, z) = \nabla\Phi(x, y, z) = \vec{i}\,\frac{\partial \Phi}{\partial x} + \vec{j}\,\frac{\partial \Phi}{\partial y} + \vec{k}\,\frac{\partial \Phi}{\partial z} \qquad \text{Cartesian coordinate}$$

Divergence

$$\text{div }\vec{F}(x, y, z) = \nabla\vec{F}(x, y, z) = \frac{\partial F_x}{\partial x} + \frac{\partial F_y}{\partial y} + \frac{\partial F_z}{\partial z}$$

Curl

$$\overrightarrow{\text{curl }}\vec{F}(x, y, z) = \nabla \wedge \vec{F}(x, y, z) = \vec{i}\left(\frac{\partial F_z}{\partial y} - \frac{\partial F_y}{\partial z}\right) + \vec{j}\left(\frac{\partial F_x}{\partial z} - \frac{\partial F_z}{\partial x}\right) + \vec{k}\left(\frac{\partial F_y}{\partial x} - \frac{\partial F_x}{\partial y}\right)$$

Laplacian operator

$$\Delta\Phi(x, y, z) = \nabla \times \nabla\Phi(x, y, z) = \text{div }\overrightarrow{\text{grad }}\Phi(x, y, z)$$

$$= \frac{\partial^2 \Phi}{\partial x^2} + \frac{\partial^2 \Phi}{\partial y^2} + \frac{\partial^2 \Phi}{\partial z^2} \qquad \text{Laplacian of scalar}$$

$$\Delta\vec{F}(x, y, z) = \nabla \times \nabla\vec{F}(x, y, z) = \vec{i}\,\Delta F_x + \vec{j}\Delta F_y + \vec{k}\,\Delta F_z \qquad \text{Laplacian of vector}$$

Polar coordinates

$$\overrightarrow{\text{grad}}\,\Phi(r,\,\theta,\,z) = \left(\frac{\partial\Phi}{\partial r};\ \frac{1}{r} \times \frac{\partial\Phi}{\partial\theta};\ \frac{\partial\Phi}{\partial z} \right)$$

$$\text{div}\,\overrightarrow{F}(r,\,\theta,\,z) = \frac{1}{r} \times \left(\frac{\partial(r \times F_r)}{\partial r} + \frac{\partial F_\theta}{\partial\theta} + r \times \frac{\partial F_z}{\partial z} \right)$$

$$\overrightarrow{\text{curl}}\,\overrightarrow{F}(r,\,\theta,\,z) = \left(\frac{1}{r}\frac{\partial F_z}{\partial\theta} - \frac{\partial F_\theta}{\partial z};\ \frac{\partial F_r}{\partial z} - \frac{\partial F_z}{\partial r};\ \frac{1}{r}\frac{\partial(r \times F_\theta)}{\partial r} - \frac{1}{r}\frac{\partial F_r}{\partial\theta} \right)$$

$$\Delta\Phi(r,\,\theta,\,z) = \frac{1}{r}\frac{\partial}{\partial r}\left(r\frac{\partial\Phi}{\partial r} \right) + \frac{1}{r^2}\frac{\partial^2\Phi}{\partial\theta^2} + \frac{\partial^2\Phi}{\partial z^2}$$

Operator relationship

Gradient

$$\overrightarrow{\text{grad}}(f + g) = \overrightarrow{\text{grad}}\,f + \overrightarrow{\text{grad}}\,g$$

$$\overrightarrow{\text{grad}}(f \times g) = g \times \overrightarrow{\text{grad}}\,f + f \times \overrightarrow{\text{grad}}\,g$$

where f and g are scalars.

Divergence

$$\text{div}\left(\overrightarrow{F} + \overrightarrow{G} \right) = \text{div}\,\overrightarrow{F} + \text{div}\,\overrightarrow{G}$$

$$\text{div}\left(f \times \overrightarrow{F} \right) = f \times \text{div}\,\overrightarrow{F} + \overrightarrow{F} \times \overrightarrow{\text{grad}}\,f$$

where f is a scalar.

$$\text{div}\left(\overrightarrow{F} \wedge \overrightarrow{G} \right) = \overrightarrow{G} \times \overrightarrow{\text{curl}}\,\overrightarrow{F} - \overrightarrow{F} \times \overrightarrow{\text{curl}}\,\overrightarrow{G}$$

Curl

$$\overrightarrow{\text{curl}}\left(\overrightarrow{F} + \overrightarrow{G} \right) = \overrightarrow{\text{curl}}\,\overrightarrow{F} + \overrightarrow{\text{curl}}\,\overrightarrow{G}$$

$$\overrightarrow{\text{curl}}\left(f \times \overrightarrow{F} \right) = f \times \overrightarrow{\text{curl}}\,\overrightarrow{F} - \overrightarrow{G} \wedge \overrightarrow{\text{grad}}\,f$$

where f is a scalar.

$$\overrightarrow{\text{curl}}\left(\overrightarrow{\text{grad}}\,f \right) = 0$$

$$\text{div}\left(\overrightarrow{\text{curl}}\,\overrightarrow{F} \right) = 0$$

Laplacian

$$\Delta f = \text{div} \; \overrightarrow{\text{grad}} \; f$$

$$\Delta \overrightarrow{F} = \overrightarrow{\text{grad}} \; \text{div} \; \overrightarrow{F} - \overrightarrow{\text{curl}} \left(\overrightarrow{\text{curl}} \; \overrightarrow{F} \right)$$

where f is a scalar.

$$\Delta(f + g) = \Delta f + \Delta g$$

$$\Delta \left(\overrightarrow{F} + \overrightarrow{G} \right) = \Delta \overrightarrow{F} + \Delta \overrightarrow{G}$$

$$\Delta(f \times g) = g \times \Delta f + f \times \Delta g + 2 \times \overrightarrow{\text{grad}} \; f \times \overrightarrow{\text{grad}} \; g$$

where f and g are scalars.

A1.3.4 Trigonometric functions

Definitions

The basic definitions may be stated in terms of right-angled triangle geometry:

$$\sin(x) = \frac{\text{opposite}}{\text{hypotenuse}}$$

$$\cos(x) = \frac{\text{adjacent}}{\text{hypotenuse}}$$

$$\tan(x) = \frac{\sin(x)}{\cos(x)} = \frac{\text{opposite}}{\text{adjacent}}$$

$$\cot(x) = \frac{\cos(x)}{\sin(x)} = \frac{\text{adjacent}}{\text{opposite}}$$

The power-series expansion of these functions are:

$$\sin(x) = x - \frac{x^3}{3!} + \frac{x^5}{5!} - \frac{x^7}{7!} + \cdots \qquad \text{for any value of } x$$

$$\cos(x) = 1 - \frac{x^2}{2!} + \frac{x^4}{4!} - \frac{x^6}{6!} + \cdots \qquad \text{for any value of } x$$

$$\tan(x) = x + \frac{x^3}{3} + \frac{2}{15}x^5 + \frac{17}{315}x^7 + \cdots \qquad \text{for } -\pi/2 < x < \pi/2$$

$$\cot(x) = \frac{1}{x} - \frac{x}{3} - \frac{1}{45}x^3 - \frac{2}{945}x^5 + \cdots \qquad \text{for } 0 < \text{Abs}(x) < \pi$$

where $n! = n \times (n - 1) \times (n - 2) \times \cdots \times 1$.

Relationships

$$\tan(x) = \frac{\sin(x)}{\cos(x)}$$

$$\cot(x) = \frac{1}{\tan(x)}$$

$$\sin^2(x) + \cos^2(x) = 1$$

$$\frac{1}{\sin^2(x)} - \cot^2(x) = 1$$

$$\frac{1}{\cos^2(x)} - \tan^2(x) = 1$$

$$\sin(-x) = -\sin(x)$$

$$\cos(-x) = \cos(x)$$

$$\tan(-x) = -\tan(x)$$

$$\cot(-x) = -\cot(x)$$

$$\sin(x) = \cos\left(\frac{\pi}{2} - x\right) \qquad \text{for } 0 < x < \pi/2$$

$$\cos(x) = \sin\left(\frac{\pi}{2} - x\right) \qquad \text{for } 0 < x < \pi/2$$

$$\tan(x) = \cot\left(\frac{\pi}{2} - x\right) \qquad \text{for } 0 < x < \pi/2$$

$$\cot(x) = \tan\left(\frac{\pi}{2} - x\right) \qquad \text{for } 0 < x < \pi/2$$

$$\sin(x + y) = \sin(x)\cos(y) + \cos(x)\sin(y)$$

$$\sin(x - y) = \sin(x)\cos(y) - \cos(x)\sin(y)$$

$$\cos(x + y) = \cos(x)\cos(y) - \sin(x)\sin(y)$$

$$\cos(x - y) = \cos(x)\cos(y) + \sin(x)\sin(y)$$

$$\tan(x + y) = \frac{\tan(x) + \tan(y)}{1 - \tan(x)\tan(y)}$$

$$\tan(x - y) = \frac{\tan(x) - \tan(y)}{1 + \tan(x)\tan(y)}$$

$$\cot(x+y) = \frac{\cot(x)\cot(y) - 1}{\cot(x) + \cot(y)}$$

$$\cot(x-y) = \frac{\cot(x)\cot(y) + 1}{\cot(x) - \cot(y)}$$

$$\sin(x) + \sin(y) = 2\sin\left(\frac{x+y}{2}\right)\cos\left(\frac{x-y}{2}\right)$$

$$\sin(x) - \sin(y) = 2\cos\left(\frac{x+y}{2}\right)\sin\left(\frac{x-y}{2}\right)$$

$$\cos(x) + \cos(y) = 2\cos\left(\frac{x+y}{2}\right)\cos\left(\frac{x-y}{2}\right)$$

$$\cos(x) - \cos(y) = -2\sin\left(\frac{x+y}{2}\right)\sin\left(\frac{x-y}{2}\right)$$

$$\sin(x)\sin(y) = \frac{1}{2}(\cos(x-y) - \cos(x+y))$$

$$\cos(x)\cos(y) = \frac{1}{2}(\cos(x-y) + \cos(x+y))$$

$$\sin(x)\cos(y) = \frac{1}{2}(\sin(x-y) + \sin(x+y))$$

Derivatives

$$d(\sin(x)) = \cos(x)dx$$

$$d(\cos(x)) = -\sin(x)dx$$

$$d(\tan(x)) = \frac{1}{\cos^2(x)}dx$$

$$d(\cot(x)) = \frac{-1}{\sin^2(x)}dx$$

$$d\left(\frac{1}{\cos(x)}\right) = \frac{\tan(x)}{\cos(x)}dx$$

$$d\left(\frac{1}{\sin(x)}\right) = \frac{-\cot(x)}{\sin(x)}dx$$

Inverse trigonometric functions

The inverse trigonometric functions are expressed as $\sin^{-1}$, $\cos^{-1}$, $\tan^{-1}$ and $\cot^{-1}$. The power-series expansion of these functions are:

$$\sin^{-1}(x) = x + \frac{1}{2} \times \frac{x^3}{3} + \frac{1 \times 3}{2 \times 4} \times \frac{x^5}{5} + \frac{1 \times 3 \times 5}{2 \times 4 \times 6} \times \frac{x^7}{7} + \cdots \qquad \text{for } -1 < x < 1$$

$$\cos^{-1}(x) = \frac{\pi}{2} - \left(x + \frac{1}{2} \times \frac{x^3}{3} + \frac{1 \times 3}{2 \times 4} \times \frac{x^5}{5} \right.$$
$$\left. + \frac{1 \times 3 \times 5}{2 \times 4 \times 6} \times \frac{x^7}{7} + \cdots \right) \qquad \text{for } -1 < x < 1$$

$$\tan^{-1}(x) = x - \frac{x^3}{3} + \frac{x^5}{5} - \frac{x^7}{7} + \cdots \qquad \text{for } -1 < x < 1$$

$$\cot^{-1}(x) = \frac{\pi}{2} - \left(x - \frac{x^3}{3} + \frac{x^5}{5} - \frac{x^7}{7} + \cdots \right) \qquad \text{for } -1 < x < 1$$

The following relationships can be established:

$$\sin^{-1}(-x) = -\sin^{-1}(x)$$

$$\cos^{-1}(-x) = \pi - \cos^{-1}(x)$$

$$\tan^{-1}(-x) = -\tan^{-1}(x)$$

$$\cot^{-1}(-x) = \pi - \cot^{-1}(x)$$

$$\sin^{-1}(x) + \cos^{-1}(x) = \frac{\pi}{2}$$

$$\tan^{-1}(x) + \cot^{-1}(x) = \frac{\pi}{2}$$

$$\sin^{-1}\left(\frac{1}{x}\right) + \cos^{-1}\left(\frac{1}{x}\right) = \frac{\pi}{2}$$

$$\sin^{-1}(x) = \tan^{-1}\left(\frac{1}{\sqrt{1 - x^2}}\right)$$

$$\cot^{-1}(x) = \frac{\pi}{2} - \tan^{-1}(x^2)$$

$$\sin^{-1}(x) + \sin^{-1}(y) = \sin^{-1}\left(x\sqrt{1 - y^2} + y\sqrt{1 - x^2} \right)$$

$$\sin^{-1}(x) - \sin^{-1}(y) = \sin^{-1}\left(x\sqrt{1 - y^2} - y\sqrt{1 - x^2} \right)$$

$$\cos^{-1}(x) + \cos^{-1}(y) = \cos^{-1}\left(xy - \sqrt{1-x^2}\,\sqrt{1-y^2} \right)$$

$$\cos^{-1}(x) - \cos^{-1}(y) = \cos^{-1}\left(xy + \sqrt{1-x^2}\,\sqrt{1-y^2} \right)$$

$$\tan^{-1}(x) + \tan^{-1}(y) = \tan^{-1}\left(\frac{x+y}{1-xy} \right)$$

$$\tan^{-1}(x) - \tan^{-1}(y) = \tan^{-1}\left(\frac{x-y}{1+xy} \right)$$

Derivatives

$$d(\sin^{-1}(x)) = \frac{1}{\sqrt{1-x^2}}\,dx \qquad \text{for } -\pi/2 < \sin^{-1}(x) < \pi/2$$

$$d(\cos^{-1}(x)) = \frac{-1}{\sqrt{1-x^2}}\,dx \qquad \text{for } 0 < \cos^{-1}(x) < \pi$$

$$d(\tan^{-1}(x)) = \frac{1}{1+x^2}\,dx \qquad \text{for } -\pi/2 < \tan^{-1}(x) < \pi/2$$

$$d(\cot^{-1}(x)) = \frac{-1}{1+x^2}\,dx \qquad \text{for } 0 < \sin^{-1}(x) < \pi$$

A1.3.5 Hyperbolic functions

Definitions

There are six hyperbolic functions that are comparable to the trigonometric functions. They are designated by adding the letter h to the trigonometric abbreviations: sinh, cosh, tanh, coth, sech and csch.

The basic definitions may be stated in terms of exponentials:

$$\sinh(x) = \frac{1}{\text{csch}(x)} = \frac{e^x - e^{-x}}{2}$$

$$\cosh(x) = \frac{1}{\text{sech}(x)} = \frac{e^x + e^{-x}}{2}$$

$$\tanh(x) = \frac{1}{\coth(x)} = \frac{e^x - e^{-x}}{e^x + e^{-x}}$$

The power-series expansion of these functions are:

$$\sinh(x) = x + \frac{x^3}{3!} + \frac{x^5}{5!} + \frac{x^7}{7!} + \cdots \qquad \text{for any value of } x$$

$$\cosh(x) = 1 + \frac{x^2}{2!} + \frac{x^4}{4!} + \frac{x^6}{6!} + \cdots \qquad \text{for any value of } x$$

$$\tanh(x) = x - \frac{x^3}{3} + \frac{2}{15}x^5 - \frac{17}{315}x^7 + \cdots \qquad \text{for } -\pi/2 < x < \pi/2$$

$$\coth(x) = \frac{1}{x} + \frac{x}{3} - \frac{1}{45}x^3 + \frac{2}{945}x^5 + \cdots \qquad \text{for } 0 < \text{Abs}(x) < \pi$$

where $n! = n \times (n - 1) \times (n - 2) \times \cdots \times 1$.

Relationships

$$\sinh(-x) = -\sinh(x)$$

$$\cosh(-x) = \cosh(x)$$

$$\cosh^2(x) - \sinh^2(x) = 1$$

$$1 - \tanh^2(x) = \text{sech}^2(x)$$

$$1 - \coth^2(x) = -\text{csch}^2(x)$$

$$\sinh(x + y) = \sinh(x)\cosh(y) + \cosh(x)\sinh(y)$$

$$\sinh(x - y) = \sinh(x)\cosh(y) - \cosh(x)\sinh(y)$$

$$\cosh(x + y) = \cosh(x)\cosh(y) + \sinh(x)\sinh(y)$$

$$\cosh(x - y) = \cosh(x)\cosh(y) - \sinh(x)\sinh(y)$$

$$\tanh(x + y) = \frac{\tanh(x) + \tanh(y)}{1 + \tanh(x)\tanh(y)}$$

$$\tanh(x - y) = \frac{\tanh(x) - \tanh(y)}{1 - \tanh(x)\tanh(y)}$$

$$\sinh(2x) = 2\sinh(x)\cosh(x)$$

$$\cosh(2x) = \cosh^2(x) + \sinh^2(x) = 2\cosh^2(x) - 1 = 1 + 2\sinh^2(x)$$

$$\tanh(2x) = \frac{2\tanh(x)}{1 + \tanh^2(x)}$$

$$\sinh\left(\frac{x}{2}\right) = \sqrt{\frac{1}{2}(\cosh(x) - 1)}$$

$$\cosh\left(\frac{x}{2}\right) = \sqrt{\frac{1}{2}(\cosh(x) + 1)}$$

$$\tanh\left(\frac{x}{2}\right) = \frac{\cosh(x) - 1}{\sinh(x)} = \frac{\sinh(x)}{\cosh(x) + 1}$$

Derivatives

$$d(\sinh(x)) = \cosh(x)\,dx$$

$$d(\cosh(x)) = \sinh(x)\,dx$$

$$d(\tanh(x)) = \text{sech}^2(x)\,dx$$

$$d(\coth(x)) = -\text{csch}^2(x)\,dx$$

$$d(\text{sech}(x)) = -\text{sech}(x)\tanh(x)\,dx$$

$$d(\text{csch}(x)) = -\text{csch}(x)\coth(x)\,dx$$

Inverse hyperbolic functions

The inverse hyperbolic functions are expressible in terms of logarithms.

$$\sinh^{-1}(x) = \ln\left(x + \sqrt{x^2 + 1}\right) \qquad \text{for any value of } x$$

$$\cosh^{-1}(x) = \ln\left(x + \sqrt{x^2 - 1}\right) \qquad \text{for } x > 1$$

$$\tanh^{-1}(x) = \frac{1}{2}\ln\left(\frac{1 + x}{1 - x}\right) \qquad \text{for } x^2 < 1$$

$$\coth^{-1}(x) = \frac{1}{2}\ln\left(\frac{x + 1}{x - 1}\right) \qquad \text{for } x^2 > 1$$

$$\text{sech}^{-1}(x) = \ln\left(\frac{1}{x} + \sqrt{\frac{1}{x^2} - 1}\right) \qquad \text{for } x < 1$$

$$\text{csch}^{-1}(x) = \ln\left(\frac{1}{x} + \sqrt{\frac{1}{x^2} + 1}\right) \qquad \text{for any value of } x$$

The following relationships can be established:

$$\sinh^{-1}(-x) = \sinh^{-1}(x)$$

$$\tanh^{-1}(-x) = -\tanh^{-1}(x)$$

$$\coth^{-1}(-x) = -\coth^{-1}(x)$$

$$\coth^{-1}(x) = \tanh^{-1}\left(\frac{1}{x}\right)$$

Derivatives

$$d(\sinh^{-1}(x)) = \frac{dx}{\sqrt{x^2 + 1}}$$

$$d(\cosh^{-1}(x)) = \frac{dx}{\sqrt{x^2 - 1}}$$

$$d(\tanh^{-1}(x)) = \frac{dx}{1 - x^2} = d(\coth^{-1}(x))$$

$$d(\operatorname{sech}^{-1}(x)) = \frac{-dx}{x\sqrt{1 - x^2}}$$

$$d(\operatorname{csch}^{-1}(x)) = \frac{-dx}{x\sqrt{1 + x^2}}$$

A1.3.6 Complex numbers

Definition

A complex number z consists of two distinct scalar (i.e. real) parts a and b, and is written in the form:

$$z = a + ib$$

where $i = \sqrt{-1}$. The first part a is called the real part and the second part b is called the imaginary part of the complex number.

The modulus (i.e. absolute value) of a complex number designated r and defined as:

$$r = \sqrt{x^2 + y^2}$$

The argument θ of the complex number is the position vector measured from the positive x-axis in an anticlockwise direction:

$$\theta = \tan^{-1}\left(\frac{y}{x}\right)$$

An alternative mode of expressing a complex number is:

$$z = x + iy = r(\cos\theta + i\sin\theta) = r\,e^{i\theta}$$

Properties

Various operations involving complex numbers are:

- addition: $z_1 + z_2 = (x_1 + x_2) + i(y_1 + y_2)$
- multiplication: $z_1 z_2 = r_1 r_2 e^{i(\theta_1 + \theta_2)}$
- power: $z^n = r^n e^{in\theta}$
- division: $z_1/z_2 = r_1/r_2 e^{i(\theta_1 + \theta_2)}$
- multiplication by i: $iz = -y + ix = r\,e^{i(\theta + \pi/2)}$
- logarithm: $\ln(z) = \ln(r) + i\theta$

Conjugate number

If a complex number is: $z = x + iy$, the conjugate number is defined as:

$$\bar{z} = x - iy = r\,e^{i(\theta - \pi/2)}$$

The main properties of a conjugate are:

$$z\bar{z} = r^2 \qquad \text{(real)}$$

$$z + \bar{z} = 2x \qquad \text{(real)}$$

$$z - \bar{z} = 2iy \qquad \text{(imaginary)}$$

A1.3.7 Polynomial equations

Presentation

A polynomial equation of degree n is:

$$a_0 + a_1 x + a_2 x^2 + \cdots + a_n x^n = 0$$

This polynomial equation (with real coefficients) has at least $(n - 2)$ real solutions.

Polynomial equation of degree two

A polynomial equation of degree two has the form:

$$a_0 + a_1 x + a_2 x^2 = 0$$

And it has two, one or zero real solutions depending upon the sign of the discriminant:

$$\Delta = a_1^2 - 4a_2 a_0$$

There is no real solution for $\Delta < 0$, one solution for $\Delta = 0$ and two real solutions for $\Delta > 0$:

$$x_1 = \frac{-a_1 + \sqrt{\Delta}}{2a_2}$$

$$x_2 = \frac{-a_1 - \sqrt{\Delta}}{2a_2}$$

Polynomial equation of degree three

A polynomial equation of degree three has at least one real solution.

$$a_0 + a_1 x + a_2 x^2 + a_3 x^3 = 0$$

Denoting A_1 and A_2 the following variables:

$$A_1 = \frac{1}{3}\frac{a_1}{a_3} - \frac{1}{9}\left(\frac{a_2}{a_3}\right)^2$$

$$A_2 = \frac{1}{6}\frac{a_2 a_1}{a_3^2} - \frac{1}{2}\frac{a_0}{a_3} - \frac{1}{27}\left(\frac{a_2}{a_3}\right)^3$$

the discriminant of the polynomial equation of degree three is:

$$\Delta = A_1^3 + A_2^2$$

If $\Delta > 0$ there is one real solution. If $\Delta = 0$ there are two real solutions and there are three real solutions when $\Delta < 0$.

Note

In practice, if an obvious solution is known (i.e. $x = -c_0$), the polynomial equation may be rewritten as:

$$a_0 + a_1 x + a_2 x^2 + a_3 x^3 = (x + c_0)(b_2 x^2 + b_1 x + b_0) = 0$$

The following relationships hold:

$$b_2 = a_3$$
$$c_0 b_2 + b_1 = a_2$$
$$c_0 b_1 + b_0 = a_1$$
$$c_0 b_0 = a_0$$

The solutions of the polynomial equation of degree three are $\{x = -c_0\}$ and the solutions of the polynomial equation of degree two:

$$b_2 x^2 + b_1 x + b_0 = 0$$

---

* * *

A1.4 ALTERNATE DEPTHS IN OPEN CHANNEL FLOW

A1.4.1 Presentation

Considering an open channel flow in a rectangular channel, the continuity and Bernoulli equations state:

$$q = Vd \tag{A1.16}$$

$$H - z_o = d + \frac{V^2}{2g} \tag{A1.17}$$

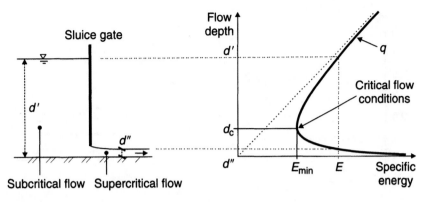

Fig. A1.1 Definition sketch: alternate depths across a sluice gate.

where q is the discharge per unit width, d is the flow depth, V is the flow velocity, H is the mean total head, z_0 is the bed elevation and g is the gravity constant. The Bernoulli equation (A1.17) is developed assuming a flat horizontal channel, hydrostatic pressure distribution and uniform velocity distribution.

For a given specific energy ($E = H - z_0$) and flow rate q_w, the system of two equations (A1.16) and (A1.17) has:

- no solution (i.e. no flow) for: $H - z_0 < 3/2 \sqrt[3]{q^2/g}$,

- one solution (i.e. critical flow conditions) for: $H - z_0 = 3/2 \sqrt[3]{q^2/g}$,   or
- two (meaningful) solutions.

In the latter case, the two possible flow depths d' and d'' are called alternate depths (Fig. A1.1). One corresponds to a subcritical flow (i.e. $Fr = q/\sqrt{gd^3} < 1$) and the other to a supercritical flow.

A1.4.2 Discussion

For a given specific energy and known discharge, the Bernoulli equation yields:

$$\left(\frac{d}{d_c}\right)^3 - \frac{H - z_0}{d_c}\left(\frac{d}{d_c}\right)^2 + \frac{1}{2} = 0 \qquad (A1.18)$$

where $d_c = \sqrt[3]{q^2/g}$ is the critical flow depth. Equation (A1.18) is a polynomial equation of degree 3 in terms of the dimensionless flow depth d/d_c and it has three real solutions for $(H - z_0)/d_c \geq 3/2$.

The solutions of equation (A1.18) (for $(H - z_0)/d_c \geq 3/2$) are:

$$\frac{d^{(1)}}{d_c} = \frac{H - z_0}{d_c}\left(\frac{1}{3} + \frac{2}{3}\cos\left(\frac{\Gamma}{3}\right)\right) \qquad \text{Subcritical flow} \qquad (A1.19a)$$

$$\frac{d^{(2)}}{d_c} = \frac{H - z_0}{d_c}\left(\frac{1}{3} + \frac{2}{3}\cos\left(\frac{\Gamma}{3} + \frac{2\pi}{3}\right)\right) \qquad \text{Negative solution} \qquad (A1.19b)$$

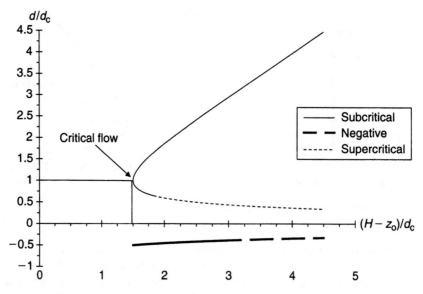

Fig. A1.2 Flow depth solutions as functions of $(H - z_0)/d_c$.

$$\frac{d^{(3)}}{d_c} = \frac{H - z_0}{d_c}\left(\frac{1}{3} + \frac{2}{3}\cos\left(\frac{\Gamma}{3} + \frac{4\pi}{3}\right)\right) \qquad \text{Supercritical flow} \qquad \text{(A1.19c)}$$

where

$$\cos\Gamma = 1 - \frac{27}{4}\left(\frac{d_c}{H - z_0}\right)^3$$

Equations (A1.19a) and (A1.19c) are plotted as functions of $(H - z_0)/d_c$ in Fig. A1.2. Note that equation (A1.19a) may be rewritten in terms of the Froude number as:

$$Fr = \left(\frac{d}{d_c}\right)^{-3/2} \qquad \text{(A1.20)}$$

It yields:

$$Fr^{(1)} = \left(\frac{H - z_0}{d_c}\left(\frac{1}{3} + \frac{2}{3}\cos\left(\frac{\Gamma}{3}\right)\right)\right)^{-3/2} \qquad \text{Subcritical flow} \qquad \text{(A1.21a)}$$

$$Fr^{(2)} = \left(\frac{H - z_0}{d_c}\left(\frac{1}{3} + \frac{2}{3}\cos\left(\frac{\Gamma}{3} + \frac{2\pi}{3}\right)\right)\right)^{-3/2} \qquad \text{Complex number} \qquad \text{(A1.21b)}$$

$$Fr^{(3)} = \left(\frac{H - z_0}{d_c}\left(\frac{1}{3} + \frac{2}{3}\cos\left(\frac{\Gamma}{3} + \frac{4\pi}{3}\right)\right)\right)^{-3/2} \qquad \text{Supercritical flow} \qquad \text{(A1.21c)}$$

Part 2 Introduction to Sediment Transport in Open Channels

Introduction to sediment transport in open channels

6.1 INTRODUCTION

Waters flowing in natural streams and rivers have the ability to scour channel beds, to carry particles (heavier than water) and to deposit materials, hence changing the bed topography. This phenomenon (i.e. *sediment transport*) is of great economical importance: e.g. to predict the risks of scouring of bridges, weirs and channel banks; to estimate the siltation of a reservoir upstream of a dam wall; to predict the possible bed form changes of rivers and estuaries (e.g. Plates 18, 19 and 20).

Numerous failures resulted from the inability of engineers to predict sediment motion: e.g. bridge collapse (pier foundation erosion), formation of sand bars in estuaries and navigable rivers, destruction of banks and levees.

In this chapter, we shall introduce the basic concepts of sediment transport in open channels.

6.2 SIGNIFICANCE OF SEDIMENT TRANSPORT

6.2.1 Sediment transport in large alluvial streams

The transport of sediment is often more visible in mountain streams, torrents and creeks (e.g. Fig. 6.1). However, larger rivers are also famous for their capacity to carry sediment load: e.g. the Nile River, the Mississippi River and the Yellow River.

The Nile River is about 6700 km long and the mean annual flow is about $97.8 \times 10^9 \, \text{m}^3$ (i.e. $3100 \, \text{m}^3/\text{s}$). Large floods take place each year from July to October with maximum flow rates over $8100 \, \text{m}^3/\text{s}$, during which the river carries a large sediment load. In Ancient Egypt, peasants and farmers expect the Nile floods to deposit fertile sediments in the 20 km wide flood plains surrounding the main channel.

The Mississippi River is about 3800 km long. Its mean annual discharge is about $536 \times 10^9 \, \text{m}^3$ (or $17\,000 \, \text{m}^3/\text{s}$). On average, the river transports about $7000 \, \text{kg/s}$ of sediment in the Gulf of Mexico. During flood periods much larger sediment-transport rates are observed. Mcmath (1883) reported maximum daily sediment-transport rates of $4.69 \times 10^9 \, \text{kg}$ (i.e. $54\,300 \, \text{kg/s}$) in the Mississippi River at Fulton (on 10 July 1880) and over $5.97 \times 10^9 \, \text{kg}$ (or $69\,110 \, \text{kg/s}$) in the Missouri River[1] at St Charles (on 3 July 1879). During the 1993 flood, the river at the Nebraska Station carried $11.9 \times 10^9 \, \text{kg}$ ($4580 \, \text{kg/s}$) of sediment load during the month of July. Between 12 and 28 July 1993, the Missouri River at Kansas City scoured the bed by 4.5 m and the sediment load reached $8700 \, \text{kg/s}$ between 26 June and 14 September 1993 at the Hermann gauging site (Bhowmik, 1996).

The Yellow River (or Huang Ho River) flows 5460 km across China and its catchment area is $745\,000 \, \text{km}^2$. The annual mean (water) flow is $48.3 \times 10^9 \, \text{m}^3$ (i.e. $1530 \, \text{m}^3/\text{s}$) and the average

---

[1] The Missouri River is the longest tributary (3726 km long) of the Mississippi River and it is sometimes nicknamed 'Big Muddy' because of the amount of sediment load.

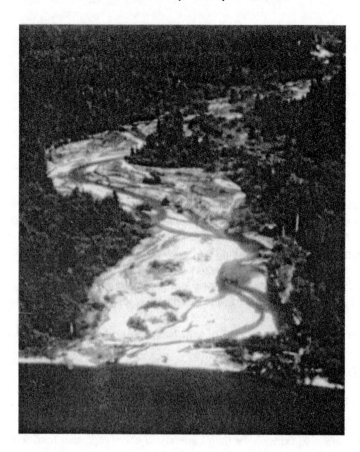

Fig. 6.1 Aerial view of Camas Creek flowing into the North Fork of the Flathead River, Montana, USA (12 July 1981) (by P. Carrara, with permission of US Geological Survey, Ref. 356ct).

annual sediment load is 1.6 billion tons which comes primarily from the Middle reach regions. In the lower reach the river bed has been subjected to numerous changes, and the location of the river estuary has varied by as much as 600 km over the past 3000 years.

6.2.2 Failures caused by sediment-transport processes

Moore Creek dam, Tamworth, Australia

The Moore Creek dam was completed in 1898 to supply water to the town of Tamworth, NSW (325 km North of Sydney) (Fig. 6.2). At the time of construction, the 18.6 m high dam was designed with advanced structural features: i.e. thin single-arch wall (7.7 m thick at base and 0.9 m thick at crest), vertical downstream wall and battered upstream face made of Portland cement concrete. The volume of the reservoir was 220 000 m^3 and the catchment area is 51 km^2.

Between 1898 and 1911, the reservoir was filled with 85 000 m^3 of sediment.[2] In 1924 the dam was out of service because the reservoir was fully silted (25 years after its construction). The dam is still standing today, although Moore Creek dam must be considered as an engineering failure. It failed because the designers did not understand the basic concepts of sediment transport nor the catchment erosion mechanisms.

---

[2] For the period 1898–1911 observations suggested that most of the siltation took place during the floods of February 1908 and January 1910.

(a)

(b)

Fig. 6.2 Photographs of the Moore Creek dam: (a) old photograph (shortly after completion), (b) recent photograph (14 June 1997).

Old Quipolly dam, Werris Creek, Australia

Completed in 1932, the Old Quipolly dam[3] is a concrete single-arch dam: it is 19 m high with a
184 m long crest (1 m thickness at crest) (Fig. 6.3). The reservoir was built to supply water to the

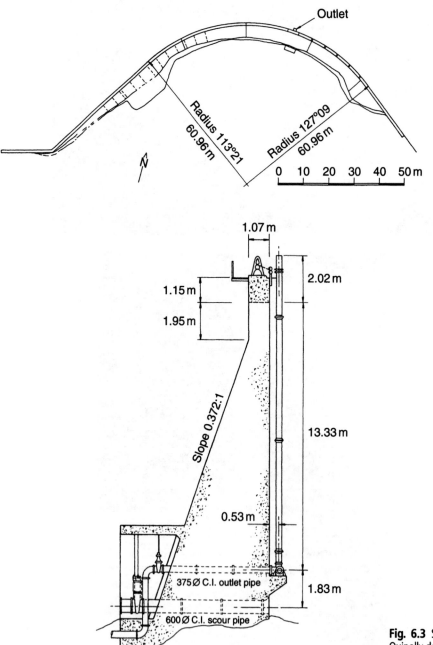

Fig. 6.3 Sketch of the Old
Quipolly dam.

---

[3] Also known as Coeypolly Creek dam No. 1 (International Commission on Large Dams, 1984). A second dam was built
in 1955, 3 km downstream of the old dam.

town of Werris Creek, NSW. The catchment area is $70\,km^2$ and the original storage capacity was $860\,000\,m^3$.

In 1941, $130\,000\,m^3$ of sediment had accumulated. In 1943 the siltation volume amounted to $290\,000\,m^3$. By 1952, more than half of the initial storage had disappeared. The reservoir was closed in 1955. Nowadays the reservoir is fully silted. The dam is still standing and the wall is in excellent condition (author's inspection in 1997). The reservoir presently acts as a sediment trap for the new Quipolly dam built in 1955 and located 3 km downstream.

The Old Quipolly dam is nevertheless another engineering failure caused by reservoir sedimentation. It is probably the most extreme siltation record documented in Australia. Indeed, several dams have also suffered siltation in South-East Australia (Table 6.1).

Mount Isa railway bridges, Queensland, Australia

In the early 1960s, the reconstruction of the railway line to Mount Isa (mining town in the Australian desert) required the construction of several bridges across creeks. Several bridges failed because of erosion of the pier foundations: e.g. bridge 235 across Julia Creek, bridge across Eastern creek (Fig. 6.4) and bridge across Corella Creek (Nonda).

The bridges collapsed because of the inability of (overseas) engineers to understand the local hydrology and associated sediment motion.

Shihmen dam, Taiwan

The Shihmen dam (Taoyuan County, Taiwan ROC) is a 133 m high dam built between 1958 and 1964. The maximum reservoir capacity was more than $60\,000\,000\,m^3$ and the catchment area is $763\,km^2$.

Although the dam was inaugurated in 1964, the reservoir began filling in May 1963. In September 1963, $20\,000\,000\,m^3$ of silt accumulated during cyclonic floods (typhoon Gloria). Between 1985 and 1995, the reservoir was dredged and over $10\,000\,000\,m^3$ of sediment was removed. But, during that period, the volume of sediment flowing into the reservoir amounted to about $15\,000\,000\,m^3$ (over 10 years). More than 100 sediment-trap dams were built upstream of Shihmen reservoir to trap incoming sediment. In 1996 only one sediment-trap dam was still functioning, all the others having filled up. It is believed that the maximum depth of water in the Shihmen reservoir was less than 40 m in 1997.

The Shihmen reservoir (Fig. 6.5) was designed to operate for at least 70 years. Thirty years after completion, it has, in fact, become a vast sediment trap with an inappropriate storage capacity to act as a flood control or water supply reservoir (Chang, 1996).

Discussion

A spectacular accident was the Kaoping river bridge failure on 27 September 2002 in Taiwan. Located between Kaohsiung and PingTung city, the 1 km long bridge failed because of scour at one pier abutment. The bridge had been in operation for about 22 years. Illegal gravel dredging upstream was suspected to be one of the causes of failure. A 100 m long bridge section dropped taking 18 vehicles with it, but there were fortunately no fatalities. Witnesses described a rumbling sound as the four-lane bridge broke and fell into the river. Recent references in bridge failures and scour include Hamill (1999) and Melville and Coleman (2000).

Table 6.1 Examples of reservoir siltation in South-East Australia

Dam (1)	Year of completion (2)	Catchment area (km²) (3)	Original reservoir capacity (m³) (4)	Cumulative siltation volume (m³) (5)	Date of record (6)
Corona dam, Broken Hill NSW, Australia	1890?	15	120 000[†]	120 000 Fully silted	1910?
Stephens Creek dam, Broken Hill NSW, Australia	1892[a]	510	24 325 000	1 820 000 2 070 000 4 500 000 De-silting*	1907 1912 1958 1971
Junction Reefs dam, Lyndhurst NSW, Australia	1896	–	200 000[†]	Fully silted	1997
Moore Creek dam, Tamworth NSW, Australia	1898	51	220 000[†]	85 000 ~ 200 000 Fully silted	1911 1924 1985
Koorawatha dam, Koorawatha NSW, Australia	1901	–	40 500	Fully silted	1997
Gap weir, Werris Creek NSW, Australia	1902	160	[†]	Fully silted	1924
Cunningham Creek dam, Harden NSW, Australia	1912	820	–[†]	216 000 258 600 379 000 522 600 758 000	1916 1917 1920 1922 1929
Illalong Creek dam, Binalong NSW, Australia	1914	130	260 000[†]	75% siltation	1997
Umberumberka dam, Broken Hill NSW, Australia	1915	420	13 197 000	3 600 000 4 500 000 5 013 000 5 700 000	1941 1961 1964 1982
Korrumbyn Creek dam, Murwillumbah NSW, Australia	1919	3	27 300[†]	Fully silted	1985
Borenore Creek dam, Orange NSW, Australia	1928[b]	22	230 000	150 000	1981
Tenterfield Creek dam, Tenterfield NSW, Australia	1930[c]	38	1 500 000	110 000	1951
Old Quipolly dam, Werris Creek NSW, Australia	1932	70	860 000[†]	130 000 290 000 ~ 430 000 Fully silted	1941 1943 1952 1985
Inverell dam, Inverell NSW, Australia	1939	600	153 000 000[†]	Fully silted	1982
Illawambra dam, Bega Valley NSW, Australia	1959	19	4500[†]	3000 Fully silted	1976 1985

Notes: [a] Spillway crest raised in 1909; [b] dam raised in 1943; [c] dam raised in 1974; [†] fully silted nowadays; * extensive de-silting in 1971.

6.3 TERMINOLOGY

Classical (clear water) hydraulics is sometimes referred to as *fixed-boundary hydraulics*. It can be applied to most man-made channels (e.g. concrete-lined channels, rock tunnels and

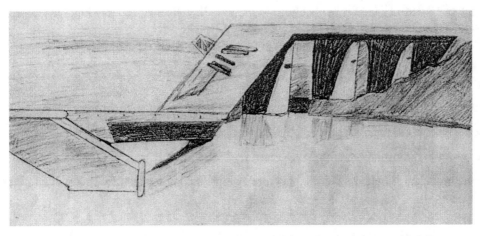

Fig. 6.4 Sketch of a collapsed railway bridge across Eastern Creek (after an original photograph). Collapse caused by pier scour.

Fig. 6.5 The Shihmen reservoir during a dry season in summer 1993 (courtesy of H.C. Pu and Sinorama).

rockfill-protected channels) and to some extent to grassed waterways. However, most natural streams are characterized by some sediment transport: i.e. the channel boundaries are movable. *Movable-boundary hydraulics* applies to streams with gravel or sand beds, estuaries (i.e. silt or sand beds), sandy coastlines and man-made canals in earth, sand or gravel.

Movable-boundary hydraulics is characterized by variable-boundary roughness and variable-channel dimensions. Strong interactive processes between the water flow and the bed force changes to take place.

6.4 STRUCTURE OF THIS SECTION

In Part 2, the lecture material is regrouped as a series of definitions (Chapter 7), the basic concepts of sediment motion (Chapters 8 and 9), the calculations of the sediment-transport capacity

(Chapters 10 and 11) and the applications to natural alluvial streams, including the concepts of erosion, accretion and bed form motion (Chapter 12).

Further examples of reservoir sedimentation are discussed in Appendix A2.1.

6.5 EXERCISES

Name two rivers which are world famous for their sediment-transport capacity.

Discuss recent bridge failures caused by river scour.

Does sediment transport occur in fixed-boundary hydraulics?

Are the boundary roughness and channel dimensions fixed in movable-boundary hydraulics?

Sediment transport and sediment properties

7

7.1 BASIC CONCEPTS

7.1.1 Definitions

Sediment transport is the general term used for the transport of material (e.g. silt, sand, gravel and boulders) in rivers and streams.

The transported material is called the *sediment load*. Distinction is made between the *bed load* and the *suspended load*. The bed load characterizes grains *rolling along the bed* while suspended load refers to grains *maintained in suspension* by turbulence. The distinction is, however, sometimes arbitrary when both loads are of the same material.

Note

The word 'sediment' refers commonly to fine materials that settles to the bottom. Technically, however, the term sediment transport includes the transport of both fine and large materials (e.g. clay, silt, gravel and boulders).

7.1.2 Bed formation

In most practical situations, the sediments behave as a non-cohesive material (e.g. sand and gravel) and the fluid flow can distort the bed into various shapes. The bed form results from the drag force exerted by the bed on the fluid flow as well as the sediment motion induced by the flow onto the sediment grains. This interactive process is complex.

In a simple approach, the predominant parameters which affect the bed form are the bed slope, the flow depth and velocity, the sediment size and particle fall velocity. At low velocities, the bed does not move. With increasing flow velocities, the inception of bed movement is reached and the sediment bed begins to move. The basic bed forms which may be encountered are the *ripples* (usually of heights less than 0.1 m), *dunes, flat bed, standing waves* and *antidunes*. At high flow velocities (e.g. mountain streams and torrents), chutes and step-pools may form. They consist of succession of chutes and free-falling nappes (i.e. supercritical flow) connected by pools where the flow is usually subcritical.

The typical bed forms are summarized in Fig. 7.1 and Table 7.1. In Table 7.1, column 3 indicates the migration direction of the bed forms. Ripples and dunes move in the downstream direction. Antidunes and step-pools are observed with supercritical flows and they migrate in the upstream flow direction. Field observations are illustrated in Fig. 7.2.

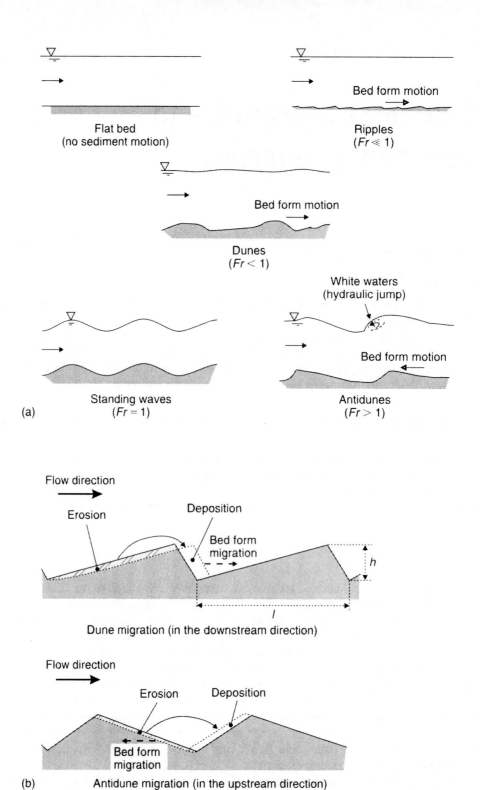

Fig. 7.1 Bed forms in movable boundary hydraulics: (a) typical bed forms and (b) bed form motion.

Table 7.1 Basic bed forms in alluvial channels (classification by increasing flow velocities)

Bed form (1)	Flow (2)	Bed form motion (3)	Comments (4)
Flat bed	No Flow (or $Fr \ll 1$)	NO	No sediment motion
Ripples	$Fr \ll 1$	D/S	Three-dimensional forms; observed also with air flows (e.g. sand ripples in a beach caused by wind)
Dunes	$Fr < 1$	D/S	Three-dimensional forms; sand dunes can also be caused by wind
Flat bed	$Fr \leq 1$	NO	Observed also with wind flow
Standing waves	$Fr = 1$	NO	Critical flow conditions; bed standing waves in phase with free-surface standing waves
Antidunes	$Fr > 1$	U/S	Supercritical flow with tumbling flow and hydraulic jump upstream of antidune crests
Chute-pools	$Fr > 1$	U/S	Very active antidunes
Step-pools	$Fr > 1$	–	Cascade of steps and pools; steps are often caused by rock bed

References: Henderson (1966) and Graf (1971).
Notes: D/S = in downstream flow direction; Fr = Froude number; U/S = in upstream flow direction.

(a)

(b)

Fig. 7.2 Field observations of bed forms: (a) Great Sand Dunes National Monument, Alamosa and Saguache Counties CO in April 1966 (courtesy of Dr Lou Maher) – dune bed forms carved by Aeolian action. (b) Sand dunes in Cape River after the flood, 270km North of Clermont QLD, Australia (8 July 1983) (courtesy of Mrs J. Hacker).

(c)

(d)

(e)

Fig. 7.2 (c) Standing waves at Takatoyo beach (Japan) in June 1999 – flow from bottom right to top left. (d) Gravel antidune at Brigalow Bend, Burdekin River, Australia in August 1995 (courtesy of Dr C. Fielding). Bed forms left after a peak flow of $8000 \, m^3/s$ – flow from the left to the right. (e) Gravel antidune at Brigalow Bend, Burdekin River, Australia in August 1995 (courtesy of Dr C. Fielding). Bed forms left after a peak flow of $8000 \, m^3/s$ – looking downstream, with an observation trench digged in the foreground.

Notes

1. Both ripples and dunes are observed with wind-blown sand and in open channel flow (Fig. 7.2a and b).
2. Note that, in alluvial rivers, dunes form with subcritical flow conditions only. Antidunes are associated with supercritical flow while standing waves are characteristics of near-critical flow conditions.
3. The transition between dune and standing-wave bed forms occurs with a flat bed. The flat bed is an unstable bed pattern, often observed at Froude numbers slighlty below unity: e.g. $Fr = 0.77$ (Kennedy, 1963), 0.83 (Laursen, 1958) and 0.57–1.05 (Julien and Rastan, 1998).

Discussion

Ripples are associated with the presence of a laminar boundary layer. Their size is independent of the flow depth d. Usually their characteristic dimensions (length l and height h) satisfy $l \ll 1000d$ and $h < 100d$.

Dunes are associated with a turbulent boundary layer. In rivers, their size is about proportional to the flow depth (see also Table 12.3). In open channels, dunes take place in subcritical flow.

With standing-wave and antidune bed forms, the free-surface profile is in phase with the bed form profile. In natural streams, antidunes and standing waves are seldom seen because the bed forms are not often preserved during the receding stages of a flood. Kennedy (1963) investigated standing-wave bed forms in laboratory while Alexander and Fielding (1997) presented superb photographs of gravel antidunes.

7.2 PHYSICAL PROPERTIES OF SEDIMENTS

7.2.1 Introduction

Distinction is made between two categories of sediment: cohesive material (e.g. clay and silt) and non-cohesive material (e.g. sand and gravel). In this section, the basic properties of both types are developed. In practice, however, we will primarily consider non-cohesive materials in this introductory course.

7.2.2 Property of single particles

The *density* of quartz and clay minerals is typically $\rho_s = 2650 \, \text{kg/m}^3$. Most natural sediments have densities similar to that of quartz.

The *relative density* of sediment particle equals:

$$s = \frac{\rho_s}{\rho} \tag{7.1}$$

where ρ is the fluid density. For a quartz particle in water, $s = 2.65$ but a quartz particle in air has a relative density, $s = 2200$.

Notes

1. In practice, a dense particle is harder to move than a light one.
2. Let us keep in mind that heavy minerals (e.g. iron and copper) have larger values of density than quartz.
3. The relative density is dimensionless. It is also called the specific gravity. It is sometimes denoted as γ_s.

Table 7.2 Sediment size classification

Class name (1)	Size range (mm) (2)	Phi-scale (ϕ) (3)	Remarks (4)
Clay	$d_s < 0.002$ to $0.004\,\text{mm}$	$+8$ to $+9 < \phi$	
Silt	0.002 to $0.004 < d_s < 0.06\,\text{mm}$	$+4 < \phi < +8$ to $+9$	
Sand	$0.06 < d_s < 2.0\,\text{mm}$	$-1 < \phi < +4$	Silica
Gravel	$2.0 < d_s < 64\,\text{mm}$	$-6 < \phi < -1$	Rock fragments
Cobble	$64 < d_s < 256\,\text{mm}$	$-8 < \phi < -6$	Original rocks
Boulder	$256 < d_s$	$\phi < -8$	Original rocks

The most important property of a sediment particle is its characteristic size. It is termed the *diameter* or *sediment size*, and denoted d_s. In practice, natural sediment particles are not spherical but exhibit irregular shapes. Several definitions of sediment size are available:

- the sieve diameter,
- the sedimentation diameter,
- the nominal diameter.

The *sieve diameter* is the size of particle which passes through a square mesh sieve of given size but not through the next smallest size sieve: e.g. $1\,\text{mm} < d_s < 2\,\text{mm}$.

The *sedimentation diameter* is the size of a quartz sphere that settles down (in the same fluid) with the same settling velocity as the real sediment particle. The *nominal diameter* is the size of the sphere of same density and same mass as the actual particle.

The sediment size may also be expressed as a function of the sedimentological size parameter ϕ (or *Phi-scale*) defined as:

$$d_s = 2^{-\phi} \tag{7.2a}$$

or

$$\phi = -\frac{\ln(d_s)}{\ln(2)} \tag{7.2b}$$

where d_s is in mm.

A typical sediment size classification is shown in Table 7.2.

Notes
1. Large particles are harder to move than small ones.
2. Air can move sand (e.g. sand dunes formed by wind action). Water can move sand, gravel, boulders or breakwater armour blocks (weighting several tonnes).
3. The sedimentation diameter is also called the standard fall diameter.

7.2.3 Properties of sediment mixture

The density of a dry sediment mixture equals:

$$(\rho_{sed})_{dry} = (1 - Po)\rho_s \tag{7.3}$$

where *Po* is the porosity factor.

The density of wet sediment is:

$$(\rho_{sed})_{wet} = Po\rho + (1 - Po)\rho_s \qquad (7.4)$$

The porosity factor ranges basically from 0.26 to 0.48. In practice, Po is typically about 0.36–0.40.

Notes

1. The density of sediments may be expressed also as a function of the void ratio: i.e. the ratio of volume of voids (or pores) to volume of solids. The void ratio is related to the porosity as:

$$\text{Void ratio} = Po/(1 - Po)$$

2. Another characteristic of a porous medium is the permeability. For a one-dimensional flow through the pores of the sediment bed, the velocity of seepage is given by the Darcy law:

$$V = -K\frac{dH}{dx}$$

where K is the hydraulic conductivity (or coefficient of permeability, in m/s) and H is the piezometric head (in m). The hydraulic conductivity not only depends on the permeability of the soil but also on the properties of the fluid and dimensional analysis yields (Raudkivi and Callander, 1976: p. 15):

$$K = k\frac{\rho g}{\mu}$$

where k is the permeability (in m^2), ρ is the fluid density, g is the gravity constant and μ is the fluid dynamic viscosity. Typical values of the hydraulic conductivity are (Raudkivi and Callander, 1976: p. 19):

Soil type:	Fine sand	Silty sand	Silt
K (m/s):	5×10^{-4} to 1×10^{-5}	2×10^{-5} to 1×10^{-6}	5×10^{-6} to 1×10^{-7}

3. Discussion: conversion between parts per million (ppm) and kilograms per cubic metre (kg/m³).

For suspended sediment, the sediment concentration may be expressed in kg/m³ and it is calculated as the ratio of dry sediment mass to volume of water–sediment mixture. It can be expressed also as a volume concentration (dimensionless). Another unit, ppm, is sometimes used. It is defined as the ratio of the weight of sediment to the weight of the water–sediment mixture times one million.

The conversion relationships are:

$$\text{Mass concentration} = \rho_s C_s$$

$$\text{Concentration in ppm by weight} = \frac{1 \times 10^6 s C_s}{1 + (s - 1)C_s}$$

where C_s is the volumetric sediment concentration and $s = \rho_s/\rho$.

Application

Calculate the dry and wet densities of a sand mixture with a 38% porosity.

Solution

Assuming a quartz sand ($\rho_s = 2650\ \mathrm{m^3/s}$), the dry and wet densities of sediment mixture are estimated using equations (7.3) and (7.4):

$$(\rho_{\mathrm{sed}})_{\mathrm{dry}} = (1 - Po)\rho_s = 1643\ \mathrm{kg/m^3}$$

$$(\rho_{\mathrm{sed}})_{\mathrm{wet}} = Po\rho + (1 - Po)\rho_s = 2022\ \mathrm{kg/m^3}$$

7.2.4 Particle size distribution

Natural sediments are mixtures of many different particle sizes and shapes. The *particle size distribution* is usually represented by a plot of the weight percentage of total sample, which is smaller than a given size plotted as a function of the particle size. A cumulative curve fitted to data points is shown in Fig. 7.3.

The characteristic *sediment size* d_{50} is defined as the size for which 50% by weight of the material is finer. Similarly the characteristic sizes d_{10}, d_{75} and d_{90} are values of grain sizes for which 10%, 75% and 90% of the material weight is finer, respectively.

d_{50} is commonly used as the characteristic grain size and the range of particle size is often expressed in terms of a sorting coefficient S:

$$S = \sqrt{\frac{d_{90}}{d_{10}}} \tag{7.5}$$

Another descriptor is the geometric standard deviation based upon a log-normal distribution of grain sizes σ_g:

$$\sigma_g = \sqrt{\frac{d_{84}}{d_{16}}} \tag{7.6}$$

Small values of S and σ_g imply a nearly uniform sediment size distribution. A large value of S means a broad sediment size distribution.

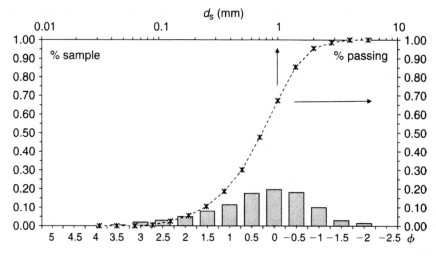

Fig. 7.3 Typical particle size distribution curve: (a) percentage sampling as a function of sedimentological size parameter ϕ (linear scale) and (b) cumulative percentage passing as a function of the particle size d_s in mm (semi-logarithmic scale).

> **Notes**
> 1. The sediment size d_{50} is called the *median grain size*.
> 2. Other definitions may be used to characterize the range of particles sizes. For example:
>
> $$\text{Gradation coefficient} = \frac{1}{2}\left(\frac{d_{84}}{d_{50}} + \frac{d_{50}}{d_{16}}\right) \quad \text{(Julien, 1995)}$$
>
> 3. The size distribution may be recorded using a settling tube (see next sections).
>
> **Comments**
> The size distribution of cohesive sediments (e.g clay and silt) may vary with the environmental conditions to which the sediments have been subjected and also the procedures which are used to determine their size distribution. In the following, we shall primarily consider non-cohesive sediments.

7.3 PARTICLE FALL VELOCITY

7.3.1 Presentation

In a fluid at rest, a suspended particle (heavier than water) falls: i.e. it has a downward (vertical) motion. The *terminal fall velocity* is the particle velocity at equilibrium, the sum of the gravity force, buoyancy force and fluid drag force being equal to zero.

In an open channel flow, the particle fall velocity is further affected by the flow turbulence and the interactions with surrounding particles.

7.3.2 Settling velocity of a single particle in still fluid

For a spherical particle settling in a still fluid, the terminal fall velocity w_o equals:

$$w_o = -\sqrt{\frac{4gd_s}{3C_d}(s-1)} \quad \text{Spherical particle} \tag{7.7}$$

where d_s is the particle diameter, C_d is the drag coefficient and $s = \rho_s/\rho$. The negative sign indicates a downward motion (for $s > 1$).

Dimensional analysis implies that the drag coefficient is a function of Reynolds number and particle shape:

$$C_d = f\left(\rho\frac{w_o d_s}{\mu}; \text{particle shape}\right) \tag{7.8}$$

where ρ and μ are the fluid density and dynamic viscosity, respectively.

At low particle Reynolds numbers ($w_o d_s/\nu < 1$), the flow around the particle is laminar. At large Reynolds numbers ($w_o d_s/\nu > 1000$), the flow around the spherical particle is turbulent and

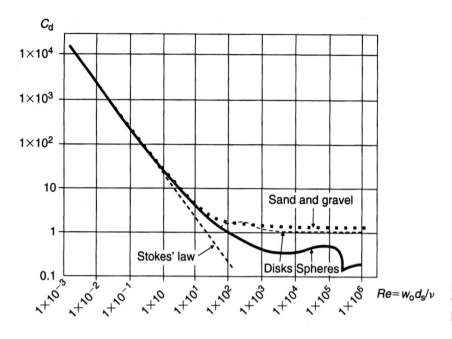

Fig. 7.4 Drag coefficient of single particle in still fluid.

Table 7.3 Computed settling velocity of sediment particles in still water (equation (7.10))

d_s (m):	5×10^{-5}	0.0001	0.0002	0.0005	0.001	0.002	0.005	0.01	0.02	0.05	0.1	0.2	0.5
w_o (m/s):	2.2×10^{-3}	8.5×10^{-3}	0.027	0.071	0.11	0.17	0.27	0.39	0.54	0.85	1.2	1.7	2.7

Notes: $s = 2.69$; w_o = terminal fall velocity of single particle in water at 20°C.

the drag coefficient is nearly constant. Typical drag coefficient values for spherical particles are presented in Fig. 7.4.

Sediment particles have irregular shapes and the drag coefficient differs from that for spherical particles. Their shape is often angular, sometimes disc shaped, and the drag coefficient can be expected to be larger than that of spheres. For sands and gravels, a simple approximation of the drag coefficient is:

$$C_d = \frac{24\mu}{\rho|w_o|d_s} + 1.5 \qquad Re < 1 \times 10^4 \tag{7.9}$$

where $|w_o|$ is the absolute value of the particle fall velocity (Fig. 7.4).

Combining equations (7.7) and (7.9), an estimate of the terminal fall velocity of a sediment particle is:

$$w_o = -\sqrt{\frac{4gd_s}{3\left(\dfrac{24\mu}{\rho|w_o|d_s} + 1.5\right)}(s-1)} \tag{7.10}$$

Computed values of settling velocity of sediment particles are reported in Table 7.3. Experimental observations are presented in Table 7.4.

Plate 1 Man-made waterways: Roman aqueduct at Tarragona, Spain in May 1997 (courtesy of Mr and Mrs Burridge). Aqueduct length: 35 km, bridge length: 249 m, bridge height: 26 m.

Plate 2 Roman aqueduct at Pont-du-Gard, France in June 1998. Aqueduct length: 49.5 km, bridge length: 275 m, bridge height: 48.8 m.

Plate 3 Photograph of large bed-forms. Dunes and runoff channels at Big Bend, Burdekin River, Australia on 20 January 1996 (courtesy of Dr C. Fielding). Bed-forms left after a peak flow of 3200 m^3/s.

(a)

(b)

Plate 4 Antidune at Big Bend, Burdekin River, Australia in January 1996 (courtesy of Dr C. Fielding). Bed-forms left after a peak flow of 3200 m^3/s. (a) Flow from the left to the right. Note the A5-size notepad on the left-hand side, used for scaling. (b) Looking downstream, tyre tracks in the foreground.

Plate 5 Photograph of man-made cascade. Overflow on the Ternay dam stepped cascade in June 1998. Dam height: 41 m (completed in 1867). Unlined rock stepped cascade. Dam refurbished in the 1980s. Small overflow: $Q \sim 2\,m^3/s$.

Plate 6 Dartmouth dam spillway (courtesy of Mr Jeffery, Goulburn-Murray Water). Dam height: 180 m (completed: 1977). Unlined rock stepped cascade. Design spillway capacity: $2755\,m^3/s$. Small overflow in 1996: $Q \sim 225\,m^3/s$.

Plate 7 When hydraulics, architecture and arts meet: Le Bosquet des Rocailles (ou Bosquet de la Salle de Bal), Jardins du Château de Versailles, France in June 1998. Built between 1680 and 1683 by Jules Mansart, sculptures by Pierre Legros and Benoît Massou, modified in 1690. The fountain includes 20 water jets and 17 cascades.

Plate 8 Hydraulic jump downstream of the Moree Weir, Moree, NSW, Australia on 16 December 1997. Flow from the left to the right.

Plate 9 Skimming flow above a stepped chute. Flow from left to right: 30-degree slope, step height: 5 cm.

(a)

(b)

Plate 10 White waters at Mount Barney Creek during a flood in February 1998, Queensland, Australia (courtesy of Dr R. Rankin). (a) Looking upstream; (b) view from the right bank, flow from the left to the right.

Plate 11 Lawn Hill Gorge in Queensland, Australia on 4 June 1984 (courtesy of Mrs J. Hacker).

Plate 12 Flood plain of the South Alligator River in March 1998 (Kakadu National Park. NT, Australia) (courtesy of Dr R. Rankin).

Plate 13 Dam hydraulics: an arch dam in a narrow valley, the Monteynard dam in June 1998. Dam height: 155 m (completed in 1962). Crest length: 210 m, design spillway capacity: 2500 m³/s.

Plate 14 Photograph of catchment hydraulics students' activity. Undergraduate students inspecting the Gold Creek dam stepped spillway on 11 September 2003.

Plate 15 Students visiting a large culvert (design flow: 220 m³/s) in eastern Brisbane on 11 September 2003.

Plate 16 Tidal bore of the Dordogne river, France on 27 September 2000. Surfers and kayacks riding the undular bore front.

Table 7.4 Terminal settling velocity of sediment particle in still water (observations)

d_s^a (mm) (1)	w_o^a (m/s) (2)	$w_o d_s/\nu$ (3)	C_d^a (4)	Comment (5)
0.089	0.005	0.44	55	Sand grains
0.147	0.013	1.9	15	
0.25	0.028	7.0	6	
0.42	0.050	21	3	
0.76	0.10	75	1.8	
1.8	0.17	304	1.5	

Notes: w_o = terminal fall velocity of single particle in water at 20°C; [a]data from Engelund and Hansen (1972).

Application

For a spherical particle, derive the basic relationship between the settling velocity, drag coefficient and particle characteristics.

Solution

Considering a spherical particle in still water, the force acting on the settling particle are:

- the drag force $0.5 C_d \rho w_o^2 A_s$,
- the weight force $\rho_s g v_s$,
- the buoyant force F_b,

where A_s is the area of the particle in the z-direction, z being the vertical axis positive upwards, C_d is the drag coefficient, g is the gravity constant, ρ is the water density, ρ_s is the particle density, w_o is the terminal fall velocity in still water and v_s is the volume of the particle.

The buoyant force on a submerged body is the difference between the vertical component of pressure force on its underside and the vertical pressure component on its upperside. To illustrate the concept of buoyancy, let us consider a diver in a swimming pool. The pressure force exerted on the diver equals the weight of water above him/her. As the pressure below him/her is larger than that immediately above, a reaction force (i.e. the buoyant force) is applied to the diver in the vertical direction. The buoyant force counteracts the pressure force and equals the weight of displaced liquid. For a sediment particle subjected to a hydrostatic pressure gradient, the buoyant force equals:

$$F_b = \rho g v_s$$

assuming a constant pressure gradient over the particle height d_s.

In the force balance, the drag force is opposed to the particle motion direction and the buoyant force is positive (upwards). At equilibrium the balance of the forces yields:

$$+\frac{1}{2} C_d \rho w_o^2 A_s - \rho_s g v_s + \rho g v_s = 0$$

The settling velocity equals:

$$w_o^2 = \frac{2g}{C_d A_s}\left(\frac{\rho_s}{\rho} - 1\right) v_s$$

Note that the settling velocity is negative (i.e. downwards).

Spherical particle

For a spherical particle, the cross-sectional area and volume of the particle are, respectively:

$$A_s = \frac{\pi}{4} d_s^2$$

$$v_s = \frac{\pi}{6} d_s^3$$

And the settling velocity becomes:

$$w_o = -\sqrt{\frac{4 g d_s}{3 C_d}\left(\frac{\rho_s}{\rho} - 1\right)} \qquad \text{Spherical particle}$$

Remarks

Note that, on Earth, the buoyant force is proportional to the liquid density ρ and to the gravity acceleration g. The buoyancy is larger in denser liquids: e.g. a swimmer floats better in the water of the Dead Sea than in fresh water. In gravitationless water (e.g. waterfall) the buoyant force is zero.

Comments

1. *Drag coefficient of spherical particles*

   Considering the flow around a sphere, the flow is everywhere laminar at very small Reynolds numbers ($Re < 1$). The flow of a viscous incompressible fluid around a sphere was solved by Stokes (1845, 1851). His results imply that the drag coefficient equals:

   $$C_d = \frac{24}{Re} \qquad \text{Laminar flow around a spherical particle (Stokes' law)}$$

   which is defined in terms of the particle size and settling velocity: $Re = \rho w_o d_s / \mu$. Stokes' law is valid for $Re < 0.1$.

   In practice, the re-analysis of a large number of experimental data with spherical particles that were unaffected by sidewall effects yielded (Brown and Lawler, 2003):

   $$C_d = \frac{24}{Re}(1 + 0.150\, Re^{0.681}) + \frac{0.407}{1 + (8710/Re)} \qquad Re < 2 \times 10^5$$

   At large Reynolds numbers ($Re > 1000$), the flow around the spherical particle is turbulent and the drag coefficient is nearly constant:

   $$C_d \approx 0.5 \qquad \text{Turbulent flow around a spherical particle } (1 \times 10^3 < Re < 1 \times 10^5)$$

2. *Drag coefficient of natural sediment particles*

   For natural sand and gravel particles, experimental values of drag coefficient were measured by Engelund and Hansen (1967) (Table 7.4). Their data are best fitted by:

   $$C_d = \frac{24}{Re} + 1.5$$

   based on experimental values obtained for sand and gravel ($Re < 1 \times 10^4$)
   Another empirical correlation was recently proposed (Cheng, 1997):

   $$C_d = \left(\left(\frac{24}{Re}\right)^{2/3} + 1\right)^{3/2}$$

   based on experimental values for natural sediment particles ($Re < 1 \times 10^4$)

3. George Gabriel Stokes (1819–1903) was a British physicist and mathematician, known for his studies of the behaviour of viscous fluids (Stokes, 1845, 1851).

4. *Terminal settling velocity of spherical particles*

   Gibbs *et al.* (1971) provided an empirical formula for the settling velocity of spherical particle ($50\,\mu\text{m} < d_s < 5\,\text{mm}$). Their results can be expressed in SI units as:

$$w_o = 10\frac{-30\nu + \sqrt{900\nu^2 + gd_s^2(s-1)(0.003869 + 2.480d_s)}}{0.011607 + 7.4405d_s} \qquad \text{Spherical particles}$$

   where ν is the kinematic viscosity of the fluid (i.e. $\nu = \mu/\rho$).

   Brown and Lawler (2003) proposed an empirical formula for spherical particles obtained with experiments unaffected by sidewall effects:

$$w_o\sqrt[3]{\frac{\rho}{g\mu(s-1)}} = \left(\left(\frac{18}{d_*^2}\right)^{0.898\left(\frac{0.936d_*+1}{d_*+1}\right)} + \left(\frac{0.317}{d_*}\right)^{0.449}\right)^{-1.114} \qquad \begin{array}{l}\text{Spherical particles}\\(Re < 2 \times 10^5)\end{array}$$

$$w_o\sqrt[3]{\frac{\rho}{g\mu(s-1)}} = \frac{d_*^2(22.5 + d_*^{2.046})}{0.0258\,d_*^{4.046} + 2.81d_*^{3.046} + 18\,d_*^{2.046} + 405} \qquad \begin{array}{l}\text{Spherical particles}\\(Re < 4 \times 10^3)\end{array}$$

   where d_* is the dimensionless particle number defined as:

$$d_* = d_s\sqrt[3]{\frac{(s-1)g}{\nu^2}}$$

5. *Terminal settling velocity of sand particles*

   For naturally worn quartz sands (0.063–1 mm sizes) settling in water, Jimenez and Madsen (2003) proposed a simple formula:

$$\frac{w_o}{\sqrt{(s-1)gd_s}} = \frac{1}{0.954 + \dfrac{5.12}{(d_s/4\nu)\sqrt{(s-1)gd_s}}} \qquad \text{Worn quartz particles } (0.063 < d_s < 1\text{ mm})$$

   where d_s is the nominal diameter or equivalent volume sphere diameter.

6. Note that the temperature affects the fall velocity as the fluid viscosity changes with temperature.

Discussion: settling velocity of sediment particles

Observed values of terminal settling velocity are reported in Table 7.4. First, note that large-size particles fall faster than small particles. For example, the terminal fall velocity of a 0.01 mm particle (i.e. silt) is about 0.004 cm/s while a 10 mm gravel settle at about 34 cm/s. Practically, fine particles (e.g. clay and silt) settle in a laminar flow motion ($w_o d_s/\nu < 1$) while large particles (e.g. gravel and boulder) fall in a turbulent flow motion ($w_o d_s/\nu > 1000$).

Further, at the limits, the fall velocity and the particle size satisfy:

$$w_o \propto d_s^2 \qquad \text{Laminar flow motion } (w_o d_s/\nu < 1) \tag{7.11a}$$

$$w_o \propto \sqrt{d_s} \qquad \text{Turbulent flow motion } (w_o d_s/\nu > 1000) \tag{7.11b}$$

Application

The size distribution of a sandy mixture is recorded using a settling tube in which the settling time in water over a known settling distance is measured. The results are:

Settling rate (cm/s)	32	16	10.7	8	5.33	4	3.2	2.7
Mass (g)	0.4	1.20	4.31	7.00	15.98	31.60	38.12	42.59

Settling rate (cm/s)	2.29	2	1.78	1.6	1.45	1.3	1.23	1.14
Mass (g)	44.80	48.04	49.07	49.60	49.81	49.87	49.95	50.00

Deduce the median grain size and the sorting coefficient $S = \sqrt{d_{90}/d_{10}}$.

Solution

First, we deduce the median settling velocity $(w_o)_{50}$ and the settling velocities $(w_o)_{10}$ and $(w_o)_{90}$ for which 10% and 90% by weight of the material settle faster, respectively. The equivalent sedimentation diameters can then be determined.

For the measured data, it yields $(w_o)_{50} = 4.56$ cm/s, $(w_o)_{10} = 10.0$ cm/s and $(w_o)_{90} = 2.27$ cm/s. The equivalent sedimentation diameter may be deduced from equation (7.7):

$$d_s = \frac{3C_d}{4g(s-1)} w_o^2$$

in which the drag coefficient may be estimated using equation (7.9):

$$C_d = \frac{24\mu}{\rho|w_o|d_s} + 1.5 \qquad \text{For sand and gravels}$$

For a known settling velocity, the equivalent sedimentation diameter may be derived from the above equations by iterative calculations or using Table 7.3.

The equivalent sedimentation diameters for the test are $d_{50} = 0.31$ mm, $d_{10} = 0.18$ mm, $d_{90} = 0.83$ mm and the sorting coefficient is $S = 2.15$.

Remarks

Fine particles settle more slowly than heavy particles. As a result, the grain sizes d_{10} and d_{90} are deduced, respectively, from the settling velocities $(w_o)_{90}$ and $(w_o)_{10}$ for which 90% and 10% by weight of the material settle faster, respectively.

7.3.3 Effect of sediment concentration

The settling velocity of a single particle is modified by the presence of surrounding particles. Experiments have shown that thick homogeneous suspensions have a slower fall velocity than that of a single particle. Furthermore, the fall velocity of the suspension decreases with increasing volumetric sediment concentration. This effect, called *hindered settling*, results from the interaction between the downward fluid motion induced by each particle on the surrounding fluid and the return flow (i.e. upward fluid motion) following the passage of a particle. As a particle settles down, a volume of fluid equal to the particle volume is displaced upwards. In thick sediment suspension, the drag on each particle tends to oppose to the upward fluid displacement.

> **Notes**
> 1. The fall velocity of a suspension w_s may be estimated as:
>
> $$w_s = (1 - 2.15C_s)(1 - 0.75C_s^{0.33})w_o \quad \text{Fall velocity of a fluid–sediment suspension}$$
>
> where w_o is the terminal velocity of a single particle and C_s is the volumetric sediment concentration. Van Rijn (1993) recommended this formula, derived from experimental work for sediment concentration up to 35%.
> 2. It must be noted that a very dense cloud of particles settling in clear water may fall faster than an individual particle. A very dense cloud tends to behave as a large particle rather than as a suspension.

7.3.4 Effect of turbulence on the settling velocity

In turbulent flows, several researchers discussed the effects of turbulence upon the sediment settling velocity. Graf (1971) presented a comprehensive review of the effects of turbulence on suspended-solid particles. Nielsen (1993) suggested that the fall velocity of sediment particles increases or decreases depending upon turbulence intensity, the particle density, and the characteristic length scale and time scale of the turbulence. Although the subject is not yet fully understood, it is agreed that turbulence may drastically affect the particle settling motion.

7.4 ANGLE OF REPOSE

Considering the stability of a single particle in a horizontal plane, the threshold condition (for motion) is achieved when the centre of gravity of the particle is vertically above the point of contact. The critical angle at which motion occurs is called the *angle of repose* ϕ_s.

The angle of repose is a function of the particle shape and, on a flat surface, it increases with angularity. Typical examples are shown in Fig. 7.5. For sediment particles, the angle of repose ranges usually from 26° to 42°. For sands, ϕ_s is typically between 26° and 34°. Van Rijn (1993) recommended to use more conservative values (i.e. larger values) for the design of stable channels (Table 7.5).

> **Note**
> For a two-dimensional polygon, the angle of repose equals 180° divided by the number of sides of the polygon. For example: $\phi_s = 60°$ for an equilateral triangle and $\phi_s = 0$ for a circle.

Angle of repose

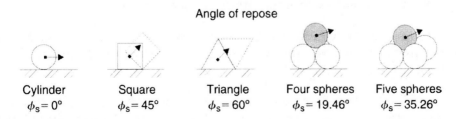

Cylinder	Square	Triangle	Four spheres	Five spheres
$\phi_s = 0°$	$\phi_s = 45°$	$\phi_s = 60°$	$\phi_s = 19.46°$	$\phi_s = 35.26°$

Fig. 7.5 Examples of angle of repose.

Table 7.5 Angle of repose for stable channel design

d_s (mm) (1)	ϕ_s: rounded particles (degrees) (2)	ϕ_s: angular particles (degrees) (3)	Comments (4)
<1	30	35	
5	32	37	Gravel
10	35	40	Gravel
50	37	42	Gravel
>100	40	45	Cobble and boulder

Note: Silicate material, see van Rijn (1993).

7.5 LABORATORY MEASUREMENTS

7.5.1 Particle size distribution

In laboratory, particle size distributions may be determined by direct measurements or indirect methods. Direct measurements include immersion and displacement volume measurements, direct measurements of particle diameter, semi-direct measurements of particle sizes using sieves.

Indirect methods of particle size measurements relate fall velocity measurements to particle size. Most common methods are the visual accumulation tube (VAT), the bottom withdrawal tube (BWT) and the pipette. The former (VAT) is used only for sands. The last two methods are used only for silts and clays.

7.5.2 Concentration of suspended sediments

Suspended sediment concentrations may be measured from 'representative' samples of the sediment-laden flow. The sampling techniques may be instantaneous sampling, point sampling or depth-integrated sampling. Graf (1971) and Julien (1995) reviewed various techniques.

7.6 EXERCISES

Bed forms

What is the difference between ripples and dunes? Can dune formation occur with wind-blown sands?

In what direction do antidunes migrate? Can antidunes be observed with wind-blown sands?

In a natural stream, can dunes form in supercritical flows?

Considering a natural stream, the water discharge is $6.4 \, \text{m}^2/\text{s}$ and the observed flow depth is 4.2 m. What is the most likely type of bed form with a movable bed?

In a natural stream, the flow velocity is 4.1 m/s and the observed flow depth is 0.8 m. What is the most likely type of bed form with a movable bed? In what direction will the bed forms migrate (i.e. upstream or downstream)?

Considering a 2.3 m wide creek, the water discharge is $1.5 \, \text{m}^3/\text{s}$ and the observed flow depth is 0.35 m. What is the most likely type of bed form with a movable bed? In what direction will the bed forms migrate (i.e. upstream or downstream)?

Sediment properties

The dry density of a sand mixture is $1655\,kg/m^3$. Calculate (and give units) (a) the sand mixture porosity and (b) the wet density of the mixture.

The characteristic grain sizes of a sediment mixture are $d_{10} = 0.1\,mm$, $d_{50} = 0.55\,mm$, $d_{90} = 1.1\,mm$. Indicate the type of sediment material. Calculate (a) the sedimentological size parameters ϕ corresponding to d_{10}, d_{50} and d_{90}, (b) the sorting coefficient and (c) the dry sediment mixture density.

Settling velocity

Considering a spherical particle (density ρ_s) settling in still water, list all the forces acting on the particle at equilibrium. Write the motion equation for the settling particle at equilibrium. Deduce the analytical expression of the settling velocity.

For a 0.03 mm diameter sphere, calculate the settling velocity in still water at 20°C. (a) Use Fig. 7.4 to calculate the drag coefficient. (b) Compare your result with the correlation of Brown and Lawler (2003).

Solution: (a) $w_o = 0.80\,mm/s$ (Fig. 7.4) and (b) $w_o = 0.86\,mm/s$ (Brown and Lawler's correlation).

Considering a 0.9 mm sphere of density $\rho_s = 1800\,kg/m^3$ settling in water, calculate the settling velocity. *Use Fig. 7.4 to calculate the drag coefficient.*

Solution: $w_o = 0.041\,m/s$.

Considering a 1.1 mm sediment particle settling in water at 20°C, calculate the settling velocity using (a) the formula of Gibbs *et al.* (1971) and (b) the semi-analytical expression derived by Chanson (1999):

$$w_o = 10\frac{-30v + \sqrt{900v^2 + gd_s^2(\mathbf{s}-1)(0.003869 + 2.480d_s)}}{0.011607 + 7.4405d_s} \qquad \text{Gibbs } et\ al.\ (1971)$$

$$w_o = -\sqrt{\frac{4gd_s}{3\left(\dfrac{24\mu}{\rho|w_o|d_s} + 1.5\right)}(\mathbf{s}-1)} \qquad \text{Chanson (1999)}$$

Calculate the fall velocity of a 0.95 mm quartz particle using the correlation of Jimenez and Madsen (2003).

Solution: $w_o = 0.1\,m/s$.

The size distribution of a sandy mixture is recorded using a settling tube in which the settling time in water over a known settling distance is measured. The results are:

Settling rate (cm/s)	35	12.0	10.0	8.0	5.0	4.0	3.0	2.8
Mass (g)	1.22	2.50	9.20	15.10	25.98	55.60	72.81	81.09

Settling rate (cm/s)	2.39	2.1	1.92	1.8	1.65	1.5	1.43	1.34
Mass (g)	89.80	93.01	96.03	96.90	98.01	99.28	99.90	100.00

Deduce the median grain size and the sorting coefficient $S = \sqrt{d_{90}/d_{10}}$

Considering a light cloud of sediment particles settling in still water, the particle size is 0.45 mm and the volumetric sediment concentration is 5%. Calculate (a) the terminal velocity of a single particle and (b) the fall velocity of the suspension.

Inception of sediment motion – occurrence of bed load motion

8

8.1 INTRODUCTION

In this chapter, the conditions for the onset of sediment transport are reviewed. First, the threshold of bed load motion is described: under what conditions do sediment particles start to roll along the bed (Plate 18)? In another section the inception of sediment suspension will be presented.

Note

In practice, bed load transport starts occurring at lower velocities than sediment suspension, for a given bed geometry and particle size distribution.

Accurate estimate of the onset of sediment motion is required: (1) to prevent erosion of channel bed, (2) to predict the risks of scouring below foundations (e.g. bridge pier foundations) and (3) to select rock armour material.

8.2 HYDRAULICS OF ALLUVIAL STREAMS

8.2.1 Introduction

In most cases, river and stream flows behave as steady flows and uniform equilibrium flow conditions are reached. Application of the momentum equation provides an expression of the mean flow velocity V (at equilibrium):

$$V = \sqrt{\frac{8g}{f}} \sqrt{\frac{D_H}{4} \sin \theta} \qquad \text{Uniform equilibrium flow} \qquad (8.1)$$

where f is the Darcy friction factor, θ is the bed slope and D_H is the hydraulic diameter.

For alluvial streams, the knowledge of the mean flow velocity is insufficient to accurately predict the occurrence of sediment motion and associated risks of scouring. The knowledge of the velocity profile and more specifically of the velocity next to the channel bed is required.

8.2.2 Velocity distributions in turbulent flows

Considering a turbulent flow along a 'smooth' boundary, the flow field can be divided into three regions (Fig. 8.1): the *inner wall region* (i.e. 'viscous sub-layer') next to the wall where the

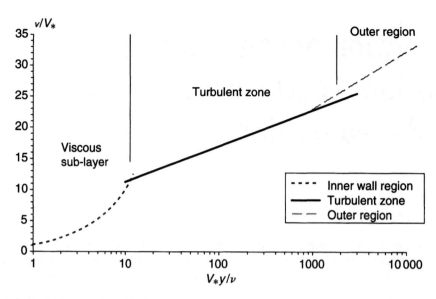

Fig. 8.1 Velocity distribution in turbulent flows.

turbulent stress is negligible and the viscous stress is large, the *outer region* where the turbulent stress is large and the viscous stress is small, and an overlap region sometimes called the *turbulent zone*.

The thickness of the inner wall region is about $(10\nu/V_*)$, where ν is the fluid kinematic viscosity (i.e. $\nu = \mu/\rho$) and V_* is the shear velocity.

Notes

1. The shear velocity is defined as: $V_* = \sqrt{\tau_0/\rho}$ where τ_0 is the mean bed shear stress and ρ is the fluid density.

2. In uniform equilibrium flow down an open channel, the average shear velocity equals:

$$V_* = \sqrt{g\frac{D_H}{4}\sin\theta} \qquad \text{Uniform equilibrium flow}$$

where D_H is the hydraulic diameter and θ is the bed slope.

Application: velocity profile in a turbulent boundary layer

For a turbulent boundary layer flow along a smooth boundary with zero pressure gradient, the velocity distribution follows (e.g. Schlichting, 1979):

$$\frac{v}{V_*} = \frac{V_*y}{\nu} \qquad \text{Viscous sublayer: } \frac{V_*y}{\nu} < 5$$

$$\frac{v}{V_*} = \frac{1}{K}\ln\left(\frac{V_*y}{\nu}\right) + D_1 \qquad \text{Turbulent zone: } 30\text{–}70 < \frac{V_*y}{\nu} \text{ and } \frac{y}{\delta} < 0.1\text{–}0.15$$

$$\frac{V_{max} - v}{V_*} = -\frac{1}{K}\ln\left(\frac{y}{\delta}\right) \qquad \text{Outer region: } \frac{y}{\delta} > 0.1\text{–}0.15$$

where v is the velocity at a distance y measured normal to the boundary, δ is the boundary layer thickness, V_{max} is the maximum velocity at the outer edge of the boundary layer (i.e. free-stream velocity), K is the von Karman constant (K = 0.40) and D_1 is a constant (D_1 = 5.5; Schlichting, 1979).

The turbulent zone equation is called the logarithmic profile or the 'law of the wall'. The outer region equation is called the 'velocity defect law' or 'outer law'. Coles (1956) extended the logarithmic profile to the outer region by adding a 'wake law' term to the right-handside term:

$$\frac{v}{V_*} = \frac{1}{K}\ln\left(\frac{V_* y}{\nu}\right) + D_1 + \frac{\Pi}{K}Wa\left(\frac{y}{\delta}\right) \qquad \text{Turbulent zone and outer region: } 30\text{-}70 < \frac{V_* y}{\nu}$$

where Π is the wake parameter, and Wa is Coles' wake function, originally estimated as (Coles, 1956):

$$Wa\left(\frac{y}{\delta}\right) = 2\sin^2\left(\frac{\pi}{2}\frac{y}{\delta}\right)$$

Roughness effects
Surface roughness has an important effect on the flow in the wall-dominated region (i.e. inner wall region and turbulent zone). Numerous experiments showed that, for a turbulent boundary layer along a rough plate, the 'law of the wall' follows:

$$\frac{v}{V_*} = \frac{1}{K}\ln\left(\frac{V_* y}{\nu}\right) + D_1 + D_2 \qquad \text{Turbulent zone: } \frac{y}{\delta} < 0.1\text{-}0.15$$

where D_2 is a function of the roughness height, of roughness shape and spacing (e.g. Schlichting, 1979; Schetz, 1993). For smooth turbulent flows, D_2 equals zero.

In the turbulent zone, the roughness effect (i.e. $D_2 < 0$) implies a 'downward shift' of the velocity distribution (i.e. law of the wall). For large roughness, the laminar sub-layer (i.e. inner region) disappears. The flow is said to be 'fully rough' and D_2 may be estimated as (Schlichting, 1979: p. 620):

$$D_2 = 3 - \frac{1}{K}\ln\left(\frac{k_s V_*}{\nu}\right) \qquad \text{Fully rough turbulent flow}$$

where k_s is the equivalent roughness height. After transformation, the velocity distribution in the turbulent zone for fully rough turbulent flow becomes:

$$\frac{v}{V_*} = \frac{1}{K}\ln\left(\frac{y}{k_s}\right) + 8.5 \qquad \text{Turbulent zone: } \frac{y}{\delta} < 0.1\text{-}0.15 \text{ for fully rough turbulent flow}$$

Discussion
In a turbulent boundary layer, the velocity distribution may be approximated by a power law function:

$$\frac{v}{V_{max}} = \left(\frac{y}{\delta}\right)^{1/N}$$

where N is a function of the boundary roughness. Buschmann and Gad-el-Hak (2003) argued advantages of both power law and logarithmic law.

For smooth turbulent flows, $N = 7$ (e.g. Schlichting, 1979). For uniform equilibrium flow in open channels, $N = K\sqrt{8/f}$, where f is the Darcy friction factor (Chen, 1990).

8.2.3 Velocity profiles in alluvial streams

Most river flows are turbulent and the velocity profile is fully developed in rivers and streams: i.e. the boundary layer thickness equals the flow depth.

For an alluvial stream, the bed roughness effect might be substantial (e.g. in a gravel-bed stream) (Plate 18) and, hence, the complete velocity profile (and the bottom shear stress) is affected by the ratio of the sediment size to the inner wall region thickness: i.e. $d_s/(10\nu/V_*)$.

If the sediment size is small compared to the sub-layer thickness (i.e. $V_* d_s/\nu < 4$–5), the flow is *smooth turbulent*. If the sediment size is much larger than the sub-layer thickness (i.e. $V_* d_s/\nu > 75$–100), the flow is called *fully rough turbulent*. For 4–5 $< V_* d_s/\nu < 75$–100, the turbulent flow regime is a *transition* regime.

Note

The law of flow resistance is related to the type of turbulent flow regime. For smooth turbulent flows, the friction factor is independent of the boundary roughness and it is a function of the mean flow Reynolds number only.

For fully rough turbulent flows, the flow resistance is independent of the Reynolds number and the friction factor is a function only of the relative roughness height.

Applications

1. For a wide river, calculate the bed shear stress and shear velocity for a steady uniform flow with a 1.5 m depth and a bed slope of 0.0003.

Solution

In uniform equilibrium flows, the shear velocity equals:

$$V_* = \sqrt{g\frac{D_H}{4}\sin\theta} \qquad \text{Uniform equilibrium flow}$$

where D_H is the hydraulic diameter. For wide channel the hydraulic diameter is about four times the flow depth d and the shear velocity becomes:

$$V_* \approx \sqrt{gd\sin\theta} \qquad \text{Uniform equilibrium flow}$$

For the given flow conditions, $V_* = 0.066$ m/s and the mean bed shear stress equals $\tau_o = 4.40$ Pa.

2. Considering an alluvial stream, the longitudinal bed slope is 0.001 and the river width is about 65 m. The flow in the river is fully rough turbulent. Velocity measurements were performed at various distance from the bed and showed that the velocity profile is approximately logarithmic. The data are:

Elevation y (m):	0.02	0.03	0.04	0.05	0.07	0.10	0.20	0.30	0.40	0.50	0.60
Velocity v (m/s):	1.32	1.47	1.54	1.63	1.75	1.85	2.10	2.23	2.33	2.41	2.47

Plot the velocity profile as $v = f(\ln(y))$. Deduce the shear velocity, the characteristic grain size and the flow rate.

In a first approximation, the characteristic grain size may be assumed to be the equivalent roughness height.

Solution

First, plot the velocity distribution.

For a fully rough turbulent flow, the logarithmic velocity profile may be estimated as (Schlichting, 1979: p. 619):

$$\frac{v}{V_*} = \frac{1}{K}\ln\left(\frac{y}{k_s}\right) + 8.5 \qquad \text{Turbulent zone and fully rough turbulent flow}$$

where v is the local velocity, k_s is the equivalent roughness height and K is the von Karman constant (K = 0.4).

The relationship $v = f(\ln(y))$ is a straight line, and the slope and constant of the linear relationship are, respectively:

$$\text{Slope: } \frac{V_*}{K}$$

$$\text{Constant: } V_*\left(8.5 - \frac{\ln(k_s)}{K}\right)$$

For the experiment data, the slope and constant of the relationship $v = f(\ln(y))$ give the shear velocity $V_* = 0.135$ m/s and the equivalent roughness height $k_s = 0.012$ m.

The estimate of the flow rate requires the knowledge of the mean flow velocity and flow depth. Both characteristics may be deduced from the shear velocity and the equivalent roughness height.

In uniform equilibrium flow and for a wide river, the shear velocity and flow depth are related by:

$$V_* \approx \sqrt{gd\sin\theta} \qquad \text{Uniform equilibrium flow}$$

Hence the flow depth equals $d = 1.86$ m. Note that the wide channel assumption is justified as $d \ll 65$ m (channel width). Furthermore, the mean flow velocity V and the shear velocity V_* are related by:

$$\frac{V}{V_*} = \sqrt{\frac{8}{f}}$$

where f is the Darcy friction factor which is a function of the relative roughness height: $k_s/D_H \approx k_s/4d$ (for a wide channel). Using the Moody diagram or the Colebrook–White formula, $f = 0.022$ for $k_s/D_H = 0.0016$ and the mean velocity equals $V = 2.6$ m/s. The total flow rate is $Q = 310$ m^3/s.

8.2.4 Forces acting on a sediment particle

For an open channel flow with a movable bed, the forces acting on each sediment particle are (Fig. 8.2):

- the gravity force $\rho_s g v_s$,
- the buoyancy force $F_b = \rho g v_s$,
- the drag force $C_d \rho A_s V^2/2$,
- the lift force $C_L \rho A_s V^2/2$,
- the reaction forces of the surrounding grains,

where v_s is the volume of the particle, A_s is a characteristic particle cross-sectional area, C_d and C_L are the drag and lift coefficients, respectively, and V is a characteristic velocity next to the channel bed.

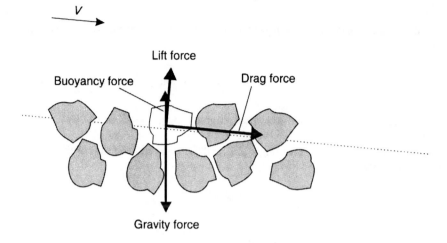

Fig. 8.2 Forces acting on a sediment particle (inter-granular forces not shown for the sake of clarity).

The gravity force and the buoyancy force act both in the vertical direction while the drag force acts in the flow direction and the lift force in the direction perpendicular to the flow direction (Fig. 8.2). The inter-granular forces are related to the grain disposition and packing.

Notes

1. The drag force and the lift force on an object are, respectively, the integration of the longitudinal component and normal component of the pressure distribution acting on the body. These expressions are often rewritten in terms of the mean flow velocity, fluid density and dimensionless coefficients: i.e. the drag and lift coefficients.
2. For an ideal-fluid flow, the drag and lift forces acting on a sphere in an uniform flow are zero (i.e. d'Alembert's paradox).
3. The pressure distribution on a grain is related to the flow pattern around the particle. In turbulent flows, several processes may take place at the particle wall: e.g. wake and flow separation (Fig. 8.3).

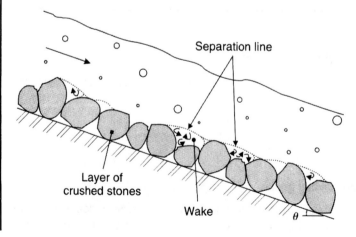

Fig. 8.3 Sketch of turbulent flow patterns around alluvial bed elements.

Hence the pressure distribution around the particle differs from the hydrostatic pressure distribution.

4. For alluvial streams, the characteristic velocity near the sediment bed is the shear velocity V_*.
5. Warning: in many alluvial channels, the shear stress (hence V_*) is not uniform around the wetted perimeter and iniertia force takes place wherever the fluid is accelerated.

8.3 THRESHOLD OF SEDIMENT BED MOTION

8.3.1 Introduction

The term *threshold of sediment motion* describes the flow conditions and boundary conditions for which the transport of sediment starts to occur. The threshold of sediment motion cannot be defined with an exact (absolute) precision but most experimental observations have provided reasonably accurate and consistent results.

8.3.2 Simple dimensional analysis

The relevant parameters for the analysis of sediment transport threshold are: the bed shear stress τ_0, the sediment density ρ_s, the fluid density ρ, the grain diameter d_s, the gravity acceleration g and the fluid viscosity μ:

$$f_1(\tau_0,\ \rho,\ \rho_s,\ \mu,\ g,\ d_s) = 0 \qquad (8.2a)$$

In dimensionless terms, it yields:

$$f_2\left(\frac{\tau_0}{\rho g d_s};\ \frac{\rho_s}{\rho};\ \frac{d_s\sqrt{\rho \tau_0}}{\mu}\right) = 0 \qquad (8.2b)$$

Discussion

The ratio of the bed shear stress to fluid density is homogeneous (in units) with a velocity squared. Introducing the *shear velocity* V_* defined as:

$$V_* = \sqrt{\frac{\tau_0}{\rho}} \qquad (8.3)$$

equation (8.2a) can be transformed as:

$$f_3\left(\frac{V_*}{\sqrt{g d_s}};\ \frac{\rho_s}{\rho};\ \rho\frac{d_s V_*}{\mu}\right) = 0 \qquad (8.2c)$$

The first term is a form of Froude number. The second is the relative density (also called specific gravity). The last term is a Reynolds number defined in terms of the grain size and shear velocity. It is often denoted as Re_* and called the *shear Reynolds number* or *particle Reynolds number*.

Note that the shear Reynolds number was previously introduced in relation to the velocity distribution in turbulent flows.

Notes

1. The boundary shear stress equals:

$$\tau_o = C_d \frac{1}{2} \rho V^2$$

where C_d is the skin friction coefficient and V is the mean flow velocity. In open channel flow, it is common practice to use the Darcy friction factor f which is related to the skin friction coefficient by $f = 4C_d$. It yields:

$$\tau_o = \frac{f}{8} \rho V^2$$

2. The shear velocity is a measure of shear stress and velocity gradient near the boundary. It may be rewritten as:

$$\frac{V_*}{V} = \sqrt{\frac{f}{8}}$$

where V is the mean flow velocity and f is the Darcy friction factor.

3. In uniform equilibrium flow down an open channel, the shear velocity equals:

$$V_* = \sqrt{g \frac{D_H}{4} \sin \theta} \qquad \text{Uniform equilibrium flow}$$

where D_H is the hydraulic diameter and θ is the bed slope. And the mean boundary shear stress equals:

$$\tau_o = \rho g \frac{D_H}{4} \sin \theta \qquad \text{Uniform equilibrium flow}$$

For a wide rectangular channel, it yields:

$$V_* \approx \sqrt{gd \sin \theta} \quad \text{Uniform equilibrium flow in wide rectangular channel}$$

$$\tau_o \approx \rho g d \sin \theta \quad \text{Uniform equilibrium flow in wide rectangular channel}$$

8.3.3 Experimental observations

Particle movement occurs when the moments of the destabilizing forces (i.e. drag, lift and buoyancy), with respect to the point of contact, become larger than the stabilizing moment of the weight force. The resulting condition is a function of the angle of repose (e.g. van Rijn, 1993: p. 4.1). Plates 18 and 19 illustrate examples of extreme bed load motion in natural streams.

Experimental observations highlighted the importance of the stability parameter τ_* (which may be derived from dimensional analysis) defined as:

$$\tau_* = \frac{\tau_o}{\rho(s-1)gd_s} \tag{8.4}$$

A critical value of the stability parameter may be defined at the inception of bed motion: i.e. $\tau_* = (\tau_*)_c$. Shields (1936) showed that $(\tau_*)_c$ is primarily a function of the shear Reynolds number $V_* d_s / \nu$ (Fig. 8.4). Bed load motion occurs for:

$$\tau_* > (\tau_*)_c \qquad \text{Bed load motion} \tag{8.5}$$

In summary: the initiation of bed load transport occurs when the bed shear stress τ_0 is larger than a critical value:

$$(\tau_o)_c = \rho(s-1)gd_s(\tau_*)_c$$

Experimental observations showed that $(\tau_*)_c$ is primarily a function of $(d_s V_*/\nu)$ (Fig. 8.4).

Notes
1. The stability parameter τ_* is called commonly the *Shields parameter* after A.F. Shields who introduced it first (Shields, 1936). It is a dimensionless parameter. Note that the Shields parameter is sometimes denoted by θ.
2. The stability parameter may be rewritten as:

$$\tau_* = \frac{V_*^2}{(s-1)gd_s} = \frac{\tau_o}{\rho(s-1)gd_s}$$

3. $(\tau_*)_c$ is commonly called the critical Shields parameter.
4. Kennedy (1995) and Buffington (1999) related the story of Albert Frank Shields.

Comments
First let us note that, for given fluid and sediment properties (i.e. ν, ρ, s) and given bed shear stress τ_0, the Shields parameter τ_* decreases with increasing sediment size: i.e. $\tau_* \propto 1/d_s$. Hence, for given flow conditions, sediment motion may occur for small particle sizes while no particle motion occurs for large grain sizes.

It is worth noting also that, for sediment particles in water (Fig. 8.4b), the Shields diagram exhibits different trends corresponding to different turbulent flow regimes:

- the smooth turbulent flow $(Re_* < 4\text{–}5)$ $0.035 < (\tau_*)_c,$
- the transition regime $(4\text{–}5 < Re_* < 75\text{–}100)$ $0.03 < (\tau_*)_c < 0.04,$
- the fully rough turbulent flow $(75\text{–}100 < Re_*)$ $0.03 < (\tau_*)_c < 0.06.$

For fully rough turbulent flows, the critical Shields parameter $(\tau_*)_c$ is nearly constant, and the critical bed shear stress for bed load motion becomes linearly proportional to the sediment size:

$$(\tau_o)_c \propto d_s \qquad \text{Fully rough turbulent flow} \tag{8.6}$$

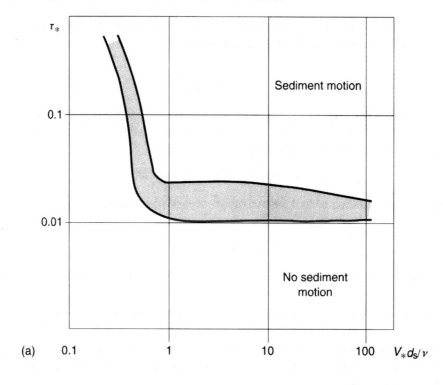

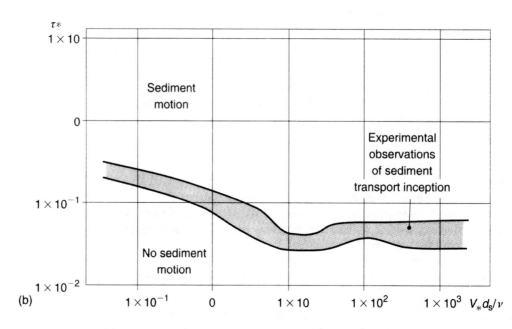

Fig. 8.4 Threshold of bed load motion (Shields diagram): (a) Shields parameter as a function of the particle Reynolds number for sand in air; (b) Shields parameter as a function of the particle Reynolds number for sediment in water (experimental data reviewed by Yalin and Karahan (1979)).

Notes

1. On the Shields diagram (Fig. 8.4), the Shields parameter and the particle Reynolds number are both related to the shear velocity and the particle size. Some researchers proposed a modified diagram: τ_* as a function of a particle parameter $d_* = (Re^2_*/\tau_*)^{1/3}$.

2. The critical Shields parameter may be estimated as (Julien, 1995):

$$(\tau_*)_c = 0.5 \tan \phi_s \qquad \text{for } d_* < 0.3$$

$$(\tau_*)_c = 0.25 d_*^{-0.6} \tan \phi_s \qquad \text{for } 0.3 < d_* < 19$$

$$(\tau_*)_c = 0.013 d_*^{0.4} \tan \phi_s \qquad \text{for } 19 < d_* < 50$$

$$(\tau_*)_c = 0.06 \tan \phi_s \qquad \text{for } 50 < d_*$$

where ϕ_s is the angle of repose and d_* is the dimensionless particle parameter defined as:

$$d_* = d_s \sqrt[3]{\frac{(s-1)g}{\nu^2}}$$

and ν is the fluid kinematic viscosity.

Applications

1. Considering a stream with a flow depth of 1.7 m and a bed slope $\sin \theta = 0.002$, indicate whether a 5 mm gravel bed will be subjected to bed load motion. Find out what is the critical particle size for bed load motion in the stream.

Solution

Assuming a wide rectangular channel and assuming that the stream flow is nearly uniform equilibrium, the shear velocity equals:

$$V_* = \sqrt{g \frac{D_H}{4} \sin \theta} \approx \sqrt{gd \sin \theta} = 0.18 \text{ m/s}$$

where d is the flow depth. The mean bed shear stress equals $\tau_o = 33.3$ Pa.

For a 5 mm gravel bed, the Shields parameter is:

$$\tau_* = \frac{V_*^2}{(s-1)gd_s} = 0.41$$

assuming $s = 2.65$. And the particle Reynolds number equals 910. For these values, Fig. 8.4b indicates bed load motion.

To estimate the critical sediment size for bed load motion in the stream, let us assume a gravel bed (i.e. fully rough turbulent flow). For a rough turbulent flow, the critical sediment size for initiation of bed motion satisfies:

$$(\tau_o)_c = (\tau_*)_c (\rho_s - \rho)gd_s = \tau_o = 33.3 \text{ Pa}$$

Assuming $(\tau_*)_c = 0.06$, it yields $d_s = 34$ mm and $Re_* = 6200$. Note that the value of Re_* corresponds to a fully rough turbulent flow and hence equation (8.6) may be used to estimate the critical particle size.

Conclusion: the largest particle size subjected to bed load motion in the stream is 34 mm. For $d_s > 34$ mm the particles will not move.

2. During a storm, the wind blows over a sandy beach (0.5 mm sand particle). The wind boundary layer is about 100 m high at the beach and the free-stream velocity at the outer edge of the wind boundary layer is 35 m/s. Estimate the risk (or not) of beach sand erosion.

Solution
First we must compute the bed shear stress or the shear velocity.

The Reynolds number $V\delta/\nu$ of the boundary layer is 2.3×10^8. The flow is turbulent. As the sand diameter is very small compared to the boundary layer thickness, let us assume that the boundary layer flow is smooth turbulent. In a turbulent boundary layer along a smooth boundary, the mean bed shear stress equals:

$$\tau_\mathrm{o} = 0.0225\rho V^2 \left(\frac{\nu}{V\delta} \right)^{1/4}$$

where ρ is the fluid density, ν is the kinematic viscosity, δ is the boundary layer thickness and V is the free-stream velocity (at the outer edge of the boundary layer) (e.g. Schlichting, 1979: p. 637; Streeter and Wylie, 1981: p. 215).

For the beach, it yields $\tau_\mathrm{o} = 0.27$ Pa. And the Shields parameter equals $\tau_* = 0.021$. The particle Reynolds number equals $Re_* = 16$.

For these values, Fig. 8.4a indicates that the flow conditions are at the onset of sediment motion. Sediment transport on the beach will be small.

Discussion
An example of destructive Aeolian transport is the Aral Sea. Following the shrinkage of the sea, 20 000 km^2 are now exposed. During dust storm, massive amounts of salt from the exposed sea bed are moved hundreds of kilometres away and ruin crop production.

8.3.4 Discussion

Several parameters may affect the inception of bed load motion: particle size distribution, bed slope, bed forms and material cohesion (van Rijn (1993) presented a comprehensive review).

(a) The particle size distribution has an effect when the size range is wide because the fine particles will be shielded by the larger particles. After an initial erosion of the fine particles, the coarser particles will form an armour layer preventing further erosion. This process is called *bed armouring*.

(b) On steep channels (Fig. 8.5), the bed slope assists in destabilizing the particles and bed motion occurs at lower bed shear stresses than in flat channels. At the limit, when the bed slope becomes larger than the repose angle (i.e. $\theta > \phi_\mathrm{s}$), the grains roll even in the absence of flow: i.e. the bed slope is unstable.

(c) When bed forms (e.g. ripples and dunes) develop, the critical bed shear stress for initiation of bed motion becomes different from that for a flat bed. Indeed, the bed shear stress above a bed form includes a skin friction component plus a form drag component which is related to the non-uniform pressure distribution in the flow surrounding the bed form (see Chapter 12).

(d) For clay and silty sediment beds, the cohesive forces between sediment particles may become important. This causes a substantial increase of the bed resistance to scouring.

A related effect is the existence of seepage through the sediment bed. When the seepage pressure forces exerted on the sediment particles become larger than the submerged weight, the grains will be subjected to some motion. This process is called *bed fluidization*.

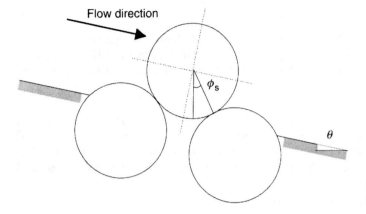

Fig. 8.5 Angle of repose for granular material on steep slope.

Notes

1. For steep channels, van Rijn (1993) suggested to define the Shields parameter as: $[(\sin(\phi_s - \theta)/\sin\phi_s)\tau_*]$, where ϕ_s is the angle of repose, θ is the bed slope and τ_* is the Shields parameter for flat channels.

2. Recently Chiew and Parker (1994) demonstrated that the critical shear velocity for bed motion on steep slopes must be estimated as:

$$\sqrt{\cos\phi_s \left(1 - \frac{\tan\phi_s}{\tan\theta}\right)}\, V_*$$

where V_* is the critical shear velocity for a flat bed. Note that the above formula is also valid for adverse slope ($\theta < 0$).

Application: bed fluidization

Considering a seepage flow through a particle bed, the forces exerted on each particle (at equilibrium) are the reaction forces of the surrounding grains, the submerged weight (i.e. weight minus buoyancy) and the pressure force induced by the seepage flow.

At the initiation of fluidization, the submerged weight equals the pressure force component in the vertical direction. For a spherical particle, it yields:

$$-(\rho_s - \rho)gv_s - \frac{dP}{dz}v_s = 0 \qquad \text{Bed fluidization condition}$$

where z is the vertical axis positive upwards, g is the gravity constant, ρ is the water density, ρ_s is the particle density and v_s is the volume of the particle. The vertical pressure gradient can be derived from Darcy's law:

$$\frac{dP}{dz} = -\rho g\left(1 + \frac{V_z}{K}\right)$$

where K is the hydraulic conductivity (in m/s) and V_z is the vertical component of the seepage velocity.

Bed fluidization occurs for:

$$V_z \geq K(s - 1) \qquad \text{Bed fluidization}$$

8.4 EXERCISES

Considering a smooth turbulent boundary layer, experimental observations suggest that the shear velocity is about 0.21 m/s and the boundary layer thickness equals 0.18 m. Calculate: (a) viscous sub-layer thickness and (b) the range of validity of the outer region velocity profile (i.e. 'velocity defect law'). (The fluid is air at 20°C.)

Considering a stream with a flow depth of 2 m and a bed slope $\sin \theta = 0.0003$, the median grain size of the channel bed is 2.5 mm. Calculate: (a) mean flow velocity, (b) shear velocity, (c) bed shear stress, (d) Reynolds number, (e) shear Reynolds number and (f) Shields parameter. (g) Will sediment motion take place? (Assume that the equivalent roughness height of the channel bed equals the median grain size.)
   Solution: Yes (bed load motion and saltation).

The wind blows over a sandy beach. The wind boundary layer is 200 m high and the free-stream velocity (i.e. at outer edge of boundary layer) is 42 m/s. Deduce the critical particle size for beach sand erosion.

Considering a stream with a flow depth of 0.95 m and a bed slope $\sin \theta = 0.0019$: (a) indicate whether a 15 mm gravel bed will be subjected to bed load motion and (b) find out the critical particle size for bed load motion in the stream.
   Solution: Particles with $d_{50} = 15$ mm will move: bed load motion and saltation. Bed load motion occurs for $d_s < 25$–30 mm.

A stream carries a discharge of 81 m³/s. The channel is 42 m wide and the longitudinal bed slope is 15 m/km. The bed consists of a mixture of fine sands ($d_{50} = 0.9$ mm). Assume that uniform equilibrium flow conditions are achieved. (a) Compute the flow depth. (b) Compute the flow velocity. (c) For such a flow, where would you locate a control: i.e. upstream or downstream? (Justify your answer.) (d) Predict the occurrence of bed motion. (e) If bed motion occurs, what is the type of bed form?

Considering the natural stream ($\sin \theta = 0.0009$ and $d_{50} = 3$ mm) the flow rate per unit width is 12 m²/s. Will the channel bed sediments be subjected to sediment motion. In the affirmative, indicate whether the sediment motion is bed load, saltation or suspension. (Assume that the equivalent roughness height of the channel bed equals the median grain size.)
   Solution: Yes (bed load motion).

A stream carries a discharge of 8 m³/s. The channel cross-section is trapezoidal with a 2.2 m bottom width and 1V:3H sidewall slopes. The longitudinal bed slope is 15 m per km. The channel bottom and sidewall consist of a mixture of fine sands ($d_{50} = 0.3$ mm). Assume that uniform equilibrium flow conditions are achieved. (a) Compute the flow depth. (b) Compute the flow velocity. (c) For such a flow, where would you locate a control: i.e. upstream or downstream? (Justify your answer.) (d) Predict the occurrence of bed motion. If bed motion occurs, what is the type of bed form?
   Assume that the equivalent roughness height equals the median sediment size.
   Solution: (a) $d = 0.44$ m, (b) $V = 5.1$ m/s, (c) upstream because the flow is supercritical, (d) bed load motion occurs. Bed form: antidunes or step-pools (depending upon channel bed type).

Inception of suspended-load motion

9

9.1 PRESENTATION

In this chapter, the inception of sediment suspension is presented. As discussed in the previous chapter, the inception of sediment motion is related to the shear velocity (or to the bed shear stress). Considering a given channel and bed material, no sediment motion is observed at very-low bed shear stress until τ_o exceeds a critical value. For τ_o larger than the critical value, bed-load motion takes place. The grain motion along the bed is not smooth, and some particles bounce and jump over the others. With increasing shear velocities, the number of particles bouncing and rebounding increases until the cloud of particles becomes a suspension.

The onset of sediment suspension is not a clearly defined condition.

Note
When particles are moving along the bed with jumps and bounces, the mode of sediment transport is sometimes called *saltation* (Fig. 10.1 later).

9.2 INITIATION OF SUSPENSION AND CRITICAL BED SHEAR STRESS

Considering a particle in suspension, the particle motion in the direction normal to the bed is related to the balance between the particle fall velocity component ($w_o \cos \theta$) and the turbulent velocity fluctuation in the direction normal to the bed. Turbulence studies (e.g. Hinze, 1975; Schlichting, 1979) suggested that the turbulent velocity fluctuation is of the same order of magnitude as the shear velocity.

With this reasoning, a simple criterion for the initiation of suspension (which does not take into account the effect of bed slope) is:

$$\frac{V_*}{w_o} > \text{Critical value} \tag{9.1a}$$

Several researchers proposed criterion for the onset of suspension (Table 9.1). In a first approximation, suspended sediment load occurs for:

$$\frac{V_*}{w_o} > 0.2 \text{ to } 2 \tag{9.1b}$$

Table 9.1 Criterion for suspended-load motion

Reference (1)	Criterion for suspension (2)	Remarks (2)
Bagnold (1966)	$\dfrac{V_*}{w_0} > 1$	As given by van Rijn (1993)
van Rijn (1984b)		Deduced from experimental investigations
	$\dfrac{V_*}{w_0} > \dfrac{4}{\sqrt[3]{\dfrac{(s-1)g}{\nu^2}d_s}}$	For $1 < \sqrt[3]{\dfrac{(s-1)g}{\nu^2}}d_s \leq 10$ where $d_s = d_{50}$
	$\dfrac{V_*}{w_0} > 0.4$	For $\sqrt[3]{\dfrac{(s-1)g}{\nu^2}}d_s > 10$ where $d_s = d_{50}$
Raudkivi (1990)	$\dfrac{V_*}{w_0} > 0.5$	Note: *rule of thumb* (Raudkivi, 1990: p. 142)! Inception of suspension (i.e. saltation)
	$\dfrac{V_*}{w_0} > 1.2$	Dominant suspended load (i.e. suspension)
Julien (1995)	$\dfrac{V_*}{w_0} > 0.2$	Turbulent water flow over rough boundaries. Inception of suspension
	$\dfrac{V_*}{w_0} > 2.5$	Dominant suspended load
Sumer *et al.* (1996)	$\dfrac{V_*^2}{(s-1)gd_s} > 2$	Experimental observations in sheet flow. Sediment size: $0.13 < d_s < 3\,\text{mm}$

Notes: $s = \rho_s/\rho$; V_* = shear velocity; w_0 = terminal settling velocity.

Application

Considering a stream with a flow depth of 3.2 m and a bed slope of $\sin\theta = 0.001$, indicate whether a 3 mm gravel bed will be subjected to sediment motion. In the affirmative, indicate whether the sediment motion is bed load, saltation or suspension.

Solution

Assuming a wide rectangular channel and assuming that the stream flow is nearly uniform equilibrium, the shear velocity equals:

$$V_* = \sqrt{g\frac{D_H}{4}\sin\theta} \approx \sqrt{gd\sin\theta} = 0.18 \text{ m/s}$$

where d is the flow depth.

For a 3 mm gravel bed, the Shields parameter is:

$$\tau_* = \frac{V_*^2}{(s-1)gd_s} = 0.67$$

assuming $s = 2.65$. The particle Reynolds number Re_* equals 540. For these values, the Shields diagram predicts sediment motion (see Chapter 8).

Let us estimate the settling velocity. For a single particle in still fluid, the fall velocity equals:

$$w_o = -\sqrt{\dfrac{4gd_s}{3\left(\dfrac{24\mu}{\rho\,|\,w_o\,|\,d_s} + 1.5\right)}(s-1)}$$

It yields $w_o = 0.20\,\text{m/s}$. Hence the ratio V_*/w_o is about unity.

Sediment transport takes place with bed load and saltation. The flow conditions are near the initiation of sediment suspension (equation (9.1b)).

9.3 ONSET OF HYPERCONCENTRATED FLOW

9.3.1 Definition

Hyperconcentrated flows are sediment-laden flows with large suspended sediment concentrations (i.e. typically more than 1% in volume). Spectacular hyperconcentrated flows are observed in the Yellow River basin (China) with volumetric concentrations larger than 8%: i.e. the sediment mass flow rate being more than 25% of the water mass flow rate.

Hyperconcentrated flows exhibit different flow properties from clear-water flows. For example, the fluid viscosity can no longer be assumed to be that of water (see Chapter 11). In practice, numerous researchers observed that the properties of hyperconcentrated flows differ from those of 'classical' sediment-laden flows when the volumetric sediment concentration exceeds 1–3%.

Notes
1. A 'hyperconcentrated flow' is a suspension flow with large volume sediment concentration (i.e. over 1–3%). Sometimes the term might also include 'mud flow' and 'debris flow', but the rheology of these flows differs significantly from that of hyperconcentrated flows. In mud and debris flow, the volumetric sediment concentration might be larger than 50%!
2. Mud is a sticky mixture of fine solid particles and water (i.e. soft wet earth).
3. Debris comprise mainly of large boulders, rock fragments, gravel-sized to clay-sized material, tree and wood material that accumulate in creeks. Debris flow characterizes sediment-laden flow in which roughly more than half of the solid volume is coarser than sand. They are often encountered in steep mountain areas: e.g. western parts of Canada, East coast of Taiwan and Austria.
4. Hyperconcentrated flows behave as non-Newtonian fluids. The relationship between the fluid shear stress and the velocity gradient is non-linear. The rheology of hyperconcentrations is currently the topic of active research (see also Chapter 11).

9.3.2 Discussion

Hyperconcentrated flows are often associated with severe land degradation in the catchment. Two dramatic examples are illustrated in Fig. 9.1 (Plate 21). In each catchment, hyperconcentrated flows occur during floods.

> **Notes**
> 1. In the Yellow River basin (China), the hyperconcentrated flows are characterized by up to 10% of the sediment material being clay and the remaining materials consisting of silt and fine sand.
> 2. In the Durance basin (France), hyperconcentrated sediment concentrations may exceed 15% (in volume), particularly in black marl[1] catchments. Meunier (1995) observed sediment generation by miniature debris flows in small size parts of the slope.

(a)

(b)

Fig. 9.1 Examples of extreme land erosion associated with hyperconcentrated suspension flows: (a) 'Moon walk' in Kaohsiung County, Taiwan ROC (September 1995). (b) Draix Catchment, France (June 1998). Black marl basin in the Durance catchment.

[1]Also called badlands (in English), 'Marnes Noires' or 'Terres Noires' (in French).

9.4 EXERCISES

During a storm, the wind blows over a beach (0.2 mm sand particles). The velocity at the outer edge of the boundary layer is 110 m/s and the boundary layer is 45 m thick. Calculate: (a) shear velocity, (b) settling velocity of 0.2 mm sand particle in air and (c) occurrence of suspended-load motion. (Assume that the equivalent roughness height of the channel bed equals the median grain size.)

Considering a stream with a flow depth of 2.3 m and a bed slope of $\sin\theta = 0.002$, the median grain size of the channel bed is 1.1 mm. Calculate: (a) mean flow velocity, (b) shear velocity, (c) settling velocity, (d) occurrence of bed-load motion and (e) occurrence of suspension. (Assume that the equivalent roughness height of the channel bed equals the median grain size.)

A stream carries a discharge of 8 m³/s. The channel cross-section is trapezoidal with a 2.2 m bottom width and 1V:3H sidewall slopes. The longitudinal bed slope is 15 m/km. The channel bottom and sidewall consist of a mixture of fine sands ($d_{50} = 0.3$ mm).

Predict the occurrence of sediment suspension. *Assume that the equivalent roughness height equals the median sediment size. Assume that uniform equilibrium flow conditions are achieved.*

Solution: This exercise is identical to one in Chapter 8. The flow is supercritical ($V = 5.1$ m/s, $d = 0.044$ m) and sediment suspension occurs because $V_*/w_o > 1$.

Sediment transport mechanisms: 10
1. Bed-load transport

10.1 INTRODUCTION

When the bed shear stress exceeds a critical value, sediments are transported in the form of bed load and suspended load. For bed-load transport, the basic modes of particle motion are *rolling* motion, *sliding* motion and *saltation* motion (Fig. 10.1).

In this chapter, formulations to predict the bed-load transport rate are presented. Figure 10.2 shows natural stream subjected to significant bed-load transport.

Notes
1. Saltation refers to the transport of sediment particles in a series of irregular jumps and bounces along the bed (Fig. 10.1a).
2. In this section, predictions of bed-load transport are developed for plane bed. Bed form motion and bed form effects on bed-load transport are not considered in this section (see Chapter 12).

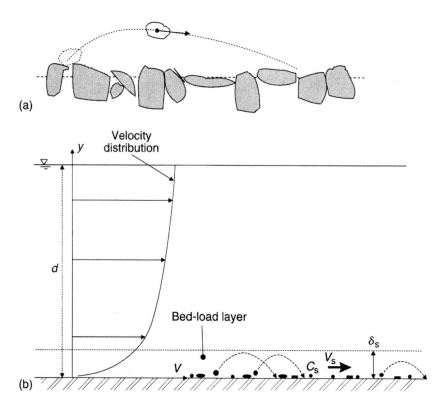

Fig. 10.1 Bed-load motion: (a) sketch of saltation motion and (b) definition sketch of the bed-load layer.

Fig. 10.2 Bed-load transport in natural streams: (a) South Korrumbyn Creek NSW, Australia looking upstream (18 August 2002). Stream bed upstream of the Korrumbyn Creek weir, fully silted today (Appendix A2.1). (b) Cedar Creek QLD, Australia looking downstream (9 December 1992) (courtesy of Mrs J. Hacker). Note the coarse material left after the flood.

Fig. 10.2 (c) Hayagawa river, Japan upstream of Nishiyama dam on November 1998 looking upstream. Note the bed-load material and the low flow channel showing the gravel bed. (d) Hayagawa river, Japan near the confluence with the Fujigawa river in November 1998 looking downstream. Note the gravel bed and on-going dredging (digger on bottom right of picture).

Definitions

The sediment transport rate may be measured by weight (units: N/s), by mass (units: kg/s) or by volume (units: m³/s). In practice the sediment transport rate is often expressed by metre width and is measured either by mass or by volume. These are related by:

$$\dot{m}_s = \rho_s q_s \qquad (10.1)$$

where $\dot{m}_s$ is the mass sediment flow rate per unit width, q_s is the volumetric sediment discharge per unit width and ρ_s is the specific mass of sediment.

10.2 EMPIRICAL CORRELATIONS OF BED-LOAD TRANSPORT RATE

10.2.1 Introduction

Bed-load transport occurs when the bed shear stress τ_o exceeds a critical value $(\tau_o)_c$. In dimensionless term, the condition for bed-load motion is:

$$\tau_* > (\tau_*)_c \qquad \text{Bed-load transport} \qquad (10.2)$$

where τ_* is the Shields parameter (i.e. $\tau_* = \tau_o/(\rho(s-1)gd_s)$) and $(\tau_*)_c$ is the critical Shields parameter for initiation of bed-load transport (Fig. 8.4).

10.2.2 Empirical bed-load transport predictions

Many researchers attempted to predict the rate of bed-load transport. The first successful development was proposed by P.F.D. du Boys in 1879. Although his model of sediment transport was incomplete, the proposed relationship for bed-load transport rate (Table 10.1) proved to be in good agreement with a large amount of experimental measurements.

Table 10.1 Empirical and semi-empirical correlations of bed-load transport

Reference (1)	Formulation (2)	Range (3)	Remarks (4)
Boys (1879)	$q_s = \lambda \tau_o(\tau_o - (\tau_o)_c)$ $\lambda = \dfrac{0.54}{(\rho_s - \rho)g}$ Schoklitsch (1914) $\lambda \propto d_s^{-3/4}$ Straub (1935)	 $0.125 < d_s < 4\,\text{mm}$	λ was called the characteristic sediment coefficient Laboratory experiments with uniform grains of various kinds of sand and porcelain Based upon laboratory data
Schoklitsch (1930)	$q_s = \lambda'(\sin\theta)^k(q - q_c)$ $q_c = 1.944 \times 10^{-2}d_s(\sin\theta)^{-4/3}$	$0.305 < d_s < 7.02\,\text{mm}$	Based upon laboratory experiments
Shields (1936)	$\dfrac{q_s}{q} = 10\dfrac{\sin\theta}{s}\dfrac{\tau_o - (\tau_o)_c}{\rho g(s-1)d_s}$	$1.06 < s < 4.25$ $1.56 < d_s < 2.47\,\text{mm}$	
Einstein (1942)	$\dfrac{q_s}{\sqrt{(s-1)gd_s^3}} =$ $2.15\exp\!\left(-0.391\dfrac{\rho(s-1)gd_s}{\tau_o}\right)$	$\dfrac{q_s}{\sqrt{(s-1)gd_s^3}} < 0.4$ $1.25 < s < 4.25$ $0.315 < d_s < 28.6\,\text{mm}$	Laboratory experiments. Weak sediment transport formula for sand mixtures Note: $d_s \approx d_{35}$ to d_{45}.

(continued)

Table 10.1 (*continued*)

Reference (1)	Formulation (2)	Range (3)	Remarks (4)
Meyer-Peter (1949, 1951)	$\dfrac{\dot{m}^{2/3}\sin\theta}{d_s} - 9.57(\rho g(s-1))^{10/9} =$ $0.462(s-1)\dfrac{\left(\rho g(\dot{m}_s)^2\right)^{2/3}}{d_s}$	$1.25 < s < 4.2$	Laboratory experiments. Uniform grain size distribution
	$\dfrac{q_s}{\sqrt{(s-1)gd_s^3}} =$ $\left(\dfrac{4\tau_0}{\rho(s-1)gd_s} - 0.188\right)^{3/2}$		Laboratory experiments. Particle mixtures Note: $d_s \approx d_{50}$
Einstein (1950)	Design chart $\dfrac{q_s}{\sqrt{(s-1)gd_s^3}} = f\left(\dfrac{\rho(s-1)gd_s}{\tau_0}\right)$	$\dfrac{q_s}{\sqrt{(s-1)gd_s^3}} < 10$ $1.25 < s < 4.25$ $0.315 < d_s < 28.6\,\text{mm}$	Laboratory experiments. For sand mixtures Note: $d_s \approx d_{35}$ to d_{45}
Schoklitsch (1950)	$\dot{m}_s = 2500(\sin\theta)^{3/2}(q-qc)$ $q_c = 0.26(s-1)^{5/3}d_{40}^{3/2}(\sin\theta)^{-7/6}$		Based upon laboratory experiments and field measurements (Danube and Aare rivers)
Nielsen (1992)	$\dfrac{q_s}{\sqrt{(s-1)gd_s^3}} =$ $\left(\dfrac{12\tau_0}{\rho(s-1)gd_s} - 0.05\right)\sqrt{\dfrac{\tau_0}{\rho(s-1)gd_s}}$	$1.25 < s < 4.22$ $0.69 < d_s < 28.7\,\text{mm}$	Re-analysis of laboratory data

Notes: $\dot{m}$: mass water flow rate per unit width; $\dot{m}_s$: mass sediment flow rate per unit width; q: volumetric water discharge; q_s: volumetric sediment discharge per unit width; $(\tau_0)_c$: critical bed shear stress for initiation of bed load.

Subsequently, numerous researchers proposed empirical and semi-empirical correlations. Some are listed in Table 10.1. Graf (1971) and van Rijn (1993) discussed their applicability. The most notorious correlations are the Meyer-Peter and Einstein formulae.

Notes
1. P.F.D. du Boys (1847–1924) was a French hydraulic engineer. In 1879, he proposed a bed-load transport model, assuming that sediment particles move in sliding layers (Boys, 1879).
2. Professor H.A. Einstein was the son of Albert Einstein.

Discussion
The correlation of Meyer-Peter (1949, 1951) has been popular in Europe. It is considered most appropriate for wide channels (i.e. large width to depth ratios) and coarse material. The Einstein's (1942, 1950) formulations derived from physical models of grain saltation and they have been widely used in America. Both the Meyer-Peter and Einstein correlations give close results (e.g. Graf, 1971: p. 150), usually within the accuracy of the data (Fig. 10.3).

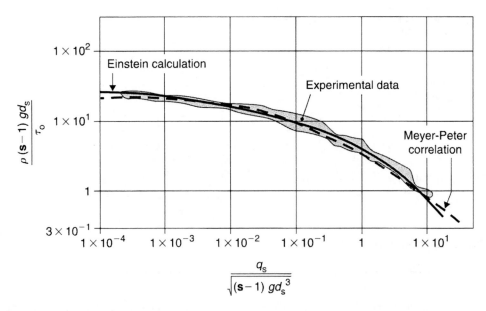

Fig. 10.3 Bed-load transport rate: comparison between Meyer-Peter formula, Einstein calculation and laboratory data (Meyer-Peter *et al.*, 1934; Gilbert, 1914; Chien, 1954).

It must be noted that empirical correlations should not be used outside of their domain of validity. For example, Engelund and Hansen (1972) indicated explicitly that Einstein's (1950) bed-load transport formula differs significantly from experimental data for large amounts of bed load (i.e. $q_s/\sqrt{(s-1)gd_s^3} > 10$).

10.3 BED-LOAD CALCULATIONS

10.3.1 Presentation

Bed-load transport is closely associated with inter-granular forces. It takes place in a thin region of fluid close to the bed (sometimes called *bed-load layer* or saltation layer) (Figs 10.1 and 10.4). Visual observations suggest that the bed-load particles move within a region of less than 10–20 particle-diameter height.

During the bed-load motion, the moving grains are subjected to hydrodynamic forces, gravity force and inter-granular forces. Conversely the (submerged) weight of the bed load is transferred as a normal stress to the (immobile) bed grains. The normal stress σ_e exerted by the bed load on the immobile bed particles is called the *effective stress* and it is proportional to:

$$\sigma_e \propto \rho(\mathbf{s}-1)g\cos\theta C_s \delta_s \tag{10.3}$$

where δ_s is the bed-load layer thickness, C_s is the volumetric concentration of sediment in the bed-load layer and θ is the longitudinal bed slope.

The normal stress increases the frictional strength of the sediment bed and the boundary shear stress applied to the top layer of the immobile grains becomes:

$$\tau_o = (\tau_o)_c + \sigma_e \tan\phi_s \tag{10.4}$$

where $(\tau_o)_c$ is the critical bed shear stress for initiation of bed load and ϕ_s is the angle of repose.

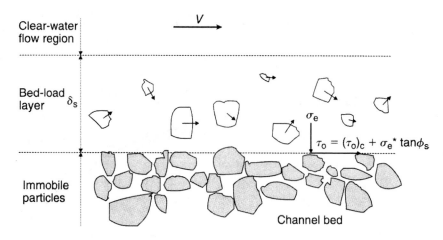

Fig. 10.4 Sketch of bed-load motion at equilibrium.

Notes
1. The concept of effective stress and associated bed shear stress (as presented above) derives from the work of Bagnold (1956, 1966).
2. For sediment particles, the angle of repose ranges usually from 26° to 42° and hence $0.5 < \tan\phi_s < 0.9$. For sands, it is common to choose $\tan\phi_s \approx 0.6$.

10.3.2 Bed-load transport rate

The bed-load transport rate per unit width may be defined as:

$$q_s = C_s \delta_s V_s \tag{10.5}$$

where V_s is the average sediment velocity in the bed-load layer (Fig. 10.1b).

Physically the transport rate is related to the characteristics of the bed-load layer: its mean sediment concentration C_s, its thickness δ_s which is equivalent to the average saltation height measured normal to the bed (Figs 10.1 and 10.4) and the average speed V_s of sediment moving along the plane bed.

Notes
1. A steady sediment transport in the bed-load layer is sometimes called a (no suspension) *sheet-flow*.
2. Note that the volumetric sediment concentration has a maximum value. For rounded grains, the maximum sediment concentration is 0.65.

10.3.3 Discussion

Several researchers proposed formulae to estimate the characteristics of the bed-load layer (Table 10.2). Figure 10.5 presents a comparison between two formulae. Overall the results are not very consistent. In practice there is still great uncertainty on the prediction of bed-load transport.

Note that the correlations of van Rijn (1984a) are probably more accurate to estimate the saltation properties (i.e. C_s, δ_s/d_s and V_s/V_*) (within their range of validity).

Table 10.2 Bed-load transport rate calculations

Reference (1)	Bed-load layer characteristics (2)	Remarks (3)
Fernandez-Luque and van Beek (1976)	$$\frac{V_s}{V_*} = 9.2\left(1 - 0.7\sqrt{\frac{(\tau_*)_c}{\tau_*}}\right)$$	Laboratory data: $1.34 \leqslant s \leqslant 4.58$ $0.9 \leqslant d_s \leqslant 3.3\,\text{mm}$ $0.08 \leqslant d \leqslant 0.12\,\text{m}$
Nielsen (1992)	$C_s = 0.65$ $$\frac{\delta_s}{d_s} = 2.5(\tau_* - (\tau_*)_c)$$ $$\frac{V_s}{V_*} = 4.8$$	Simplified model
Van Rijn (1984a, 1993)	$$C_s = \frac{0.117}{d_s}\left(\frac{\nu^2}{(s-1)g}\right)^{1/3}\left(\frac{\tau_*}{(\tau_*)_c} - 1\right)$$ $$\frac{\delta_s}{d_s} = 0.3\left(d_s\left(\frac{(s-1)g}{\nu^2}\right)^{1/3}\right)^{0.7}\sqrt{\frac{\tau_*}{(\tau_*)_c} - 1}$$ $$\frac{V_s}{V_*} = 9 + 2.6\log_{10}\left(d_s\left(\frac{(s-1)g}{\nu^2}\right)^{1/3}\right) - 8\sqrt{\frac{(\tau_*)_c}{\tau_*}}$$ $$C_s = \frac{0.117}{d_s}\left(\frac{\nu^2}{(s-1)g}\right)^{1/3}\left(\frac{\tau_*}{(\tau_*)_c} - 1\right)$$ $$\frac{\delta_s}{d_s} = 0.3\left(d_s\left(\frac{(s-1)g}{\nu^2}\right)^{1/3}\right)^{0.7}\sqrt{\frac{\tau_*}{(\tau_*)_c} - 1}$$ $$\frac{V_s}{V_*} = 7$$	For $\dfrac{\tau_*}{(\tau_*)_c} < 2$ and $d_s = d_{50}$ Based on laboratory data $0.2 \leqslant d_s \leqslant 2\,\text{mm}$ $d > 0.1\,\text{mm}$ $Fr < 0.9$ $d_s = d_{50}$ Based on laboratory data $0.2 \leqslant d_s \leqslant 2\,\text{mm}$ $d > 0.1\,\text{m}$ $Fr < 0.9$

Notes: V_*: shear velocity; $(\tau_*)_c$: critical Shields parameter for initiation of bed load.

Notes

1. The calculations detailed in Table 10.2 apply to flat channels (i.e. $\sin\theta < 0.001$–0.01) and in absence of bed forms (i.e. plane bed only).
2. For steep channels several authors showed a strong increase of bed-load transport rate. It is believed that the longitudinal bed slope affects the transport rate because the threshold conditions (i.e. initiation of bed load) are affected by the bed slope, the sediment motion is changed with steep bed slope and the velocity distribution near the bed is modified.

Discussion

The prediction of bed-load transport rate is *not* an accurate prediction. One researcher (van Rijn, 1984a) stated explicitly that:

"the overall inaccuracy […] may not be less than a factor 2"

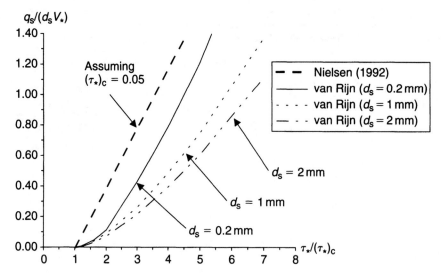

Fig. 10.5 Dimensionless bed-load transport rate $q_s/(d_s V_*)$ as a function of the dimensionless Shields parameter $\tau_*/(\tau_*)_c$ (Table 10.2).

10.4 APPLICATIONS

10.4.1 Application No. 1

The bed-load transport rate must be estimated for the Danube river (Central Europe) at a particular cross-section. The known hydraulic data are flow rate of about $530 \, \mathrm{m^3/s}$, flow depth of $4.27 \, \mathrm{m}$ and bed slope being about 0.0011. The channel bed is a sediment mixture with a median grain size of $0.012 \, \mathrm{m}$ and the channel width is about $34 \, \mathrm{m}$.

Predict the sediment-load rate using the Meyer-Peter correlation, the Einstein formula, and equation (10.5) using both Nielsen and van Rijn coefficients.

First calculations

Assuming a wide channel, the mean shear stress and shear velocity equals, respectively:

$$\tau_o = \rho g d \sin \theta = 998.2 \times 9.81 \times 4.27 \times 0.0011 = 46.0 \ \mathrm{Pa}$$

$$V_* = \sqrt{gd \sin \theta} = 0.215 \ \mathrm{m/s}$$

The Shields parameter equals:

$$\tau_* = \frac{\tau_o}{\rho(s-1)gd_s} = \frac{46.0}{998.2 \times 1.65 \times 9.81 \times 0.012} = 0.237$$

assuming $s = 2.65$ (quartz particles). And the particle Reynolds number is:

$$Re_* = \frac{V_* d_s}{\nu} = \frac{0.215 \times 0.012}{1.007 \times 10^{-6}} = 2558$$

For these flow conditions (τ_*, Re_*), the Shields diagram predicts sediment motion:

$$\tau_* = 0.237 > (\tau_*)_c \approx 0.05$$

Note that V_*/w_o is small and the flow conditions are near the initiation of suspension. In first approximation, the suspended sediment transport will be neglected.

Approach No. 1: Meyer-Peter correlation

For the hydraulic flow conditions, the dimensionless parameter used for the Meyer-Peter formula is:

$$\frac{\rho(s-1)gd_s}{\tau_o} = \frac{998.2 \times 1.65 \times 9.81 \times 0.012}{46} = 4.215$$

Application of the Meyer-Peter formula (for a sediment mixture) leads to:

$$\frac{q_s}{\sqrt{(s-1)gd_s^3}} = 0.66$$

Hence: $q_s = 0.0035\,\text{m}^2/\text{s}$.

Approach No. 2: Einstein function

For the hydraulic flow conditions, the dimensionless parameter used for the Einstein formula is:

$$\frac{\rho(s-1)gd_{35}}{\tau_o}$$

In the absence of information on the grain size distribution, it will be assumed that $d_{35} \approx d_{50}$. It yields:

$$\frac{\rho(s-1)gd_{35}}{\tau_o} \approx 4.215$$

For a sediment mixture, the Einstein formula gives:

$$\frac{q_s}{\sqrt{(s-1)gd_s^3}} = 0.85$$

Hence: $q_s = 0.0045\,\text{m}^2/\text{s}$.

Approach No. 3: bed-load calculation (equation (10.5))

The bed-load transport rate per unit width equals:

$$q_s = C_s \delta_s V_s \tag{10.5}$$

Using Nielsen's (1992) simplified model, it yields:

$$q_s = 0.65 \times 2.5(\tau_* - (\tau_*)_c)d_s 4.8V_*$$
$$= 0.65 \times 2.5 \times (0.237 - 0.05) \times 0.012 \times 4.8 \times 0.215 = 0.0038\,\text{m}^2/\text{s}$$

With the correlation of van Rijn (1984a), the saltation properties are:

$$C_s = 0.00145$$

$$\frac{\delta_s}{d_s} = 31.6$$

$$\frac{V_s}{V_*} = 7$$

and the sediment transport rate is: $q_s = 0.00083\,\text{m}^2/\text{s}$.

Summary

	Meyer-Peter	Einstein	Equation (10.5) (Nielsen)	Equation (10.5) (van Rijn)
Q_s (m³/s)	0.119	0.153	0.128	0.0281 (?)
q_s (m²/s)	0.0035	0.0045	0.0038	0.00083 (?)
Mass sediment rate (kg/s)	314	405	339	74.6 (?)

Four formulae were applied to predict the sediment transport rate by bed load. Three formulae give reasonably close results. Let us review the various formulae.

Graf (1971) commented that the Meyer-Peter formula 'should be used carefully at [...] high mass flow rate', emphasizing that most experiments with large flow rates used by Meyer-Peter *et al.* (1934) were performed with light sediment particles (i.e. lignite breeze, s = 1.25). Graf stated that one advantage of the Meyer-Peter formula is that 'it has been tested with large grains'.

The Einstein formula has been established with more varied experimental data than the Meyer-Peter formula. And the present application is within the range of validity of the data (i.e. $q_s \sqrt{(s-1)gd_s^3} = 0.85 \ll 10$).

Equation (10.5) gives reasonably good overall results using Nielsen's (1992) simplified model. In the present application, the grain size (0.012 m) is large compared to the largest grain size used by van Rijn (1984a) to validate his formulae (i.e. $d_s \leq 0.002$ m). Hence it is understandable that equation (10.5) with van Rijn's formulae is inaccurate (and not applicable).

For the present application, it might be recommended to consider the Meyer-Peter formula which was developed and tested in Europe.

Note
All bed-load formulae would predict the maximum bed-load transport rate for a stream in equilibrium. This transport capacity may or may not be equal to the actual bed load if the channel is subjected to degradation or aggradation (see Chapter 12).

10.4.2 Application No. 2

A wide stream has a depth of 0.6 m and the bed slope is 0.0008. The bed consists of a mixture of heavy particles ($\rho_s = 2980$ kg/m³) with a median particle size $d_{50} = 950$ μm.

Compute the bed-load transport rate using the formulae of Meyer-Peter and Einstein, and equation (10.5) for uniform equilibrium flow conditions.

First calculations
Assuming a wide channel, the mean shear stress and shear velocity equals, respectively:

$$\tau_o = \rho g d \sin \theta = 998.2 \times 9.80 \times 0.6 \times 0.0008 = 4.70 \text{ Pa}$$

$$V_* = \sqrt{gd \sin \theta} = 0.069 \text{ m/s}$$

The Shields parameter equals:

$$\tau_* = \frac{\tau_o}{\rho(s-1)gd_s} = \frac{4.70}{998.2 \times 1.98 \times 9.80 \times 0.00095} = 0.255$$

The particle Reynolds number is:

$$Re_* = \frac{V_* d_s}{\nu} = \frac{0.069 \times 0.00095}{1.007 \times 10^{-6}} = 65.1$$

For these flow conditions (τ_*, Re_*), the Shields diagram predicts sediment motion:

$$\tau_* = 0.255 > (\tau_*)_c \approx 0.05$$

Note that V_*/w_o is less than 0.7. In first approximation, the suspended sediment transport is negligible.

Approach No. 1: Meyer-Peter correlation

For the hydraulic flow conditions, the dimensionless parameter used for the Meyer-Peter formula is:

$$\frac{\rho(s-1)gd_s}{\tau_o} = \frac{998.2 \times 1.98 \times 9.80 \times 0.00095}{4.7} = 3.91$$

Application of the Meyer-Peter formula (for a sediment mixture) leads to:

$$\frac{q_s}{\sqrt{(s-1)gd_s^3}} = 0.76$$

Hence: $q_s = 9.82 \times 10^{-5}\,\text{m}^2/\text{s}$.

Approach No. 2 : Einstein function

For the hydraulic flow conditions, the Einstein formula is based on the d_{35} grain size. In the absence of information on the grain size distribution, it will be assumed that $d_{35} \approx d_{50}$.

For a sediment mixture, the Einstein formula gives:

$$\frac{q_s}{\sqrt{(s-1)gd_s^3}} \approx 1$$

Hence: $q_s = 1.29 \times 10^{-4}\,\text{m}^2/\text{s}$.

Approach No. 3: bed-load calculation (equation (10.5))

The bed-load transport rate per unit width equals:

$$q_s = C_s \delta_s V_s \tag{10.5}$$

Using Nielsen's (1992) simplified model, it yields:

$$q_s = 0.65 \times 2.5(\tau_* - (\tau_*)_c)d_s\, 4.8V_*$$
$$= 0.65 \times 2.5 \times (0.255 - 0.05) \times 0.00095 \times 4.8 \times 0.069 = 1.05 \times 10^{-4}\,\text{m}^2/\text{s}$$

With the correlation of van Rijn (1984a), the saltation properties are:

$$C_s = 0.019$$

$$\frac{\delta_s}{d_s} = 5.848$$

$$\frac{V_s}{V_*} = 7$$

and the sediment transport rate is: $q_s = 0.5 \times 10^{-4}\,\text{m}^2/\text{s}$.

Summary

	Meyer-Peter	Einstein	Equation (10.5) (Nielsen)	Equation (10.5) (van Rijn)
q_s (m²/s)	0.98×10^{-5}	1.2×10^{-4}	1.05×10^{-4}	0.5×10^{-4}
Mass sediment rate (kg/s/m)	0.29	0.38	0.31	0.15

All the formulae give consistent results (within the accuracy of the calculations!).

For small-size particles (i.e. $d_{50} < 2\,\text{mm}$), the formulae of van Rijn are recommended because they were validated with over 500 laboratory and field data.

Note, however, that 'the overall inaccuracy of the predicted (bed-load) transport rates may not be less than a factor 2' (van Rijn, 1984a: p. 1453).

10.4.3 Application No. 3

The North Fork Toutle river flows on the north-west slopes of Mount St. Helens (USA), which was devastated in May 1980 by a volcanic eruption. Since 1980 the river has carried a large volume of sediments.

Measurements were performed on 28 March 1989 at the Hoffstadt Creek bridge. At that location the river is 18 m wide. Hydraulic measurements indicated that the flow depth was 0.83 m, the depth-averaged velocity was 3.06 m/s and the bed slope was $\sin\theta = 0.0077$. The channel bed is a sediment mixture with a median grain size of 15 mm and $d_{84} = 55\,\text{mm}$.

Predict the sediment-load rate using the Meyer-Peter correlation, the Einstein formula and equation (10.5) using both Nielsen and van Rijn coefficients.

First calculations

Assuming a wide channel ($d = 0.83\,\text{m} \ll 18\,\text{m}$), the mean shear stress and shear velocity equals, respectively:

$$\tau_o = \rho g d \sin\theta = 998.2 \times 9.81 \times 0.83 \times 0.0077 = 62.6 \text{ Pa}$$

$$V_* = \sqrt{gd \sin\theta} = 0.25 \text{ m/s}$$

The Shields parameter equals:

$$\tau_* = \frac{\tau_o}{\rho(s-1)gd_s} = 0.258$$

assuming $s = 2.65$ (quartz particles) and using $d_s = d_{50}$. And the particle Reynolds number is:

$$Re_* = \frac{V_* d_s}{\nu} = 3725$$

For these flow conditions (τ_*, Re_*), the Shields diagram predicts sediment motion:

$$\tau_* = 0.258 > (\tau_*)_c \approx 0.05$$

Note that V_*/w_o is small ($V_*/w_o \approx 0.5$) and the flow conditions are near the initiation of suspension. In first approximation, the suspended sediment load will be neglected.

Approach No. 1: Meyer-Peter correlation

For the hydraulic flow conditions, the dimensionless parameter used for the Meyer-Peter formula is:

$$\frac{\rho(s-1)gd_s}{\tau_0} = 3.87$$

using $d_s = d_{50}$. Application of the Meyer-Peter formula (for a sediment mixture) would lead to:

$$\frac{q_s}{\sqrt{(s-1)gd_s^3}} = 0.78$$

Hence: $q_s = 0.0057\,\mathrm{m^2/s}$.

Approach No. 2: Einstein function

For the hydraulic flow conditions, the dimensionless parameter used for the Einstein formula is:

$$\frac{\rho(s-1)gd_{35}}{\tau_0}$$

In absence of information on the grain size distribution, we assume that $d_{35} \approx d_{50}$.

For a sediment mixture, the Einstein formula gives:

$$\frac{q_s}{\sqrt{(s-1)gd_s^3}} \sim 40$$

But note that the flow conditions are outside of the range of validity of the formula. That is, the Einstein formula should not be used.

Approach No. 3: bed-load calculation (equation (10.5))

The bed-load transport rate per unit width equals:

$$q_s = C_s \delta_s V_s \qquad (10.5)$$

Using Nielsen's (1992) simplified model, it yields:

$$q_s = 0.65 \times 2.5(\tau_* - (\tau_*)_c)d_s\,4.8V_* = 0.0061\,\mathrm{m^2/s}$$

With the correlation of van Rijn (1984a), the saltation properties are:

$$C_s = 0.00129$$

$$\frac{\delta_s}{d_s} = 38.97$$

$$\frac{V_s}{V_*} = 7$$

and the sediment transport rate is: $q_s = 0.0013\,\mathrm{m^2/s}$.

Summary

	Meyer-Peter	Einstein	Equation (10.5) (Nielsen)	Equation (10.5) (van Rijn)	Data
Q_s (m³/s)	0.10	N/A	0.11	0.024	
q_s (m²/s)	0.0057	N/A	0.0061	0.0023	
Mass sediment rate (kg/s)	274	N/A	290	63	205.2

Pitlick (1992) described in depth the field study performed at the Hoffstadt Creek bridge on the North Fork Toutle river. On 28 March 1989, the main observations were:

$$d = 0.83 \, \text{m}, \ V = 3.06 \, \text{m/s}, \ \sin\theta = 0.0077, \ f = 0.054, \ \tau_0 = 63 \, \text{N/m}^2, \ C_s = 0.031,$$
$$\dot{m}_s = 11.4 \, \text{kg/m/s}$$

The channel bed was formed in dunes (up to 0.16 m high).

Discussion

First, let us note that two methods of calculations are incorrect: Einstein formula and equation (10.5) using van Rijn's correlation. The flow conditions and sediment characteristics are outside of the range of applicability of Einstein's formula as $q_s/\sqrt{(s-1)gd_s^3} > 10$. The grain size (0.015 m) is larger than the largest grain size used by van Rijn (1984a) to validate his formulae (i.e. $d_s \leqslant 0.002$ m). Secondly it is worth noting that the Meyer-Peter formula and equation (10.5) using Nielsen's correlations give reasonable predictions.

This last application is an interesting case: it is well documented. The river flow is characterized by heavy bed-load transport and the bed forms are a significant feature of the channel bed.

10.5 EXERCISES

Considering a 20 m wide channel, the bed slope is 0.00075 and the observed flow depth is 3.2 m. The channel bed is sandy ($d_s = 50.008$ m). Calculate (a) mean velocity, (b) mean boundary shear stress, (c) shear velocity, (d) Shields parameter and (e) occurrence of bed-load motion. If bed-load motion occurs, calculate: (f) bed-load layer sediment concentration, (g) bed-load layer thickness, (h) average sediment velocity in bed-load layer and (i) bed-load transport rate. (Assume that the equivalent roughness height of the channel bed equals the median grain size. Use Nielsen simplified model.)

Considering a wide channel, the discharge is 20 m²/s, the observed flow depth is 4.47 m and the bed slope is 0.001. The channel bed consists of a sand mixture ($d_{50} = 0.011$ m). Calculate the bed-load transport rate using: (a) Meyer-Peter correlation, (b) Einstein function, (c) Nielsen simplified model and (d) van Rijn correlations. (Assume that the equivalent roughness height of the channel bed equals the median grain size.)

A 25 m wide channel has a bed slope of 0.0009. The bed consists of a mixture of light particles ($\rho_s = 2350 \, \text{kg/m}^3$) with a median particle size $d_{50} = 1.15$ mm. The flow rate is 7.9 m³/s. Calculate the bed-load transport rate at uniform equilibrium flow conditions using the formulae of Meyer-Peter, Einstein, Nielsen and van Rijn. (Assume that the equivalent roughness height of the channel bed equals the median grain size.)

In a section of the river Fujigawa, the bed slope is roughly 0.01. The bed consists of a mixture of gravels with median particle size $d_{50} = 18$ cm. For a flow rate of 24 m²/s, calculate the bed-load transport rate at uniform equilibrium flow conditions using the formula of Nielsen. (Assume that the equivalent roughness height of the channel bed equals the median grain size.)

   Solution: $d = 3.31$ m, $V_* = 0.55$ m/s, $\tau_* = 0.10$, $q_s = 0.042$ m²/s.

   Remark: The Fujigawa river is infamous for massive sediment loads during floods (Fig. 10.6).

Fig. 10.6 Fujigawa river, Japan on 2 November 2001 looking upstream.

Sediment transport mechanisms: 11
2. Suspended-load transport

11.1 INTRODUCTION

Sediment suspension can be described as the motion of sediment particles during which the particles are surrounded by fluid. The grains are maintained within the mass of fluid by turbulent agitation without (frequent) bed contact. Sediment suspension takes place when the flow turbulence is strong enough to balance the particle weight (i.e. $V_*/w_o > 0.2$–2, Chapter 9).

The amount of particles transported by suspension is called the *suspended load*. In this section, the mechanisms of suspended-load transport are described first. Then methods to predict the suspended-load transport rate are presented.

Notes
1. In natural streams suspension is more likely to occur with fine particles (e.g. silt and fine sands). Figure 11.1 illustrates an example of suspension material left after the flood.
2. The term *wash load* describes an inflow of fine particles in suspension which do not interact with the bed material (bed load and suspension) and remains in suspension (Fig. 12.1).

Fig. 11.1 Suspension material left after the flood in 1963 (by E.D. McKee, with permission of US Geological Survey, Ref. 79ct). Silt deposit in the flood plain of the Colorado River, Grand Canyon National Park, USA. Note the shrinkage cracks.

Mechanisms of suspended-load transport

The transport of suspended matter occurs by a combination of advective turbulent diffusion and convection. *Advective diffusion* characterizes the random motion and mixing of particles through the water depth superimposed to the longitudinal flow motion. In a stream with particles heavier than water, the sediment concentration is larger next to the bottom and turbulent diffusion induces an upward migration of the grains to region of lower concentrations. A time-averaged balance between settling and diffusive flux derives from the continuity equation for sediment matter:

$$D_s \frac{dc_s}{dy} = -w_o c_s \tag{11.1}$$

where c_s is the local sediment concentration at a distance y measured normal to the channel bed, D_s is the sediment diffusivity and w_o is the particle settling velocity.

Sediment motion by *convection* occurs when the turbulent mixing length is large compared to the sediment distribution length scale. Convective transport may be described as the entrainment of sediments by very-large scale vortices: e.g. at bed drops, in stilling basins and hydraulic jumps (Fig. 11.2).

Notes

1. The advective diffusion process is sometimes called gradient diffusion.
2. Equation (11.1) derives from Fick's law. It is valid at equilibrium: i.e. when the turbulent diffusion balances exactly the particle settling motion.
3. The sediment diffusivity D_s is also called sediment mixing coefficient or sediment diffusion coefficient.

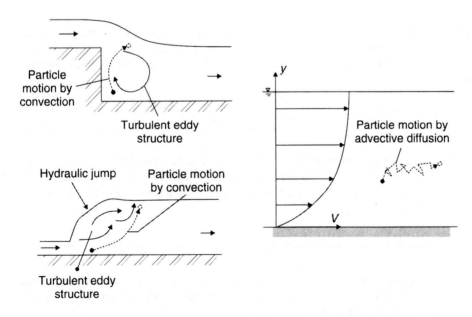

Fig. 11.2 Suspended-sediment motion by convection and diffusion processes.

11.2 ADVECTIVE DIFFUSION OF SEDIMENT SUSPENSION

11.2.1 Introduction

For a constant sediment diffusion coefficient, the solution of the continuity equation for sediment (equation (11.1)) is:

$$c_s = (C_s)_{y=y_s} \exp\left(-\frac{w_o}{D_s}(y - y_s)\right) \qquad \text{Constant diffusivity coefficient} \qquad (11.2)$$

where $(C_s)_{y=y_s}$ is the sediment concentration at a reference location $y = y_s$. Equation (11.2) is valid for uniform turbulence distribution at steady-state conditions: e.g. it can be applied to suspension in tanks with oscillating and rotating stirring devices.

In natural (flowing) streams, the turbulence is generated by boundary friction: it is stronger close to the channel bed than near the free surface. Hence, the assumption of constant sediment diffusivity (i.e. D_s = constant) is not realistic and equation (11.2) should not be applied to natural streams.

11.2.2 Sediment concentration in streams

In flowing streams the sediment diffusivity may be assumed to be nearly equal to the turbulent diffusion coefficient (i.e. the 'eddy viscosity'). The eddy viscosity is a coefficient of momentum transfer and it expresses the transfer of momentum from points where the momentum per unit volume (ρv) is high to points where it is lower. It is also called the turbulent mixing coefficient.

In open channel flows, the eddy viscosity and hence the sediment diffusivity D_s may be estimated as:

$$D_s \approx KV_*(d - y)\frac{y}{d} \qquad (11.3)$$

where d is the flow depth, V_* is the shear velocity and K is the von Karman constant (K = 0.4).

For a parabolic diffusivity law (equation (11.3)), the integration of the continuity equation for sediment (equation (11.1)) gives the distribution of sediment concentration across the flow depth:

$$c_s = (C_s)_{y=y_s}\left(\frac{(d/y) - 1}{(d/y_s) - 1}\right)^{w_o/(KV_*)} \qquad (11.4)$$

where $(C_s)_{y=y_s}$ is the reference sediment concentration at the reference location $y = y_s$.

Equation (11.4) was first developed by Rouse (1937) and it was successfully verified with laboratory and field data (e.g. Vanoni, 1946). Typical plots are presented in Fig. 11.3.

Notes
1. The dimensionless number (w_o/KV_*) is sometimes called the *Rouse number* after H. Rouse.
2. Equation (11.4) is not valid very close to the wall. It predicts an infinite bed concentration (i.e. at $y = 0$). Typically it is used for $y > y_s$.

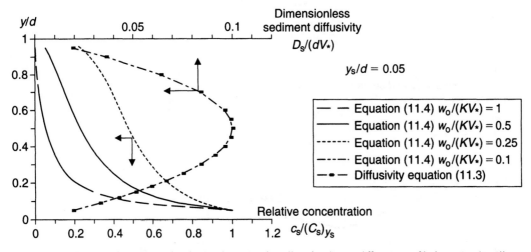

Fig. 11.3 Sediment concentration distribution (equation (11.4)) and sediment diffusivity profile (equation (11.3)).

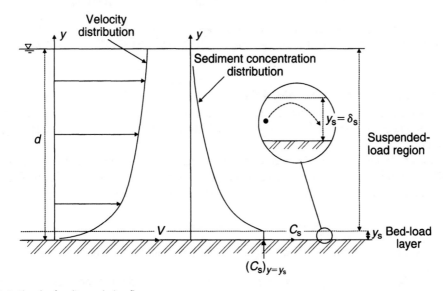

Fig. 11.4 Sketch of sediment-laden flows.

11.2.3 Discussion

Sediment motion and flow regions in sediment-laden flows

In a sediment-laden flow, the basic flow regions consist of:

- the bed-load layer next to the channel bottom (if $\tau_* > (\tau_*)_c$),
- a suspension region above (if sediment suspension occurs) (Fig. 11.4).

The type of sediment motion is a function of the sediment properties, bed slope and flow conditions. The results may be summarized in a modified Shields diagram shown in Fig. 11.5. Figure 11.5 presents the Shields parameter τ_* as a function of a dimensionless particle parameter $d_* = d_s \sqrt[3]{(s-1)g/\nu^2}$. The critical Shields parameter for initiation of sediment motion is plotted in solid line and a criterion for initiation of suspension (i.e. $V_*/w_o = 1$) is also shown.

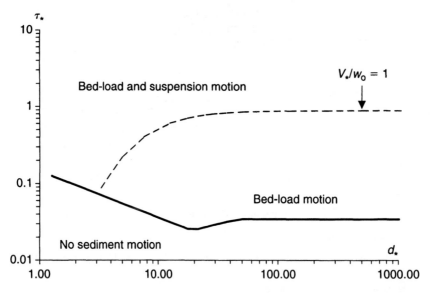

Fig. 11.5 Threshold of sediment motion (bed-load and suspension): Shields parameter $\tau_* = \tau_o/(\rho g(s-1)d_s)$ as a function of the dimensionless particle parameter $d_* = d_s \sqrt[3]{(s-1)g/\nu^2}$ (for water at 20°C and $s = 2.69$).

Reference concentration and elevation

Several researchers showed that the accurate prediction of sediment concentration distributions (i.e. equation (11.3)) depends critically upon the selection of the reference concentration $(C_s)_{y=y_s}$ and elevation y_s.

Very close to the bed (i.e. in the bed-load layer), sediment motion occurs by bed-load transport (rolling, sliding and saltation). Suspension takes place above the bed-load layer (Fig. 11.4). Hence a logical choice for y_s is the outer edge of the bed-load layer:

$$y_s = (\delta_s)_{bl} \qquad \text{Bed-load layer thickness} \qquad (11.5)$$

and the reference concentration must be taken as the bed-load layer sediment concentration:

$$(C_s)_{y=y_s} = (C_s)_{bl} \qquad \text{Average bed-load layer sediment concentration} \qquad (11.6)$$

where the subscript bl refers to the bed-load layer properties (Table 10.2) (van Rijn, 1984b).

Sediment diffusivity

Although equation (11.4) was successfully compared with numerous data, its derivations implies a parabolic distribution of the sediment diffusivity (i.e. equation (11.3)). The re-analysis of model and field data (Anderson, 1942; Coleman, 1970) shows that the sediment diffusivity is not zero next the free surface (as predicted by equation (11.3)). The distribution of sediment diffusion coefficient is better estimated by a semi-parabolic profile:

$$D_s \approx K V_*(d-y)\frac{y}{d} \qquad y/d < 0.5 \qquad (11.7a)$$

$$D_s \approx 0.1 V_* d \qquad \text{Outer region } (y/d > 0.5) \qquad (11.7b)$$

Table 11.1 Dimensionless sediment diffusivities D_s/dV_* in suspended-sediment flows and for salt diffusion in open channel flows

Reference (1)	D_s/dV_* (2)	Comments (3)
Sediment-laden flows		
Lane and Kalinske (1941)[a]	0.067[b]	
Anderson (1942)[c]	0.08–0.3[d]	Enoree river (USA). $0.9 < d < 1.52\,\text{m}$
Coleman (1970)	0.05–0.4[d]	Model data $q_w \leqslant 0.2\,\text{m}^2/\text{s}$
Matter diffusion in open channel		
Elder (1959)	0.228	Model data; lateral (transverse) diffusion of permanganate solution; $d = 0.01–0.15\,\text{m}$
Fischer *et al.* (1979)	0.067	Vertical mixing
	0.15	Transverse mixing in straight channels; average of experimental measurements ($0.09 \leqslant D_s/V_*d \leqslant 0.26$)
	0.4–0.8	Transverse mixing in curved channels and irregular sides; experimental data ($0.36 \leqslant D_s/V_*d \leqslant 3.4$)

Notes: [a] as reported in Graf (1971); [b] depth-averaged value; [c] see also Coleman (1970); [d] sediment diffusion coefficient in the outer flow region ($y/d > 0.5$).

Comments

1. The above developments (equations (11.3) and (11.7)) imply that the diffusion of sediment particles has the same properties as the diffusion of small coherent fluid structures (i.e. $D_s = \nu_T$).

   In practice, the sediment diffusivity D_s does not equal exactly the turbulent mixing coefficient (i.e. eddy viscosity). First, the diffusion of solid particles differs from the diffusion of fluid particles. Secondly, the solid sediment particles may interfere with the flow turbulence and, hence, the presence of solid particles might affect the turbulent mixing coefficient. Finally, the assumption of $D_s = \nu_T$ does not hold well for all grain sizes.

2. Table 11.1 summarizes a number of experimental data from which values of dimensionless diffusivity $D_s/(dV_*)$ were estimated. The results are compared with values of dimensionless diffusivity for salt diffusion in streams.

3. A recent study (Chanson, 1997) developed an analogy between the diffusion of suspended particles and that of entrained air bubbles in open channel supercritical flows. Interestingly, the analysis showed the same order of magnitude for the mean (dimensionless) diffusion coefficient in both types of flows (air–water and sediment–water flows).

11.3 SUSPENDED-SEDIMENT TRANSPORT RATE

11.3.1 Presentation

Considering sediment motion in an open channel (Fig. 11.4), the suspended-load transport rate equals:

$$q_s = \int_{\delta_s}^{d} c_s V \, \mathrm{d}y \tag{11.8}$$

where q_s is the volumetric suspended-load transport rate per unit width, c_s is the sediment concentration (equation (11.4)), V is the local velocity at a distance y measured normal to the channel bed, d is the flow depth and δ_s is the bed-load layer thickness.

Notes

1. Equation (11.8) implies that the longitudinal velocity component of the suspended load equals that of the water flow in which the particles are convected.
2. In practice, field measurements are often presented as:

$$q_s = \int_{\delta_s}^{d} c_s v \, dy = (C_s)_{mean} \, q = (C_s)_{mean} \, Vd$$

where q is the water discharge per unit width, V is the mean flow velocity, d is the flow depth and $(C_s)_{mean}$ is the depth-averaged sediment concentration (or mean suspended-sediment concentration).

11.3.2 Calculations

The suspended-load transport rate may be computed using equation (11.8) in which c_s is computed using equation (11.4), and δ_s and $(C_s)_{bl}$ are deduced from the bed-load layer characteristics (Chapter 10):

$$c_s = (C_s)_{y=y_s} \left(\frac{(d/y) - 1}{(d/y_s) - 1} \right)^{w_o/(KV_*)} \tag{11.4}$$

$$(C_s)_{bl} = \frac{0.117}{d_s} \left(\frac{\nu^2}{(s-1)g} \right)^{1/3} \left(\frac{\tau_*}{(\tau_*)_c} - 1 \right) \quad \text{(van Rijn, 1984a)} \tag{11.6}$$

$$(\delta_s)_{bl} = 0.3 d_s \left(d_s \left(\frac{(s-1)g}{\nu^2} \right)^{1/3} \right)^{0.7} \sqrt{\frac{\tau_*}{(\tau_*)_c} - 1} \quad \text{(van Rijn, 1984a)} \tag{11.7}$$

where $\tau_* = \tau_o/(\rho(s-1)gd_s)$, $(\tau_*)_c$ is the critical Shields parameter for bed-load initiation, d_s is the sediment size, τ_o is the bed shear stress, $s = \rho_s/\rho$ and ν is the kinematic viscosity of the fluid.

For a rough-turbulent flow (e.g. gravel-bed streams), the velocity distribution may be estimated as:

$$\frac{V}{V_*} = \frac{1}{K} \ln\left(\frac{y}{k_s} \right) + 8.5 \tag{11.9}$$

where V_* is the shear velocity, K is the von Karman constant (K = 0.4) and k_s is the equivalent roughness height.

Notes

1. The sediment concentration in the bed-load layer has an upper limit of 0.65 (for rounded grains). Equation (11.6) should be used in the form:

$$(C_s)_{bl} = \text{Min}\left[\frac{0.117}{d_s} \left(\frac{\nu^2}{(s-1)g} \right)^{1/3} \left(\frac{\tau_*}{(\tau_*)_c} - 1 \right); 0.65 \right] \tag{11.6b}$$

2. Equation (11.9) was obtained for the turbulent zone (i.e. $y/\delta < 0.1$–0.15) of a fully rough-turbulent boundary layer. In fully developed open channel flow, it is valid for $y/d < 0.1$–0.15.
3. Some researchers (e.g. Graf, 1971; Julien, 1995) proposed to extend equation (11.9) (in first approximation) to the entire flow field. Such an approximation is reasonable with heavy particles as most of the sediment flux (i.e. $C_s V$) occurs in the lowest flow region (i.e. in or close to the turbulent zone).
4. For (fully developed) turbulent open channel flows, the velocity distribution may also be expressed as:

$$\frac{V}{V_{max}} = \left(\frac{y}{d}\right)^{1/N}$$

where V_{max} is the free-surface velocity. By continuity, the free-surface velocity equals:

$$V_{max} = \frac{N+1}{N}\frac{q}{d} = \frac{N+1}{N}V_* \sqrt{\frac{8}{f}} \qquad \text{For a wide rectangular channel}$$

and q is the discharge per unit width and f is the Darcy friction factor.

For uniform equilibrium flow, Chen (1990) showed that the exponent N is related to the Darcy friction factor:

$$N = K\sqrt{\frac{8}{f}}$$

11.3.3 Application

A mixture of fine sands ($d_{50} = 0.1\,\text{mm}$) flows on a steep slope ($\sin\theta = 0.03$). The water discharge is $q = 1.5\,\text{m}^2/\text{s}$:

(a) Predict the occurrence of bed-load motion and/or suspension.
(b) If bed-load motion occurs, estimate the characteristics of the bed-load layer using the formulae of van Rijn (1984a) and calculate the bed-load transport rate.
(c) If suspension occurs, plot the sediment concentration distribution and velocity distribution on a same graph (plot v/V_* and c_s as functions of y/d).
(d) Calculate the suspended-load transport rate.
(e) Calculate the total sediment transport rate.

Note: Perform the calculations for a flat bed and assuming uniform equilibrium flow. The bed roughness height k_s will be assumed to be the median grain size.

Solution

First, the basic flow properties (mean velocity, flow depth and Darcy friction factor) must be computed. Then the shear velocity and mean bottom shear stress are deduced. Finally, the fall velocity of the sediment particle is calculated.

The mean flow velocity, flow depth and Darcy friction factor are the solution of a system of three equations (with three unknowns): the continuity equation (for water), the momentum equation for the uniform equilibrium flow and the friction factor formula. These are:

$$Q = VA \qquad\qquad\qquad [\text{C}]$$

$$V = \sqrt{\frac{8g}{f}} \sqrt{\frac{D_H}{4} \sin \theta} \qquad\qquad \text{[M]}$$

$$f = \mathscr{F}\left(\frac{VD_H}{\nu}; \frac{k_s}{D_H}\right) \qquad \text{Darcy friction factor calculation}$$

where V is the mean flow velocity, D_H is the hydraulic diameter, Q is the water discharge and A is the cross-sectional area of the flow (in plane normal to flow direction).

For a wide rectangular channel and assuming turbulent flow, the above equations become:

$$q \approx Vd \qquad \text{Wide rectangular channel}$$

$$V \approx \sqrt{\frac{8g}{f}} \sqrt{d \sin \theta} \qquad \text{Uniform equilibrium flow in wide rectangular channel}$$

$$\frac{1}{\sqrt{f}} = -2.0 \log_{10}\left(\frac{k_s}{3.71 D_H} + \frac{2.51\nu}{VD_H \sqrt{f}}\right) \qquad \text{Turbulent flow (Colebrook–White formula)}$$

Iterative calculations yield:

$$V = 6.62 \text{ m/s}$$

$$f = 0.012$$

$$d = 0.227 \text{ m}$$

The shear velocity and bed shear stress are:

$$V_* = V\sqrt{\frac{f}{8}} = 0.258 \,\text{m/s}$$

$$\tau_0 = \rho V_*^2 = 66.6 \,\text{Pa}$$

For a 0.1 mm sediment particle, the settling velocity is:

$$w_0 = 0.0081 \,\text{m/s}$$

(a) Let us predict the occurrence of bed-load motion using the Shields diagram. The Shields parameter and particle Reynolds number are:

$$\tau_* = \frac{\tau_0}{\rho(s - 1)gd_s} = 41.15$$

$$Re_* = \frac{V_* d_s}{\nu} = 25.66$$

The Shields diagram indicates $(\tau_*)_c \approx 0.08$. Hence bed-load motion occurs (because $\tau_* > (\tau_*)_c$).

Now let us consider the occurrence of suspension. The ratio V_*/w_0 equals 32. Hence suspension takes place.

(b) The characteristics of the bed-load layer are computed using the formulae of van Rijn (1984a):

$$(C_s)_{bl} = Min\left(\frac{0.117}{d_s}\left(\frac{\nu^2}{(s-1)g} \right)^{1/3}\left(\frac{\tau_*}{(\tau_*)_c} - 1 \right); 0.65 \right) \qquad (11.6b)$$

$$(\delta_s)_{bl} = 0.3d_s\left(d_s\left(\frac{(s-1)g}{\nu^2} \right)^{1/3} \right)^{0.7}\sqrt{\frac{\tau_*}{(\tau_*)_c} - 1} \qquad (11.7)$$

Calculations yield:

$$(C_s)_{bl} = 0.65 \qquad (\delta_s)_{bl} = 0.0013\,m$$

The bed-load transport rate equals:

$$(q_s)_{bl} = C_s\delta_s V_s \qquad (11.5)$$

Using the formulae of Nielsen (1992), the transport rate per unit width equals:

$$(q_s)_{bl} = 0.0083\,m^2/s$$

(c) The flow Reynolds number VD_H/ν equals 1.5×10^3 and the shear Reynolds number of the flow $k_s V_*/\nu$ equals 25.7. The flow is turbulent in the transition (between smooth and fully rough). Hence, the above calculation of the Darcy friction factor is correct. But the entire velocity distribution cannot be estimated by equation (11.9) (valid for fully rough-turbulent flow only). Instead, the velocity distribution may be described by:

$$\frac{V}{V_{max}} = \left(\frac{y}{d} \right)^{1/N}$$

where V_{max} is the free-surface velocity:

$$V_{max} = \frac{N+1}{N}V_*\sqrt{\frac{8}{f}} = 7.3\ m/s \qquad \text{For a wide rectangular channel}$$

For uniform equilibrium flow, Chen (1990) showed that the exponent N is related to the Darcy friction factor:

$$N = K\sqrt{\frac{8}{f}} = 10.2$$

The volumetric sediment concentration distribution can be estimated using the Rouse profile:

$$c_s = (C_s)_{bl}\left(\frac{(d/y) - 1}{(d/(\delta_s)_{bl}) - 1} \right)^{w_0/(KV_*)} \qquad (11.4)$$

where K is the von Karman constant (K = 0.4).

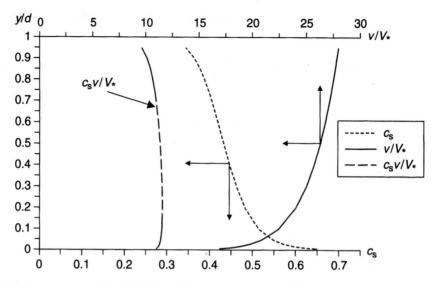

Fig. 11.6 Sediment concentration distribution, dimensionless velocity profile v/V_* and dimensionless suspended-sediment flux $c_s v/V_*$.

Figure 11.6 presents the distributions of sediment concentration c_s, of dimensionless velocity v/V_* and of the dimensionless suspended-sediment flux $c_s v/V_*$.
(d) The suspended-load transport rate is calculated as:

$$q_s = \int_{y_s}^{d} c_s v \, dy \qquad (11.8)$$

Numerical integration gives:

$$q_s = 0.62 \text{ m}^2\text{/s} \qquad \text{Suspension transport rate}$$

(e) The total sediment transport rate equals:

$$q_s = 0.625 \text{ m}^2\text{/s}$$

Note that most sediment transport occur by suspension. The contribution of bed-load transport is negligible.

11.4 HYPERCONCENTRATED SUSPENSION FLOWS

11.4.1 Presentation

The last example showed a sediment-laden flow with a large amount of suspension. The depth-averaged sediment concentration was about 30%. With such an amount of sediment:

- Are the (sediment–water) fluid properties the same as those of clear water?
- What are the interactions between suspended sediment and turbulence?
- Can the flow properties be computed as for those of clear-water flows (e.g. Chézy equation and Colebrook–White formula)?

Applications

Wan and Wang (1994) described numerous examples of hyperconcentrated flows in the Yellow River basin (China).

During one hyperconcentrated flood event, the Yellow River bed was eroded by 8.8 m in 72 h (Longmen, Middle reach of Yellow River, August 1970, $Q = 13\,800\,\mathrm{m^3/s}$). During another flood event, a wash load (sediment concentration: $964\,\mathrm{kg/m^3}$) was transported over 50 km in narrow and deep canals.

'Clogging' of open channels might also result during the recession of a hyperconcentrated flood. A whole river might stop flowing during sometimes (e.g. Heihe river, July 1967).

11.4.2 Fluid properties

In a homogeneous suspension, the basic fluid properties (e.g. density and viscosity) differ from that of clear water. The density of a particle–water mixture equals:

$$\rho_{\mathrm{mixt}} = \rho(1 + C_{\mathrm{s}}(\mathbf{s} - 1)) \tag{11.10}$$

where ρ is the water density, C_{s} is the volumetric sediment concentration and $\mathbf{s} = \rho_{\mathrm{s}}/\rho$. The dynamic viscosity of the mixture equals:

$$\mu_{\mathrm{mixt}} = \mu(1 + 2.5C_{\mathrm{s}}) \tag{11.11}$$

where μ is the water dynamic viscosity (Einstein, 1906, 1911).

Note

More sophisticated formulations of the mixture viscosity were proposed. For example:

$$\mu_{\mathrm{mixt}} = \mu(1 + 2.5C_{\mathrm{s}} + 6.25C_{\mathrm{s}}^{2}) \qquad \text{(Graf, 1971)}$$

$$\mu_{\mathrm{mixt}} = \mu(1 - 1.4C_{\mathrm{s}})^{-2.5} \qquad \text{(Wan and Wang, 1994)}$$

Rheology

For a Newtonian fluid (e.g. clear water), the shear stress acting in a direction (e.g. x-direction) is proportional to the velocity gradient in the normal direction (e.g. y-direction):

$$\tau = \mu \frac{\mathrm{d}v}{\mathrm{d}y} \tag{11.12}$$

where μ is the dynamic viscosity of the fluid (see Chapter 1).

Hyperconcentrated suspensions do not usually satisfy equation (11.12). They are called non-Newtonian fluids. For homogenous suspension of fine particles, the constitutive law of the fluid is:

$$\tau = \tau_1 + \mu_{\mathrm{mixt}} \frac{\mathrm{d}v}{\mathrm{d}y} \tag{11.13}$$

where μ_{mixt} is the dynamic viscosity of the sediment–water mixture and τ_1 is a constant, called the *yield stress*.

Notes

1. Equation (11.12) is also called Newton's law of viscosity.
2. Fluids satisfying equation (11.13) are called *Bingham plastic fluids*.
3. The Bingham plastic model is well suited to sediment-laden flows with large concentrations of fine particles (Wan and Wang, 1994; Julien, 1995).
4. The yield stress of hyperconcentrated mixture is generally formulated as a function of sediment concentration. Julien (1995) proposed:

$$\tau_1 \approx 0.1 \exp(3(C_s - 0.05)) \qquad \text{For sands}$$

$$\tau_1 \approx 0.1 \exp(23(C_s - 0.05)) \qquad \text{For silts (70\%) and clays (30\%)}$$

$$\tau_1 \approx 0.1 \exp(23(C_s - 0.05)) \qquad \text{For silts (70\%) and clays (30\%)}$$

11.4.3 Discussion

The characteristics of hyperconcentrated flows are not yet fully understood. It is clear that the fluid and flow properties of hyperconcentration cannot be predicted as those of clear-water flows. Furthermore, the interactions between bed-load and suspended-load motions are not known, and there are substantial differences between fine-particle hyperconcentrated flows and large-particle debris flows.

Currently, the topic is under active research investigations. Wan and Wang (1994) presented a solid review of the Chinese expertise, in particular with fine-particle hyperconcentrations. Wang (2002) discussed examples of flow instabilities: e.g. river clogging, 'paving way' process and lava flows. Takahashi (1991) reviewed the Japanese knowledge in large-particle flows and mountain debris flows. Meunier (1995) described a series of field investigations in France, including studies of the catchment hydrology, soil erosion, hydraulics and sediment transport. Julien (1995) presented a brief summary of American experience.

Practically, hyperconcentrated flows must be closely monitored. They could lead to severe soil erosion, disastrous reservoir siltation and channel clogging with associated upstream flooding. For example, the reservoir shown in Plate 21 became fully silted in less than 2 years, despite the presence of upstream debris dams in the catchment. Figure 11.7 presents a small-debris flow fan in a large-debris flow channel (Inset).

11.5 EXERCISES

Sediment suspension is observed in a hydraulic jump stilling basin. Does sediment motion occur by advective diffusion or convection?

Solution: convection.

A natural stream has a bed slope of 0.008 and the observed flow depth is 0.87 m. The bed consists of very-fine sands ($d_{50} = 0.18$ mm). Experimental observations indicate that the bed material is entrained in suspension and that the sediment concentration at a distance: $y_s = 0.1$ m from the bed is 15%. (a) Calculate the Rouse number $w_o/(KV_*)$. (b) Plot the sediment concentration distribution between $y_s = 0.1$ mm and the free surface. (Assume that the sediment concentration distribution follows the Rouse profile.)

Fig. 11.7 River fan in a mini-debris flow in the Osawa-gawa debris flood plain on 1 November 2001. The Osawa channel is located downstream of a major fault on the Western slope of Mt Fuji (in background of photograph).

Fine sands ($d_{50} = 0.2\,\text{mm}$) in suspension flow down a 3 m wide steep mountain river ($\sin\theta = 0.032$). The normal depth equals 0.23 m. Calculate (a) water discharge, (b) mean boundary shear stress, (c) shear velocity, (d) settling velocity and (e) Shields parameter. (f) Predict the occurrence of bed-load motion and/or suspension. If sediment motion occurs, calculate (g) bed-load layer sediment concentration, (h) bed-load layer thickness, (i) average sediment velocity (in bed-load layer), (j) bed-load transport rate, (k) suspension transport rate and (l) total sediment transport rate. (Assume that the sediment concentration distribution follows the Rouse profile.)

Sediment transport capacity and total sediment transport

12

12.1 INTRODUCTION

In this chapter we review the calculations of the total sediment transport rate. They are based upon previous results (Chapters 10 and 11) obtained for flat bed, and they can be used to predict the maximum sediment transport capacity. The effects of sediment erosion, accretion and bed-form motion are then discussed. Later the effects of bed forms on flow resistance are presented.

12.2 TOTAL SEDIMENT TRANSPORT RATE (SEDIMENT TRANSPORT CAPACITY)

12.2.1 Presentation

The total sediment discharge is the total volume of sediment particles in motion per unit time. It includes the sediment transport by bed-load motion and by suspension as well as the wash load (Fig. 12.1).

Numerous formulae were proposed to predict the bed-load transport (Chapter 10), the suspension discharge (Chapter 11) and the total transport rate (e.g. Table 12.1). Despite significant progresses during the past decades, none of these can accurately predict the 'real' sediment motion in natural streams.

Most importantly, let us remember that existing formulae (e.g. Chapters 10 and 11, Table 12.1) predict only the *sediment transport capacity* of a known bed sediment mixture: i.e. the maximum sediment transport rate. They neither take into account the sediment inflow nor erosion and accretion.

Discussion

Horn *et al.* (1999) performed field measurements of sediment transport rates during flood events in Queensland rivers (Australia). They compared their data with classical correlations and found that none of them were suitable to the investigated streams, either overpredicting or underpredicting the sediment transport rates by nearly one order of magnitude either way. The findings emphasize the limitations of these formulae.

12.2.2 Calculation of the sediment transport capacity

The sediment transport capacity (i.e. maximum total sediment transport rate) equals the sum of the bed-load transport rate and suspension transport rate:

$$q_s = (q_s)_{bl} + (q_s)_s \qquad (12.1)$$

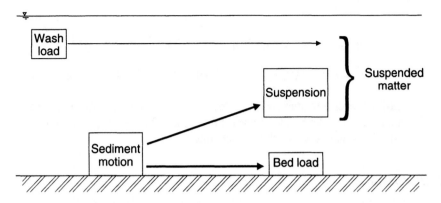

Fig. 12.1 Sediment transport classification.

Table 12.1 Empirical and semi-empirical correlations of total sediment transport

Reference (1)	Formulation (2)	Range (3)	Remarks (4)
Einstein (1950)	$q_s = (q_s)_{bl}\left[1 + I_1 \ln\left(\dfrac{30d}{d_s}\right) + I_2\right]$ where I_1 and I_2 are two integrals deduced from design charts	Small suspension transport rate	
Engelund and Hansen (1967)	$q_s = 0.4f\dfrac{\tau_o}{\rho}\sqrt{\dfrac{d_s}{(s-1)g}}$	$0.19 < d_s < 0.93\,\text{mm}$	Based on laboratory data by Guy *et al.* (1966) ($B = 2.44\,\text{m}$). Validated for $d_{50} > 0.15\,\text{mm}$ and $\sqrt{d_{90}/d_{10}} < 1.6$
Graf (1971)	$\dfrac{q_s}{q} = 10.39\dfrac{VD_H/4}{\sqrt{(s-1)gd_{50}^3}}\left(\dfrac{(s-1)d_{50}}{\sin\theta D_H/4}\right)^{-2.52}$	$0.1 < \dfrac{(s-1)d_{50}}{\sin\theta(D_H/4)} < 15$	Based on experimental data in open channel and pipe flows
van Rijn (1984c)	$\dfrac{q_s}{q} = 0.005\left(\dfrac{V - (V)_c}{\sqrt{(s-1)gd_{50}}}\right)^{2.4}\left(\dfrac{d_{50}}{d}\right)^{1.2}$ $+ 0.012\left(\dfrac{V - (V)_c}{\sqrt{(s-1)gd_{50}}}\right)^{2.4}\left(\dfrac{d_{50}}{d}\right)$ $\times\left(d_{50}\sqrt[3]{\dfrac{(s-1)g}{v^2}}\right)^{-0.6}$ $(V)_c = 0.19d_{50}^{0.1}\log_{10}\left(\dfrac{D_H}{d_{90}}\right)$ for $0.1 < d_{50} < 0.5\,\text{mm}$ $(V)_c = 8.5d_{50}^{0.6}\log_{10}\left(\dfrac{D_H}{d_{90}}\right)$ for $0.5 < d_{50} < 2\,\text{mm}$	$0.1 < d_s < 2\,\text{mm}$ $d_{84}/d_{16} = 4$ $1 < d < 20\,\text{m}$ $0.5 < V < 2.5\,\text{m/s}$	Regression analysis of calculations (at 15°C)

Notes: B: channel width; D_H: hydraulic diameter; q_s: total volumetric sediment discharge per unit width.

where:

$$(q_s)_{bl} = (C_s)_{bl}(\delta_s)_{bl}(V_s)_{bl} \qquad \text{Bed-load transport rate} \qquad (12.2)$$

$$(q_s)_s = \int_{\delta_s}^{d} C_s V \, dy = (C_s)_{mean} q \qquad \text{Suspended-sediment transport rate} \qquad (12.3)$$

and δ_s is the bed-load layer thickness, d is the flow depth, q is the water discharge per unit width and the subscript bl refers to the bed-load layer. Equations (12.2) and (12.3) have been detailed in Chapters 10 and 11, respectively (for flat-bed channels).

12.2.3 Discussion

Equations (12.1)–(12.3) give an estimate of the (maximum) sediment transport capacity of the flow for a given channel configuration and for a flat movable bed. 'Reasonable' predictions might be obtained for straight prismatic channels with relatively wide cross-section formed with uniform bed material, but not in natural streams.

Limitations of the calculations include:

* non-uniformity of the flow and presence of secondary currents,
* bends, channel irregularities, bank shape, formation of bars and presence of bed forms (e.g. Figs 7.2 and 12.2),
* change of flow regime (e.g. change in bed slope),
* transition regime (e.g. between dunes and flat bed, Fig. 12.5).

12.3 EROSION, ACCRETION AND SEDIMENT BED MOTION

12.3.1 Presentation

Erosion or *accretion* of the channel bed is not only a function of the sediment transport capacity (equation (12.1)) but also of the inflow conditions (i.e. discharge and sediment inflow). Let us consider some examples.

(1) Change of bed slope
When the bed slope decreases, the mean velocity in the downstream reach is lower and hence the lower reach has a smaller sediment transport capacity than the upstream reach for a given discharge and channel cross-sectional shape. As a result, the sediment inflow is larger than the sediment transport capacity and accretion usually takes place in the downstream reach.

(2) Increase of channel width
When the channel width increases, lower flow velocities take place in the downstream section: the sediment transport capacity is smaller in the downstream reach and accretion may take place. This geometry is well suited to the design of settling basins: e.g. at the intake of an irrigation channel, a settling basin will trap sediment materials to prevent siltation of the irrigation system.

(3) Change in flood flow
A change in river discharge (e.g. receding flood) affects the flow properties: e.g. flow depth, mean velocity and boundary shear stress. Often erosion is observed during the rising stage of a flood and accretion occurs during the falling stage (e.g. Fig. 12.2). As a result, massive sediment movement may be observed: e.g. formation of sand bars in rivers, foundation scouring (houses and bridge piers) and shift of the main river course.

Fig. 12.2 Sediment deposits after floods: (a) Fujigawa river, Japan, on 2 November 2001. Looking upstream at the sediment material left after the last floods. (b) Cattle Creek at Gargett bridge on August 1977. Large bank of gravel left after the flood (courtesy of Mrs J. Hacker). Major tributary of Pioneer River, Inland of Mackay QLD, Australia.

Erosion and accretion must be predicted using the continuity equation for the bed sediment material.

Note

Accretion is the increase of channel bed elevation resulting from the accumulation of sediment deposits. It is also called deposition and aggradation.

12.3.2 Continuity principle for bed material

By continuity the longitudinal change in sediment transport rate, $\partial q_s/\partial s$, must be equal to the change in bed elevation, taking into account the porosity of the bed material. Considering a small control

Fig. 12.2 (c) Flinders River on 11 June 1984. Sediment deposit on bank after the flood (courtesy of Mrs J. Hacker). At Molesworth QLD, Australia. River subject to very larger floods during tropical cyclones. (d) Mirani Road Bridge looking upstream on August 1977 (courtesy of Mrs J. Hacker). Debris of old bridge piers in the foreground.

volume (Fig. 12.3a), the continuity equation for sediment states that the change in sediment transport rate Δq_s during a time Δt equals:

$$\Delta q_s \Delta t = -(1 - Po)\Delta z_o \cos \theta \Delta s \qquad (12.4a)$$

where Po is the bed porosity, Δz_o is the change in bed elevation during the time Δt, Δs is the length of the control volume in the flow direction and θ is the longitudinal bed slope.

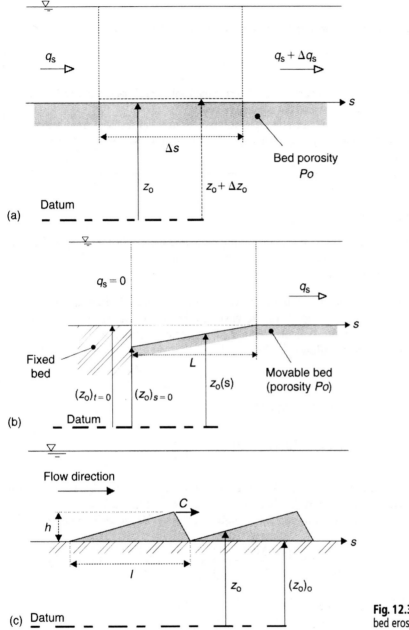

Fig. 12.3 Sediment transport and bed erosion/accretion.

Equation (12.4a) may be re-arranged as follows:

$$\frac{\Delta q_s}{\Delta s} = -(1 - Po)\frac{\Delta z_o}{\Delta t}\cos\theta \qquad (12.4b)$$

An increase in sediment transport rate (i.e. $\Delta q_s/\Delta s > 0$) implies a drop in bed elevation and hence erosion. Conversely a decrease in sediment transport rate (i.e. $\Delta q_s/\Delta s < 0$) implies accretion.

Notes
1. The continuity equation for bed material may be rewritten in a differential form:

$$\frac{\partial q_s}{\partial s} = -(1 - Po)\cos\theta\frac{\partial z_o}{\partial t} \tag{12.4c}$$

assuming no lateral sediment inflow and where s is the longitudinal co-ordinate in the flow direction.
2. The term $(1 - Po)$ is sometimes called the *packing factor*.
3. For the calculation of erosion and accretion, the volumes of material deduced from sediment transport rates must be computed taking into account the porosity (i.e. the voids) of the sediment bed (i.e. equation (12.4)).

12.3.3 Applications

Practical applications of the continuity equation for movable bed materials are presented below: a movable bed downstream of a fixed bed and the migration of dunes above a fixed bed (Fig. 12.3), the erosion rate of a stream reach.

(A) Considering a movable bed downstream of a fixed bed, the channel is horizontal and the bed is initially flat (i.e. $t = 0$). Let us compute the downstream sediment discharge q_s assuming a movable bed erosion such that the bed elevation $z_o(s, t)$ varies linearly with time and that $q_s(s = 0, t) = 0$ (i.e. no sediment inflow).

Solution
The elevation of the movable bed ($s > 0$) equals:

$$z_o(s, t) = (z_o)_{t=0} - \Delta(t)\left(1 - \frac{s}{L}\right)t$$

where t is the time, s is the longitudinal co-ordinate, $\Delta(t) = (z_o)_{t=0} - (z_o)_{s=0}$, $(z_o)_{t=0}$ is the initial bed elevation and $(z_o)_{s=0}$ is the movable bed elevation at $s = 0$ (Fig. 12.3b).

The integration of the continuity equation for movable bed material (equation (12.4)) yields:

$$q_s(s, t) = \int_0^s \frac{\partial q_s}{\partial s} ds = \int_0^s (1 - Po)\frac{\Delta(t)}{t}\left(1 - \frac{s}{L}\right)ds = \frac{(1 - Po)\Delta(t)}{t}s\left(1 - \frac{s}{2L}\right)$$

At $s = L$, the sediment transport rate equals:

$$q_s(L, t) = \frac{(1 - Po)\Delta(t)L}{t}$$

Note that, in this particular case, the calculations are independent of the flow conditions because the variations of the bed elevation $z_o(s, t)$ are defined.

(B) Considering the migration of two-dimensional bed forms with constant shape (Fig. 12.3c), let us compute the local sediment transport rate and the mean sediment transport rate assuming no sediment suspension. We shall assume that the bed forms migrate over a horizontal fixed bed.

Solution

If the bed forms move at a constant velocity C without any change of shape, the channel bed elevation may be described as:

$$z_o = (z_o)_o + F(s - Ct)$$

where s is the longitudinal co-ordinate, t is the time, $(z_o)_o$ is the fixed-bed elevation and F is a function describing the bed-form shape.

The integration of the continuity equation for sediment material (equation (12.4)) yields:

$$q_s(s, t) = (1 - Po) \, C \, F(s - Ct)$$

assuming no sediment suspension.

Discussion

For dune bed forms (Fig. 12.3c, $C > 0$), the above result implies that $q_s(s, t)$ is zero at the troughs and it is maximum at the dune crest. This indicates that the bed shear stress at the surface of the dunes (i.e. effective shear stress) varies from a negligible value at the trough (i.e. less then $(\tau_o)_c$) to a maximum value at the crest. For anti-dunes (Fig. 12.4) (Plate 22), the same equation may apply but the bed-form velocity C is negative. Hence $q_s(s, t)$ is maximum at the wave trough and minimum at the wave crest.

Overall, the mean sediment transport rate (i.e. bed-load transport) of two-dimensional triangular dune bed forms (Fig. 12.3c) equals:

$$q_s = \frac{1}{2}(1 - Po) \, C \, h$$

Fig. 12.4 Photograph of anti-dune flows: active anti-dunes in a rip channel leading into breaking waves, Terasawa beach (Japan) on 13 October 2001.

Note that this result has a similar form as equation (12.2). In alluvial streams, however, the dune pattern is often very irregular, three dimensional and continuously changing.

Tsubaki et al. (1953) reported experimental measurements of dune migration in prototype channels. During one run, the flow conditions were: $d = 0.467$ m, $V = 0.765$ m/s, coarse sand ($d_{50} = 1.26$ mm). They observed that $h = 0.0826$ m and $C = 1.01 \times 10^{-3}$ m/s, and measured the sediment transport rate $q_s = 5.22 \times 10^{-5}$ m^2/s. For the experiment, the above calculations yield:

$$\frac{1}{2}(1 - Po)Ch \approx 0.5(1 - 0.4) \times 1.01 \times 10^{-3} \times 0.0826 = 2.5 \times 10^{-5} \text{ m}^2/\text{s}$$

That is, the calculation gives the order of magnitude of the sediment transport rate, although the developments assumed triangular bed forms while Tsubaki et al. (1953) observed non-triangular bed forms.

(C) A 1 km long reach of a river has a longitudinal slope $\sin\theta = 0.0014$. The water discharge is 8.5 m^2/s and the mean sediment concentration of the inflow is 0.25%. The sediment transport capacity of the reach is 0.13 m^2/s. Deduce the rate of erosion (or accretion) of the channel bed for the 1 km long reach.

Solution
Let us consider the 1 km long section of a river with a bed slope $\theta = 0.08°$. The sediment inflow is:

$$(q_s)_o = [(C_s)_{mean}]_o\, q = 0.0025 \times 8.5 = 0.021 \text{ m}^2/\text{s}$$

The maximum sediment transport rate of the reach (i.e. the sediment transport capacity) is larger than the sediment inflow:

$$q_s = 0.13 \text{ m}^2/\text{s} > (q_s)_o = 0.021 \text{ m}^2/\text{s}$$

Hence, bed erosion will take place along the reach. The continuity equation for the reach is:

$$\frac{\Delta q_s}{\Delta s} = \frac{q_s - (q_s)_o}{L} = -(1 - Po)\frac{\Delta z_o}{\Delta t}\cos\theta = -(1 - Po)\frac{\partial z_o}{\partial t}\cos\theta$$

assuming an uniform erosion rate along the river reach and no lateral sediment inflow, and where L is the reach length ($L = 1000$ m). It yields:

$$\frac{\partial z_o}{\partial t} = -\frac{1}{(1 - 0.4) \times 0.9999}\frac{0.13 - 0.021}{1000} = -0.00018 \text{ m/s}$$

assuming a 40% porosity of the movable bed.

In summary, the channel bed is eroded and the bed elevation drops at a rate of 0.65 m per h (15.7 m per day!). The erosion rate will remain constant until the modifications of the channel bed profile change the sediment transport capacity of the reach.

12.4 SEDIMENT TRANSPORT IN ALLUVIAL CHANNELS

12.4.1 Introduction

The above developments are based upon results (Chapters 10 and 11) obtained with flat bed. In practice, bed forms are observed (Chapter 7, Figs 7.1 and 7.2).

Movable hydraulics is characterized by a dual interaction between bed forms and flow resistance. The type of bed form and the sediment transport rate are both related to the discharge which is, in turn, a function of the bed-form resistance.

12.4.2 Influence of bed forms on flow resistance

In an alluvial channel, the boundary friction is related to the skin friction (or grain-related friction) and to the form losses caused by the bed forms. Figure 12.5 presents a typical relationship between the mean boundary shear stress and the flow velocity. It indicates that the effect of bed forms is particularly substantial with ripples and dunes.

In alluvial streams the mean bed shear stress τ_0 may be divided into:

$$\tau_0 = \tau_0' + \tau_0'' \tag{12.5}$$

where τ_0' is the skin friction shear stress and τ_0'' is the form-related shear stress.

Skin friction shear stress
The *skin friction shear stress* equals:

$$\tau_0' = \frac{f}{8}\rho V^2 \tag{12.6}$$

where ρ is the fluid density, V is the mean flow velocity and f is the Darcy–Weisbach friction factor. For turbulent flows, the friction factor may be calculated using the Colebrook–White formula:

$$\frac{1}{\sqrt{f}} = -2.0 \log_{10}\left(\frac{k_s}{3.71D_H} + \frac{2.51}{Re\sqrt{f}}\right) \qquad \text{Turbulent flow} \tag{12.7}$$

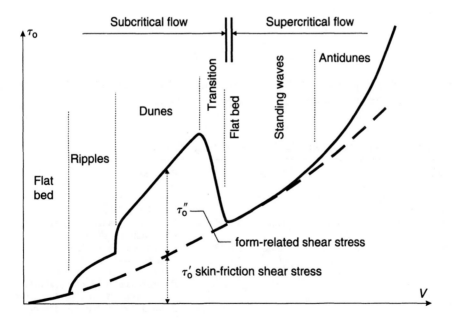

Fig. 12.5 Bed shear stress as a function of the mean flow velocity.

where D_H is the hydraulic diameter, $Re = VD_H/\nu$, ν is the fluid kinematic viscosity and k_s is the equivalent roughness height, which is related to the sediment size (Table 12.2).

The skin friction shear stress τ_0' is hence a function of the grain size d_s, fluid properties (ρ, ν) and flow characteristics (V, D_H).

Notes
1. τ_0' is also called the *effective bed shear stress*.
2. It must be emphasized that there is no definite correlation between the grain size and the equivalent roughness height (e.g. Table 12.2). Usually k_s is larger than the median grain size and a first approximation might be: $k_s = 2d_{90}$.
3. In practice, the shear velocity (and hence the friction factor) might be deduced from velocity distribution measurements (on flat bed). Another method consists in deducing the friction factor from the energy slope (i.e. from the slope of the total head line).

Bed-form shear stress

The *bed-form shear stress* τ_0'' is related to the fluid pressure distribution on the bed form and to the form loss. Fundamental experiments in irregular open channels (Kazemipour and Apelt, 1983; Fig. 12.6) showed that the form losses could account for up to 92% of the total loss.

Next to the surface of a bed form, the fluid is accelerating immediately upstream of the bed-form crest and decelerating downstream of the crest, associated sometimes with separation and recirculation (Fig. 12.7). The form loss may be crudely analysed as a sudden expansion downstream of the bed-form crest. For a two-dimensional dune element, it yields:

$$\tau_0'' = \frac{1}{2}\rho V^2 \frac{h^2}{ld} \tag{12.8}$$

where h and l are, respectively, the height and length of the dune (Fig. 12.7). Experimental observations of dune dimensions are summarized in Table 12.3.

Table 12.2 Equivalent roughness height k_s of sediment materials (flat-bed channel)

Reference (1)	k_s (2)	Range (3)	Remarks (4)
Engelund (1966)	$2.5d_{65}$		Experimental data
Kamphuis (1974)	1.5–$2.5d_{90}$	$d/d_{90} > 10$, $B/d > 3$	Laboratory data ($B = 0.38$ m); $d_{84}/d_{16} \sim 1.73$
	$\approx 2d_{90}$		Large values of d/d_{90}
Maynord (1991)	$2d_{90}$	$d/d_{90} > 2$, $B/d \geqslant 5$, $Re_* \geqslant 100$	Re-analysis of laboratory data: $0.0061 < d_{50} < 0.076$ m $0.007 < d_{90} < 0.071$ m
van Rijn (1984c)	$3d_{90}$		Re-analysis of flume and field data
Sumer *et al.* (1996)	$1.5d_{50}$–$21.1d_{50}$		Experimental data in sheet flow ($B = 0.3$ m); $0.00013 < d_{50} < 0.003$ m; $1 \leqslant d_{84}/d_{16} \leqslant 2.3$

Notes: B: channel width; k_s: equivalent roughness height (grain friction related).

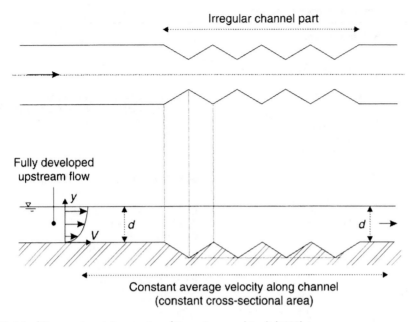

Fig. 12.6 Sketch of the experimental apparatus of Kazemipour and Apelt (1983).

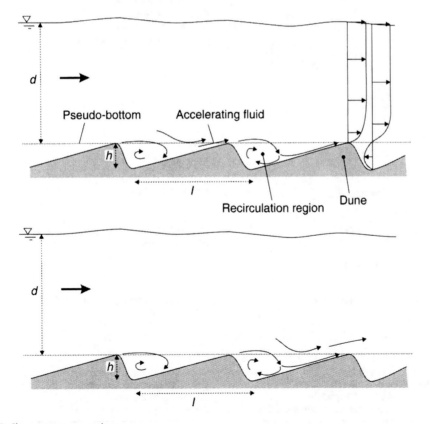

Fig. 12.7 Flow pattern over a dune.

Table 12.3 Characteristic dimensions of bed forms

Reference (1)	Bed-form dimensions (2)	Comments (3)
Dunes		
Tsubaki *et al.* (1953)	$3.4 \leqslant \dfrac{l}{d} \leqslant 10.5$ $0.083 \leqslant \dfrac{h}{d} \leqslant 0.23$	Irrigation canals: $0.8 \leqslant B \leqslant 3.8\,\text{m}$ $1.03 \leqslant d_{50} \leqslant 2.28\,\text{mm}$, $0.108 \leqslant d \leqslant 0.467\,\text{m}$, $0.28 \leqslant Fr \leqslant 0.53$, $8 \times 10^{-4} \leqslant \sin\theta \leqslant 0.0019$
Laursen (1958)	$\dfrac{l}{d} = -7.1$	Model data: $d_{50} = 0.1\,\text{mm}$, $0.076 \leqslant d \leqslant 0.305\,\text{m}$, $0.28 \leqslant Fr \leqslant 0.47$, $4 \times 10^{-4} \leqslant \sin\theta \leqslant 0.0018$
Yalin (1964)	$\dfrac{h}{d} = \dfrac{1}{6}\left(1 - \dfrac{(\tau_o)_c}{\tau_o}\right)$ $\dfrac{l}{d} \approx 5$	Based on model and field data: $0.137 \leqslant d_{50} \leqslant 2.45\,\text{mm}$ $0.013 \leqslant d \leqslant 28.2\,\text{m}$ $1 \times 10^{-5} \leqslant \sin\theta \leqslant 0.014$
Engelund and Hansen (1967)	$\dfrac{l}{h} = \dfrac{1.88}{f}$	Based on laboratory data by Guy *et al.* (1966)
van Rijn (1984c)	$\dfrac{h}{d} = 0.11\left(\dfrac{d_{50}}{d}\right)^{0.3}$ $\times \left[1 - \exp\left(-0.5\left(\dfrac{\tau_o}{(\tau_o)_c} - 1\right)\right)\right]$ $\times \left[25 - \left(\dfrac{\tau_o}{(\tau_o)_c} - 1\right)\right]$ $\dfrac{l}{d} = 7.33$	Based on model and field data: $0.19 \leqslant d_{50} \leqslant 3.6\,\text{mm}$ $0.11 \leqslant d \leqslant 16\,\text{m}$ $0.34 \leqslant V \leqslant 1.55\,\text{m/s}$ *Note:* τ_o and $(\tau_o)_c$ are effective bed shear stresses
Standing waves		
Tison (1949)	$\dfrac{h}{d} = 0.2$ $\dfrac{h}{l} \approx 0.09\text{–}0.14$	Laboratory data: $d_{50} = 0.25\,\text{mm}$ $0.035 \leqslant d \leqslant 0.05\,\text{m}$ $0.42 \leqslant Fr \leqslant 0.68$
Kennedy (1963)	$\dfrac{l}{d} = 2\pi Fr^2$ $\dfrac{h}{l} \approx 0.014$	Ideal fluid flow calculations validated with laboratory data
Chanson (2000)	$\dfrac{h}{d} = 1.31(1 - Fr)$ $\dfrac{l}{d} \approx 11.1 Fr^{3.5}$ (i.e. $3.4 < \dfrac{l}{d} < 8.3$)	Based upon undular jump free-surface data: $0.7 \leqslant Fr \leqslant 0.9$
Anti-dunes Alexander and Fielding (1997)	$l = 8\text{–}19\,\text{m}$ $h = 0.25\text{–}1\,\text{m}$ $h/l \leqslant 0.13$ typically (observed maximum h/l: 0.28)	Gravel anti-dunes in Burdekin river (Australia) for $Q = 7700\,\text{m}^3/\text{s}$ and $3200\,\text{m}^3/\text{s}$: $0.15 \leqslant d_{50} \leqslant 2.8\,\text{mm}$ $2 \leqslant d \leqslant 4\,\text{m}$

Notes: f: Darcy friction factor; h: bed-form height; l: bed-form length (Fig. 12.7).

Notes

1. At a sudden channel expansion, the head loss equals:

$$\Delta H = k \frac{V_1^2}{2g} \left(1 - \frac{A_1}{A_2} \right)^2 \qquad \text{Borda–Carnot formula}$$

where A_1 and A_2 are, respectively, the upstream and downstream cross-sectional areas of the flow ($A_1 < A_2$), V_1 is the upstream flow velocity and k is a head loss coefficient, function of the ratio A_1/A_2. For A_1/A_2 close to unity, $k = 1$. This formula is called the Borda–Carnot formula after the French scientists J.C. de Borda (1733–1799) and S. Carnot (1796–1832).

2. The Borda–Carnot formula may be applied to a wide rectangular channel with two-dimensional dunes (Fig. 12.7):

$$\Delta H = k \frac{V^2}{2g} \left(1 - \frac{1}{1 + h/d} \right)^2$$

where V is the mean flow velocity at the bed-form crest and d is the flow depth measured from the bed-form crest (Fig. 12.7). For $h/d \ll 1$, it becomes:

$$\Delta H \approx \frac{V^2}{2g} \left(\frac{h}{d} \right)^2 \qquad h \ll d$$

That is, the head loss along a dune element (length l) equals:

$$\Delta H \approx \left(\frac{h^2 D_H}{d^2 l} \right) \frac{l}{D_H} \frac{V^2}{2g} \qquad h \ll d$$

For a wide channel, it yields equation (12.8) (see Chapter 4).

3. Equation (12.8) was first applied to alluvial streams by Engelund (1966).

4. Several researchers investigated the characteristic sizes of bed forms; van Rijn (1993) presented a comprehensive review. Typical dimensions of dune bed form are presented in Table 12.3.

5. For steep mountain streams with step-pool bed forms, flow resistance predominantly form drag. Recent experiments under controlled flow conditions yielded:

$$\sqrt{\frac{8}{f}} = 3.86 \ln \left(\frac{d}{\text{Std}(z_o)} \right) - 1.19$$

$$1.5 < h/\text{Std}(z_o) < 10, \quad 0.02 \leqslant S_o \leqslant 0.10, \quad 0.015 < d_{50} < 0.037 \text{ m}$$

where d is the average water depth and $\text{Std}(z_o)$ is the standard deviation of the bed elevations (Aberle and Smart, 2003). For the largest water depths the data tended to $f \sim 0.11$–0.22 that is close to flow resistance data in skimming flow down stepped chute and form drag modelling (Chanson *et al.*, 2002) (Section 17.3 and Appendix A4.1).

Important

The bed-load transport must be related to the effective shear stress (skin friction shear stress) only and not to the form roughness. In natural rivers, the Shields parameter and bed-load layer characteristics must be calculated using the skin friction bed shear stress.

In other words, the onset of sediment motion must be predicted using the Shields parameter defined as:

$$\tau_* = \frac{\tau_0'}{\rho g(s-1)d_s}$$

Comments

All characteristic parameters of bed-load transport and bed-load layer must be computed in terms of the *effective bed shear stress* τ_0'. This includes the bed-load layer thickness $(\delta_s)_{bl}$, sediment concentration in bed-load layer $(C_s)_{bl}$, bed-load velocity $(V_s)_{bl}$ and bed-load transport rate $(q_s)_{bl}$.

For the suspended material, the main parameters of the sediment concentration distribution $c_s(y)$ and velocity distribution $v(y)$ are related to the *total bed shear stress*. That is, the Rouse number (w_o/KV_*), the shear velocity V_* and the mean flow velocity V are calculated in terms of the mean boundary shear stress $\tau_0(\tau_0 = \tau_0' + \tau_0'')$.

12.4.3 Design chart

The above considerations have highlighted the complexity of the interactions between sediment transport, bed form and flow properties.

For pre-design calculations, Engelund and Hansen (1967) developed a design chart which regroups all the relevant parameters (Fig. 12.8):

- a Froude number $(Vd)/\sqrt{g(s-1)d_s^3}$,
- a dimensionless sediment transport rate $q_s/\sqrt{g(s-1)d_s^3}$,
- the bed slope $\sin\theta$,
- the dimensionless flow depth d/d_s,
- the bed form.

Figure 12.8 takes into account the effect of bed forms and it may be used also to predict the type of bed form. In the region for which dunes occur, only two of the first four relevant parameters are required to deduce the entire flow properties.

Notes

1. The design chart of Engelund and Hansen (1967) was developed for fully rough turbulent flows. It was validated with the experimental data of Guy *et al.* (1966).
2. The calculations are not valid for ripple bed forms with $d_sV_*/\nu < 12$. The chart does not predict the presence of ripple bed forms.
3. For the calculation of erosion and accretion, the sediment volumes, deduced from the sediment transport rates, must be computed taking into account the porosity (i.e. the voids) of the sediment bed.
4. Engelund and Hansen (1967) used the median grain size $d_s = d_{50}$ for their design chart (Fig. 12.8).

12.5 APPLICATIONS

12.5.1 Presentation

Practically, the most important variables in designing alluvial channels are the water discharge Q, the sediment transport rate Q_s and the sediment size d_s.

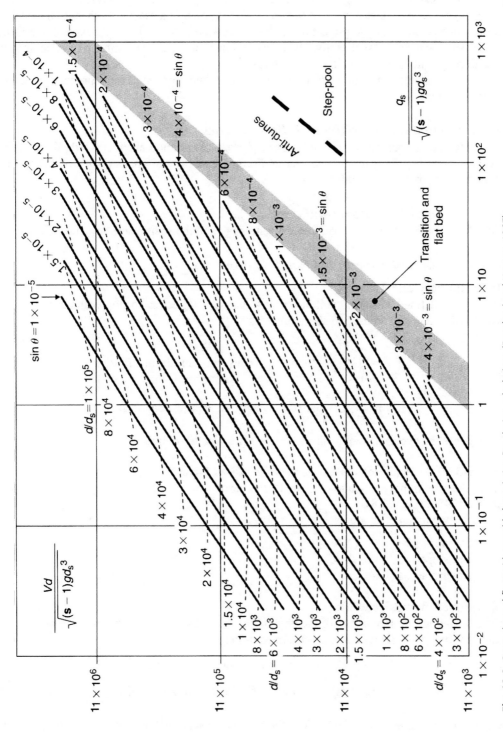

Fig. 12.8 Open channel flow with movable boundaries. Pre-design calculations (Engelund and Hansen, 1967).

The complete calculations of the flow properties and sediment transport are deduced by an iterative process. In a first stage, simplified design charts (e.g. Engelund and Hansen, 1967, Fig. 12.8) may be used to have a 'feel' of the type of bed forms. In a second stage, complete calculations must be developed to predict the hydraulic flow conditions (V, d), the type of bed forms and the sediment transport capacity. Then the continuity equation for sediment material may be used to assess the rate of erosion (or accretion).

Complete calculations (for a prismatic section of alluvial channel) must proceed in several successive steps.

(A) Determination of the channel characteristics

These include the bed slope $\sin\theta$, the shape of cross-section and channel width, the movable bed properties (ρ_s, Po, particle size distribution), eventually the presence of bed forms (e.g. resulting from past flood events).

(B) Selection of the inflow conditions

The inflow conditions are the total water discharge Q and the sediment inflow (i.e. $(C_s)_{mean}$).

(C) Calculations of sediment-laden flow properties

[1] Preliminary calculations

Assuming a flat bed, uniform equilibrium flow and $k_s = d_s$, deduce a first estimate of the flow properties (V and d) and the bed shear stress τ_o:

$$V = \sqrt{\frac{8g}{f}}\sqrt{\frac{D_H}{4}\sin\theta} \qquad \text{Uniform equilibrium flow}$$

$$\tau_o = \rho g \frac{D_H}{4}\sin\theta \qquad \text{Uniform equilibrium flow}$$

Then assess the occurrence of sediment motion (i.e. Shields diagram).

[2] Pre-design calculations to assess the type of bed form

Using the preliminary calculations, apply the results to the design chart of Engelund and Hansen (1967) (Fig. 12.8).

The known parameters are usually:

- the Froude number $(Vd)/\sqrt{g(s-1)d_{50}^3}$,
- the bed slope $\sin\theta$,
- the dimensionless ratio d/d_{50}.

[3] Complete flow calculations

The calculations are iterative until the type of bed form, the bed resistance and the flow conditions satisfy the continuity and momentum equations:

$$Q = VA \qquad\qquad\qquad\qquad\qquad\qquad\qquad\qquad \text{[C]}$$

$$\tau_o = \rho g \frac{D_H}{4}\sin\theta \qquad \text{Uniform equilibrium flow} \qquad\qquad \text{[M]}$$

where V is the mean flow velocity, A is the flow cross-sectional area and D_H is the hydraulic diameter.

(a) Select a mean flow velocity V.
(b) Compute the skin friction shear stress τ_o':

$$\tau_o' = \frac{f}{8}\rho V^2 \qquad \text{Skin friction shear stress}$$

where the roughness height might be assumed $k_s = 2d_{90}$.
(c) In presence of bed forms, calculate the bed form shear stress:

$$\tau_o'' = \frac{1}{2}\rho V^2 \frac{h^2}{ld} \qquad \text{Bed-form shear stress}$$

where h and l are bed-form height and length (Table 12.3).
(d) Compare $\tau_o' + \tau_o''$ and τ_o

If $(\tau_o' + \tau_o'') > \tau_o$, select a lower flow velocity to satisfy the momentum equation. Iterative calculations proceed until $[(\tau_o' + \tau_o'') = \tau_o]$ within a reasonable accuracy.

[4] Sediment transport calculations
First the sediment transport capacity is calculated:

$$Q_s = (Q_s)_{bl} + (Q_s)_s$$

Then the continuity equation for the sediment material is applied to the channel reach to predict erosion or accretion.

Comments
A series of calculations is valid for a prismatic section of the channel. Each time that one channel characteristics changes (e.g. slope, sediment size and cross-sectional shape), the calculations must be performed again.

12.5.2 Application No. 1

An alluvial stream in which the flow depth is 1.8 m has a longitudinal bed slope of $\sin\theta = 0.001$. The characteristics of the bed material are: $d_{50} = 0.8$ mm and $d_{90} = 2.1$ mm. Predict the occurrence of sediment motion, the type of bed form and the mean flow velocity.

Solution
The calculations will proceed in three steps:

[1] Preliminary calculations (assuming flat bed and $k_s = d_{50}$).
[2] Determination of type of bed form (using Fig. 12.8).
[3] Complete calculations (taking into account the grain roughness and the type of bed form).

Step [1]: Assuming a wide channel and uniform equilibrium flow, the bed shear stress equals:

$$\tau_o = \rho g \frac{D_H}{4}\sin\theta \approx \rho gd \sin\theta = 17.6 \text{ Pa} \qquad \text{Uniform equilibrium flow} \qquad \text{[M]}$$

Assuming a flat bed and $k_s = d_{50}$, the mean velocity is:

$$V = \sqrt{\frac{8g}{f}}\sqrt{\frac{D_H}{4}\sin\theta} \qquad \text{Uniform equilibrium flow and flat bed}$$

The (iterative) calculations give: $V = 3.36$ m/s and $f = 0.0125$.
    The Shield parameter equals:

$$\tau_* = \frac{\tau_o}{\rho(s-1)gd_{50}} = 1.36$$

and the particle Reynolds number ($d_{50}V_*/\nu$) equals 106. For these values, the Shields diagram (Fig. 8.4) predicts *sediment motion*.

Step [2]: Using the preliminary results, let us use the design chart of Engelund and Hansen (1967) (Fig. 12.8). The known parameters are:

- $(Vd)/\sqrt{g(s-1)d_{50}^3} = 6.7 \times 10^4$,
- the bed slope: $\sin\theta = 0.001$,
- the dimensionless flow depth: $d/d_{50} = 2.25 \times 10^3$.

Figure 12.8 indicates that the bed form is either transition (from dune to flat bed) or flat bed. Note that the preliminary calculations (Step [1]) predicted a flow Froude number: $Fr = 0.8$. According to Table 7.1 and Fig. 7.1, the type of bed form is the *flat bed*.

Step [3]: The complete flow calculations are developed. The total bed shear stress must satisfy the momentum equation:

$$\tau_o = \tau_o' + \tau_o'' \approx \rho gd \sin\theta = 17.6 \text{ Pa}$$

where τ_o' is the skin friction shear stress and τ_o'' is the form-related shear stress. The skin friction shear stress equals:

$$\tau_o' = \frac{f}{8}\rho V^2$$

assuming $k_s = 2d_{90}$ and the bed-form shear stress is calculated as:

$$\tau_o'' = \frac{1}{2}\rho V^2 \frac{h^2}{ld}$$

For a flat bed, $h = 0$ and $\tau_o'' = 0$. Hence it yields:

$$\tau_o = \tau_o' + 0 = 17.6 \text{ Pa}$$

Application of the continuity and momentum equations gives the mean flow velocity, friction factor and Reynolds number: $V = 2.84$ m/s and $Re = 5 \times 10^6$ (fully rough turbulent flow).
    Note: the mean velocity (computed using τ_o) differs from the value deduced from the preliminary calculations (Step [1]).

12.5.3 Application No. 2

An alluvial stream in which the flow depth is 1.35 m has a longitudinal bed slope of $\sin\theta = 0.0002$. Characteristics of the sediment mixture are: $d_{50} = 1.5$ mm and $d_{90} = 2.5$ mm. Predict the sediment transport capacity (taking into account the bed form).

Solution
The calculations proceed in four steps.

Step [1]: Assuming a wide channel and uniform equilibrium flow, the bed shear stress equals:

$$\tau_0 = \rho g \frac{D_H}{4} \sin \theta \approx \rho g d \sin \theta = 2.64 \text{ Pa} \qquad \text{Uniform equilibrium flow} \qquad [\text{M}]$$

Assuming a flat bed and $k_s = d_{50}$ (i.e. $k_s/D_H \approx 2.8 \times 10^{-4}$), the mean velocity is:

$$V = \sqrt{\frac{8g}{f}} \sqrt{\frac{D_H}{4}} \sin \theta \qquad \text{Uniform equilibrium flow and flat bed}$$

The (iterative) calculations give: $V = 1.19 \text{ m/s}$ and $f = 0.015$.
    The Shield parameter equals:

$$\tau_* = \frac{\tau_0}{\rho(s-1)gd_{50}} = 0.109$$

and the particle Reynolds number ($d_{50}V_*/\nu$) equals 77. For these values, the Shields diagram (Fig. 8.4) predicts *sediment motion*.

Step [2]: Using the preliminary results, let us use the design chart of Engelund and Hansen (1967) (Fig. 12.8). The known parameters are:

- $(Vd)/\sqrt{g(s-1)d_{50}^3} = 6.89 \times 10^3$,
- the bed slope: $\sin \theta = 2 \times 10^{-4}$,
- the dimensionless flow depth: $d/d_{50} = 6.75 \times 10^3$.

Figure 12.8 indicates that the bed forms are *dunes*. Note that Fig. 12.8 suggests that $q_s \sim 0.05 \sqrt{(s-1)gd_s^3}$.

Step [3]: The complete flow calculations are developed. The total bed shear stress must satisfy the momentum equation:

$$\tau_0 = \tau_0' + \tau_0'' \approx \rho g d \sin \theta = 2.64 \text{ Pa} \qquad [\text{M}]$$

where τ_0' is the skin friction shear stress and τ_0'' is the form-related shear stress. The skin friction shear stress equals:

$$\tau_0' = \frac{f}{8} \rho V^2$$

assuming $k_s = 2d_{90}$ and the bed form shear stress is calculated as:

$$\tau_0'' = \frac{1}{2} \rho V^2 \frac{h^2}{ld}$$

where the dune dimensions may be calculated using the correlations of van Rijn (1984c):

$$\frac{h}{d} = 0.11 \left(\frac{d_{50}}{d}\right)^{0.3} \left(1 - \exp\left(-0.5\left(\frac{\tau_0}{(\tau_0)_c} - 1\right)\right)\right)\left(25 - \left(\frac{\tau_0}{(\tau_0)_c} - 1\right)\right)$$

$$\frac{l}{d} = 7.33$$

The calculations are iterative:

1. assume V,
2. compute h, l, τ_0' and τ_0'',

3 check the momentum equation by comparing $(\tau_0' + \tau_0'')$ and τ_0 (=2.64 Pa),
4. if $(\tau_0' + \tau_0'') > \tau_0$, select a smaller flow velocity V; if $(\tau_0' + \tau_0'') < \tau_0$, select a larger flow velocity,
5. repeat the calculations until $(\tau_0' + \tau_0'') = \tau_0$ with an acceptable accuracy.

Complete calculations give: $V = 0.9\,\mathrm{m/s}$, $\tau_0' = 1.95\,\mathrm{Pa}$, $\tau_0'' = 0.70\,\mathrm{Pa}$, $h = 0.15\,\mathrm{m}$ and $l = 9.9\,\mathrm{m}$.

Step [4]: The sediment transport capacity is the sum of the bed-load transport rate and suspension transport rate.

The bed-load transport rate (see Chapter 10) is calculated using flow properties based upon the effective shear stress τ_0': i.e. V_*', τ_*', $(\tau_*)_c'$, $(C_s)_{bl}$, $(d_s)_{bl}$ and $(V_s)_{bl}$.

Complete calculations confirm bed-load motion and the bed-load transport rate equals $2.48 \times 10^{-5}\,\mathrm{m^2/s}$.

The suspension transport rate (see Chapter 11) must be computed using the overall bed shear stress τ_0. For the present application, the ratio V_*/w_o equals 0.36, where $V_* = \sqrt{\tau_0/\rho} = 0.051\,\mathrm{m/s}$.

The flow conditions are near the onset of suspension and the suspension transport rate may be assumed negligible. Overall the sediment transport capacity of the alluvial stream with dune bed forms is: $q_s = 2.48 \times 10^{-5}\,\mathrm{m^2/s}$ (i.e. $0.066\,\mathrm{kg/s/m}$).

12.6 EXERCISES

Continuity equation for sediment material

A 5 km long reach of a river has a longitudinal slope 1.4 m per km and the channel bed consists of fine sands. The river is 110 m wide and the flow rate is a water discharge of $920\,\mathrm{m^3/s}$. The mean sediment concentration of the inflow is 0.22%. Preliminary calculations indicate that the sediment transport capacity of the reach is $0.0192\,\mathrm{m^2/s}$. (a) Calculate the rate of erosion (or accretion) of the channel bed for the 5 km long reach. (b) Compute the change in bed elevation (positive upwards) during 1 day.

Considering a 300 m long reach of a stream (bed slope 0.0004), the inflow sediment transport rate is $0.00034\,\mathrm{m^2/s}$. During a 2 h flood event, the river bed was scoured by 0.5 m in average over the reach. Compute the sediment transport capacity of the reach during the 2 h event. Assume a sand and gravel river bed.

Total sediment transport capacity

An alluvial stream in which the flow depth is 2.2 m has a bed slope of 0.0015. The bed material characteristics are: $d_{50} = 0.95\,\mathrm{mm}$, $d_{90} = 3.2\,\mathrm{mm}$ and $\rho_s = 2480\,\mathrm{kg/m^3}$. Predict (a) occurrence of sediment motion, (b) type of bed form, (c) mean flow velocity and (d) total sediment transport capacity. (Take into account the bed form. Assume $k_s = 2d_{90}$.)

Considering a 1500 m reach of an alluvial channel (85 m wide), the median grain size of the movable bed is 0.5 mm, $d_{90} = 3.15\,\mathrm{mm}$ and the longitudinal bed slope is $\sin\theta = 0.00027$. The observed flow depth is 1.24 m and the mean sediment concentration of the inflow is 2.1%. Calculate (a) discharge, (b) effective boundary shear stress, (c) type of bed form, (d) bed-form shear stress, (e) total shear stress, (f) total sediment transport capacity and (g) rate of erosion (or accretion). (Assume uniform equilibrium flow conditions. Take into account the bed form and use the design chart of Engelund and Hansen (1967) to predict the type of bed form. Assume $k_s = 2d_{90}$.)

Part 2 Revision exercises

REVISION EXERCISE NO. 1

Considering a 52 m wide stream with a flow depth of 0.85 m and a bed slope of $\sin \theta = 0.012$, the median grain size of the channel bed is 0.35 mm. (a) Compute the water discharge. (b) Predict the occurrence of bed-load motion and/or suspension. (c) If bed-load motion occurs, calculate the bed-load transport rate. (d) If suspension occurs, plot the sediment concentration distribution (Rouse concentration distribution) and velocity distribution on the same graph (plot v/V_* and C_s as functions of y/d). (e) Calculate the suspended-load transport rate. (f) Calculate the total sediment transport rate.

Note: Perform the calculations for a flat bed and assuming a uniform equilibrium flow. The bed roughness height k_s will be assumed to be the median grain size.

REVISION EXERCISE NO. 2

A channel of trapezoidal cross-section (bottom width 2 m and sidewall slope 1V:3H) has a longitudinal slope of 0.012. The channel bed and sloping sidewall consist of a mixture of fine sands ($d_s = 0.2$ mm). The flow rate is $Q = 4$ m³/s and normal flow conditions are achieved. (a) Compute the flow depth. (b) Will bed-load motion take place? (c) Will sediment suspension occur? (d) Calculate the total boundary shear stress. (e) What is the type of bed forms? (Assume uniform equilibrium flow conditions. Take into account the bed form and use the design chart of Engelund and Hansen (1967) to predict the type of bed form. Assume $k_s = d_s$. The fluid is water at 20°C and the sediment grains are quartz particles.)

REVISION EXERCISE NO. 3

Considering a 50 m reach of an alluvial channel (25 m wide), the bed material characteristics are: $d_{50} = 0.5$ mm, $d_{90} = 1.95$ mm and $\sin \theta = 0.00025$. The water discharge is 16.5 m³/s and the mean sediment concentration of the inflow is 0.8%. Predict the rate of erosion (or accretion) that will take place. (Assume uniform equilibrium flow conditions. Take into account the bed form and use the design chart of Engelund and Hansen (1967) to predict the type of bed form. Assume $k_s = 2d_{90}$.)

REVISION EXERCISE NO. 4

An artificial canal carries a discharge of 25 m³/s. The channel cross-section is trapezoidal and symmetrical, with a 0.5 m bottom width and 1V:2.5H sidewall slopes. The longitudinal bed slope is 8.5 m/km. The channel bottom and sidewall consist of a mixture of fine sands ($d_{50} = 0.3$ mm).

A gauging station is set at a bridge crossing the waterway. The observed flow depth, at the gauging station, is 2.2 m. (a) Compute the flow velocity, Darcy friction factor, boundary shear stress and shear velocity. (b) At the gauging station, is the flow laminar, smooth turbulent, transition turbulent or fully rough turbulent? (c) Compute the critical flow depth. (d) From where is the flow controlled? (e) Predict the occurrence of bed motion and the type of bed-load motion. (f) If bed motion occurs, what is the likely type of bed form? (g) If bed-load motion occurs, compute the bed-load transport rate (using Nielsen's formula).

Use the Moody and Shields diagrams. Assume that the equivalent roughness height equals the median sediment size.

REVISION EXERCISE NO. 5

An artificial canal carries a discharge of $25\,\text{m}^3/\text{s}$. The channel cross-section is trapezoidal and symmetrical, with a 0.5 m bottom width and 1V:2.5H sidewall slopes. The longitudinal bed slope is 8.5 m/km. The channel bottom and sidewall consist of a mixture of fine sands ($d_s = 0.3\,\text{mm}$).

A gauging station is set at a bridge crossing the waterway. The observed flow depth, at the gauging station, is 2.2 m. Compute (a) the flow velocity, (b) the Darcy friction factor, (c) the boundary shear stress, (d) the critical flow depth. (e) From where is the flow controlled? (f) Predict the occurrence of sediment motion and the type of sediment-load motion. (g) If bed motion occurs, what is the likely type of bed form. (h) Calculate the sediment load transport rate.

Assume that the equivalent roughness height equals the median sediment size.

Solution: (c) $\tau_o = 5\,\text{Pa}$; (d) $d_c = 1.73\,\text{m}$ (trapezoidal channel, see Chapter 3, Section 3.4); (f) $\tau_* = 1.1$, $V_*/w_o = 0.5$, hence bed-load motion; (g) Bed forms: dunes to flat bed; (h) $q_s = 0.00017\,\text{m}^2/\text{s}$ (Nielsen's formula).

Appendix to Part 2

A2.1 SOME EXAMPLES OF RESERVOIR SEDIMENTATION

A2.1.1 Introduction

Natural streams and rivers have the ability to scour channel beds and to carry a large amount of solid particles. When a dam is built across a river, it often acts as a sediment trap. After several years of use, a reservoir might become full of sediments and cease to provide water storage.

In this appendix, some examples of reservoir siltation are described.

A2.1.2 Reservoir siltation in Australia

The Umberumberka dam, Broken Hill, NSW

Built between 1913 and 1915, the Umberumberka dam was completed in 1915 to supply water to the town of Broken Hill, NSW (900 km west-north-west of Sydney). The 41 m high dam was designed as a concrete arched dam. The original volume of the reservoir was 13 197 000 m^3 and the catchment area is 420 km^2 (Wasson and Galloway, 1986).

The dam is located on the Umberumberka Creek, at a scarp immediately downstream of an alluvial fan. The average rainfall over the catchment is about 220 mm (for the period 1939–1972). The dam traps almost all the incoming sediment material. Estimates of the volume of sediment trapped in the reservoir have been made regularly since the dam construction.

The volume of trapped sediment is summarized in Table A2.1 and compared with other dams in the same region (i.e. Corona dam and Stephens Creek dam). Note that most siltation occurred between 1915 and 1941 and the rate of sedimentation has since decreased drastically (Wasson and Galloway, 1986). The Umberumberka reservoir is still in use today.

The Korrumbyn Creek dam, Murwillumbah, NSW

The Korrumbyn Creek dam was built between 1917 and 1918, and completed in late 1918 to supply water to the town of Murwillumbah, NSW (150 km south of Brisbane). The 14.1 m high

Table A2.1 Reservoir siltation in the Broken Hill region (north-east area of New South Wales, Australia)

Dam (1)	Year of completion (2)	Catchment area (km^2) (3)	Original reservoir capacity (m^3) (4)	Cumulative siltation volume (m^3) (5)	Date of record (6)
Corona dam, Broken Hill, NSW, Australia	1890?	15	120 000[a]	120 000 Fully silted	1910?
Stephens Creek dam, Broken Hill, NSW, Australia	1892	510	24 325 000	4 500 000 De-silting[b]	1958 1971
Umberumberka dam, Broken Hill, NSW, Australia	1915	420	13 197 000	3 600 000	1941
				4 500 000	1961
				5 013 000	1964
				5 700 000	1982

Notes: [a] Fully silted nowadays, [b] extensive de-silting in 1971.

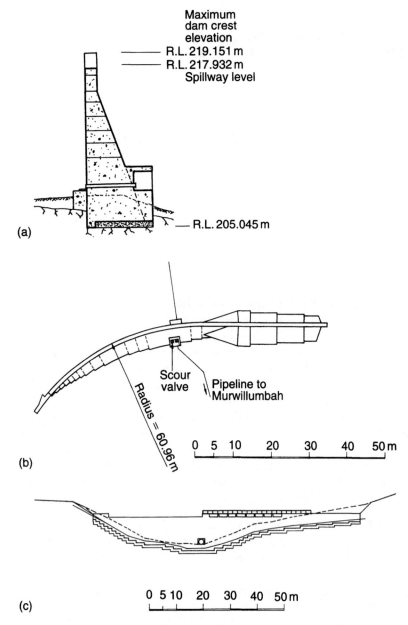

Fig. A2.1 Sketch of the Korrumbyn Creek dam after original drawings: (a) cross-section of the wall at the bottom outlets, (b) top view and (c) view from downstream.

dam was designed with advanced structural features: i.e. single arch wall (1.1 m thick at crest, 5.2 m thick at base and 61 m radius in plan) with the left-abutment tangent of gravity cross-section, made of Portland cement concrete (Fig. A2.1). The cross-section of the arch wall has a vertical upstream wall and a battered downstream face. The dam was equipped with two bottom outlets (one pipe outlet valve and one scour valve) and with an overfall spillway (at elevation 217.9 m R.L.). The volume of the reservoir was 27 300 m³ and the catchment area is only 3 km² (Chanson and James, 1998).

Fig. A2.2 Recent photographs of the Korrumbyn Creek dam taken on 25 April 1997: (a) view of the subtropical forest occupying the reservoir – note the dam wall in the bottom right corner and (b) dam wall and outlet system.

The reservoir was rapidly abandoned because a log jammed the scour pipe entrance during a flood and could not be removed: i.e. the dam could no longer be scoured to remove the sediments. Prior to the incident, the water level used to drop during dry periods and the water would turn green and foul as it warmed up, making it unfit for use. Further the catchment is very steep (bed slope > 7.6°) and the channel bed contains a wide range of sediment materials. As a result the reservoir silted up very rapidly by bed load. For these reasons the life of the reservoir was very short (less than 20 years).

The dam still stands today, the reservoir being occupied by an overgrown tropical rainforest (Fig. A2.2).

Table A2.2 Siltation of Cunningham Creek and Illalong Creek reservoirs

Dam (1)	Year of completion (2)	Catchment area (km²) (3)	Original reservoir capacity (m³) (4)	Cumulative siltation volume (m³) (5)	Date of record (6)
Cunningham Creek dam, Harden, NSW, Australia	1912	820	–	216 000	1916
				258 600	1917
				379 000	1920
				522 600	1922
				758 000	1929
				Fully silted	1998
Illalong Creek dam, Binalong, NSW, Australia	1914	130	2.6×10^5	Fully silted	1998

Table A2.3 Examples of extreme reservoir sedimentation rates

Reservoir (1)	Sedimentation rate (m³/km²/year) (2)	Study period (3)	Catchment area (km²) (4)	Annual rainfall (mm) (5)
Asia				
Wu-Sheh (Taiwan) (S)	10 838	1957–1958	205	–
	9959	1959–1961	205	–
	7274	1966–1969	205	–
Shihmen (Taiwan) (S)	4366	1958–1964	763	>2000
Tsengwen (Taiwan) (S)	6300	1973–1983	460	3000
Muchkundi (India)	1165	1920–1930?	67	–
North Africa				
El Ouldja (Algeria) (W)	7960 (F)	1948–1949	1.1	1500
El Fodda (Algeria) (W)	5625 (F)	1950–1952	800	555
	3060 (F)	1932–1948	800	555
Hamiz (Algeria) (W)	1300	1879–1951	139	–
El Gherza (Algeria)	615	1951–1967	1300	35
	577	1986–1992	1300	35
North America				
Sweetwater (USA)	10 599	1894–1895	482	240
White Rock (USA)	570	1923–1928	295	870
Zuni (USA) (*)	546	1906–1927	1290	250–400
Roosevelt (USA)	438	1906–1925	14 900	–
Europe				
Saignon (France) (*)	25 714	1961–1963	3.5	–
Saifnitz (Austria) (*)	6820	1876	4	–
Monte Reale (Italy) (*)	1927	1904–1905	436	–
Wetzmann (Australia) (*)	1852	1883–1884	324	–
Pont-du-Loup (France) (*)	1818	1927–1928	750	–
Pontebba (Austria)	1556	1862–1880	10	–
Lavagnina (Italy)	784	1884–1904	26	1800
Roznov (Poland) (S)	398	1958–1961	4885	600
Cismon (Italy)	353	1909–1919	496	1500
Abbeystead (UK) (*)	308	1930–1948	49	1300–1800
Porabka (Poland) (S)	288	1958–1960	1082	600
Australia				
Quipolly (*)	1143	1941–1943	70	686
Pykes Creek	465	1911–1945	125	–
Umberumberka	407	1961–1964	420	220
Corona (*)	400	1890–1910	15	–
Eildon	381	1939–1940	3885	–
Moore Creek (*)	174	1911–1924	51	674
Pekina Creek (*)	174	1911–1944	136	340–450
Korrumbyn Creek (*)	*1400 (?)*	1918–1924?	3	1699

Notes: (S) summer rainfall climate; (W) winter rainfall climate; (F) important flushing; (*) fully silted today; (–) data not available.
Reference: Chanson (1998).

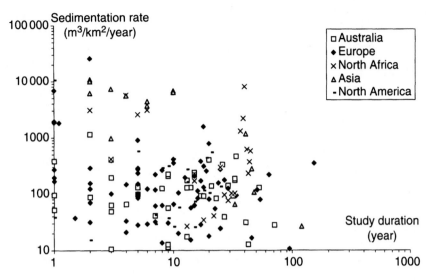

Fig. A2.3 Extreme reservoir sedimentation rates.

The Cunningham Creek dam, Harden, NSW

The Cunningham Creek dam, located near Harden, NSW, was completed in 1912 to supply water to the railway. The 13 m high dam is a curved arched dam and the catchment area is 850 km^2. The reservoir was over 1.3 km long and about 110 m wide, and the reservoir area at dam crest level is 150 000 m^2.

The dam was equipped with a bottom outlet (pipe outlet), and a crane was installed at the dam crest to facilitate the outlet operation and to assist in clearing the outlet (Hellström, 1941).

The reservoir silted up very rapidly (Table A2.2) and it was well documented (Hellström, 1941). Attempts to flush sediments were not successful. During the first 10 years of operation, the silt was deposited at a nearly constant rate of 49 000 m^3 per annum (60 m^3/km^2/year). The dam was nearly full (~90%) by 1929. The reservoir was filled primarily by suspended load. Another railway dam (Illalong Creek dam), located less than 30 km from Cunningham Creek dam, suffered a similar sedimentation problem (Table A2.2) (Chanson, 1998).

A2.1.3 Extreme reservoir siltation

Reservoir sedimentation results from soil erosion in the catchment (wind, rainfall and man-made erosion), surface runoff, sediment transport in the creeks, streams and rivers, and sediment trapping in the reservoir. Altogether reservoir sedimentation is a very complex process. Practically, the primary consequence is the reduction of the reservoir capacity and of its economical and strategic impact.

Extreme reservoir sedimentation has been observed all around the world (Table A2.3). Usually the sedimentation rate is defined as: (sediment inflow − sediment flushing dredging)/year/km^2 of catchment. Prototype observations are shown in Fig. A2.3 where all the reported data must be considered as extreme events. In some cases, extreme siltation rates might lead to the reservoir siltation and its disuse (Table A2.3).

Part 3 Hydraulic Modelling

Summary of basic hydraulic principles   13

13.1 INTRODUCTION

For design engineers, it is essential to predict accurately the behaviour of hydraulic structures under the design conditions (e.g. maximum discharge capacity), the operation conditions (e.g. gate operation) and the emergency situations (e.g. probable maximum flood for a spillway). During design stages, the engineers need a reliable prediction 'tool' to compare the performances of various design options.

In fluid mechanics and hydraulics, many problems are insoluble by theory or by reference to empirical data. For example, Plates 23 and 24 illustrate two extreme flow situations: a dry channel (Plate 23) and the same waterway in flood (Plate 24). In practice, a proper design of hydraulic structures derives from a model: a model being defined as a system which operates similarly to other systems and which provides accurate prediction of other systems' characteristics.

In hydraulics, three types of models are commonly used: analytical models, physical models and computational models. Analytical models are theoretical solutions of the fundamental principles within a framework of basic assumptions: e.g. the method of characteristics for unsteady open channel flows (Chapter 17). Numerical models are computer softwares which solve the basic fluid mechanics equations: e.g. numerical integration of the backwater equation to predict the longitudinal free-surface profile of a gradually varied open channel flow (Chapter 15). Their application is restricted, however, to simple flow situations and boundary conditions for which the basic equations can be numerically integrated and are meaningful. The calibration and validation of computational models are extremely difficult, and most computer models are applicable only in very specific range of flow conditions and boundary conditions. Most often, physical models must be used, including for the validation of computational models (Chapter 14).

In this section, we consider open channels in which water flows with a free surface. First, the basic principles are summarized. Then physical and numerical modelling of hydraulic flows are discussed.

13.2 BASIC PRINCIPLES

13.2.1 Introduction

The basic fluid mechanics principles are the continuity equation (i.e. conservation of mass), the momentum principle (or conservation of momentum) and the energy equation. A related principle is the Bernoulli equation which derives from the motion equation (e.g. Section 2.2.3, and Liggett (1993)).

13.2.2 Basic equations in open channel flows

In an open channel flow the free surface is always a streamline and it is all the time at a constant absolute pressure (usually atmospheric). The driving force of the fluid motion is gravity and the fluid is water in most practical situations.

For an incompressible fluid (e.g. water), for zero lateral inflow (i.e. Q constant) and for steady flow conditions, the *continuity equation* yields:

$$Q = V_1 A_1 = V_2 A_2 \tag{13.1}$$

where Q is the total discharge (i.e. volume discharge), V is the mean flow velocity across the cross-section A and the subscripts 1 and 2 refer to the upstream and downstream locations, respectively.

The *momentum equation* states that the change of momentum flux equals the sum of all the forces applied to the fluid. Considering an arbitrary control volume, it is advantageous to select a volume with a control surface perpendicular to the flow direction, denoted by s. For a steady and incompressible flow, the momentum equation gives:

$$F_s = \rho Q (V_{s2} - V_{s1}) \tag{13.2}$$

where F_s is the resultant of the forces in the s-direction, ρ is the fluid density and $(V_{si})_{i=1,2}$ is the velocity component in the s-direction. In hydraulics, the *energy equation* is often written as:

$$H_1 = H_2 + \Delta H \tag{13.3}$$

where H is the mean total head and ΔH is the head loss.

Notes
Along any streamline, the total head is defined as:

$$H = \frac{v^2}{2g} + z + \frac{P}{\rho g} \qquad \text{Total head along a streamline}$$

where v, z, P are the local fluid velocity, altitude and pressure, respectively.

Assuming a hydrostatic pressure distribution, the mean total head (in a cross-section normal to the flow direction) equals:

$$H = \alpha \frac{V^2}{2g} + z_o + d \cos \theta \qquad \text{Mean total head}$$

where V is the mean flow velocity (i.e. $V = Q/A$), z_o is the bed elevation (positive upwards), θ is the bed slope and α is the kinetic energy correction coefficient (i.e. Coriolis coefficient).

13.2.3 The Bernoulli equation

The Bernoulli equation derives from the Navier–Stokes equation. The derivation is developed along a *streamline*, for a *frictionless* fluid (i.e. no viscous effect), for the volume force potential (i.e. gravity) being independent of the time and for a *steady* and *incompressible* flow.

Along a streamline, the Bernoulli equation implies:

$$\frac{P}{\rho} + gz + \frac{V^2}{2} = \text{constant} \tag{13.4}$$

Application: smooth transition

The Bernoulli equation may be applied to smooth transitions (e.g. a step and a weir) for a steady, incompressible and frictionless flow. Assuming a hydrostatic pressure distribution and an uniform velocity distribution, the continuity equation and Bernoulli equation applied to a horizontal channel give:

$$Q = VA = \text{constant}$$

$$H = z_\text{o} + d + \frac{V^2}{2g} = \text{constant}$$

13.3 FLOW RESISTANCE

In open channel flows, bottom and sidewall friction resistance can be neglected over a short transition[1] as a first approximation. But the approximation of frictionless flow is no longer valid for long channels. Considering a water supply canal extending over several kilometres, the bottom and sidewall friction retards the fluid, and at equilibrium the friction force counterbalances exactly the weight force component in the flow direction.

The laws of flow resistance in open channels are essentially the same as those in closed pipes (Henderson, 1966). The head loss ΔH over a distance Δs (along the flow direction) is given by the Darcy equation:

$$\Delta H = f \frac{\Delta s}{D_\text{H}} \frac{V^2}{2g} \tag{13.5}$$

where f is the Darcy coefficient[2], V is the mean flow velocity and D_H is the hydraulic diameter or equivalent pipe diameter.

In open channels and assuming hydrostatic pressure distribution, the energy equation may be conveniently rewritten as:

$$d_1 \cos\theta_1 + z_\text{o1} + \alpha_1 \frac{V_1^2}{2g} = d_2 \cos\theta_2 + z_\text{o2} + \alpha_2 \frac{V_2^2}{2g} + \Delta H \tag{13.6}$$

where the subscripts 1 and 2 refer to the upstream and downstream cross-section of the control volume, d is the flow depth measured normal to the channel bottom and V is the mean flow velocity.

---

[1] In open channel hydraulics, the flow below a sluice gate, a hydraulic jump, the flow above a weir, or at an abrupt drop or rise may be considered as short transitions.

[2] Also called the Darcy–Weisbach friction factor or head loss coefficient.

Notes

1. The hydraulic diameter is defined as:

$$D_H = 4\frac{\text{cross-sectional area}}{\text{wetted perimeter}} = \frac{4A}{P_w}$$

The hydraulic diameter is also called the equivalent pipe diameter.

2. The Darcy friction factor of open channel flows can be estimated as for pipe flows:

$$f = \frac{64}{Re} \qquad \text{Laminar flow } (Re < 2000)$$

$$\frac{1}{\sqrt{f}} = -2.0\log_{10}\left(\frac{k_s}{3.71D_H} + \frac{2.51}{Re\sqrt{f}}\right)$$

Turbulent flow (Colerook–Whites formula: Colebrook, 1939)

where k_s is the equivalent roughness height and Re is the Reynolds number: $Re = \rho V D_H/\mu$. A less-accurate formula may be used to initialize the calculation with the Colebrook–White formula:

$$f = 0.1\left(1.46\frac{k_s}{D_H} + \frac{100}{Re}\right)^{1/4}$$

Turbulent flow (Altsul's formula: Idelchik, 1969, 1986)

Physical modelling of hydraulics

14.1 INTRODUCTION

Definition: the physical hydraulic model

A physical model is a scaled representation of a hydraulic flow situation. Both the boundary conditions (e.g. channel bed and sidewalls), the upstream flow conditions and the flow field must be scaled in an appropriate manner (Fig. 14.1).

Physical hydraulic models are commonly used during *design* stages to *optimize* a structure and to *ensure a safe operation* of the structure. Furthermore, they have an important role to assist non-engineering people during the '*decision-making*' process. A hydraulic model may help the decision-makers to visualize and to picture the flow field, before selecting a 'suitable' design.

In civil engineering applications, a physical hydraulic model is usually a smaller-size representation of the prototype (i.e. the *full-scale* structure) (e.g. Fig. 14.2). Other applications of model studies (e.g. water treatment plant and flotation column) may require the use of models larger than the prototype. In any case the model is investigated in a laboratory under controlled conditions.

Discussion

Hydraulic modelling cannot be disassociated from the basic theory of fluid mechanics. To be efficient and useful, experimental investigations require theoretical guidance which derives primarily from the basic principles (see Chapter 13) and the theory of similarity (see next section).

In the present section, we will consider the physical modelling of hydraulic flows: i.e. the use of laboratory models (with controlled flow conditions) to predict the behaviour of prototype flow situations.

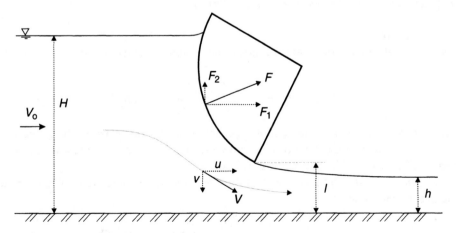

Fig. 14.1 Basic flow parameters.

(a1)   (a2)

Fig. 14.2 Example of physical models: (a) Stepped spillway, Les Olivettes dam (Vailhan, France). Dam height: 36 m, design flow: 290 m³/s (1-in-1000 year flood), $\theta = 53°$, step height: 0.6 m, roller compacted concrete dam for flood mitigation and irrigation. (a1) Prototype stepped spillway in March 2003 (courtesy of Mr and Mrs Chanson). View from the left bank. (a2) Two-dimensional spillway model (1:10 scale) for $q = 0.025$ m²/s (courtesy of Mr Paul Royet).

14.2 BASIC PRINCIPLES

In a physical model, the flow conditions are said to be similar to those in the prototype if the model displays similarity of form (*geometric similarity*), similarity of motion (*kinematic similarity*) and similarity of forces (*dynamic similarity*).

Basic-scale ratios

Geometric similarity implies that the ratios of prototype characteristic lengths to model lengths are equal:

$$L_r = \frac{l_p}{l_m} = \frac{d_p}{d_m} = \frac{H_p}{H_m} \qquad \text{Length} \qquad (14.1)$$

where the subscripts p and m refer to prototype (full scale) and model parameters, respectively, and the subscript r indicates the ratio of prototype-to-model quantity. Length, area and volume are the parameters involved in geometric similitude.

Kinematic similarity implies that the ratios of prototype characteristic velocities to model velocities are the same:

$$V_r = \frac{v_p}{v_m} = \frac{(V_1)_p}{(V_1)_m} = \frac{(V_2)_p}{(V_2)_m} \qquad \text{Velocity} \qquad (14.2)$$

(b1)

(b2)

Fig. 14.2 (b) Breakwater jetty for the Changhua Reclamation area, along the north-west coastline of Taiwan ROC (January 1994). (b1) Prototype breakwater jetty. (b2) Model breakwater jetty in a wave flume (Tainan Hydraulic Laboratory).

Dynamic similarity implies that the ratios of prototype forces to model forces are equal:

$$F_r = \frac{(F_1)_p}{(F_1)_m} = \frac{(F_2)_p}{(F_2)_m} \qquad \text{Force} \qquad (14.3)$$

Work and power are other parameters involved in dynamic similitude.

Notes

1. Geometric similarity is not enough to ensure that the flow patterns are similar in both model and prototype (i.e. kinematic similarity).
2. The combined geometric and kinematic similarities give the prototype-to-model ratios of time, acceleration, discharge and angular velocity.

Subsequent-scale ratios

The quantities L_r, V_r and F_r (defined in equations (14.1)–(14.3)) are the basic-scale ratios. Several scale ratios can be deduced from equations (14.1) to (14.3):

$$M_r = \rho_r L_r^3 \qquad \text{Mass} \tag{14.4}$$

$$t_r = \frac{L_r}{V_r} \qquad \text{Time} \tag{14.5}$$

$$Q_r = V_r L_r^2 \qquad \text{Discharge} \tag{14.6}$$

$$P_r = \frac{F_r}{L_r^2} \qquad \text{Pressure} \tag{14.7}$$

where ρ is the fluid density. Further scale ratios may be deduced in particular flow situations.

Application

In open channel flows, the presence of the free surface means that gravity effects are important. The Froude number ($Fr = V/\sqrt{gL}$) is always significant. Secondary scale ratios can be derived from the constancy of the Froude number[1] which implies:

$$V_r = \sqrt{L_r} \qquad \text{Velocity}$$

Other scale ratios are derived from the Froude similarity (e.g. Henderson, 1966):

$$Q_r = V_r L_r^2 = L_r^{5/2} \qquad \text{Discharge}$$

$$F_r = \frac{M_r L_r}{T_r^2} = \rho_r L_r^3 \qquad \text{Force}$$

$$P_r = \frac{F_r}{L_r^2} = \rho_r L_r \qquad \text{Pressure}$$

14.3 DIMENSIONAL ANALYSIS

14.3.1 Basic parameters

The basic relevant parameters needed for any dimensional analysis (Fig. 14.1) may be grouped into the following groups:

(a) Fluid properties and physical constants (see Appendix A1.1). These consist of the density of water ρ (kg/m^3), the dynamic viscosity of water μ (N s/m^2), the surface tension of air and water σ (N/m), the bulk modulus of elasticity of water E_b (Pa) and the acceleration of gravity g (m/s^2).

---

[1] It is assumed that the gravity acceleration is identical in both model and prototype.

(b) Channel (or flow) geometry. These may consist of the characteristic length(s) L (m).
(c) Flow properties. These consist of the velocity(ies) V (m/s) and the pressure difference(s) ΔP (Pa).

14.3.2 Dimensional analysis

Taking into account all basic parameters, dimensional analysis yields:

$$\mathcal{F}_1(\rho, \mu, \sigma, E_b, g, L, V, \Delta P) = 0 \tag{14.8}$$

There are eight basic parameters and the dimensions of these can be grouped into three categories: mass (M), length (L) and time (T). The Buckingham Π-theorem (Buckingham, 1915) implies that the quantities can be grouped into five ($5 = 8 - 3$) independent dimensionless parameters:

$$\mathcal{F}_2\left(\frac{V}{\sqrt{gL}};\ \frac{\rho V^2}{\Delta P};\ \frac{\rho VL}{\mu};\ \frac{V}{\sqrt{\dfrac{\sigma}{\rho L}}};\ \frac{V}{\sqrt{\dfrac{E_b}{\rho}}}\right) = 0 \tag{14.9a}$$

$$\mathcal{F}_2(Fr;\ Eu;\ Re;\ We;\ Ma) = 0 \tag{14.9b}$$

The first ratio is the Froude number Fr, characterizing the ratio of the inertial force to gravity force. Eu is the Euler number, proportional to the ratio of inertial force to pressure force. The third dimensionless parameter is the Reynolds number Re which characterizes the ratio of inertial force to viscous force. The Weber number We is proportional to the ratio of inertial force to capillary force (i.e. surface tension). The last parameter is the Sarrau–Mach number, characterizing the ratio of inertial force to elasticity force.

Notes
1. The Froude number is used generally for scaling free-surface flows, open channels and hydraulic structures. Although the dimensionless number was named after William Froude (1810–1879), several French researchers used it before: e.g. Bélanger (1828), Dupuit (1848), Bresse (1860) and Bazin (1865a). Ferdinand Reech (1805–1880) introduced the dimensionless number for testing ships and propellers in 1852, and the number should be called the Reech–Froude number.
2. Leonhard Euler (1707–1783) was a Swiss mathematician and physicist, and a close friend of Daniel Bernoulli.
3. Osborne Reynolds (1842–1912) was a British physicist and mathematician who expressed first the 'Reynolds number' (Reynolds, 1883).
4. The Weber number characterizing the ratio of inertial force over surface tension force was named after Moritz Weber (1871–1951), German Professor at the Polytechnic Institute of Berlin.
5. The Sarrau–Mach number is named after Professor Sarrau who first highlighted the significance of the number (Sarrau, 1884) and E. Mach who introduced it in 1887. The Sarrau–Mach number was once called the Cauchy number as a tribute to Cauchy's contribution to wave motion analysis.

Discussion
Any combination of the dimensionless numbers involved in equation (14.9) is also dimensionless and may be used to replace one of the combinations. It can be shown that one parameter can be

replaced by the Morton number $Mo = (g\mu^4)/(\rho\sigma^3)$, also called the liquid parameter, since:

$$Mo = \frac{We^3}{Fr^2 Re^4} \tag{14.10}$$

The Morton number is a function only of fluid properties and gravity constant. For the same fluids (air and water) in both model and prototype, Mo is a constant (i.e. $Mo_p = Mo_m$).

14.3.3 Dynamic similarity

Traditionally model studies are performed using geometrically similar models. In a geometrically similar model, true dynamic similarity is achieved if and only if each dimensionless parameters (or Π-terms) has the same value in both model and prototype:

$$Fr_p = Fr_m; \quad Eu_p = Eu_m; \quad Re_p = Re_m; \quad We_p = We_m; \quad Ma_p = Ma_m \tag{14.11}$$

Scale effects will exist when one or more Π-terms have different values in the model and prototype.

Practical considerations
In practice, hydraulic model tests are performed under controlled flow conditions. The pressure difference ΔP may be usually controlled. This enables ΔP to be treated as a dependent parameter. Further compressibility effects are small in clear-water flows[2] and the Sarrau–Mach number is usually very small in both model and prototype. Hence, dynamic similarity in most hydraulic models is governed by:

$$\frac{\Delta P}{\rho V^2} = \mathscr{F}_3\left(\frac{V}{\sqrt{gL}}; \frac{\rho V L}{\mu}; \frac{V}{\sqrt{\frac{\sigma}{\rho L}}} \right) \tag{14.12a}$$

$$Eu = \mathscr{F}_3(Fr; Re; We) \qquad \text{Hydraulic model tests} \tag{14.12b}$$

There are a multitude of phenomena that might be important in hydraulic flow situations: e.g. viscous effects, surface tension and gravity effect. The use of the same fluid on both prototype and model prohibits simultaneously satisfying the Froude, Reynolds and Weber number scaling criteria (equation (14.12)) because the Froude number similarity requires $V_r = \sqrt{L_r}$, the Reynolds number scaling implies that $V_r = 1/L_r$ and the Weber number similarity requires: $V_r = 1\sqrt{L_r}$.

In most cases, only the most dominant mechanism is modelled. Hydraulic models commonly use water and/or air as flowing fluid(s). In *fully enclosed flows* (e.g. pipe flows), the pressure

[2] This statement is not true in air–water flows (e.g. free-surface aerated flows) as the sound celerity may decrease down to about 20 m/s for 50% volume air content (e.g. Cain, 1978; Chanson, 1997).

losses are basically related to the Reynolds number *Re*. Hence, a Reynolds number scaling is used: i.e. the Reynolds number is the same in both model and prototype. In *free-surface flows* (i.e. flows with free surface), gravity effects are always important and a Froude number modelling is used (i.e. $Fr_m = Fr_p$) (e.g. Fig. 14.2).

Discussion

When inertial and surface tension forces are dominant, a Weber number similarity must be selected. Studies involving air entrainment in flowing waters (i.e. white waters), de-aeration in shaft or bubble plumes are often based upon a Weber number scaling.

The Euler number is used in practice for the scaling of models using air rather than water: e.g. hydraulic models in wind tunnels, or a manifold system with water flow which is scaled at a smaller size with an air flow system.

14.3.4 Scale effects

Scale effects may be defined as the distortions introduced by effects (e.g. viscosity and surface tension) other than the dominant one (e.g. gravity in free-surface flows). They take place when one or more dimensionless parameters (see Section 14.3.3) differ between model and prototype.

Scale effects are often small but they are not always negligible altogether. Considering an overflow above a weir, the fluid is subjected to some viscous resistance along the invert. However the flow above the crest is not significantly affected by resistance, the viscous effects are small and the discharge–head relationship can be deduced as for ideal-fluid flow.

In free-surface flows, the gravity effect is dominant. If the same fluid (i.e. water) is used in both model and prototype, it is impossible to keep both the Froude and Reynolds numbers in model and full scale. Indeed it is elementary to show that a Froude similitude implies $(Re)_r = L_r^{3/2}$, and the Reynolds number becomes much smaller in the model than in the prototype (if $L_r < 1$).

Note that different fluids may be used to have the same Reynolds and Froude numbers in model and prototype but this expedient is often neither practical nor economical.

Some examples of scale effects

Example No. 1
Considering the drag exerted on two-dimensional bodies, Fig. 14.3 shows the effects of the Reynolds number on the drag coefficient. Dynamic similarity (equation (14.12)) requires the drag coefficient to be the same in model and prototype. If the Reynolds number is smaller in the model than at full scale (in most practical cases), Fig. 14.3 suggests that the model drag coefficient would be larger than that of the prototype and dynamic similarity could not be achieved. Moreover, the drag force comprises the form drag and the surface drag (i.e. skin friction). In small-size models, the surface drag might become predominant, particularly if the model flow is not fully rough turbulent or the geometrical scaling of roughness height is not achievable.

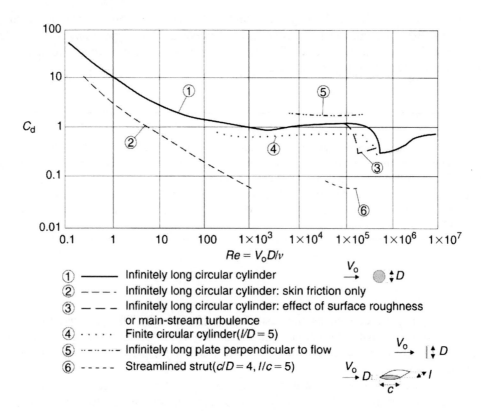

Fig. 14.3 Drag coefficient on two-dimensional bodies.

In practice, an *important rule* in model studies is that the model Reynolds number Re_m should be kept as large as possible, so that the model flow is fully rough turbulent (if prototype flow conditions are fully rough turbulent).

Example No. 2

Another example is the effect of corner radius on the drag force on two-dimensional bodies (Fig. 14.4). Figure 14.4 shows significant differences in the Reynolds number–drag coefficient relationships depending upon the relative radius r/D. When the corner radius on prototype is small and L_r is large, it is impossible to have the same ratio of corner radius to body size in model and prototype because the model cannot be manufactured with the required accuracy. In such cases, the drag force is not scaled adequately.

Example No. 3

A different example is the flow resistance of bridge piers. Henderson (1966) showed that the resistance to flow of normal bridge pier shapes is such that the drag coefficient is about or over unity, implying that the form drag is a significant component of the total drag. In such a case, the viscous effects are relatively small and dynamic similarity is achievable, provided that the model viscous effects remain negligible.

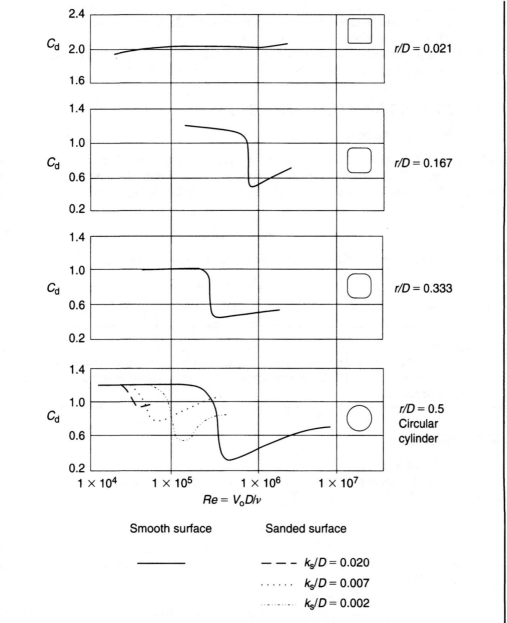

Fig. 14.4 Effect of corner radius and surface roughness on the drag coefficient of two-dimensional bodies.

Discussion

If scale effects would become significant in a model, a smaller prototype-to-model scale ratio L_r should be considered to minimize the scale effects. For example, in a 100:1 scale model of an open channel, the gravity effect is predominant but viscous effects might be significant. A geometric-scale ratio of 50:1 or 25:1 may be considered to reduce or eliminate viscous scale effects.

Another example is the entrainment of air bubbles in free-surface flows. Gravity effects are predominant but it is recognized that surface tension scale effects can take place for $L_r > 10$–20 (or $L_r < 0.05$–0.1) (e.g. Wood, 1991; Chanson, 1997).

At the limit, no scale effect is observed only at full scale (i.e. $L_r = 1$) as all the Π-terms (equation (14.11)) have the same values in the prototype and model when $L_r = 1$.

14.4 MODELLING FULLY ENCLOSED FLOWS

14.4.1 Reynolds models

Fully enclosed flow situations include pipe flows, turbomachines and valves. For such flow situations, viscosity effects on the solid boundaries are important. Physical modelling is usually performed with a Reynolds similitude: i.e. the Reynolds number is kept identical in model and prototype:

$$Re_p = Re_m \tag{14.13}$$

If the same fluid is used in model and prototype, equation (14.13) implies:

$$V_r = 1/L_r \qquad \text{Reynolds similitude}$$

For $L_r > 1$, the model velocity must be larger than that in the prototype.

Discussion

For example, if the model scale is 10:1 (i.e. $L_r = 10$), the velocity in the model must be 10 times that in the prototype. By using a different fluid in the mode, the ratio (μ_r/ρ_r) becomes different from unity and V_m can be reduced.

14.4.2 Discussion

Flow resistance in pipe flows

For pipe flows, the Darcy equation relates the pressure losses to the pipe geometry (diameter D and length L) and to the flow velocity V:

$$\Delta P = f \frac{L}{D} \frac{\rho V^2}{2} \tag{14.14}$$

where f is the Darcy–Weisbach friction factor. After transformation and combining with equation (14.10), it leads:

$$\frac{fL}{2D} = Eu = \mathcal{F}_4(Fr; Re; We; Ma; \ldots) \tag{14.15}$$

In pipe flows, gravity and surface tension have no effect on the pressure losses. For steady liquid flows, the compressibility effects are negligible. The roughness height k_s is, however, an additional characteristic length. For an uniformly distributed roughness, equation (14.15) becomes:

$$\frac{fL}{2D} = \mathcal{F}_4\left(Re; \frac{k_s}{D}\right) \tag{14.16}$$

Equation (14.16) expresses the dimensionless relationship between friction losses in pipes, the Reynolds number and relative roughness. An illustration is the Moody diagram (Fig. 14.5).

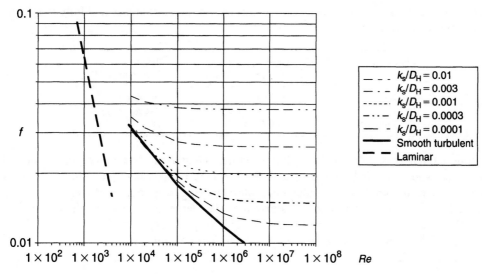

Fig. 14.5 Friction factor versus Reynolds number in pipe flows.

Skin friction and form drag

Considering the drag on a body (e.g. Figs 14.3 and 14.4), the pressure losses associated with the modification of the flow field caused by the presence of the body are usually expressed in terms of the drag force F_d on the body. The Euler number is rewritten as $Eu = F_d/\rho V^2 A$, where A is the projection of the body in the plane normal to the flow direction. F_d/A is equivalent to the pressure loss ΔP.

Equations (14.10) and (14.15) may be combined to relate the drag coefficient C_d to the Π-terms:

$$C_d = \frac{Eu}{2} = \frac{F_d}{(\rho/2)V^2 A} = \mathcal{F}_5(Fr;\ Re;\ We;\ Ma; \dots) \qquad (14.17)$$

In equation (14.17), the Reynolds number Re is related to the *skin friction drag* due to viscous shear as well as to *form drag* resulting from the separation of the flow streamlines from the body.

14.4.3 Practical considerations in modelling fully enclosed flows

The flow regime in pipes is either laminar or turbulent. In industrial applications, it is commonly accepted that the flow becomes turbulent for Reynolds numbers larger than 2000–3000, the Reynolds number being defined in terms of the equivalent pipe diameter D and of the mean flow velocity V.

For turbulent flows, the flow regime can be sub-divided into three categories: smooth, transition and fully rough. Each category can be distinguished as a function of a shear Reynolds number defined as:

$$Re_* = \rho \frac{V_* k_s}{\mu} \qquad (14.18)$$

where V_* is the shear velocity. The transition between smooth turbulence and fully rough turbulence is approximately defined as:

Flow situation Ref.	Open channel flow (Henderson, 1966)	Pipe flow (Schlichting, 1979)
Smooth turbulent	$Re_* < 4$	$Re_* < 5$
Transition	$4 < Re_* < 100$	$5 < Re_* < 75$
Fully rough turbulent	$100 < Re_*$	$75 < Re_*$

Dynamic similarity of fully enclosed flows implies the same resistance coefficient in model and in prototype. This can be achieved with the Reynolds number being the same in model and prototype, or with both flows in model and prototype being fully rough turbulent (Fig. 14.5).

If the full-scale flow is turbulent, it is extremely *important* to ensure that the model flow is also turbulent. Laminar and turbulent flows exhibit very important basic differences. In most cases, *turbulent flows should not be scaled* with laminar flow models.

Note

The Reynolds number can be kept constant by changing the flowing fluid. For example, the atmospheric wind flow past a tall building could be modelled in a small-size water tunnel at the same Reynolds number.

14.5. MODELLING FREE-SURFACE FLOWS

14.5.1 Presentation

In free-surface flows (e.g. rivers and wave motion), gravity effects are predominant. Model-prototype similarity is performed usually with a Froude similitude:

$$Fr_p = Fr_m \qquad (14.19)$$

If the gravity acceleration is the same in model and prototype, a Froude number modelling implies:

$$V_r = \sqrt{L_r} \qquad \text{Froude similitude}$$

Note that the model velocity is less than that in the prototype for $L_r > 1$ and the time scale equals $t_r = \sqrt{L_r}$.

Remarks

A Froude number modelling is typically used when friction losses are small and the flow is highly turbulent: e.g. spillways, overflow weirs and flow past bridge piers. It is also used in studies involving large waves: e.g. breakwater or ship models.[3]

A main concern is the potential for scale effects induced by viscous forces. Scale effects caused by surface tension effects are another concern, in particular when free-surface aeration (i.e. air entrainment) takes place.

---

[3] The testing of ship models is very specialized. Interestingly, F. Reech and W. Froude were among the firsts to use the Froude similitude for ship modelling.

14.5.2 Modelling hydraulic structures and wave motion

In hydraulic structures and for wave motion studies (Fig. 14.2), the gravity effect is usually predominant in prototype. The flow is turbulent, and hence viscous and surface tension effects are negligible in prototype if the flow velocity is reasonably small. In such cases a Froude similitude must be selected.

The most economical strategy is:

1. to choose a geometric-scale ratio L_r such as to keep the model dimensions small,
2. to ensure that the model Reynolds number Re_m is large enough to make the flow turbulent at the smallest test flows.

14.5.3 Modelling rivers and flood plains

In river modelling, gravity effects and viscous effects are basically of the same order of magnitude. For example, in uniform equilibrium flows (i.e. normal flows), the gravity force component counterbalances exactly the flow resistance and the flow conditions are deduced from the continuity and momentum equations.

In practice, river models are scaled with a Froude similitude (equation (14.19)) and viscous scale effects must be minimized. The model flow must be turbulent, and possibly fully rough turbulent with the same relative roughness as for the prototype:

$$Re_m > 5000 \tag{14.20}$$

$$(k_s)_r = L_r \tag{14.21}$$

where the Reynolds number is defined in terms of the hydraulic diameter (i.e. $Re = \rho V D_H / \mu$).

Distorted models

A distorted model is a physical model in which the geometric scale is different between each main direction. For example, river models are usually designed with a larger scaling ratio in the horizontal directions than in the vertical direction: $X_r > Z_r$. The scale distortion does not seriously distort the flow pattern and it usually gives good results.

A classical example of distorted model is that of the Mississippi river built by the US Army Corps of Engineers. The Mississippi basin is about $3\,100\,000\,km^2$ and the river is nearly $3800\,km$ long. An outdoor model was built with a scale of 2000:1. If the same scaling ratio was applied to both the vertical and horizontal dimensions, prototype depths of about $6\,m$ would imply model depths of about $3\,mm$. With such small-flow depths, surface tension and viscous effects would be significant. The Mississippi model was built, in fact, with a distorted scale: $Z_r = 100$ and $X_r = 2000$. Altogether the model size is about $1.5\,km$ per $2\,km$!

A distorted model of rivers is designed with a Froude similitude:

$$Fr_p = Fr_m \tag{14.19}$$

where the Froude number scaling ratio is related to the vertical-scale ratio:

$$Fr_r = \frac{V_r}{\sqrt{Z_r}} \tag{14.22}$$

As for an undistorted model, the distorted model flow must be turbulent (equation (14.20)), and preferably fully rough turbulent with the same relative roughness as for the prototype:

$$(k_s)_r = Z_r \tag{14.23}$$

The Froude similitude (equation (14.22)) implies:

$$V_r = \sqrt{Z_r} \qquad \text{Velocity} \tag{14.24}$$

$$Q_r = V_r X_r Z_r = Z_r^{3/2} X_r \qquad \text{Discharge} \tag{14.25}$$

$$t_r = \frac{X_r}{V_r} = \frac{X_r}{\sqrt{Z_r}} \qquad \text{Time} \tag{14.26}$$

$$(\tan \theta)_r = \frac{Z_r}{X_r} \qquad \text{Longitudinal bed slope} \tag{14.27}$$

where θ is the angle between the channel bed and the horizontal.

With a distorted-scale model, it is possible to select small physical models (i.e. X_r large). In addition to the economical and practical benefits, distorted models have also the following advantages compared to non-distorted models:

- the flow velocities and turbulence in the model are larger (equation (14.24));
- the time scale is reduced (equation (14.26));
- the model Reynolds number is larger, improving the prototype-to-model dynamic similarity;
- the larger vertical scale (i.e. $Z_r < X_r$) allows a greater accuracy on the flow depth measurements.

Discussion

Practically it is recommended that the model distortion (i.e. the ratio X_r/Z_r) should be less than 5–10. Some disadvantages of distorted models may be mentioned for completeness: the velocity directions are not always reproduced correctly, and some observers might be distracted unfavourably by the model distortion leading to inaccurate or incorrect judgements.

Movable-bed models

Movable-bed hydraulic models are some of the most difficult types of models and they often give unsatisfactory results.

The primary difficulty is to scale both the sediment movement and the fluid motion. Furthermore, the bed roughness becomes a function of the bed geometry and of the sediment transport. Early movable-bed model studies on the River Mersey (England) and Seine River (France) in the 1880s showed that the time scale governing the fluid flow differs from the time scale governing sediment motion (see Appendix A3.1).

A detailed analysis of sediment transport modelling is developed in Appendix A3.1. Several authors (e.g. Henderson, 1996: pp. 497–508; Graf, 1971: pp. 392–398) also discussed various methods for 'designing' a movable-bed model.

The most important point is the need to verify and to calibrate a movable-bed model before using it as a prediction tool.

14.5.4 Resistance scaling

The modelling of flow resistance is not a simple matter. Often the geometric similarity of roughness height and spacing is not enough. For example, it is observed sometimes that the model does not reproduce the flow patterns in the prototype because the model is too 'smooth' or too 'rough'. In some cases (particularly with large-scale ratio L_r), the model flow is not as turbulent as the prototype flow. A solution is to use roughness elements (e.g. mesh, wire and vertical rods) to enhance the model flow turbulence, hence to simulate more satisfactorily the prototype flow pattern.

Another aspect is the scaling of the resistance coefficient. The flow resistance can be described in terms of the Darcy friction factor or an empirical resistance coefficient (e.g. Chézy or Gauckler–Manning coefficients).

In uniform equilibrium flows, the momentum equation implies:

$$V_r = \sqrt{L_r} = \sqrt{\frac{(D_H)_r (\sin \theta)_r}{f_r}} \tag{14.28}$$

For an undistorted model, a Froude similitude (equation (14.19) and (14.28)) implies that the model flow resistance will be similar to that in prototype:

$$f_r = 1 \tag{14.29}$$

Most prototype flows are fully rough turbulent and the Darcy friction factor is primarily a function of the relative roughness.

Another approach is based upon the Gauckler–Manning coefficient. The Chézy equation implies that, in gradually varied and uniform equilibrium flows, the following scaling relationship holds:

$$V_r = \sqrt{L_r} = \frac{1}{(n_{\text{Manning}})_r} ((D_H)_r)^{2/3} \sqrt{(\sin \theta)_r} \tag{14.30}$$

For an undistorted scale model, equation (14.30) becomes:

$$(n_{\text{Manning}})_r = L_r^{1/6} \tag{14.31}$$

Equation (14.31) indicates that the notion of complete similarity is applied both to the texture of the surface and to the shape of its general outline (Henderson, 1966). In practice, the lowest achievable value of n_{Manning} is about 0.009–0.010 s/m$^{1/3}$ (i.e. for glass). With such a value, the prototype resistance coefficient $(n_{\text{Manning}})_p$ and the Gauckler–Manning coefficient similarity $(n_{\text{Manning}})_r$ could limit the maximum geometrical similarity ratio L_r. If L_r is too small (typically less than 40), the physical model might neither be economical nor convenient.

In summary, a physical model (based upon a Froude similitude) has proportionally more resistance than the prototype. If the resistance losses are small (e.g. at a weir crest), the resistance-scale effects are not considered. In the cases of river and harbour modelling, resistance is significant. The matter may be solved using distorted models.

Distorted models

With a distorted-scale model, equations (14.28) and (14.30) become, respectively:

$$V_r = \sqrt{Z_r} = \sqrt{\frac{(D_H)_r (\sin \theta)_r}{f_r}} \tag{14.32}$$

$$V_r = \sqrt{Z_r} = \frac{1}{(n_{\text{Manning}})_r} ((D_H)_r)^{2/3} \sqrt{(\sin \theta)_r} \tag{14.33}$$

For a wide channel (i.e. $(D_H)_r = Z_r$) and a flat slope (i.e. $(\sin\theta)_r = (\tan\theta)_r$), the scaling of flow resistance in distorted models implies:

$$f_r = \frac{Z_r}{X_r} \qquad \text{Wide channel and flat slope} \qquad (14.34)$$

$$(n_{\text{Manning}})_r = \frac{Z_r^{2/3}}{\sqrt{X_r}} \qquad \text{Wide channel and flat slope} \qquad (14.35)$$

Discussion

In practice $Z_r/X_r < 1$ and equation (14.34) would predict a model friction factor larger than that in the prototype. But equation (14.35) could imply a model resistance coefficient larger or smaller than that in the prototype depending upon the ratio $Z_r^{2/3}/\sqrt{X_r}$!

14.6 DESIGN OF PHYSICAL MODELS

14.6.1 Introduction

Before building a physical model, the engineers must have the appropriate topographic and hydrological field information. Then the type of model must be selected and a question arises:

'Which is the dominant effect: e.g. viscosity, gravity and surface tension?'

14.6.2 General case

In the general case, the engineer must choose a proper geometric scale. The selection procedure is an iterative process:

1. Select the smallest geometric-scale ratio L_r to fit within the constraints of the laboratory.
2. For L_r, and for the similitude criterion (e.g. Froude or Reynolds), check if the discharge can be scaled properly in model, based upon the maximum model discharge $(Q_m)_{max}$.

For L_r and the similitude criterion, is the maximum model discharge large enough to model the prototype flow conditions?

---

3. Check if the flow resistance scaling is achievable in the model.

Is it possible to achieve the required f_m or $(n_{\text{Manning}})_m$ in the model?

---

4. Check the model Reynolds number Re_m for the smallest test flow rate.

For $(Q_p)_{min}$, what are the flow conditions in model: e.g. laminar or turbulent, smooth turbulent or fully rough-turbulent? If the prototype flow is turbulent, model flow conditions must be turbulent (i.e. typically $Re_m > 5000$).

---

5. Choose the convenient scale.

When a simple physical model is not feasible, more advanced modelling techniques can be used: e.g. a two-dimensional model (e.g. spillway flow) and a distorted-scale model (e.g. river flow).

14.6.3 Distorted-scale models

For a distorted-scale model, the engineer must select two (or three) geometric scales. The model design procedure is again an iterative process:

1. Select the smallest horizontal-scale ratio X_r to fit within the constraints of the laboratory.
2. Determine the possible range of vertical-scale Z_r such as:
   - the smallest-scale $(Z_r)_1$ is that which gives the limit of the discharge scaling ratio, based upon the maximum model discharge $(Q_m)_{max}$;
   - the largest-scale $(Z_r)_2$ is that which gives the feasible flow resistance coefficient (i.e. feasible f_m or $(n_{Manning})_m$);
   - check the distortion ratio X_r/Z_r (X_r/Z_r should be <5–10).
3. Check the model Reynolds number Re_m for the smallest test flow rate. This might provide a new largest vertical-scale ratio $(Z_r)_3$:
   - check the distortion ratio X_r/Z_r.
4. Select a vertical-scale ratio which satisfies: $(Z_r)_1 < Z_r < Min[(Z_r)_2, (Z_r)_3]$. If this condition cannot be satisfied, a smaller horizontal-scale ratio must be chosen:
   - check the distortion ratio X_r/Z_r; in practice it is recommended that X_r/Z_r should be <5–10.
5. Choose the convenient scales (X_r, Z_r).

14.7 SUMMARY

Physical hydraulic modelling is a design technique used by engineers to optimize the structure design, to ensure the safe operation of the structure and/or to facilitate the decision-making process.

In practice most hydraulic models are scaled with either a Froude or a Reynolds similitude: i.e. the selected dimensionless number is the same in model and prototype (i.e. full scale).

The most common fluids are air and water. Free-surface flow modelling is most often performed with the same fluid (e.g. water) in full scale and model. Fully enclosed flow modelling might be performed with water in prototype and air in model. The selection of fluid in model and prototype fixes the density-scale ratio ρ_r.

Table 14.1 summarizes the scaling ratios for the Froude and Reynolds similitudes.

Table 14.1 Scaling ratios for Froude and Reynolds similitudes (undistorted model)

Parameter	Unit	Scale ratio with		
		Froude law[a]	Froude law[a] (distorted model)	Reynolds law
(1)	(2)	(3)	(4)	(5)
Geometric properties				
Length	m	L_r	X_r, Z_r	L_r
Area	m^2	L_r^2	–	L_r^2
Kinematic properties				
Velocity	m/s	$\sqrt{L_r}$	$\sqrt{Z_r}$	$1/L_r \times \mu_r/\rho_r$
Discharge per unit width	m^2/s	$L_r^{3/2}$	$Z_r^{3/2}$	μ_r/ρ_r
Discharge	m^3/s	$L_r^{5/2}$	$Z_r^{3/2}X_r$	$L_r\mu_r/\rho_r$
Time	s	$\sqrt{L_r}$	$X_r/\sqrt{Z_r}$	$L_r^2\rho_r/\mu_r$
Dynamic properties				
Force	N	$\rho_r L_r^3$	–	μ_r^2/ρ_r
Pressure	Pa	$\rho_r L_r$	$\rho_r Z_r$	$\mu_r^2/\rho_r \times 1/L_r^2$
Density	kg/m^3	ρ_r	ρ_r	ρ_r
Dynamic viscosity	Pa s	$L_r^{3/2}\sqrt{\rho_r}$	–	μ_r
Surface tension	N/m	L_r^2	–	$\mu_r^2/\rho_r \times 1/L_r$

Note: [a]Assuming identical gravity acceleration in model and prototype.

14.8 EXERCISES

A butterfly valve is to be tested in laboratory to determine the discharge coefficient for various openings of the disc. The prototype size will be 2.2 m in diameter and it will be manufactured from cast steel with machined inside surfaces (roughness height estimated to be about 0. 5 mm). The maximum discharge to be controlled by the valve is 15 m³/s. The laboratory model is at a 5:1 scale model.

(a) What surface condition is required in the model? What model discharge is required to achieve complete similarity with the prototype, if water is used in both? (b) Can these conditions be achieved? (c) If the maximum flow available for model tests is 200 l/s, could you accurately predict prototype discharge coefficients from the results of the model tests?

Summary sheet

(a) $(k_s)_m =$	$Q_m =$
(b) Yes/No	Reasons:
(c) $Re_p =$	Re_m
Discussion:	

The inlet of a Francis turbine is to be tested in laboratory to determine the performances for various discharges. The prototype size of the radial flow rotor will be inlet diameter = 0.6 m, width = 0.08 m and inlet cross-flow area = $\pi \times$ diameter $\times$ width. It will be manufactured from cast steel with machined inside surfaces (roughness height estimated to be about 0.3 mm). The maximum discharge to be turbined (by the Francis wheel) is 1.4 m³/s. The laboratory model is at a 5:1 scale model.

(a) What surface condition is required in the model? What model discharge is required to achieve complete similarity with the prototype, if water is used in both? (b) Can these conditions be achieved? (Compute the minimum required model total head and flow rate. Compare these with the pump performances of a typical hydraulic laboratory: $H \sim$ 10 m and $Q \sim$ 100 l/s.) (c) If the maximum flow available for model tests is 150 l/s, would you be able to accurately predict prototype performances from the results of the model tests? (Justify your answer.)

Summary sheet

(a) $(k_s)_m =$	$Q_m =$
(b) Yes/No	Reasons:
(c) $Re_p =$	Re_m
Discussion:	

An overflow spillway is to be designed with an uncontrolled crest followed by a stepped chute and a hydraulic jump dissipator. The maximum spillway capacity will be 4300 m³/s. The width of the crest, chute and dissipation basin will be 55 m. A 50:1 scale model of the spillway is to be built. Discharges ranging between the maximum flow rate and 10% of the maximum flow rate are to be reproduced in the model.

(a) Determine the maximum model discharge required. (b) Determine the minimum prototype discharge for which negligible-scale effects occur in the model. (Comment on your result.) (c) What will be the scale for the force ratio?

Laboratory tests indicate that operation of the basin may result in unsteady wave propagation downstream of the stilling basin with model wave amplitude of about 0.05 m and model wave period of 47 s. Calculate (d) the prototype wave amplitude (e) the prototype wave period.

Summary sheet

(a) Maximum Q_m =	
(b) Minimum Q_p =	Why?
(c) Force$_r$ =	
(d) A_p =	
(e) t_p =	

A 35.5:1 scale model of a concrete overfall spillway and stilling basin is to be built. The prototype discharge will be 200 m³/s and the spillway crest length is 62 m (a) Determine the maximum model discharge required and the minimum prototype discharge for which negligible-scale effects occur in the model. (b) In tests involving baffle blocks for stabilizing the hydraulic jump in the stilling basin, the force measured on each block was 9.3 N. What is the corresponding prototype force? (c) The channel downstream of the stilling basin is to be lined with rip-rap (angular blocks of rock) approximately 650 mm in size. The velocity measured near the rip-rap is as low as 0.2 m/s. Check whether the model Reynolds number is large enough for the drag coefficient of the model rocks to be the same as in the prototype. What will be the scale for the force ratio?

Summary sheet

(a) Maximum Q_m = Minimum Q_p =	Why?
(b) Force$_p$ =	
(c) Re_m = Force ratio =	Comment:

A sluice gate will be built across a 25 m wide rectangular channel. The maximum prototype discharge will be 275 m³/s and the channel bed will be horizontal and concrete lined (i.e. smooth). A 35:1 scale model of the gate is to be built for laboratory tests.

(a) What similitude should be used? (b) Calculate model width. (c) Calculate maximum model flow rate.

For one particular gate opening and flow rate, the *laboratory* flow conditions are upstream flow depth of 0.2856 m and downstream flow depth of 0.0233 m. (d) Compute the model discharge. State the basic principle(s) involved. (e) Compute the model force acting on the sluice gate. State the basic principle(s) involved. (f) What will be the corresponding prototype discharge and force on gate? (g) What will be the scale for the force ratio?

Gate operation may result in unsteady flow situations. If a prototype gate operation has the following characteristics: gate opening duration = 15 min, initial discharge = 180 m³/s and new discharge = 275 m³/s, calculate the following: (h) Gate opening duration. (i) Discharges to be used in the model tests.

Summary sheet

(a) Similitude	Why?
(b) B_m =	
(c) $(Q_m)_{max}$ =	
(d) Q_m =	Principle(s):
(e) F_m =	Principle(s):
(f) Q_p = $Force_p$ =	
(g) $Force_r$ =	
(h) t_m =	
(i) Q_m = (before)	Q_m = (after)

A hydraulic jump stilling basin, equipped with baffle blocks, is to be tested in laboratory to determine the dissipation characteristics for various flow rates. The maximum prototype discharge will be 220 m³/s and the rectangular channel will be 10 m wide. (Assume the channel bed to be horizontal and concrete lined, i.e. smooth.) A 40:1 scale model of the stilling basin is to be built. Discharges ranging between the maximum flow rate and 10% of the maximum flow rate are to be reproduced in the model.

(a) What similitude should be used? (Justify your selection.) (b) Determine the maximum model discharge required. (c) Determine the minimum prototype discharge for which negligible-scale effects occur in the model. (Comment your result.)

For one particular inflow condition, the *laboratory* flow conditions are upstream flow depth of 0.019 m, upstream flow velocity of 2.38 m/s and downstream flow depth of 0.122 m (d) Compute the model force exerted on the baffle blocks. (State the basic principle(s) involved.) (e) What is the direction of force in (d) i.e. upstream or downstream? (f) What will be the corresponding prototype force acting on the blocks? (g) Compute the prototype head loss.

Operation of the basin may result in unsteady wave propagation downstream of the stilling basin. (h) What will be the scale for the time ratio?

Tests will be made on a model sea wall of 1/18 prototype size. (a) If the prototype wave climate is wave period = 12 s, wavelength = 20 m and wave amplitude = 2.1 m, what wave period, wavelength and wave amplitude should be used in the model tests? (b) If the maximum force exerted by a wave on the model sea wall is 95 N, what corresponding force will be exerted on the prototype?

Summary sheet

(a) $t_m =$ $L_m =$	
$A_m =$	
(b) Force$_p =$	

A fixed bed model is to be built of a certain section of a river. The maximum full-scale discharge is 2750 m³/s, the average width of the river is 220 m and the bed slope is 0.16 m/km. The Gauckler–Manning coefficient for the prototype is estimated at 0.035 s/m$^{1/3}$. Laboratory facilities limit the scale ratio to 200:1 and maximum model discharge is 45 l/s. Note that the smoothest model surface feasible has a Gauckler–Manning coefficient of about 0.014 s/m$^{1/3}$. Discharges ranging between the maximum flow rate and 15% of the maximum flow rate are to be reproduced in the model.

Determine the acceptable maximum and minimum values of the vertical-scale ratio Z_r. Select a suitable scale for practical use, and calculate the corresponding model values of the Gauckler–Manning coefficient, maximum discharge and normal depth (at maximum discharge). (It can be assumed that the river channel is wide (i.e. $D_H \sim 4d$) on both model and prototype for all flows. Assume that uniform equilibrium flow is achieved in model and prototype.)

Summary sheet

Minimum $Z_r =$	Why?
Maximum $Z_r =$	Why?
Alternative Maximun $Z_r =$	Why?
Allowable range for $Z_r =$	
Your choice for $Z_r =$	Why?
Corresponding values of $Q_r =$	$(n_{Manning})_r =$
$(n_{Manning})_m =$	Maximum $Q_m =$
$d_m =$	At maximum flow rate

An artificial concrete channel model is to be built. Laboratory facilities limit the scale ratio to 50:1 and maximum model discharge is 50 l/s. The maximum full-scale discharge is 150 m³/s, the cross-section of the channel is approximately rectangular (50 m bottom width) and the bed slope is 0.14 m per km. (*Note*: The roughness height of the prototype is estimated as 3 mm while the smoothest model surface feasible has a Darcy friction factor of about $f = 0.03$.) Discharges ranging between the maximum flow rate and 10% of the maximum flow rate are to be reproduced in the model.

For an undistorted model. (a) What would be the model discharge at maximum full-scale discharge? (b) What would be the Darcy coefficient of the model flow? (c) What would be the Darcy coefficient of the prototype channel? (d) Comment and discuss your findings. (Assume normal flow conditions.)

A distorted model is to be built. (e) Determine the acceptable maximum and minimum values of the vertical-scale ratio Z_r. (f) Select a suitable scale for practical use. (g) Calculate the corresponding model values of Darcy coefficient. (h) Calculate the corresponding model values of maximum discharge. (i) Calculate the corresponding model values of normal depth (at maximum discharge).

A fixed bed model is to be made of a river with a surface width of 80 m. The Gauckler–Manning coefficient for the river is estimated at $0.026 \, \text{s/m}^{1/3}$. Scale ratios of $X_r = 150$ and $Z_r = 25$ have been selected: (a) Find the required model values of the Gauckler–Manning coefficient corresponding to prototype depths of water of 2.0 and 5.0 m, if the cross-sectional shape is assumed to be rectangular. (b) What material would you recommend to use in the laboratory model for a prototype depth of 2.0 m?

A broad-crested weir is to be designed to operate with critical flow above its crest and no downstream control. The maximum discharge capacity will be $135 \, \text{m}^3/\text{s}$. The weir crest will be 2.5 m above the natural channel bed and its width will be 15 m. A 20:1 scale model of the weir is to be built. Discharges ranging between the maximum flow rate and 10% of the maximum flow rate are to be reproduced in the model.

Calculate the upstream and upstream flow depths at maximum flow rate on prototype, the horizontal component of the sliding force acting on the weir at maximum flow rate.

Determine the maximum model discharge required and the minimum prototype discharge for which negligible-scale effects occur in the model.

Solution: (a) Prototype: $d_1 = 5.4 \, \text{m}$, $d_2 = 0.95 \, \text{m}$, $F = 1 \, \text{MN}$. (b) Model: $(Q_m)_{max} = 0.075 \, \text{m}^3/\text{s}$, $(Q_p)_{min} = 2.5 \, \text{m}^3/\text{s}$.

A (prototype) sluice gate is installed in a 50 m wide rectangular channel. The flow depth upstream of the gate is 3.2 m and the flow rate is $75 \, \text{m}^3/\text{s}$. The bed is horizontal and smooth.

(a) Calculate the prototype downstream flow depth, and the force acting on the gate at design flow conditions. The prototype sluice gate experienced some unusual vibrations (period: 1.5 s) *for the above flow conditions* and a 12:1 scale model of the sluice gate is to be built. (b) Determine the model flow rate, the downstream flow velocity in the model, the (expected) model vibration period and the minimum prototype discharge for which negligible-scale effects occur in this model. (*This last question is a general question, not related to the above flow conditions.*)

Solution: (a) $d_2 = 0.195 \, \text{m}$, $F = 2 \times 10^6 \, \text{N}$. (b) $Q_m = 0.150 \, \text{m}^3/\text{s}$, $(d_2)_m = 0.016 \, \text{m}$, $T_m = 0.4 \, \text{s}$, $(Q_p)_{min} = 3.9 \, \text{m}^3/\text{s}$.

Numerical modelling of steady open channel flows: backwater computations

15.1 INTRODUCTION

Numerical models of hydraulic flows are computer programs designed to solve the basic fluid mechanics equations. They are applied typically to simple flows with uncomplicated boundary conditions for which the basic equations can be numerically integrated and are meaningful. The calibration and validation of computational models are extremely difficult, and most computer models are applicable only to a very specific range of flow conditions and boundary conditions.

In the present section, we present an example of hydraulic modelling: the numerical integration of the energy equation for gradually varied flows (GVFs) in open channel. It is one of the simplest forms of numerical model: i.e. the numerical integration of a steady one-dimensional flow.

15.2 BASIC EQUATIONS

15.2.1 Presentation

Considering a steady open channel flow and along a streamline in the s-direction, the energy equation written in terms of mean total head is:

$$\frac{\partial H}{\partial s} = \frac{\partial}{\partial s}\left(z + \frac{P}{\rho g} + \frac{v^2}{2g}\right)_{\text{mean}} = -f\frac{1}{D_H}\frac{V^2}{2g} \tag{15.1a}$$

where s is the distance along the channel bed (Fig. 15.1), z is the elevation (positive upwards), P is the pressure, v is the fluid velocity along a streamline, f is the Darcy friction factor, D_H is

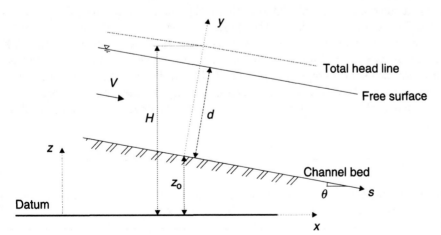

Fig. 15.1 Sketch of open channel flow.

the hydraulic diameter, V is the mean flow velocity (i.e. $V = Q/A$) and the subscript 'mean' refers to the mean total head over the cross-sectional area. In the following development, we shall consider the mean flow properties only.

The energy equation (i.e. equation (15.1)) is also called the *backwater equation*. It is usually rewritten as:

$$\frac{\partial H}{\partial s} = -S_f \tag{15.1b}$$

where S_f is the friction slope defined as:

$$S_f = \frac{\tau_0}{\rho g D_H} = f \frac{1}{D_H} \frac{V^2}{2g} \tag{15.2}$$

where τ_0 is the boundary shear stress and D_H is the hydraulic diameter.

Notes
1. S_f is called the friction slope or energy slope. That is, it is the slope of the total energy line (or total headline – THL) (Fig. 15.1).
2. Assuming a hydrostatic pressure gradient, equation (15.1) can be transformed as:

$$\frac{\partial H}{\partial s} = \frac{\partial}{\partial s}\left(z_0 + d \cos\theta + \alpha \frac{V^2}{2g}\right) = -S_f$$

   where z_0 is the bed elevation, d is the flow depth measured normal to the channel bottom, θ is the longitudinal bed slope and α is the kinetic energy correction coefficient (i.e. Coriolis coefficient).
3. Equation (15.1a) is the basic backwater equation, first developed by J.B. Bélanger (Bélanger, 1828). It is a one-dimensional model of gradually varied steady flows in open channel.
4. Further developments of the backwater equation are presented in Appendix A3.2.

15.2.2 Basic assumptions

Equation (15.1) may be applied to natural and artificial channels. It is important to remember the basic assumptions (Chapter 5).

The backwater calculations are developed assuming [H1] a non-uniform flow, [H2] a steady flow, [H3] that the flow is gradually varied and [H4] that, at a given section, the flow resistance is the same as for a uniform flow for the same depth and discharge, regardless of trends of the depth.

The GVF calculations (i.e. backwater calculations) neither apply to uniform equilibrium flows nor to unsteady flows nor to rapidly varied flows (RVFs). Furthermore, the last assumption [H4] implies that the Darcy, Chézy or Gauckler–Manning equations may be used to estimate the flow resistance, although these equations were originally developed for uniform flows only (see Section 15.2.4).

Note
In GVFs, the friction slope and Darcy coefficient are calculated as in uniform equilibrium flow (see Chapter 4) but using the local non-uniform equilibrium flow depth.

15.2.3 Applications

In natural streams, the cross-sections are irregular and variable, and the location of the free-surface elevation in GVFs is predicted using equation (15.1). Furthermore, the bed slope is usually small in natural systems (i.e. typically $\tan\theta < 0.01$). For such small slopes, the following approximation holds: $\tan\theta \approx \sin\theta \approx \theta$, where θ is in radians. The flow depth (measured normal to the channel bed) approximately equals the free-surface elevation above channel bed within 0.005%. These variables are used interchangeably.

At hydraulic structures and spillways, the flow is often supercritical. The channel slope may range from zero to steep (over 75°). Negative slopes might also take place (e.g. flip bucket). With such structures, it is extremely important to assess first whether GVF calculations (i.e. backwater calculations) can be applied or not (see Section 15.2.2). In the affirmative, it is frequent that the velocity distribution is not uniform and the pressure distribution might not be hydrostatic.

In man-made channels, simpler forms of the backwater equation are available. Assuming a hydrostatic pressure distribution, the backwater equation may be expressed in terms of the flow depth:

$$\frac{\partial d}{\partial s}(\cos\theta - \alpha Fr^2) = S_\mathrm{o} - S_\mathrm{f} \tag{15.3}$$

where $S_\mathrm{o} = \sin\theta$, α is the Coriolis coefficient (or kinetic energy correction coefficient) and Fr is the Froude number defined as:

$$Fr = \frac{Q}{\sqrt{g(A^3/B)}} \tag{15.4}$$

where A is the flow cross-sectional area and B is the open channel free-surface width.

Notes

1. If the velocity varies across the section, the mean of the velocity head $(v^2/2g)_\mathrm{mean}$ is not equal to $V^2/2g$, where the subscript 'mean' refers to the mean value over a cross-section. The ratio of these quantities is called the Coriolis coefficient, denoted as α and defined as:

$$\alpha = \frac{\int_A \rho v^3\,\mathrm{d}A}{\rho V^3 A}$$

2. In artificial channels of simple shape, equation (15.3) can be integrated analytically (e.g. Chow, 1973: pp. 237–242) or semi-analytically (e.g. Henderson, 1966: pp. 130–136; Chow, 1973: pp. 252–262).
3. Equation (15.4) is a common definition of the Froude number for a channel of irregular cross-sectional shape.
4. For a rectangular channel, the Froude number equals: $Fr = V/\sqrt{gd}$.

15.2.4 Discussion: flow resistance calculations

For channels extending over several kilometres, the boundary shear opposes the fluid motion and retards the flow. The flow resistance and gravity effects are of the same order of magnitude and it is essential to accurately predict the flow resistance. In fact, the friction force exactly equals and opposes the weight force component in the flow direction at equilibrium (i.e. normal flow conditions).

The flow resistance caused by friction is given by the Darcy equation:

$$\Delta H = f \frac{\Delta s}{D_H} \frac{V^2}{2g} \tag{15.5a}$$

where ΔH is the total head loss over a distance Δs (in the flow direction), f is the Darcy coefficient, V is the mean flow velocity and D_H is the hydraulic diameter or equivalent pipe diameter. Using the definition of the friction slope (equation (15.2)), equation (15.5a) can be rewritten as:

$$V = \sqrt{\frac{8g}{f}} \sqrt{\frac{D_H}{4} S_f} \tag{15.5b}$$

Equation (15.5b) is sometimes called the Chézy equation. It is valid for GVFs and for uniform equilibrium flows.

Discussion

Note the similarity between equations (15.1a) and (15.5a). Equation (15.5a) may be rewritten as:

$$\frac{\Delta H}{\Delta s} = f \frac{1}{D_H} \frac{V^2}{2g} \qquad \text{Head loss gradient}$$

where ΔH is the head loss (always positive).

Empirical resistance coefficients

The Chézy equation was originally introduced (in 1768) as an empirical correlation by the French engineer A. Chézy:

$$V = C_{\text{Chézy}} \sqrt{\frac{D_H}{4} S_f} \tag{15.6}$$

where $C_{\text{Chézy}}$ is the Chézy coefficient (Units: $\text{m}^{1/2}/\text{s}$), ranging typically from $30\,\text{m}^{1/2}/\text{s}$ (small rough channel) up to $90\,\text{m}^{1/2}/\text{s}$ (large smooth channel).

An empirical formulation (called the Gauckler–Manning formula) was deduced from the Chézy equation:

$$V = \frac{1}{n_{\text{Manning}}} \left(\frac{D_H}{4} \right)^{2/3} \sqrt{\sin \theta} \tag{15.7}$$

where n_{Manning} is the Gauckler–Manning coefficient (Units: $\text{s}/\text{m}^{1/3}$). The Gauckler–Manning coefficient was assumed to be a characteristic of the surface roughness alone. Such an approximation might be reasonable as long as the water is neither too shallow nor the channel too narrow. But the assumption is not strictly correct. It can be shown that, for a given channel, the Gauckler–Manning coefficient depends upon the discharge, flow depth and relative roughness.

Notes
1. The Chézy equation was introduced in 1768 by A. Chézy when designing canals for the water supply of the city of Paris. Some researchers assumed wrongly that $C_{Chézy}$ would be a constant, but the original study never implied such an assumption.
2. The Gauckler–Manning formula was proposed first in 1867 (Gauckler, 1867). In 1889, R. Manning proposed two formulae: one became known as the 'Gauckler–Manning formula' but Manning preferred to use the second formula that he presented in his paper (Manning, 1890)!

Discussion

The friction slope may be expressed in terms of the Darcy, Chézy or Gauckler–Manning coefficients:

$$S_f = f \frac{1}{D_H} \frac{V^2}{2g} \tag{15.2}$$

$$S_f = \frac{1}{C_{Chézy}^2} \frac{V^2}{D_H/4} \tag{15.8}$$

$$S_f = n_{Manning}^2 \frac{V^2}{\left(D_H/4\right)^{4/3}} \tag{15.9}$$

where V is the mean flow velocity.

In practice, equation (15.2) must be the preferred friction slope definition. In the case of complex bed forms, equations (15.8) or (15.9) might be used if some information on $C_{Chézy}$ or $n_{Manning}$ is available.

15.3 BACKWATER CALCULATIONS

15.3.1 Presentation

In the next sections, we will develop a basic case of backwater calculations. First, a general method is described. Then calculations are developed and commented. At each step, a practical case is shown. The same test case is considered: i.e. a reservoir discharging into a channel, followed by an overfall (Fig. 15.2).

15.3.2 Method

Firstly, a sketch of the longitudinal profile of the channel must be drawn. Control structures (e.g. gate and weir) must be included.

Secondly, the location of 'obvious' control sections[1] must be highlighted as at these locations the flow depth is known. Classical control sections include sluice gate, overflow gate, broad-crested weir, hydraulic jump, overfall and bottom slope change from very flat to very steep (or from very steep to very flat).

---

[1]A control section is defined as the location where critical flow occurs (see Chapter 3).

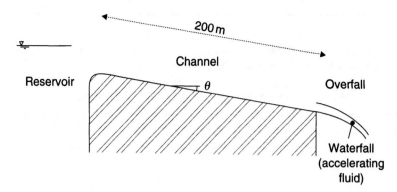

Fig. 15.2 Sketch of a long channel carrying water from a reservoir to an overfall.

Thirdly, the uniform flow properties must be computed at each position along the channel. In practice, uniform flow conditions must be re-computed after *each* change of bottom slope, total discharge, cross-sectional shape (e.g. from rectangular to trapezoidal), cross-sectional characteristics (e.g. bottom width) and boundary roughness (i.e. bottom and sidewalls).

Then plot the uniform flow properties on the longitudinal sketch of the channel. Highlight each change of flow regime (i.e. sub- to supercritical flow and super- to subcritical flow): these are control sections.

Lastly sketch the composite free-surface profiles.

This first analysis is *uppermost important.* An incorrect estimate of the composite profile may lead to wrong or false results which are neither necessarily detected by computer programs nor 'computer experts'. It is the responsibility of the hydraulic engineer to state clearly and accurately the correct boundary conditions.

Backwater calculations start from positions of known flow depth. Calculations may be performed in the upstream and downstream flow directions up to the next control sections. At a control section, the type of control affects the flow computation through the control section. That is, the Bélanger equation (i.e. momentum equation) must be applied for a hydraulic jump; the Bernoulli equation (and specific energy concept) is applied for a gate (or a weir); for a smooth change from sub- to supercritical flow (e.g. spillway crest), calculations continue downstream of the critical flow depth section.

Application

Let us consider the outflow from a reservoir of constant free-surface elevation into a long channel (200 m length). The channel ends by a free overfall (at the downstream end) (Fig. 15.2). The channel has a rectangular cross-section (width: 3.5 m). The channel bed is made of smooth concrete and the equivalent roughness height of bed and sidewalls is $k_s = 2$ mm. Assuming that the flow rate in the channel is 5 m³/s, sketch the composite profile for each of the following channel slopes:

(a) Case 1: $\theta = 0.02°$.
(b) Case 2: $\theta = 1.4°$.

Solution

From the first sketch (Fig. 15.2), we note two regions where the flow characteristics are obvious:

1. In the reservoir, the fluid is quiescent (still). The flow must be subcritical.
2. Downstream of the overfall crest, the free-falling flow is accelerating. The flow must be supercritical.

First result: there must be (at least) one control section between the reservoir and the waterfall as the flow regime must change from subcritical in the reservoir to supercritical at the overfall.

For a constant-slope (and constant-width) channel, one set of uniform flow conditions must be computed. The normal flow computations give:

(a) Case 1: $\theta = 0.02°$, $d_0 = 1.33$ m, $V_0 = 1.07$ m/s, $Fr_0 = 0.30$.
(b) Case 2: $\theta = 1.4°$, $d_0 = 0.307$ m, $V_0 = 4.65$ m/s, $Fr_0 = 2.68$.

In Case 1, the channel flow will tend towards subcritical normal flow conditions (i.e. $Fr_0 = 0.3$) while in Case 2 the waters will accelerate towards supercritical equilibrium flow conditions ($Fr_0 = 2.7$).

Conclusions
In Case 1, the flow is subcritical all along the channel. At the downstream end of the channel, the free-falling water is accelerated and become supercritical. Critical flow conditions are observed only near the end of the channel (see discussion in Henderson, 1966: pp. 28–29 and 191–202). The complete composite profile is plotted in Fig. 15.3a. Note that the flow in the reservoir is also subcritical: i.e. there is no control section at the upstream end of the channel.

In Case 2, the channel flow is supercritical. At the channel intake, there is a change of flow regime from subcritical (in the reservoir) to supercritical. This is a control section (Fig. 15.3b). At the downstream of the channel, the flow is accelerated and remains supercritical: i.e. critical flow conditions are not observed at the overfall.

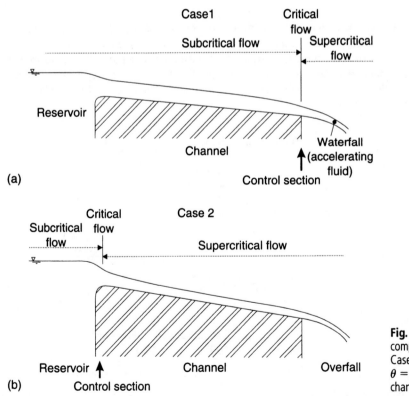

Fig. 15.3 Sketch of the composite profiles: (a) Case 1: channel slope, $\theta = 0.02°$. (b) Case 2: channel slope, $\theta = 1.4°$.

15.3.3 Calculations

Firstly, the calculations must start from positions of known flow depth. Secondly, the reader must remember that subcritical flows are controlled from downstream while supercritical flows are controlled from upstream. In practice, it is strongly advised to start backwater calculations of subcritical flow from the downstream control and to proceed in the upstream direction. Similarly, it is recommended to start backwater computations of supercritical flow from the upstream control and to proceed in the flow direction.

The calculations must be conducted as detailed in Section 5.1. An example is detailed below.

Application

Let us consider the same example: the outflow from a reservoir of constant free-surface elevation into a long channel (200 m length). The channel ends by a free overfall at the downstream end
(Fig. 15.2). The channel has a rectangular cross-section (width: 3.5 m). The equivalent roughness
height of the channel bed is k_s = 2 mm.

Assuming that the flow rate in the channel is 5 m³/s, compute the backwater profile for the following two different channel slopes:

(a) Case 1: θ = 0.02°.
(b) Case 2: θ = 1.4°.

Subsidiary question: in each case, what is the reservoir free-surface elevation above the channel
intake (crest)?

Solution

Case 1: In Case 1, a control section is located at the downstream end of the channel (Fig. 15.3a). At
that location (i.e. s = 200 m) the flow depth equals the critical flow depth d_c = 0.593 m (in first
approximation). Backwater calculations are performed from the control section in the upstream flow
direction up to the channel end. Results are summarized in Fig. 15.4. Note that, near the upstream
end of the channel, the flow depth equals 0.96 m: i.e. normal flow conditions do not take place
because the channel is too short.

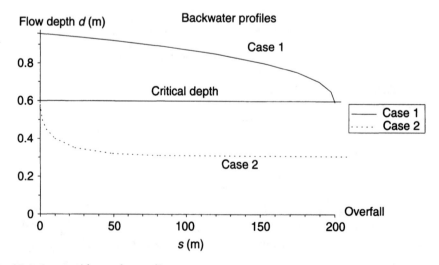

Fig. 15.4 Computed free-surface profiles.

The flow depth in the reservoir is deduced by assuming a smooth transition (i.e. frictionless) between the reservoir and the channel. The application of the Bernoulli equation indicates that the total head in the reservoir is the same as at the channel intake (where $d = 0.96$ m and $V = 1.49$ m/s). As the velocity in the reservoir is negligible, the reservoir free-surface elevation must be 1.07 m above the channel bed (at intake).

Case 2: In Case 2, the control section is located at the upstream end of the channel (Fig. 15.3b). At that location (i.e. $s = 0$ m) the flow depth equals the critical flow depth $d_c = 0.593$ m. Backwater calculations are performed from the control section into the downstream flow direction up to the downstream channel end. Results are summarized in Fig. 15.4. Near the downstream end, the flow depth equals 0.31 m: i.e. the flow is almost uniform equilibrium.

The flow depth in the reservoir is deduced by assuming a smooth flow transition (frictionless) between the reservoir and the channel. The application of the Bernoulli equation indicates that the total head in the reservoir is the same as at the channel intake (where $d = d_c = 0.593$ m and $V = V_c = 2.4$ m/s). As the velocity in the reservoir is negligible, the free-surface elevation above the channel bed (at intake) equals 0.889 m.

15.3.4 Comments

The above application is a simple case with only one control section and two distinctive boundary conditions (upstream reservoir and downstream overfall) (Fig. 15.3). Usually, backwater profiles include more control sections. Particular controls might be required to apply the Bernoulli equation (e.g. gate) or the momentum equation (e.g. hydraulic jump) to continue the calculations. If the longitudinal channel profile includes a relatively long prismatic channel, uniform equilibrium flow conditions may also be attained.

Remember that uniform equilibrium flows, hydraulic jumps and control sections are some form of singularity. Altogether backwater calculations are a difficult exercise, and they require practice and experience. Note that, in the previous applications (Figs. 15.2–15.4), two regions of RVF exist, although the channel flow is gradually varied: at the channel intake and at the overfall. In the region of RVF, the backwater calculations cannot accurately predict the flow properties.

15.4 NUMERICAL INTEGRATION

15.4.1 Introduction

The backwater equation (equation (15.1a)) is a non-linear equation which cannot be solved analytically, but it can be integrated numerically. One of the most common integration methods is the standard step method.

The method developed herein (i.e. step method/depth calculated from distance) is used by several numerical models (e.g. HEC, Table 15.1).

Comments

There are two integration step methods: the 'step method/distance calculated from depth' and the 'step method/depth calculated from distance'. Henderson (1966) discussed each one in detail.

Note that the 'step method/depth calculated from distance' (developed below and used in several softwares) assumes a mild slope with subcritical flow or steep slope with supercritical flow. It is indeterminate for critical flow conditions!

Table 15.1 Examples of computational models for open channel flow calculations

Model (1)	Description (2)	Remarks (3)
HydroChan	One-dimensional model for steady flow down flat slope (assuming hydrostatic pressure distribution and prismartic cross-section). Flow resistance computed using the Darcy friction factor (fully rough-turbulent flow), Gauckler–Manning formula or Keulegan friction coefficient.	Shareware software developed by Hydrotools.
HEC-1 and HEC-2	One-dimensional model for steady subcritical flow down flat slope (assuming hydrostatic pressure distribution). Flow resistance computed using the Gauckler–Manning formula.	Developed by USACE. Standard step method/depth calculated from distance.
WES	One-dimensional model for supercritical flow in curved channels. Flow resistance calculated using the Colebrook–White formula.	Developed by USACE-WES.
MIKE-11	One-dimensional model.	Developed by DHI.
GENFLO2D	Two-dimensional model (analytical solution of Navier–Stokes equation). Steady quasi-uniform flows in complex compound channels.	Developed by Water Solutions Mixing length turbulence model.
TELEMAC	Two-dimensional model (finite element method): Saint–Venant equations. Flow resistance computed using the Strickler formula.	Developed by SOGREAH.
TABS-2/RAM-2	Two-dimensional model (finite element method). RAM-2 is the steady flow hydraulic model.	Developed by USACE-WES.

15.4.2 Standard step method (depth calculated from distance)

The basic equation is:

$$\frac{\partial H}{\partial s} = -S_f \tag{15.1b}$$

where H is the mean total head over the cross-section and S_f is the friction slope. Assuming a hydrostatic pressure gradient, the mean total head equals:

$$H = z_o + d \cos \theta + \alpha \frac{V^2}{2g} \tag{15.10}$$

where z_o is the bed elevation, d is the flow depth measured normal to the channel bottom, θ is the channel slope, α is the Coriolis coefficient, V is the mean velocity and g is the gravity constant (Fig. 15.1).

Considering the GVF sketched in Fig. 15.5, the backwater equation can be integrated between two stations denoted $\{i\}$ and $\{i + 1\}$:

$$\frac{H_{i+1} - H_i}{s_{i+1} - s_i} \approx -\frac{1}{2}(S_{f_{i+1}} + S_{f_i}) \tag{15.11a}$$

where s is the streamwise coordinate, and the subscripts i and $i + 1$ refer to stations $\{i\}$ and $\{i + 1\}$, respectively. It yields:

$$H_{i+1} - H_i \approx -\frac{1}{2}(S_{f_{i+1}} + S_{f_i})(s_{i+1} - s_i) \tag{15.11b}$$

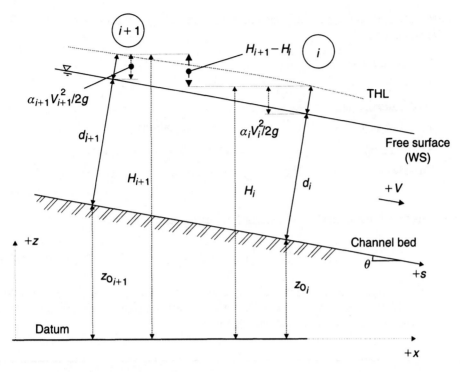

Fig. 15.5 Backwater calculations: definition sketch.

If the flow conditions at station $\{i\}$ are known, the total head at station $\{i + 1\}$ equals:

$$H_{i+1} \approx H_i - \frac{1}{2}(S_{f_{i+1}} + S_{f_i})(s_{i+1} - s_i) \qquad (15.11c)$$

Notes

1. The elevation (i.e. altitude) is positive upwards.
2. By convention, the downstream direction is the positive direction.
3. The calculations are solved iteratively because the friction slope at station $\{i + 1\}$ is unknown beforehand.

15.4.3 Computational algorithm

The calculations can be performed only if the total flow rate Q is known.

The calculation steps are:

Step 1. At station $\{i\}$ (i.e. $s = s_i$), the water level and hence the flow depth are known. The following variables can be deduced

- Cross-sectional area A_i
- Wetted perimeter P_{w_i}
- Hydraulic diameter $D_{H_i} = 4A_i/P_{w_i}$
- Flow velocity $V_i = Q/A_i$ (continuity equation)
- Friction slope S_{f_i} (equation (15.2), (15.8) or (15.9))
- Total head $H_i = z_{0_i} + d_i \cos \theta_i = \alpha_i V_i^2/2g$

Step 2. Estimate (i.e. guess) the water level (or flow depth $d_{i+1}^{(1)}$) at station $\{i + 1\}$ (i.e. $s = s_{i+1}$).
Step 3. Calculate then:

- Cross-sectional area $\quad A_{i+1}^{(1)}$
- Wetted perimeter $\quad P_{\mathrm{w}_{i+1}}^{(1)}$
- Hydraulic diameter $\quad D_{\mathrm{H}_{i+1}}^{(1)}$
- Flow velocity $\quad V_{i+1}^{(1)} \quad$ (continuity equation)
- Friction slope $\quad S_{\mathrm{f}_{i+1}}^{(1)} \quad$ (equation (15.2), (15.8) or (15.9))
- Total head $\quad H_{i+1}^{(1)} = z_{0_{i+1}} + d_{i+1}^{(1)} \cos \theta_{i+1} + \alpha_{i+1}^{(1)} (V_{i+1}^{(1)})^2/(2g)$

$$\text{(equation (15.10))}$$

Step 4. Check the calculations by calculating $H_{i+1}^{(*)}$ using equation (15.11c):

$$H_{i+1}^{(*)} \approx H_i - \frac{1}{2}(S_{\mathrm{f}_{i+1}^{(1)}} + S_{\mathrm{f}_i})(s_{i+1} - s_i) \qquad \text{(equation (15.11C))}$$

Step 5. If $H_{i+1}^{(*)} = H_{i+1}^{(1)}$ with an acceptable accuracy, the estimate of $d_{i+1}^{(1)}$ is acceptable, and $d_{i+1} = d_{i+1}^{(1)}$. And then we can proceed to the next interval. In the negative, we must repeat the Steps 2–5 until an acceptable accuracy is obtained.

Note
For a full-scale channel, an acceptable accuracy might be: $H_{i+1}^{(*)} - H_{i+1}^{(1)} < 1\,\mathrm{mm}$

15.4.4 Flood plain calculations

Considering a main channel with adjacent flood plains (Fig. 15.6), the flood plains and shallow water zones are usually much rougher than the river channel (Section 4.5.3). Assuming that the friction slope of each portion is the same, the flow resistance may be estimated from:

$$Q = \sum_{i=1}^{n} Q_i = \sum_{i=1}^{n} \sqrt{\frac{8g}{f_i}} A_i \sqrt{\frac{(D_{\mathrm{H}})_i}{4}} \sqrt{S_{\mathrm{f}}} \qquad (15.12)$$

where Q is the total discharge in the cross-section, Q_i is the flow rate in the section $\{i\}$, the subscript i refers to the section $\{i\}$ and n is the number of sections (e.g. $n = 4$ in Fig. 15.6).

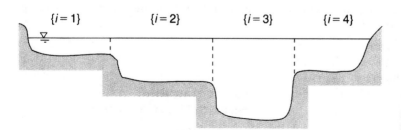

$\{i = 1\}\qquad \{i = 2\}\qquad \{i = 3\}\qquad \{i = 4\}$

Fig. 15.6 Sketch of a flood plain cross-section (with four sections).

Conversely, if the discharge, cross-sectional shape and free-surface elevation are known, the friction slope may be estimated as:

$$S_f = \frac{Q|Q|}{\left(\sum_{i=1}^{n} \sqrt{\frac{8g}{f_i}} A_i \sqrt{\frac{(D_H)_i}{4}}\right)^2}$$

(15.13)

where $|Q|$ is the magnitude of the flow rate. Equation (15.13) is general and it takes into account the flow direction. Indeed the friction slope has the same sign as the flow direction.

Notes

1. In natural channels, the total head losses include regular losses (i.e. friction) and singular losses (e.g. channel expansion). Examples of singular head losses include sudden channel expansions and contractions. The backwater calculations must be adapted to account for the singular losses along a reach of length L:

$$H_2 - H_1 = -S_f L - K\frac{V^2}{2g}$$

where K is a singular head loss coefficient which may be calculated from *reference handbook* (e.g. Idelchik, 1969, 1986).

For an abrupt sidewall expansion, the singular head loss may be estimated as:

$$\Delta H = \left(1 - \frac{W_1}{W_2}\right)^2 \frac{V_1^2}{2g} \qquad \text{Sudden sidewall expansion}$$

where W is the channel width, and the subscripts 1 and 2 refer to the upstream and downstream flow conditions. The above result is valid for $Fr_1 < 0.5$ (Henderson, 1966). For gradual expansions, the singular head loss coefficient is smaller: e.g. by tapering the sidewalls with a 14° angle to the flow direction, the head loss coefficient is about one third of the abrupt expansion situation.

In the case of channel contractions, the energy losses are smaller than for expansions. For an abrupt contraction, the singular head loss is about:

$$\Delta H = K\frac{V_2^2}{2g} \qquad \text{Sudden sidewall contraction}$$

where V_2 is the downstream flow velocity and $K \approx 0.23$–0.35 for square-edged contractions in rectangular channels and $K \approx 0.1$–0.2 for rounded-edged contractions (Henderson, 1966).

2. Under particular flow conditions, sections of the flood plain may operate under subcritical flow conditions while other sections are characterized by supercritical flows. Such situations, called mixed flow regimes, require special attention.

15.5 DISCUSSION

To solve the backwater equation (15.1), it is *essential* to determine correctly the boundary conditions: i.e. the flow conditions imposed upstream, downstream and along the channel. These boundary conditions are in practice:

(a) control devices (e.g. sluice gate, weir and reservoir) that impose flow conditions for the depth, the discharge or a relationship between the discharge and the depth;

(b) geometric characteristics (e.g. bed altitude, channel width and local roughness).

The computations proceed from a hydraulic control or from a position of known flow characteristics. Usually the integration proceeds in the upstream direction if the flow is subcritical, and in the downstream flow direction for supercritical flows.

It must be emphasized that the calculations depend critically on the assumed flow resistance coefficient. In practice, a great uncertainty applies to the assumed friction factor. Also note that the above calculations are applicable to both subcritical and supercritical flows within the GVF calculation assumptions (Section 15.2.2). The only other assumption is that of hydrostatic pressure distribution.

In practice, design engineers *must remember* that backwater calculations are not valid for RVF situations (e.g. gate and weir crest) including across a hydraulic jump. A hydraulic jump is a RVF situation. The relationship between the upstream and downstream flow conditions derives from the momentum equation (i.e. the Bélanger equation). Furthermore the backwater calculations should not be applied to unsteady flow situations.

15.6 COMPUTER MODELS

Several computer models have been developed to compute backwater profiles (e.g. Table 15.1). One example is discussed in Appendix A3.2. The simplest are one-dimensional models based on the backwater equation (i.e. equation (8.1)). Most of these use a step method/depth calculated from distance (see Section 15.4.2): e.g. HEC-1 and HEC-2. More complicated hydraulic models are based on the Saint-Venant equations (see Chapter 16) and they are two dimensional or three dimensional.

The use of computer model requires a sound understanding of the basic equations and a good knowledge of the limitations of the models. It is *uppermost important* that computer program users have a thorough understanding of the hydraulic mechanisms and of the key physical processes that take place in the system (i.e. river and catchment). The writer knows too many 'computer users' who have little idea of the model limitations.

Practically, hydraulic engineers must consider both physical and numerical modelling. Physical modelling assists in understanding the basic flow mechanisms as well as visualizing the flow patterns. In addition, physical model data may be used to calibrate and to verify a numerical model. Once validated, the computational model is a valuable tool to predict a wide range of flow conditions.

15.7 EXERCISES

A waterway has a trapezoidal cross-section with a 7 m bottom width, a vertical sidewall and a battered (1V:5H) sidewall. The longitudinal bed slope is 6 m/km. The channel bottom and sidewalls are concrete lined. The waterway carries a discharge of 52 m³/s. (a) Compute the critical flow depth. (b) Compute the normal flow depth.

A gauging station is set at a bridge crossing the waterway. The observed flow depth at the gauging station is 2.2 m. (c) From where is the flow controlled: i.e. upstream or downstream? (Justify your answer.) (d) Compute the average bed shear stress at the gauging station. (e) Upstream of the gauging station, will the flow depth be larger than, equal to or less than that at the gauging station? (Justify your answer.)

An artificial canal carries a discharge of 25 m³/s. The channel cross-section is trapezoidal and symmetrical, with a 0.5 m bottom width and 1V:2.5H sidewall slopes. The longitudinal bed slope is 8.5 m/km. The channel bottom and sidewall consist of a mixture of fine sands (k_s = 0.3 mm).

(a) What is the normal depth, boundary shear stress and shear velocity at uniform equilibrium? A gauging station is set at a bridge crossing the waterway. The *observed* flow depth, at the gauging station, is 2.2 m. (b) Calculate the Darcy friction factor, boundary shear stress and shear velocity at the gauging station. Is the flow laminar, smooth turbulent or fully rough turbulent? At the gauging station, from where is the flow controlled?

Solution:

(a) $d_o = 1.23$ m, $\tau_o = 51$ Pa.

(b) $\tau_o = 5$ Pa, $d_c = 1.7$ m (subcritical flow conditions, downstream control).

Write a calculation sheet to integrate the backwater equation in the form:

$$\frac{\partial H}{\partial s} = -S_f$$

using the standard step method ('distance calculated from depth'). In order to limit the task, use the following specifications: 1. Cross-sectional shape limited to symmetrical trapezoidal and specified bottom width, angle of side slope, elevation of bed level above datum and distance of section from a reference location (positive in downstream direction). 2. In the iterative procedure required to calculate the correct total head and distance which constitute the result of the integration, you may use any depth increment provided that the changes in friction slope remain small. (For engineering purposes, agreement would be satisfactory if $\Delta S_f/(S_o - S_f) < 5\%$ and $\Delta S_f/S_f < 5\%$.)

In order to validate your calculation sheet, use it to compute the following backwater profile: trapezoidal channel (6 m bottom width and 1V:1.5H side slope), bed slope $S_o = 0.001$, $k_s = 25$ mm and $Q = 3.84\,\text{m}^3/\text{s}$.

Calculate the free-surface elevation upstream of a gauging station, where the known depth is 1.05 m and the bed elevation is RL 779.8 m, at the following locations: $s = 0, -20, -89, -142$ m.

Prepare (1) a brief description of the calculations sheet, (2) a printout of the complete calculation sheet, (3) the results of the validation test, (4) the attached summary sheet.

Summary sheet

$Q =$

s (m)	Free-surface elevation (m RL)	V (m/s)
0		
−20 m		
−89 m		
−142 m		

Uniform equilibrium flow conditions in the channel:

$$d_o =$$
$$Fr_o =$$

Type of free-surface profile observed:

Unsteady open channel flows: 16
1. Basic equations

Summary

In this chapter, the basic equations for unsteady open channel flows are detailed. The continuity and momentum equations are developed for one-dimensional flows: i.e. the Saint-Venant equations. After introducing the basic assumptions, both integral and differential forms of the Saint-Venant equations are introduced. Then the method of characteristics is presented.

16.1 INTRODUCTION

Common examples of unsteady open channel flows include flood flows in rivers and tidal flows in estuaries, irrigation channels, headrace and tailrace channel of hydropower plants, navigation canals, stormwater systems and spillway operation. Figure 16.1(a1) and (a2) illustrates extreme unsteady flow conditions. Figure 16.1(a1) and (a2) shows the propagation of an instability wave down a stepped spillway. Children wandering on the side footpath give the scale of the phenomenon. Figure 16.1b shows a dam break wave down a flat stepped waterway.

In unsteady open channel flows (e.g. Fig. 16.1), the velocities and water depths change with time and longitudinal position. For one-dimensional applications, the relevant flow parameters (e.g. V and d) are functions of time and longitudinal distance. Analytical solutions of the basic equations are nearly impossible because of their non-linearity, but numerical techniques may provide approximate solutions for some specific cases.

Notes

1. The first major mathematical model of a river system was developed in 1953 (Stoker, 1953). J.J. Stoker, E. Isaacson and A. Troesch, from the Courant Institute, New York University, developed and implemented a numerical model for flood wave profiles in the Ohio and Mississippi rivers (Stoker, 1953; Isaacson *et al.*, 1954, 1956). This was followed by important developments in the late 1950s, in particular, by A. Preissmann and J.A. Cunge in France. Yevdjevich (1975) and Montes (1998: pp. 470–471) summarized the historical developments in numerical modelling of unsteady open channel flows.
2. The name of Vujica Yevdjevich is sometimes spelled Jevdjevich or Yevdyevich.
3. James Johnston Stoker was a professor at the Courant Institute, New York University. His book on water waves is a classical publication (Stoker, 1957).
4. Alexandre Preissmann (1916–1990) was born and educated in Switzerland. From 1958, he worked for the development of mathematical models at Sogreah in Grenoble. Born and educated in Poland, Jean A. Cunge worked in France at Sogreah and he lectured at the Hydraulics and Mechanical Engineering School of Grenoble (ENSHMG).

(a1)

(a2)

Fig. 16.1 Example of unsteady open channel flows: (a) Unsteady wave propagation in Sorpe dam spillway on 2 November 1998 (courtesy of Marq Redeker, Ruhrverband). Instability wave induced by resonance in the step pools (Chanson, 2001: pp. 293–294).

16.2 BASIC EQUATIONS

16.2.1 Presentation

The basic one-dimensional unsteady open channel flow equations are commonly called the Saint-Venant equations. They are named after the French engineer Adhémar Jean Claude Barré

Fig. 16.1 (b) Dam break wave down a stepped waterway. $\theta = 3.4°$, $W = 0.5\,m$, step height: $0.0715\,m$ and $Q = 0.040\,m^3/s$. Looking upstream at the incoming wave front.

de Saint-Venant who developed them in 1871. These equations are based upon a number of *key, fundamental assumptions*:

- [H1] the flow is one dimensional: i.e. the velocity is uniform in a cross-section and the transverse free-surface profile is horizontal;
- [H2] the streamline curvature is very small and the vertical fluid accelerations are negligible; as a result, the pressure distributions are hydrostatic;
- [H3] the flow resistance and turbulent losses are the same as for a steady uniform flow for the same depth and velocity, regardless of trends of the depth;
- [H4] the bed slope is small enough to satisfy the following approximations: $\cos\theta \approx 1$ and $\sin\theta \approx \tan\theta$;
- [H5] the water density is a constant.

With these basic hypotheses, the flow can be described at any point and any time by two variables: e.g. V and d, or Q and d, where V is the flow velocity, d is the water depth and Q is the water discharge. Hence the unsteady flow properties must be described by two equations: the conservation of mass and conservation of momentum.

Notes

1. Adhémar Jean Claude Barré de Saint-Venant (1797–1886), French engineer of the 'Corps des Ponts-et-Chaussées', developed the equation of motion of a fluid particle in terms of the shear and normal forces exerted on it (Barré de Saint-Venant, 1871a,b). His original development was introduced for both fluvial and estuarine systems.

2. The assumption [H4] is valid within 0.1% for $\theta < 2.57°$, and within 1% for $\theta < 8.12°$ where θ is the angle between the channel invert and the horizontal.
3. The five basic assumptions are valid for any channel cross-sectional shape. However, the cross-sectional shape is indirectly limited by the assumptions of one-dimensional flow, horizontal transverse free surface and hydrostatic pressure.
4. The Saint-Venant equations were further developed for fixed boundary channels. That is, sediment motion is neglected.
5. The equations of conservation of momentum and conservation of energy are equivalent if the two relevant variables (e.g. V and d) are continuous functions. At a discontinuity (e.g. a hydraulic jump), the equivalence becomes untrue. Unsteady open channel flow equations must be based upon the continuity and momentum principles, which are applicable to both the continuous and discontinuous flow situations (see Chapter 4).

16.2.2 Integral form of the Saint-Venant equations

Continuity equation

Considering the control volume defined by the cross-sections 1 and 2 (Fig. 16.2), located between $x = x_1$ and $x = x_2$, between the times $t = t_1$ and $t = t_2$, the continuity principle states that the net mass flux into the control volume equals the net mass increase of the control volume between the times t_1 and t_2. It yields:

$$\int_{t_1}^{t_2} (\rho V_1 A_1 - \rho V_2 A_2)dt + \int_{x_1}^{x_2} ((\rho A)_{t_2} - (\rho A)_{t_1})dx = 0 \qquad (16.1)$$

where ρ is the fluid density, V is the flow velocity, A is the flow cross-sectional area, the subscripts 1 and 2 refer to the upstream and downstream cross-section, respectively (Fig. 16.2), and the subscripts t_1 and t_2 refer to the instants $t = t_1$ and $t = t_2$, respectively.

Defining $Q = VA$, and dividing equation (16.1) by the density ρ, the continuity equation becomes:

$$\int_{t_1}^{t_2} (Q_1 - Q_2)dt + \int_{x_1}^{x_2} (A_{t_2} - A_{t_1})dx = 0 \qquad (16.2)$$

Momentum principle

The application of the momentum principle in the x-direction implies that the net change of momentum in the control volume between the instants t_1 and t_2 plus the rate of change of momentum flux across the volume equals the sum of the forces applied to the control volume in the x-direction. The momentum per unit volume equals ρV. The momentum flux equals $\rho V^2 A$.

The net change of momentum in the control volume equals:

$$\int_{x_1}^{x_2} ((\rho VA)_{t_1} - (\rho VA)_{t_2})dx \qquad (16.3)$$

while the rate of change of momentum flux across the volume is equal to:

$$\int_{t_1}^{t_2} ((\rho V^2 A)_1 - (\rho V^2 A)_2)dt \qquad (16.4)$$

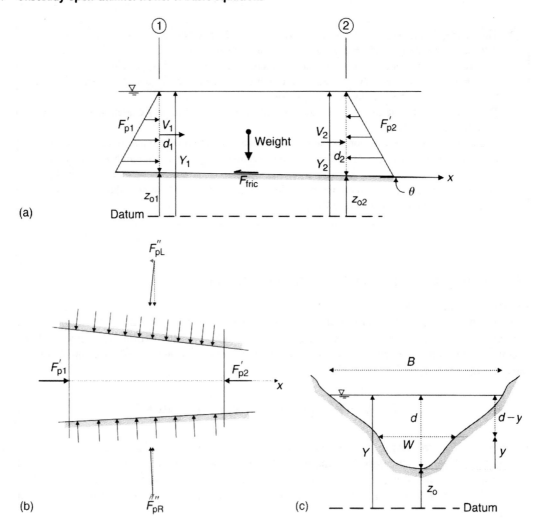

Fig. 16.2 Definition sketch: (a) side view, (b) top view and (c) cross-section.

The forces acting on the control volume contained between sections 1 and 2 are the pressure forces at sections 1 and 2, the pressure force components on the channel sidewalls if the channel width vary in the x-direction, the weight of the control volume, the reaction force of the bed (equal, in magnitude, to the weight in absence of vertical fluid motion) and the flow resistance opposing fluid motion.

The pressure forces acting on sections 1 and 2 equal:

$$\int_{t_1}^{t_2} (F'_{p1} - F'_{p2}) \, dt = \int_{t_1}^{t_2} ((\rho I_1)_1 - (\rho I_1)_2) \, dt \qquad (16.5)$$

where:

$$I_1 = \int_0^d (d - y) W \, dy$$

y is the distance measured from the bottom, d is the water depth and W is the channel width at the distance y above the bed (Fig. 16.2c).

When the channel width changes with distance, the pressure force components on the right and left sidewalls equal:

$$\int_{t_1}^{t_2} (F''_{pL} + F''_{pR})\,dt = \int_{t_1}^{t_2} \int_{x_1}^{x_2} \rho g I_2\; dx\; dt \tag{16.6}$$

where:

$$I_2 = \int_0^d (d - y)\left(\frac{\partial W}{\partial x}\right)_{y=\text{constant}} dy$$

Equation (16.6) states that an increase in wetted surface in the x-direction, the water depth d being constant, induces a positive sidewall pressure force component. The above expression of F''_p is valid only for gradual variations in cross-section. For sudden changes in cross-sectional shapes, forces other than the hydrostatic pressure force take place.

The gravity force component in the flow direction (i.e. x-direction) is:

$$\int_{t_1}^{t_2} \int_{x_1}^{x_2} \rho g A S_o\; dx\; dt \tag{16.7}$$

where the bed slope $S_o = \sin\theta$ may be approximated to: $S_o \approx -\partial z_o/\partial x$ (assumption [H4], Section 16.2.1).

Fluid motion is opposed by flow resistance and shear forces exerted on the wetted surfaces. The integration of the friction force F_{fric} between $t = t_1$ and $t = t_2$ is:

$$\int_{t_1}^{t_2} F_{\text{fric}}\; dt = \int_{t_1}^{t_2} \int_{x_1}^{x_2} \rho g A S_f\; dx\; dt \tag{16.8}$$

where S_f is the friction slope defined as:

$$S_f = \frac{4\tau_o}{\rho g D_H} \tag{16.9}$$

τ_o is the average boundary shear stress and D_H is the hydraulic diameter.

Notes
1. The basic assumptions of the Saint-Venant equations (Section 16.2.1) imply that the vertical pressure distribution is hydrostatic: i.e. $P(y) = \rho g(d - y)$ for $0 \leqslant y \leqslant d$.
2. Simple geometrical considerations give:

$$W(x,\; y = d,\; t) = B(x,\; t)$$

$$\int_0^d W\,dy = A$$

$$\frac{\partial A}{\partial y} = B$$

where the free-surface width B (Fig. 16.2) and the flow cross-sectional area A are functions of both distance x and time t.

3. The left bank is located on the left-hand side of an observer when looking downstream. Conversely the right bank is on the right-hand side of an observer looking downstream.

4. Yen (2002) discussed specifically the definitions of the friction slope. One series of definitions is based upon the momentum principle (e.g. equation (16.9)) while another is based upon the energy principle. All definitions are equivalent for uniform equilibrium flow conditions.

Combining equations (16.3)–(16.8), and dividing by the density ρ which is assumed to be constant, the momentum equation becomes:

$$\int_{x_1}^{x_2} ((VA)_{t_1} - (VA)_{t_2})dx + \int_{t_1}^{t_2} ((V^2A)_1 - (V^2A)_2)dt$$

$$= \int_{t_1}^{t_2} ((I_1)_1 - (I_1)_2)dt + \int_{t_1}^{t_2} \int_{x_1}^{x_2} gI_2 \, dx \, dt + \int_{t_1}^{t_2} \int_{x_1}^{x_2} gA(S_o - S_f)dx \, dt \quad (16.10)$$

Equation (16.10) is the cross-sectional integration of the principle of momentum conservation for one-dimensional unsteady open channel flows. It states that the net change of momentum in control volume (i.e. unsteady term) plus the rate of change of momentum flux across the volume equal the pressure forces acting on sections 1 and 2, plus the pressure force components on the right and left sidewalls, plus the gravity force component in flow direction and the friction force F_{fric} acting on the wetted surface between $t = t_1$ and $t = t_2$.

Equations (16.2) and (16.10) form a system of two equations based upon the Saint-Venant equation assumptions (Section 16.2.1). They were developed without any additional assumption and some characteristic parameter (e.g. V, A and d) might not be continuous. If one or more parameter is discontinuous, equations (16.2) and (16.10) remain valid if the Saint-Venant hypotheses are respected.

16.2.3 Differential form of the Saint-Venant equations

The differential form of the Saint-Venant equations may be derived from the integral form (Section 16.2.2) if the relevant parameters are continuous and differentiable functions with respect to x and y. The Taylor-series expansion of each parameter follows:

$$A_{t_2} = A_{t_1} + \frac{\partial A}{\partial t} \Delta t + \frac{\partial^2 A}{\partial t^2} \frac{\Delta t^2}{2} + \cdots \quad (16.11a)$$

$$Q_2 = Q_1 + \frac{\partial Q}{\partial x} \Delta x + \frac{\partial^2 Q}{\partial x^2} \frac{\Delta x^2}{2} + \cdots \quad (16.11b)$$

where $\Delta t = t_2 - t_1$ and $\Delta x = x_2 - x_1$. Neglecting the second order term, the continuity equation (equation (16.2)) yields:

$$\int_{x_1}^{x_2} \int_{t_1}^{t_2} \left(\frac{\partial A}{\partial t} + \frac{\partial Q}{\partial x} \right) dt \, dx = 0 \quad (16.12)$$

Similarly the momentum principle (equation (16.10)) becomes:

$$\int_{x_1}^{x_2}\int_{t_1}^{t_2}\left(\frac{\partial Q}{\partial t}+\frac{\partial(V^2 A)}{\partial x}\right)dt\ dx=\int_{x_1}^{x_2}\int_{t_1}^{t_2}-g\left(\frac{\partial I_1}{\partial x}-I_2-(S_o-S_f)A\right)dt\ dx \quad (16.13)$$

If equations (16.12) and (16.13) are valid everywhere in the (x, t) plane, they are also valid over an infinitely small area $(dx\ dt)$ and it yields:

$$\frac{\partial A}{\partial t}+\frac{\partial Q}{\partial x}=0 \qquad \text{Continuity equation} \qquad (16.14)$$

$$\frac{\partial Q}{\partial t}+\frac{\partial}{\partial x}(V^2 A+gI_1)=g(S_o-S_f)A+gI_2 \qquad \text{Momentum equation} \quad (16.15\text{a})$$

Based upon geometrical considerations and using Leibnitz rule, it can be proved that the terms I_1 and I_2 satisfy:

$$\frac{\partial}{\partial x}(gI_1)=gA\frac{\partial d}{\partial x}+gI_2 \qquad (16.16)$$

Replacing into the momentum equation (equation (16.15a)), it yields:

$$\frac{\partial Q}{\partial t}+\frac{\partial}{\partial x}(V^2 A)+gA\frac{\partial d}{\partial x}=gA(S_o-S_f) \qquad \text{Momentum equation} \qquad (16.15\text{b})$$

Equation (16.15b) is valid for prismatic and non-prismatic channels. That is, the pressure force contribution caused by a change in cross-sectional area at a channel expansion (or contraction) is exactly balanced by the pressure force component on the channel banks in the flow direction.

The continuity and momentum equations may be rewritten in terms of the free-surface elevation $(Y = d + z_o)$ and flow velocity V only. It yields:

$$\frac{\partial Y}{\partial t}+\frac{A}{B}\frac{\partial V}{\partial x}+V\left(\frac{\partial Y}{\partial x}+S_o\right)+\frac{V}{B}\left(\frac{\partial A}{\partial x}\right)_{d=\text{constant}}=0 \qquad (16.17)$$

$$\frac{\partial V}{\partial t}+V\frac{\partial V}{\partial x}+g\frac{\partial Y}{\partial x}+gS_f=0 \qquad (16.18\text{a})$$

The system of equations (16.17) and (16.18a) forms the basic *Saint-Venant equations*.

Notes
1. A *prismatic* channel has a cross-sectional shape independent of the longitudinal distance along the flow direction. That is, the width $W(x, y, t)$ is only a function of y and it is independent of x and t.
2. Equation (16.17) is the *continuity equation* while equation (16.18a) is called the *dynamic equation*. The Saint-Venant equations (equations (16.17) and (16.18a)) are also called *dynamic wave equations*.
3. In his original development, Barré de Saint-Venant (1871a) obtained the system of two equations:

$$\frac{\partial A}{\partial t}+\frac{\partial Q}{\partial x}=0 \qquad (16.14)$$

$$\frac{1}{g}\frac{\partial V}{\partial t} + \frac{\partial}{\partial x}\left(\frac{V^2}{2g}\right) + \frac{\partial Y}{\partial x} + S_f = 0 \tag{16.18b}$$

where Y is the free-surface elevation (i.e. $Y = d + z_o$).

4. Equation (16.18b) is dimensionless. The first term is related to the slope of flow acceleration energy line, the second term is the longitudinal slope of the kinetic energy line, the third one is the longitudinal slope of the free surface and the last term is the friction slope. In equation (16.18b), the two first terms are inertial terms. In steady flows, the acceleration term $\partial V/\partial t$ is zero.

5. The Saint-Venant equations may be expressed in several ways. For example, the continuity equation may be written as:

$$\frac{\partial A}{\partial t} + \frac{\partial Q}{\partial x} = 0 \tag{16.14}$$

$$B\frac{\partial d}{\partial t} + \frac{\partial Q}{\partial x} = 0$$

$$\frac{\partial Y}{\partial t} + \frac{A}{B}\frac{\partial V}{\partial x} + V\left(\frac{\partial Y}{\partial x} + S_o\right) + \frac{V}{B}\left(\frac{\partial A}{\partial x}\right)_{d=\text{constant}} = 0 \tag{16.17}$$

$$\frac{\partial d}{\partial t} + \frac{A}{B}\frac{\partial V}{\partial x} + V\frac{\partial d}{\partial x} + \frac{V}{B}\left(\frac{\partial A}{\partial x}\right)_{d=\text{constant}} = 0$$

$$\frac{\partial Y}{\partial t} + \frac{A}{B}\frac{\partial V}{\partial x} + V\left(\frac{\partial Y}{\partial x} + S_o\right) = 0 \tag{16.19}$$

The dynamic equation may be written as:

$$\frac{\partial V}{\partial t} + V\frac{\partial V}{\partial x} + g\frac{\partial Y}{\partial x} + gS_f = 0 \tag{16.18a}$$

$$\frac{\partial V}{\partial t} + V\frac{\partial V}{\partial x} + g\frac{\partial d}{\partial x} + g(S_f - S_o) = 0$$

$$\frac{\partial Q}{\partial t} + \frac{\partial}{\partial x}(V^2 A) + gA\frac{\partial d}{\partial x} + gA(S_f - S_o) = 0 \tag{16.15b}$$

$$\frac{1}{g}\frac{\partial V}{\partial t} + \frac{\partial}{\partial x}\left(\frac{V^2}{2g}\right) + \frac{\partial d}{\partial x} + (S_f - S_o) = 0$$

16.2.4 Flow resistance estimate

The laws of flow resistance in open channels are essentially the same as those in closed pipes, although, in open channel, the calculations of the boundary shear stress are complicated by the existence of the free surface and the wide variety of possible cross-sectional shapes (Chapters 4, 5, 13 and 15). The head loss ΔH over a distance L along the flow direction is given by the Darcy equation:

$$\Delta H = f\frac{L}{D_H}\frac{V^2}{2g} \tag{16.20}$$

where f is the Darcy–Weisbach friction factor, V is the mean flow velocity and D_H is the hydraulic diameter or equivalent pipe diameter. In gradually varied flows, it yields:

$$V = \sqrt{\frac{8g}{f}} \sqrt{\frac{D_H}{4} S_f} \qquad (16.21a)$$

where S_f is the friction slope (Chapters 4 and 5). Equation (16.21a) may be rewritten as:

$$S_f = \frac{f}{8g(D_H/4)} V|V| \qquad (16.21b)$$

The friction slope equals the square of the velocity times a non-linear term.

Discussion

In open channels, the Darcy equation (equation (16.21a)) is the only sound method to estimate the friction loss. For various reasons, empirical resistance coefficients (e.g. Chézy coefficient and Gauckler–Manning coefficient) were and are still used. Their use for man-made channels is *highly inaccurate* and improper. Most friction coefficients are completely empirical and are limited to fully rough turbulent water flows.

In natural streams, the flow resistance may be expressed in terms of the Chézy equation:

$$V = C_{\text{Chézy}} \sqrt{\frac{D_H}{4} S_f}$$

where $C_{\text{Chézy}}$ is the Chézy coefficient (Units: $m^{1/2}/s$). The Chézy equation was first introduced as an *empirical correlation*. The Chézy coefficient ranges typically from $30\,m^{1/2}/s$ (small rough channel) up to $90\,m^{1/2}/s$ (large smooth channel). Another *empirical formulation*, called the Gauckler–Manning formula, was developed for turbulent flows in rough channels. The Gauckler–Manning coefficient is an empirical coefficient found to be a characteristic of the surface roughness primarily (Chapter 4).

Liggett (1975) summarized our poor understanding of the friction loss process nicely: '*The (Chézy and Gauckler–Manning) equations express our continuing ignorance of turbulent processes*' (p. 45).

Remarks

1. James A. Liggett is an emeritus professor in fluid mechanics at Cornell University and a former editor of the Journal of Hydraulic Engineering.
2. The friction slope may be expressed as a function of the velocity as:

$$S_f = \frac{f}{8g(D_H/4)} V|V| \qquad (16.21b)$$

$$S_f = \frac{1}{C_{\text{Chézy}}^2(D_H/4)} V|V|$$

$$S_f = \frac{n_{\text{Manning}}^2}{(D_H/4)^{4/3}} V|V|$$

where $|V|$ is the magnitude of the flow velocity and $n_{Manning}$ is the Gauckler–Manning coefficient (Units: $s/m^{1/3}$). The friction has the same sign as the velocity. For a wide rectangular channel, $D_H = 4A/P_w \approx 4d$ and it yields:

$$S_f = \frac{f}{8gd}V|V| \tag{16.21c}$$

$$S_f = \frac{1}{C_{Chézy}^2 d}V|V|$$

$$S_f = \frac{n_{Manning}^2}{d^{4/3}}V|V|$$

Flood plain calculations

Considering a main channel with adjacent overbanks (Fig. 16.3), the shallow-water zones are usually much rougher than the river channel (Section 4.5.3 and Section 15.4.4). The friction slope is assumed to be identical in each portion and the flow resistance may be estimated from:

$$Q = \sum_{i=1}^{n} Q_i = \sum_{i=1}^{n} \sqrt{\frac{8g}{f_i}} A_i \sqrt{\frac{(D_H)_i}{4}} \sqrt{S_f} \tag{16.22}$$

where Q is the total discharge in the cross-section, Q_i is the flow rate in the section $\{i\}$ and the subscript i refers to the section $\{i\}$ (Fig. 16.3). If the discharge, cross-sectional shape and free-surface elevation are known, the friction slope equals:

$$S_f = \frac{Q|Q|}{\left(\sum_{i=1}^{n} \sqrt{\frac{8g}{f_i}} A_i \sqrt{\frac{(D_H)_i}{4}} \right)^2} \tag{16.23}$$

where $|Q|$ is the magnitude of the flow rate. Note that equation (16.23) does take into account the flow direction.

16.2.5 Discussion

The system of equations (16.2) and (16.10) is the integral form of the Saint-Venant equations while the system of equations (16.17) and (16.18a) is the differential form of the Saint-Venant equations.

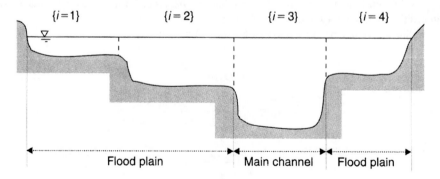

$\{i = 1\}$ $\{i = 2\}$ $\{i = 3\}$ $\{i = 4\}$

Flood plain   Main channel   Flood plain

Fig. 16.3 Sketch of a flood plain cross-section (with four sections).

The system of two equations formed by equations (16.2) and (16.10), and equations (16.17) and (16.18a), are equivalent *if and only if all* the variables and functions are continuous and differentiable. In particular, for discontinuous solutions (e.g. hydraulic jump), equations (16.2) and (16.10) are applicable but equations (16.17) and (16.18a) might not be valid.

The Saint-Venant equations were developed within very specific assumptions (Section 16.2.1). If these basic hypotheses are not satisfied, the Saint-Venant equations are not valid. For example, in an undular jump, the streamline curvature and vertical acceleration is important, and the pressure is not hydrostatic; in a mountain stream, the bed slope is not small; in sharp bends, the centrifugal acceleration may be important and the free surface is not horizontal in the radial direction.

16.3 METHOD OF CHARACTERISTICS

16.3.1 Introduction

The method of characteristics is a mathematical technique to solve a system of partial differential equations such as the Saint-Venant equations which may be expressed in terms of the water depth d:

$$\frac{\partial d}{\partial t} + \frac{A}{B}\frac{\partial V}{\partial x} + V\frac{\partial d}{\partial x} + \frac{V}{B}\left(\frac{\partial A}{\partial x}\right)_{d=\text{constant}} = 0 \tag{16.24}$$

$$\frac{\partial V}{\partial t} + V\frac{\partial V}{\partial x} + g\frac{\partial d}{\partial x} + g(S_f - S_o) = 0 \tag{16.25}$$

Equations (16.24) and (16.25) are equivalent to equations (16.17) and (16.18a).

In an open channel, a small disturbance can propagate upstream and downstream. For a rectangular channel, the celerity of a small disturbance equals $C = \sqrt{gd}$ (Chapter 3). In a channel of irregular cross-section, the celerity of a small wave is $C = \sqrt{gA/B}$, where A is the flow cross-section and B is the free-surface width (Fig. 16.2c). The celerity C characterizes the propagation of a small disturbance (i.e. small wave) relative to the fluid motion. For an observer standing on the bank, the absolute speed of the small wave is $(V + C)$ and $(V - C)$ where V is the flow velocity.

The differentiation of the celerity C with respect to time and space satisfies, respectively:

$$\frac{\partial}{\partial t}(C^2) = 2C\frac{\partial C}{\partial t} \approx g\frac{\partial d}{\partial t} \tag{16.26}$$

$$\frac{\partial}{\partial x}(C^2) = 2C\frac{\partial C}{\partial x} \approx g\frac{\partial d}{\partial x} \tag{16.27}$$

Replacing the terms $\partial d/\partial t$ and $\partial d/\partial x$ by equations (16.26) and (16.27), respectively, in the Saint-Venant equations, it yields:

$$2\frac{\partial C}{\partial t} + 2V\frac{\partial C}{\partial x} + C\frac{\partial V}{\partial x} = 0 \tag{16.28}$$

$$\frac{\partial V}{\partial t} + 2C\frac{\partial C}{\partial x} + V\frac{\partial V}{\partial x} + g(S_f - S_o) = 0 \tag{16.29}$$

The addition of the two equations gives:

$$\left(\frac{\partial}{\partial t} + (V + C)\frac{\partial}{\partial x}\right)(V + 2C) + g(S_f - S_o) = 0 \qquad (16.30)$$

while the subtraction of equation (16.29) from equation (16.28) yields:

$$\left(\frac{\partial}{\partial t} + (V - C)\frac{\partial}{\partial x}\right)(V - 2C) + g(S_f - S_o) = 0 \qquad (16.31)$$

Equations (16.30) and (16.31) may be rewritten as:

$$\frac{D}{Dt}(V + 2C) = -g(S_f - S_o) \qquad \text{Forward characteristic} \qquad (16.32a)$$

$$\frac{D}{Dt}(V - 2C) = -g(S_f - S_o) \qquad \text{Backward characteristic} \qquad (16.32b)$$

along their *respective* characteristic trajectories:

$$\frac{dx}{dt} = V + C \qquad \text{Forward characteristic C1} \qquad (16.33a)$$

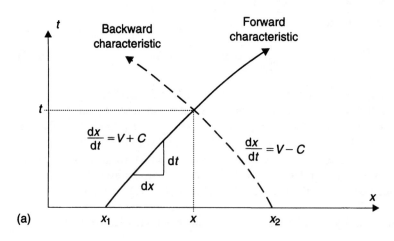

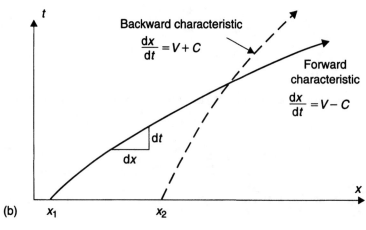

Fig. 16.4 Definition sketch of the forward and backward characteristics: (a) subcritical flow conditions and (b) supercritical flow conditions.

$$\frac{dx}{dt} = V - C \qquad \text{Backward characteristic C2} \qquad (16.33b)$$

These trajectories are called characteristic directions or characteristics of the system. For an observer travelling along the forward characteristics (Fig. 16.4, equation (16.33a)), equation (16.32a) is valid at any point. For an observer travelling on the backward characteristics (equation (16.33b)), equation (16.32b) is satisfied everywhere.

The system of four equations formed by equations (16.32) and (16.33) are equivalent to the system of equations (16.17) and (16.18a). They represent the characteristic system of equations that replaces the differential form of the Saint-Venant equations.

The families of forward (C1) and backward (C2) characteristics are shown in the (x, t) plane in Fig. 16.4 for sub- and supercritical flows.

Discussion

The absolute differential D/Dt of a scalar function $\Phi(x, t)$ equals (Appendix A1.3):

$$\frac{D\Phi}{Dt} = \frac{\partial \Phi}{\partial t} + \frac{\partial x}{\partial t}\frac{\partial \Phi}{\partial x}$$

and this may be rewritten as:

$$\frac{D\Phi}{Dt} = \frac{\partial \Phi}{\partial t} + V\frac{\partial \Phi}{\partial x}$$

Along a characteristic trajectory:

$$\frac{dx}{dt} = V \pm C$$

the absolute differential of the scalar function $\Phi(x, t)$ equals:

$$\frac{D\Phi}{Dt} = \left(\frac{\partial}{\partial t} + (V \pm C)\frac{\partial}{\partial x}\right)\Phi$$

Notes

1. The method of characteristics was derived from the work of the French mathematician Gaspard Monge (1746–1818) and it was first applied by the Belgian engineer Junius Massau (Massau, 1889, 1900) to graphically solve a system of partial differential equations. Today it is acknowledged to be the most accurate, reliable of all numerical integration techniques.

2. The forward characteristic trajectory is often called the *C1 characteristic* and positive characteristic, sometimes denoted as $C+$. The backward characteristic is often called the *C2 characteristic* and negative characteristic, sometimes denoted as $C-$.

3. For a horizontal, frictionless channel, the characteristic system of equations becomes:

$$V + 2C = \text{constant} \qquad \text{Along C1 characteristic}$$

$$V - 2C = \text{constant} \qquad \text{Along C2 characteristic}$$

The constants $(V + 2C)$ and $(V - 2C)$ are called the *Riemann invariants*.

4. If the term $g(S_f - S_o)$ is constant along the channel and independent of the time, equation (16.28) becomes:

$$\frac{D}{Dt}(V + 2C + g(S_f - S_o)t) \qquad \text{Along C1 characteristic}$$

$$\frac{D}{Dt}(V - 2C + g(S_f - S_o)t) \qquad \text{Along C2 characteristic}$$

That is, the term $(V + 2C + g(S_f - S_o)t)$ is a constant along the forward characteristic and the term $(V - 2C + g (S_f - S_o)t)$ is constant along the backward characteristic trajectory C2.

Discussion: graphical solution of the characteristic system of equations

Considering the case of a wave characterized by $S_o = S_f$, and for which the initial flow conditions (i.e. V and d) are known at each location x. Figure 16.5 illustrates four initial points D1–D4. The slope of the characteristic trajectories is known at each initial point: i.e. $dx/dt = V \pm C$. The forward and backward trajectories can be approximated by straight lines[1], which intersect at the points E1–E3 (Fig. 16.5). The characteristic equations give the velocity and celerity at the next time step. For example, at the point E1:

$$V_{E1} + 2C_{E1} = V_{D1} + 2C_{D1} \qquad \text{Along C1 characteristic}$$

$$V_{E1} - 2C_{E1} = V_{D2} - 2C_{D2} \qquad \text{Along C2 characteristic}$$

Similarly at E2 and E3, and then at F1, F2 and ultimately at G1. When the distance δx tends to zero, the graphical solution approaches the exact solution of the Saint-Venant equations.

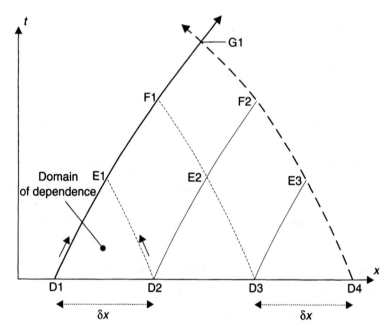

Fig. 16.5 Network of characteristic trajectories and domain of dependence.

[1]With slope $dt/dx = 1/(V + C)$ and $dt/dx = 1(V - C)$ respectively.

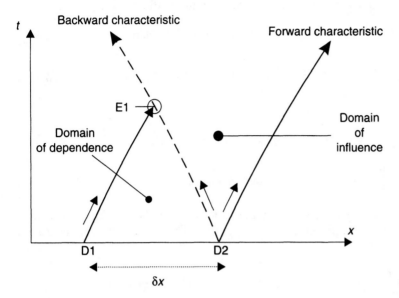

Fig. 16.6 Sketch of the region of influence.

In the example illustrated in Fig. 16.5, the flow properties at the point G1 are functions of the initial flow properties everywhere in the reach D1–D4. The interval D1–D4 is called the interval of influence. In practice, an infinite number of characteristic curves crosses the interval D1–D4 and the region D1–G1–D2 is called the *domain of dependence* for the point G1.

In the computational process (Fig. 16.6), the properties of the point D2, whose initial conditions are given, do influence a region delimited by the forward and backward characteristic trajectories through D2. This region is called the *domain of influence* (Fig. 16.6).

Notes

1. When conducting a linear interpolation from a point where the flow properties are known (e.g. D1 and D2 in Fig. 16.5), the time step δt must be selected such as the point E1 is within the domain of influence of the interval D1–D2. That is, the time step must satisfy the conditions:

$$\delta t \leq \frac{\delta x}{|V + C|} \quad \text{and} \quad \delta t \leq \frac{\delta x}{|V - C|}$$

where $|V + C|$ is the absolute value of the term $(V + C)$. This condition is sometimes called the Courant condition or Courant–Friedrichs–Lewy (CFL) condition.

16.3.2 Boundary conditions

The method of characteristics is a mathematical technique to solve the system of partial differential equations formed by the continuity and momentum equations (i.e. Saint-Venant equations):

$$\frac{\partial Y}{\partial t} + \frac{A}{B}\frac{\partial V}{\partial x} + V\left(\frac{\partial Y}{\partial x} + S_o\right) + \frac{V}{B}\left(\frac{\partial A}{\partial x}\right)_{d=\text{constant}} = 0 \tag{16.17}$$

$$\frac{\partial V}{\partial t} + V\frac{\partial V}{\partial x} + g\frac{\partial Y}{\partial x} + gS_f = 0 \tag{16.18a}$$

by replacing the above two equations by a system of four ordinary differential equations:

$$\frac{D}{Dt}(V + 2C) = -g(S_f - S_o) \qquad \text{Forward characteristic} \qquad (16.32a)$$

$$\frac{D}{Dt}(V - 2C) = -g(S_f - S_o) \qquad \text{Backward characteristic} \qquad (16.32b)$$

along respectively:

$$\frac{dx}{dt} = V + C \qquad \text{Forward characteristic C1} \qquad (16.33a)$$

$$\frac{dx}{dt} = V - C \qquad \text{Backward characteristic C2} \qquad (16.33b)$$

The shape of the characteristic trajectories is a function of the flow conditions. Figure 16.7 illustrates four examples. Figure 16.6a presents a subcritical flow. The slope of the characteristics satisfies at each point:

$$\tan \Delta_1 = \frac{dt}{dx} = \frac{1}{V + C} \qquad \text{Along the forward (C1) characteristic}$$

$$\tan \Delta_2 = \frac{dt}{dx} = \frac{1}{V - C} \qquad \text{Along the backward (C2) characteristic}$$

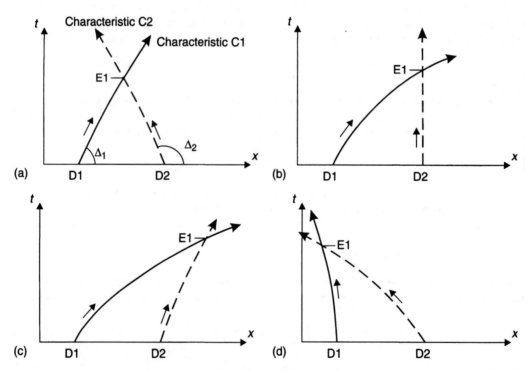

Fig. 16.7 Shape of characteristics as a function of the flow conditions: (a) subcritical flow, (b) critical flow, (c) super-critical flow and (d) supercritical flow in the negative x-direction.

where Δ is the angle between the characteristics and the x-axis (Fig. 16.7a). Figure 16.7b shows a critical flow. That is, $V = C$ and $\Delta_2 = \pi/2$. Figure 16.7c and d illustrates both supercritical flow conditions.

Figure 16.6c corresponds to a torrential flow in the positive x-direction, while Fig. 16.7d shows a supercritical flow in the negative x-direction. In a supercritical flow (Fig. 16.7c), the velocity is greater than the celerity of small disturbances: i.e. $V > C$. As a result, $\tan \Delta_2$ is positive.

Initial and boundary conditions

Assuming the simple case $g(S_f - S_o) = 0$, the characteristic system of equations is greatly simplified. That is, the Riemann invariants are constant along the characteristic trajectories:

$$V + 2C = \text{constant} \qquad \text{Along C1 characteristic}$$

$$V - 2C = \text{constant} \qquad \text{Along C2 characteristic}$$

Considering an *infinitely long reach*, the initial flow conditions ($t = 0$) are defined by two parameters (e.g. V and d). From each point D1 on the x-axis, two characteristics develop along which the Riemann invariants are known constants everywhere for $t > 0$ (Fig. 16.6). Similarly, at each point E1 in the (x, $t > 0$) plane, two characteristic trajectories intersect and the flow properties (V, d) are deduced from the initial conditions ($t = 0$). It can be mathematically proved that the Saint-Venant equations have a solution for $t > 0$ if and only if two flow conditions (e.g. V and d) are known at $t = 0$ everywhere.

Considering a *limited reach* ($x_1 < x < x_2$) with subcritical flow conditions (Fig. 16.8a), the flow properties at the point E2 are deduced from the initial flow conditions ($V(t = 0)$, $d(t = 0)$) at the points D1 and D2. At the point E1, however, only one characteristic curve intersects the boundary ($x = x_1$). This information is not sufficient to calculate the flow properties at the point E1: i.e. one additional information (e.g. V or d) is required. In other words, one flow condition must be prescribed at E1. The same reasoning applies at the point E3. In summary, in a bounded reach with subcritical flow, the solution of the Saint-Venant equations requires the knowledge of two flow parameters at $t = 0$ and one flow property at each boundary for $t > 0$.

Figure 16.8b shows a supercritical flow. At the point E2, two characteristics intersect and all the flow properties may be deduced from the initial flow conditions. But, at the upstream boundary (point E1), two flow properties are required to solve the Saint-Venant equations.

Discussion: types of boundary conditions

The initial flow conditions may include the velocity $V(x, 0)$ and water depth $d(x, 0)$, the velocity $V(x, 0)$ and the free-surface elevation $Y(x, 0)$, or the flow rate $Q(x, 0)$ and the cross-sectional area $A(x, 0)$. For a limited river section, the prescribed boundary condition(s) (i.e. at $x = x_1$ and $x = x_2$) may be the water depth $d(t)$ or the free-surface elevation $Y(t)$, the flow rate $Q(t)$ or flow velocity $V(t)$, or a relationship between the water depth and flow rate. For the example shown in Fig. 16.8a, several combinations of boundary conditions are possible: e.g. $\{d(x_1, t) \text{ and } d(x_2, t)\}$, $\{d(x_1, t) \text{ and } Q(x_2, t)\}$, $\{d(x_1, t) \text{ and } Q(x_2, t) = f(d(x_2, t))\}$, $\{Q(x_1, t) \text{ and } d(x_2, t)\}$ or $\{Q(x_1, t) \text{ and } Q(x_2, t)\}$.

Note that it is impossible to set $Q(x_1, t) = f(d(x = 0, t))$ unless critical flow conditions occur at the upstream boundary section ($x = x_1$).

For a set of boundary conditions $\{Q(x_1, t) \text{ and } Q(x_2, t)\}$, the particular case $Q(x_1, t) = Q(x_2, t)$ implies that the inflow equals the outflow. The mass of water does not change and the variations of the free surface are strongly affected by the initial conditions.

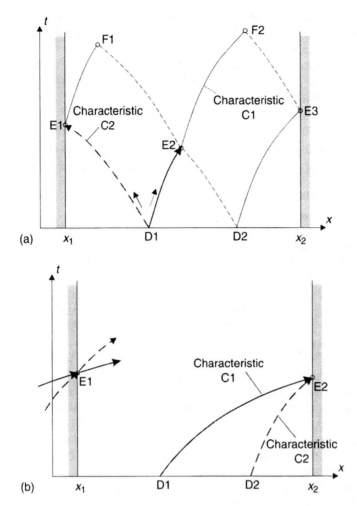

Fig. 16.8 Example of boundary conditions: (a) subcritical flow and (b) supercritical flow.

16.3.3 Application: numerical integration of the method of characteristics

The characteristic system of equations may be regarded as the conservation of basic flow properties (velocity and celerity) when viewed by observers travelling on the characteristic trajectories. At the intersection of a forward characteristic and a backward characteristic, there are four unknowns (x, t, V, C) which are the solution of the system of four differential equations. Note that such a solution is impossible at a flow discontinuity (e.g. surge front).

The characteristic system of equations may be numerically integrated. Ideally a variable grid of point may be selected in the (x, t) plane where each point (x, t) is the crossing of two characteristics: i.e. a forward characteristic and a backward characteristic. However, the variable grid method requires two types of interpolation because (1) the hydraulic and geometric properties of the channel are defined only at a limited number of sections, and (2) the computed results are found at a number of (x, t) points unevenly distributed in the domain. In practice, a fixed grid method is preferred by practitioners: i.e. the Hartree method.

The Hartree scheme uses fixed time and spatial intervals (Fig. 16.9). The flow properties are known at the time $t = (n - 1)\delta t$. At the following time step $t + \delta t$, the characteristics intersecting

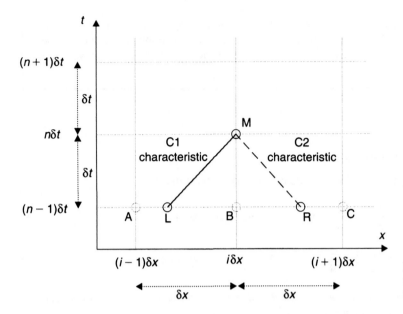

Fig. 16.9 Numerical integration of the method of characteristics by the Hartree method.

at point $M(x = i\delta x, t = n\delta t)$ are projected backward in time where they intersect the line $t = (n - 1)\delta t$ at points L and R whose locations are unknown (Fig. 16.9). Since x_L and x_R do not coincide with grid points, the velocity and celerity at points R and L must be interpolated between $x = (i - 1)\delta x$ and $x = i\delta x$, and between $x = i\delta x$ and $x = (i + 1)\delta x$, respectively. The discretization of the characteristic system of equations gives:

$$V_L + 2C_L = V_M + 2C_M + g(S_f - S_o)\delta t \qquad \text{Forward characteristic} \qquad (16.34a)$$

$$V_R - 2C_R = V_M - 2C_M + g(S_f - S_o)\delta t \qquad \text{Backward characteristic} \qquad (16.34b)$$

$$\frac{x_M - x_L}{\delta t} = V_L + C_L \qquad \text{Forward characteristic} \qquad (16.34c)$$

$$\frac{x_M - x_R}{\delta t} = V_R - C_R \qquad \text{Backward characteristic} \qquad (16.34d)$$

assuming $(S_f - S_o)$ constant during the time step δt. The system of four equations has four unknowns: V_M and C_M, and x_L and x_R, where the subscripts M, L, R refer, respectively, to points M, L and R defined in Fig. 16.9.

The stability of the solution of the method of characteristics is based upon the Courant condition. That is, the time step δt and the interval length δx must satisfy:

$$\frac{|V + C|}{\delta x/\delta t} \leqslant 1 \quad \text{and} \quad \frac{|V - C|}{\delta x/\delta t} \leqslant 1$$

where V is the flow velocity and C is the small wave celerity.

Notes

1. The *Hartree method* was named after the English physicist Douglas R. Hartree (1897–1958). The scheme is sometimes called the Hartree–Fock method after the Russian physicist V. Fock who generalized Hartree's scheme.

> 2. With the Hartree scheme, the solution is artificially smoothened by interpolation errors to esti-
> mate V_R, C_R, V_L and C_L. The errors are cumulative with time and the resulting inaccuracy limits
> the applicability of the method. In practice, finite difference methods offer many improvement in
> accuracy.

16.4 DISCUSSION

16.4.1 The dynamic equation

The differential form of the Saint-Venant equations may be written as:

$$\frac{\partial A}{\partial t} + \frac{\partial Q}{\partial x} = 0 \qquad \text{Continuity equation} \qquad (16.14)$$

$$\frac{1}{g}\left(\frac{\partial V}{\partial t} + V\frac{\partial V}{\partial x}\right) + \frac{\partial d}{\partial x} + S_f - S_o = 0 \qquad \text{Dynamic equation} \qquad (16.25b)$$

For *steady flows*, the continuity equation states $\partial Q/\partial x = 0$ (i.e. constant flow rate) in absence of
lateral inflow and the dynamic equation yields:

$$\frac{\partial d}{\partial x} + \frac{1}{g}V\frac{\partial V}{\partial x} + S_f - S_o = 0 \qquad \text{Steady flows} \qquad (16.34e)$$

Equation (16.34e) is basically the backwater equation for flat channels (Chapter 5) and it may be
rewritten as:

$$\frac{\partial d}{\partial x}\left(1 - \beta\frac{V^2}{g(A/B)}\right) = S_o - S_f \qquad \text{Steady flows} \qquad (16.34f)$$

where β is the momentum correction coefficient which is assumed as a constant (Chapter 3).
Note a major difference between the backwater equation and equation (16.34f). The former
derives from the energy equation but the latter was derived from the momentum principle.
   For steady uniform equilibrium flows, the dynamic equation yields:

$$S_f - S_o = 0 \qquad \text{Steady uniform equilibrium flows} \qquad (16.35)$$

Simplification of the dynamic wave equation
In the general case of unsteady flows, the dynamic equation (equation (16.25b)) may be simpli-
fied under some conditions, if the acceleration term $\partial V/\partial t$ and the inertial term $V\partial V/\partial x$ become
small. For example, when the flood flow velocity increases from 1 to 2 m/s in 3 h (i.e. rapid vari-
ation), the dimensionless acceleration term $(1/g)\partial V/\partial t$ equals 9.4×10^{-6}; when the velocity
increases from 1.0 to 1.4 m/s along a 10 km reach (e.g. reduction in channel width), the longitudinal slope of the kinetic energy line $(1/g)V\partial V/\partial x$ is equal to 4.9×10^{-6}. For comparison, the
average bed slope S_o of the Rhône river between Lyon and Avignon (France) is about
0.7×10^{-3} and the friction slope is of the same order of magnitude; the average bed slope and
friction slope of the Tennessee river between Watts Bar and Chickamaugo dam is 0.22×10^{-3};
during a flood in the Missouri river, the discharge increased very rapidly from 680 to 2945 m³/s,
but the acceleration term was less than 5% of the friction slope; on the Kitakami river (Japan),

Table 16.1 Simplification of the dynamic wave equation

Equation (1)	Dimensionless expression (2)	Remarks (3)
Dynamic wave equation	$\dfrac{1}{g}\left(\dfrac{\partial V}{\partial t} + V\dfrac{\partial V}{\partial x}\right) + \dfrac{\partial d}{\partial x} + S_f - S_o = 0$	Saint-Venant equation
Diffusive wave equation	$\dfrac{\partial d}{\partial x} + S_f - S_o = 0$	See Section 17.6
Kinematic wave equation	$S_f - S_o = 0$	See Section 17.5

actual flood records showed that the dimensionless acceleration and inertial terms were less than 1.5% than the term $\partial d/\partial x$ (Miller and Cunge, 1975: p. 189).

The dynamic equation may be simplified when one or more terms become negligible. Table 16.1 summarizes various forms of the dynamic wave equation which may be solved in combination with the unsteady flow continuity equation (equation (16.14)).

16.4.2 Limitations of the Saint-Venant equations

The Saint-Venant equations were developed for one-dimensional flows with hydrostatic pressure distributions, small bed slopes, constant water density, no sediment motion and assuming that the flow resistance is the same as for a steady uniform flow for the same depth and velocity.

Limitations of the Saint-Venant equation applications include two- and three-dimensional flows, shallow-water flood plains where the flow is nearly two-dimensional, undular and wavy flows, and the propagation of sharp discontinuities.

Flood plains
In two-dimensional problems involving flood propagation over inundated plains (e.g. Plate 24), the flood build-up is relatively slow, except when dykes break. The resistance terms are the dominant term in the dynamic equation. The assumption of one-dimensional flow becomes inaccurate.

Cunge (1975b) proposed an extension of the Saint-Venant equation in which flood plains are considered as storage volumes and with a system of equations somehow comparable to a network analysis.

Non-hydrostatic pressure distributions
The Saint-Venant equations are inaccurate when the flow is not one dimensional and the pressure distributions are not hydrostatic. Examples include flows in sharp bends, undular hydraulic jumps and undular tidal bores. Other relevant situations include supercritical flows with shock waves and flows over a ski jump (Chapter 3).

Remarks
1. Undular bores (and surges) are positive surge characterized by a train of secondary waves (or undulations) following the surge front. They are called Boussinesq–Favre waves in homage to the contributions of J.B. Boussinesq and H. Favre. Classical studies of undular surges include Benet and Cunge (1971) and Treske (1994).

2. With supercritical flows, a flow disturbance (e.g. change of direction and contraction) induces the development of shock waves propagating at the free surface across the channel (e.g. Ippen and Harleman, 1956). Shock waves are also called lateral shock waves, oblique hydraulic jumps, Mach waves, crosswaves and diagonal jumps.

   Shock waves induce flow concentrations. They create a local discontinuity in terms of water depth and pressure distributions.

Sharp discontinuities

The differential form of the Saint-Venant equations does not apply across sharp discontinuities. Plates 8 and 25 and Fig. 16.10 illustrate two examples: a hydraulic jump in a natural stream and a weir. At a hydraulic jump, the momentum equation must be applied across the jump front (Chapter 4) (Plate 25). At the weir crest, the Bernoulli equation may be applied (Chapters 3 and 19). If the weir becomes submerged, the structure acts as a large roughness and a singular energy loss.

Remarks

Another example of sharp transition is a sudden channel contraction. Assuming a smooth and short transition, the Bernoulli principle implies conservation of total head. Barnett (2002) discussed errors induced by the Saint-Venant equations for the triangular channel contraction.

16.4.3 Concluding remarks

The Saint-Venant equations are the unsteady flow equivalent of the backwater equation. The latter was developed for steady gradually varied flows and it is derived from the energy equation, while the Saint-Venant equations are derived from the momentum equation. Both the backwater and Saint-Venant equations are developed for one-dimensional flows, assuming a hydrostatic pressure distribution, for gradually varied flows, neglecting sediment transport and assuming that the flow resistance is the same as for uniform equilibrium flow conditions for the same

Fig. 16.10 Example of flow situations where the Saint-Venant equations are *not* applicable. Weir structure in Aichi forest park, Toyogawa catchment on 30 April 1999.

depth and velocity. The Saint-Venant equations assume further that the slope is small. Note that the backwater equation may account for the bed slope (Chapters 5 and 15) and a pressure correction coefficient may be introduced if the pressure is not hydrostatic.

Overall the assumptions behind the Saint-Venant equations are very restrictive and limit the applicability of the result (see Section 16.4.2).

16.5 EXERCISES

Write the five basic assumptions used to develop the Saint-Venant equations.

Were the Saint-Venant equations developed for movable boundary hydraulic situations?

Are the Saint-Venant equations applicable to a steep slope?
Solution: In open channel flow hydraulics, a 'steep' slope is defined when the uniform equilibrium flow is supercritical (Chapter 5). The notion of steep and mild slope is not only a function of the bed slope but also is a function of the flow resistance.

A basic assumption of the Saint-Venant equations is a bed slope small enough such that $\cos \theta \approx 1$ and $\sin \theta \approx \tan \theta \approx \theta$. It is based solely upon the invert angle with the horizontal θ. The following table summarizes the error associated with the approximation with increasing angle θ.

θ (degrees)	θ (radians)	$1 - \cos \theta$	$\sin \theta / \tan \theta$
0	0	0	1
0.5	0.008727	3.81×10^{-5}	0.999962
1	0.017453	0.000152	0.999848
2	0.034907	0.000609	0.999391
4	0.069813	0.002436	0.997564
6	0.10472	0.005478	0.994522
8	0.139626	0.009732	0.990268
10	0.174533	0.015192	0.984808
12	0.20944	0.021852	0.978148
15	0.261799	0.034074	0.965926
20	0.349066	0.060307	0.939693
25	0.436332	0.093692	0.906308

Express the differential form of the Saint-Venant equations in terms of the water depth and flow velocity. Compare the differential form of the momentum equation with the backwater equation.

What is the dynamic wave equation? From which fundamental principle does it derive?

What are the two basic differences between the dynamic wave equation and the backwater equation?
Solution: The backwater equation derives from energy considerations for steady flow motion.

Is the dynamic wave equation applicable to a hydraulic jump?
Solution: The dynamic wave equation is the differential form of the unsteady momentum equation. It might not be applicable to a discontinuity (e.g. a hydraulic jump), although the integral form of the Saint-Venant equations is (Section 16.2.2).

Is the dynamic wave equation applicable to an undular hydraulic jump or an undular surge?
Solution: Both hydraulic jump and positive surge are a flow discontinuity, and the differential form of the unsteady momentum equation might not be applicable. However, the pressure distribution is not hydrostatic beneath waves, and this includes undular jumps and surges (e.g. Chanson, 1995a, Montes and Chanson, 1998). As the Saint-Venant equations were

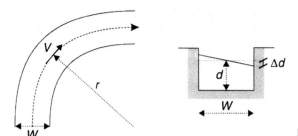

Fig. E.16.1 Sketch of a circular channel bend.

developed assuming a hydrostatic pressure distributions, they are neither applicable to undular hydraulic jump flows nor undular surges.

Considering a channel bend, estimate the conditions for which the basic assumption of quasi-horizontal transverse free surface is no longer valid. *Assume a rectangular channel of width W much smaller than the bend radius r.*

Solution: In a channel bend, the flow is subjected to a centrifugal acceleration acting normal to the flow direction and equal to V^2/r where r is the radius of curvature. The centrifugal pressure force induces a greater water depth at the outer bank than in a straight channel (Fig. E.16.1).

In first approximation, the momentum equation applied in the transverse direction yields:

$$\frac{1}{2}\rho g(d + \Delta d)^2 - \frac{1}{2}\rho g(d - \Delta d)^2 = \rho\frac{V^2}{r}dW$$

assuming $W \ll r$ and a flat horizontal channel. The rise Δd in free-surface elevation is about:

$$\Delta d \approx \frac{V^2}{2rg}W$$

The change in water depth from the inner to outer bank is less than 1% if the channel width, curvature and water depth satisfy:

$$\frac{V^2}{rg}\frac{W}{d} < 0.01$$

Remark

When the bend radius is comparable to the channel width, the change in curvature radius across the width must be taken into account. A simple development shows that, for a horizontal transverse invert, the transverse variations in channel depth and depth-averaged velocity satisfy, respectively:

$$\frac{\partial d}{\partial r} = \frac{V^2}{gr}$$

$$\frac{\partial V}{\partial r} = -\frac{V}{r}$$

where r is the radial distance measured from the centre of curvature and V is the depth-averaged velocity (Henderson, 1966: pp. 251–258). The results show that the water depth increases from the inner bank to the outer bank while the depth-averaged velocity is maximum at the inner bank. In practice, flow resistance affects the velocity field but maximum velocity is observed towards the inner bank.

Give the expression of the friction slope in terms of the flow rate, cross-sectional area, hydraulic diameter and Darcy friction factor only. Then, express the friction slope in terms of the flow rate and Chézy coefficient. Simplify both expressions for a wide rectangular channel.
Solution:

$$S_f = \frac{f}{8gd^3W^2}Q^2$$

$$S_f = \frac{Q^2}{C_{Ch\acute{e}zy}^2 d^3W^2}$$

where d is the water depth and W is the channel width.

Considering the flood plain sketched in Fig. 4.12 (Chapter 4, Chanson, 1999: p. 80), develop the expression of the friction slope in terms of the total flow rate and respective Darcy friction factors.

Considering a long horizontal channel with the fluid initially at rest, a small wave is generated at the origin at $t = 0$ (e.g. by throwing a stone in the channel). What is the wave location as a function of time for an observer standing on the bank? (Consider only the wave travelling in the positive x-direction.) What is the value of $(V + 2C)$ along the forward characteristics for an initial water depth $d_0 = 1.5\,\text{m}$?
Solution: A small disturbance will propagate with a celerity $C = \sqrt{gd_0}$. For a channel initially at rest, the wave location, for an observer standing on the bank is given by the forward characteristics trajectory:

$$\frac{dx}{dt} = V + C = 0 + \sqrt{gd_0} \qquad \text{Forward characteristics C1}$$

The integration gives:

$$x = \sqrt{gd_0}\,t$$

In a horizontal channel with the fluid initially at rest, $S_o = 0$ (horizontal channel) and $S_f = 0$ (no flow). Along the forward characteristics, $(V + 2C)$ is a constant and it is equal to 7.7 m/s.

Note that, at the wave crest, the water depth is greater than d_0, hence the wave celerity is greater than $\sqrt{gd}$. As a result, the velocity V becomes negative beneath the wave crest (Section 17.2.1).

Considering a long channel, flow measurements at two gauging stations given at $t = 0$:

	Station 1	Station 2
Location x (km)	7.1	8.25
Water depth (m)	2.2	2.45
Flow velocity (m/s)	+0.35	+0.29

In the (x, t) plane, plot the characteristics issuing from each gauging station. (*Assume straight lines.*) Calculate the location, time and flow properties at the intersection of the characteristics issuing from the two gauging stations. (*Assume $S_f = S_o = 0$.*)
Solution:
$x = 7.7\,\text{km}$, $t = 120\,\text{s}$, $V = +0.06\,\text{m/s}$ and $d = 2.34\,\text{m}$.

The flow conditions correspond to a reduction in flow rate. At $x = 7.7\,\text{km}$ and $t = 120\,\text{s}$ and $q = 0.14\,\text{m}^2/\text{s}$, compared to $q_1 = 0.77\,\text{m}^2/\text{s}$ and $q_2 = 0.71\,\text{m}^2/\text{s}$ at $t = 0$.
The analysis of flow measurements in a river reach gave:

	Station 1	Station 2
Location x (km)	11.8	13.1
Water depth (m)	0.65	0.55
Flow velocity (m/s)	+0.5	+0.55

at $t = 1\,\text{h}$. Assuming a kinematic wave (i.e. $S_o = S_f$), plot the characteristics issuing from the measurement stations *assuming straight lines*. Calculate the flow properties at the intersection of the characteristics.

Solution: For a kinematic wave problem (i.e. $S_o = S_f$), the characteristic system of equations becomes:

$$\frac{D}{Dt}(V + 2C) = -g(S_f - S_o) = 0 \quad \text{Forward characteristics}$$

$$\frac{D}{Dt}(V - 2C) = -g(S_f - S_o) = 0 \quad \text{Backward characteristics}$$

along the characteristic trajectories:

$$\frac{dx}{dt} = V + C \quad \text{Forward characteristics C1}$$

$$\frac{dx}{dt} = V - C \quad \text{Backward characteristics C2}$$

At the intersection of the forward characteristics issuing from Station 1 with the backward characteristics issuing from Station 2, the trajectory equations satisfy:

$$x = x_1 + (V_1 + C_1)t$$
$$x = x_2 + (V_2 - C_2)t$$

where x and t are the location and time of the intersection. At the intersection, the flow properties satisfy:

$$V + 2C = V_1 + 2C_1$$
$$V - 2C = V_2 - 2C_2$$

where $C = \sqrt{gd_1}$ and $C = \sqrt{gd_2}$.
The solutions of the characteristic system of equations yields: $x = 12.6\,\text{km}$, $t = 263\,\text{s}$, $V = 0.80\,\text{m/s}$, $C = 2.44\,\text{m/s}$, $d = 0.61\,\text{m}$ and $q = 0.49\,\text{m}^2/\text{s}$. As a comparison, $q_1 = 0.325\,\text{m}^2/\text{s}$ and $q_2 = 0.30\,\text{m}^2/\text{s}$ at $t = 0$. That is, the flow situation corresponds to an increase in flow rate.

Considering a supercritical flow (flow direction in the positive x-direction), how many boundary conditions are needed for $t > 0$ and where?

What is the difference between the backwater equation, diffusive wave equation, dynamic wave equation and kinematic wave equation? Which one(s) does(do) apply to unsteady flows?

Are the basic, original Saint-Venant equations applicable to the following situations: (1) flood plains, (2) mountain streams, (3) the Brisbane river in Brisbane, (4) the Mississippi river near Saint Louis and (5) an undular tidal bore?

In an irrigation canal, flow measurements at several gauging stations given at $t = 0$:

	Gauge 1	Gauge 2	Gauge 3	Gauge 4	Gauge 5
Location x (km)	1.2	1.31	1.52	1.69	1.95
Water depth (m)	0.97	0.96	0.85	0.78	0.75
Flow velocity (m/s)	+51	+49	+0.46	+0.42	+0.405

The canal has a trapezoidal cross-section with a 1 m base width and 1V:2H sidewalls.

In the (x, t) plane, plot the characteristics issuing from each gauging station. (*Assume straight lines.*) Calculate the location, time and flow properties at the intersections of the characteristics issuing from the gauging stations. Repeat the process for all the domain of dependance. (*Assume $S_f = S_o = 0$.*)
Solution: At the latest (i.e. last) intersection of the characteristic curves, $x = 1.61$ km, $t = 39$ s, $V = +0.59$ m/s and $C = 2.35$ m/s. The flow conditions correspond to an increase in flow rate. At $x = 1.61$ km, $t = 39$ s, $Q = 1.58$ m^3/s and $d = 0.933$ m, compared to $Q_1 = 1.45$ m^3/s and $Q_5 = 0.929$ m^3/s at $t = 0$.

Downstream of a hydropower plant, flow measurements in the tailwater channel indicated at $t = 0$:

	Station 1	Station 2	Station 3	Station 4
Location x (m)	95	215	310	605
Water depth (m)	0.37	0.45	0.48	0.52
Flow velocity (m/s)	+1.55	+1.44	+1.36	+1.10

The canal has a rectangular cross-section with a 9 m width.

In the (x, t) plane, plot the characteristics issuing from each gauging station. (*Assume straight lines.*) Calculate the location, time and flow properties at the intersections of the characteristics issuing from the gauging stations. Repeat the process for all the domain of dependance. (*Assume $S_f = S_o = 0$.*)
Solution: At the latest (i.e. last) intersection of the characteristic curves, $x = 421$ m, $t = 45$ s, $V = +0.59$ m/s and $C = 2.19$ m/s. The flow conditions correspond to a reduction in turbine discharge. At $x = 421$ km and $t = 45$ s and $Q = 4.28$ m^3/s, compared to $Q_1 = 5.16$ m^3/s and $Q_5 = 5.15$ m^3/s at $t = 0$.

Unsteady open channel flows: 2. Applications   17

Summary

In this chapter, the Saint-Venant equations for unsteady open channel flows are applied. Basic applications include small waves and monoclinal waves, the simple-wave problem, positive and negative surges and the dam break wave.

17.1 INTRODUCTION

In unsteady open channel flows, the velocities and water depths change with time and longitudinal position. For one-dimensional applications, the continuity and momentum equations yield the Saint-Venant equations (Chapter 16). The application of the Saint-Venant equations is limited by some *basic assumptions*: [H1] the flow is one dimensional, [H2] the streamline curvature is very small and the pressure distributions are hydrostatic, [H3] the flow resistance are the same as for a steady uniform flow for the same depth and velocity, [H4] the bed slope is small enough to satisfy $\cos\theta \approx 1$ and $\sin\theta \approx \tan\theta$ and [H5] the water density is a constant. Sediment transport is further neglected.

With these hypotheses, the flow can be characterized at any point and any time by two variables: e.g. V and Y where V is the flow velocity and Y is the free-surface elevation. The unsteady flow properties are described by a system of two partial differential equations:

$$\frac{\partial Y}{\partial t} + \frac{A}{B}\frac{\partial V}{\partial x} + V\left(\frac{\partial Y}{\partial x} + S_o\right) + \frac{V}{B}\left(\frac{\partial A}{\partial x}\right)_{d=\text{constant}} = 0 \qquad (17.1)$$

$$\frac{\partial V}{\partial t} + V\frac{\partial V}{\partial x} + g\frac{\partial Y}{\partial x} + gS_f = 0 \qquad (17.2)$$

where A is the cross-sectional area, B is the free-surface width, S_o is the bed slope and S_f is the friction slope. The Saint-Venant equations (equations (17.1) and (17.2)) cannot be solved analytically because of non-linear terms (e.g. S_f) and complicated functions (e.g. $A(d)$ and $B(d)$).

A mathematical technique to solve the system of partial differential equations formed by the Saint-Venant equations is the method of characteristics. It yields a characteristic system of equations:

$$\frac{D}{Dt}(V + 2C) = -g(S_f - S_o) \qquad \text{Forward characteristic} \qquad (17.3a)$$

$$\frac{D}{Dt}(V - 2C) = -g(S_f - S_o) \qquad \text{Backward characteristic} \qquad (17.3b)$$

along:

$$\frac{dx}{dt} = V + C \qquad \text{Forward characteristic C1} \qquad (17.4a)$$

$$\frac{dx}{dt} = V - C \qquad \text{Backward characteristic C2} \qquad (17.4b)$$

where C is celerity of a small disturbance for an observer travelling with the flow: $C = \sqrt{gA/B}$. For an observer travelling along the forward characteristic, equation (17.3a) is valid at any point. For an observer travelling on the backward characteristic, equation (17.3b) is satisfied everywhere. The system of four equations formed by equations (17.3) and (17.4) represents the characteristic system of equations that replaces the differential form of the Saint-Venant equations.

Note

Along the forward characteristic trajectory:

$$\frac{dx}{dt} = V + C$$

the absolute derivative of $(V + 2C)$ equals:

$$\frac{D}{Dt}(V + 2C) = \left(\frac{\partial}{\partial t} + (V + C)\frac{\partial}{\partial x} \right)(V + 2C) \qquad \text{Forward characteristic}$$

Conversely, the absolute derivative of $(V - 2C)$ along the backward characteristic is:

$$\frac{D}{Dt}(V - 2C) = \left(\frac{\partial}{\partial t} + (V - C)\frac{\partial}{\partial x} \right)(V - 2C) \qquad \text{Backward characteristic}$$

17.2 PROPAGATION OF WAVES

17.2.1 Propagation of a small wave

Considering a simple-wave propagating in a horizontal channel initially at rest, the wave height is Δd and the wave propagation speed (or celerity) is U (Fig. 17.1). For an observer travelling with the wave, the continuity and energy equations may be written between an upstream location and the cross-section where the wave height is maximum. Neglecting energy loss and assuming a prismatic rectangular channel, it yields:

$$Ud = (U - \Delta U)(d + \Delta d) \qquad \text{Continuity equation} \qquad (17.5)$$

$$d + \frac{U^2}{2g} = d + \Delta d + \frac{(U - \Delta U)^2}{2g} \qquad \text{Bernoulli equation} \qquad (17.6)$$

where d is the water depth in the channel initially at rest.
After transformation, the wave celerity equals:

$$U = \sqrt{\frac{2g(d + \Delta d)^2}{2d + \Delta d}} \qquad (17.7a)$$

For a small wave, the celerity of the disturbance becomes:

$$U \approx \sqrt{gd}\left(1 + \frac{3}{4}\frac{\Delta d}{d}\right) \qquad \text{Small wave in fluid at rest} \qquad (17.7b)$$

For an infinitely small disturbance, equation (17.7a) yields: $U = \sqrt{gd}$.
For a small wave propagating in an uniform flow (velocity V), the propagation speed of a small wave equals:

$$U \approx V + \sqrt{gd}\left(1 + \frac{3}{4}\frac{\Delta d}{d}\right) \qquad \text{Small wave in uniform flow} \qquad (17.8)$$

where U is the wave celerity for an observer standing on the bank and assuming that the wave propagates in the flow direction.

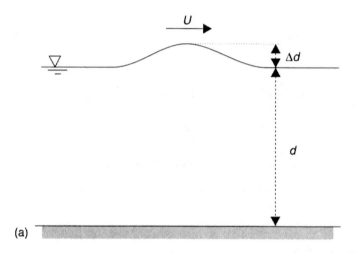

(a)

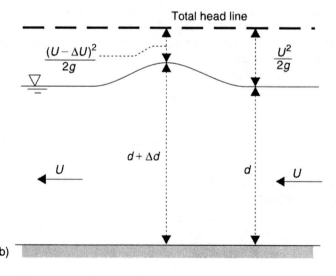

(b)

Fig. 17.1 Sketch of a small wave propagation: (a) view by an observer standing on the bank and (b) view by an observer travelling with the wave.

Notes
1. A celerity is a characteristic velocity or speed. For example, the celerity of sound is the sound speed.
2. In July 1870, Barré de Saint-Venant derived equation (17.7a) and compared it favourably with the experiments of H.E. Bazin (Barré de Saint-Venant, 1871b: p. 239).
3. Henri Emile Bazin (1829–1917) was a French engineer, member of the French 'Corps des Ponts-et-Chaussées' and later of the Académie des Sciences de Paris. He worked as an assistant of Henri P.G. Darcy at the beginning of his career.

17.2.2 Propagation of a known discharge (monoclinal wave)

Considering the propagation of a known flow rate Q_2 in a channel, the initial discharge in that channel is Q_1 prior to the passage of the monoclinal wave (Fig. 17.2). Upstream and downstream of the wave, the flow conditions are steady and the propagating wave is assumed to have a constant shape. The celerity of the wave must be greater than the downstream flow velocity. The continuity equation between sections 1 and 2 gives:

$$(U - V_1)A_1 = (U - V_2)A_2 \tag{17.9}$$

where U is the wave celerity for an observer standing on the bank, V is the flow velocity, A is the cross-sectional area and the subscripts 1 and 2 refer to the initial and new (steady) flow conditions, respectively. After transformation, it yields:

$$U = \frac{Q_2 - Q_1}{A_2 - A_1} \tag{17.10a}$$

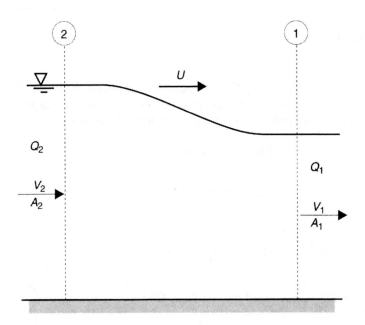

Fig. 17.2 Propagation of a known flow rate.

For a small variation in discharge, it becomes:

$$U = \frac{\partial Q}{\partial A} \tag{17.10b}$$

Using equation (17.10a), the flood discharge wave propagation is predicted as a function of the cross-sectional shape and flow velocity.

Discussion

The discharge wave celerity U is a function of the steady flow rate Q. Considering an uniform equilibrium flow in a wide rectangular channel, the discharge equals:

$$Q = \sqrt{\frac{8g}{f}} A \sqrt{d S_o}$$

The celerity of a discharge wave becomes:

$$U = \frac{\partial Q}{\partial A} = \frac{\partial Q/\partial d}{\partial A/\partial d} = \frac{(3/2)\sqrt{8g/f}\sqrt{d S_o} B}{B} = \frac{3}{2}V \qquad \text{Constant Darcy friction factor}$$

where V is the flow velocity and assuming a constant Darcy friction factor f.

Equation (17.10a) is highly dependent upon the flow resistance formula (e.g. Darcy–Weisbach, Chézy and Gauckler–Manning). For example, for a wide rectangular channel, the Gauckler–Manning formula yields:

$$U = \frac{\partial Q}{\partial A} = \frac{5}{3}V \qquad \text{Constant Gauckler–Manning friction coefficient}$$

Notes

1. The propagation wave of a known discharge is known as a *monoclinal wave*.
2. Equation (17.10a) was first developed by A.J. Seddon in 1900 and used for the Mississippi River. It is sometimes called Seddon equation or Kleitz–Seddon equation.

17.3 THE SIMPLE-WAVE PROBLEM

17.3.1 Basic equations

A simple wave is defined as a wave for which $S_o = S_f = 0$ with initially constant water depth and flow velocity. In the system of Saint-Venant equations, the dynamic equation becomes a kinematic wave equation (Section 15.4.1). The characteristic system of equations for a simple wave is:

$$\frac{D}{Dt}(V + 2C) = 0 \qquad \text{Forward characteristic}$$

$$\frac{D}{Dt}(V - 2C) = 0 \qquad \text{Backward characteristic}$$

along:

$$\frac{dx}{dt} = V + C \qquad \text{Forward characteristic C1}$$

$$\frac{dx}{dt} = V - C \qquad \text{Backward characteristic C2}$$

Basically, $(V + 2C)$ is constant along the forward characteristic. That is, for an observer moving at the absolute velocity $(V + C)$. Similarly $(V - 2C)$ is a constant along the backward characteristic. The characteristic trajectories can be plotted in the (x, t) plane (Fig. 17.3). They represent the path of observers travelling in the forward and backward characteristics. On each forward characteristic, the slope of the trajectory is $1/(V + C)$ while $(V + 2C)$ is a constant along the characteristic trajectory. Altogether the characteristic trajectories form contour lines of $(V + 2C)$ and $(V - 2C)$. For the simple-wave problem $(S_o = S_f = 0)$, a family of characteristic trajectories is a series of straight lines if any one curve of the family (i.e. C1 or C2) is a straight line (Henderson, 1966).

Initially, the flow conditions are uniform everywhere: i.e. $V(x, t = 0) = V_o$ and $d(x, t = 0) = d_o$. At $t = 0$, a disturbance (i.e. a simple wave) is introduced at the origin ($x = 0$, Fig. 17.3). The wave propagates to the right along the forward characteristic C1 with a velocity $(V_o + C_o)$ where V_o is the initial flow velocity and C_o is the initial celerity: $C_o = \sqrt{gA_o/B_o}$.[1] The disturbance must propagate into the undisturbed flow region with a celerity assumed to be C_o implying that the wave front is assumed small enough. In the (x, t) plane, the positive characteristic D1–E2 divides the flow region into a region below the forward characteristic which is unaffected by the disturbance (i.e. zone of quiet) and the flow region above where the effects of the wave are felt.

In the *zone of quiet*, the flow properties at each point may be deduced from two characteristic curves intersecting the line ($t = 0$) for $x > 0$ where $V = V_o$ and $C = C_o$. In turn, the characteristic system of equations yields: $C = C_o$ and $V = V_o$ everywhere in the zone of quiet, also called the undisturbed zone.

Considering a point E1 located on the left boundary (i.e. $x = 0$, $t > 0$), the backward characteristic issuing from this point intersects the first positive characteristic and it satisfies:

$$V_{E1} - 2C_{E1} = V_o - 2C_o \tag{17.11a}$$

Considering the forward characteristic issuing from the same point E1 (line E1–F1), the slope of the C1 characteristic satisfies:

$$\left(\frac{dt}{dx}\right)_{E1} = \frac{1}{V_{E1} + C_{E1}} = \frac{1}{3C_{E1} + V_o - 2C_o} \tag{17.11b}$$

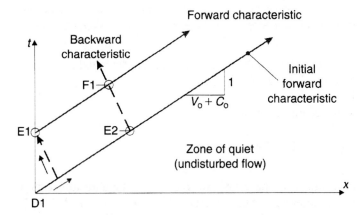

Fig. 17.3 Characteristic curves of a simple wave.

[1] For the subcritical flow sketched in Fig. 17.2. If $x = 0$ is not a boundary condition, a similar reasoning may be developed on the left side of the (x, t) plane following the negative characteristics issuing from the origin (e.g. Section 17.7).

At the boundary (i.e. Point E1), one flow parameter (V or C) is prescribed. The second flow parameter and the slope of the forward characteristic at the boundary are then deduced from equation (17.11a).

Considering a point F1 in the (x, t) plane, both forward and backward characteristic trajectories intersect (Fig. 17.3). The flow conditions at F1 must satisfy:

$$V_{F1} + 2C_{F1} = V_{E1} + 2C_{E1} \qquad \text{Forward characteristic} \qquad (17.12a)$$

$$V_{F1} - 2C_{F1} = V_o - 2C_o \qquad \text{Backward characteristic} \qquad (17.12b)$$

This linear system of equations has two unknowns V_{F1} and C_{F1}. In turn, the flow properties can be calculated everywhere in the (x, t) plane, but in the zone of quiet where the flow is undisturbed.

Although the simple-wave problem is based upon drastic assumptions, it may be relevant to a number of flow situations characterized by rapid changes such that the acceleration term $\partial V/\partial t$ is much larger than both bed and friction slopes. Examples include the rapid opening (or closure) of a gate and the dam break wave. Henderson (1966: pp. 289–294) presented a comprehensive discussion of the simple-wave problem.

Remarks
1. The forward characteristic issuing from the origin ($x = 0$, $t = 0$) is often called the *initial forward characteristic* or initial characteristic.
2. Stoker (1957) defined the wave motion under which the forward characteristics are straight as a 'simple wave'.
3. In the (x, t) plane (e.g. Fig. 17.3), the slope of a forward characteristic is:

$$\frac{dt}{dx} = \frac{1}{V(x, t) + C(x, t)}$$

17.3.2 Application

Considering a simple-wave problem where a disturbance originates from $x = 0$ at $t = 0$, the calculations must proceed in a series of successive steps. For the subcritical flow sketched in Fig. 17.3, this is:

Step 1. Calculate the initial flow conditions at $x > 0$ for $t = 0$. These include the initial flow velocity V_o and the small wave celerity $C_o = \sqrt{gA_o/B_o}$.

Step 2. Plot the initial forward characteristic.
The initial forward characteristic trajectory is:

$$t = \frac{x}{V_o + C_o} \qquad \text{Initial forward characteristic}$$

The C1 characteristic divides the (x, t) space into two regions. In the undisturbed flow region (zone of quiet), the flow properties are known everywhere: i.e. $V = V_o$ and $C = C_o$.

Step 3. Calculate the flow conditions (i.e. $V(x = 0, t_o)$ and $C(x = 0, t_o)$) at the boundary ($x = 0$) for $t_o > 0$. One flow condition is prescribed at the boundary: e.g. the flow rate is known or the water depth is given. The second flow property is calculated using the backward characteristics intersecting the boundary:

$$V(x = 0, t_o) = V_o + 2(C(x = 0, t_o) - C_o) \qquad (17.11a)$$

Step 4. Draw a family of forward characteristics issuing from the boundary ($x = 0$) at $t = t_o$. The forward characteristic trajectory is:

$$t = t_o + \frac{x}{V(x = 0, t_o) + C(x = 0, t_o)} \qquad \text{where } t_o = t(x = 0)$$

Step 5. At the required time t, or required location x, calculate the flow properties along the family of forward characteristics.

The flow conditions are the solution of the characteristic system of equations along the forward characteristics originating from the boundary (Step 4) and the negative characteristics intersecting the initial forward characteristic. It yields:

$$V(x, t) = \frac{1}{2}(V(x = 0, t_o) + 2C(x = 0, t_o) + V_o - 2C_o)$$

$$C(x, t) = \frac{1}{4}(V(x = 0, t_o) + 2C(x = 0, t_o) - V_o + 2C_o)$$

Discussion

In the solution of the simple-wave problem, Step No. 2 is important to assess the extent of the influence of the disturbance. In Steps No. 3 and 5, the celerity C is a function of the flow depth: $C = \sqrt{gA/B}$ where the flow cross-section A and free-surface width B are both functions of the water depth.

It is important to remember the basic assumptions of the simple-wave analysis. That is, a wave for which $S_o = S_f = 0$ with initially constant water depth d_o and flow velocity V_o. Furthermore, the front of the wave is assumed small enough for the initial forward characteristic trajectory to satisfy $dx/dt = V_o + C_o$.

Application

A basic application of the simple-wave method is a river discharging into the sea and the upstream extent of tidal influence onto the free surface. Considering a stream discharging into the sea, the tidal range is 0.8 m and the tidal period is 12 h 25 min. The initial flow conditions are $V = 0.3$ m/s and $d = 0.4$ m corresponding to low tide. Neglecting bed slope and flow resistance, and starting at a low tide, calculate how far upstream the river level will rise 3 h after low tide and predict the free-surface profile at $t = 3$ h.

Solution

The prescribed boundary condition at the river mouth is the water depth:

$$d(x = 0, t_o) = 0.4 + \frac{0.8}{2}\left(1 + \cos\left(\frac{2\pi}{T}t_o - \pi\right)\right)$$

where T is the tide period ($T = 44\,700$ s).

Let us select a coordinate system with $x = 0$ at the river mouth and x positive in the upstream direction. The initial conditions are $V_0 = -0.3$ m/s and $C_0 = 1.98$ m/s assuming a wide rectangular channel.

Then the flow conditions at the boundary ($x = 0$) are calculated at various times t_0:

$$C(x = 0, t_0) = \sqrt{gd(x = 0, t_0)}$$

$$V(x = 0, t_0) = 2C(x = 0, t_0) + V_0 - 2C_0 \qquad (17.11a)$$

The results are summarized below:

t_0 (s) (1)	$d(x = 0, t_0)$ (m) (2)	$C(x = 0, t_0)$ (m/s) (3)	$V(x = 0, t_0)$ (m/s) (4)	$V + C$ (m/s) (5)	$V + 2C$ (m/s) (6)
0	0.40	1.98	−0.30	1.68	3.66
4470	0.48	2.16	0.06	2.22	4.38
8940	0.68	2.57	0.89	3.46	6.04
13410	0.92	3.01	1.76	4.77	7.77
17880	1.12	3.32	2.38	5.70	9.01
22350	1.20	3.43	2.60	6.03	9.46

Next the C1 characteristic trajectories can be plotted from the left boundary ($x = 0$). As the initial C1 characteristic is a straight line, all the C1 characteristics are straight lines with the slope:

$$\frac{dt}{dx} = \frac{1}{V(x, t) + C(x, t)} = \frac{1}{V(x = 0, t_0) + C(x = 0, t_0)}$$

Results are shown in Fig. 17.4 for the first 3 h ($0 < t < 3$ h). Figure 17.4a presents three C1 characteristics. Following the initial characteristic, the effects of the tide are felt as far upstream as $x = +18.1$ km at $t = 3$ h.

At $t = 3$ h, the flow property between $x = 0$ and 18 100 m are calculated from:

$$V(x, t = 3\,h) + 2C(x, t = 3\,h) = V(x = 0, t_0) + 2C(x = 0, t_0) \qquad \text{Forward characteristic}$$

$$V(x, t = 3\,h) - 2C(x, t = 3\,h) = V_0 - 2C_0 \qquad \text{Backward characteristic}$$

where the equation of the forward characteristic is:

$$t = t_0 + \frac{x}{V(x = 0, t_0) + C(x = 0, t_0)} \qquad \text{Forward characteristic}$$

These three equations are three unknowns: $V(x, t = 3\,h)$, $C(x, t = 3\,h)$ and $t_0 = t(x = 0)$ for the C1 characteristics. The results in terms of the water depth, flow velocity and celerity at high tide are presented in Fig. 17.4b.

Remarks

- Practically, the calculations of the free-surface profile are best performed by selecting the boundary point (i.e. the time t_0), calculating the flow properties at the boundary, and then drawing the C1 characteristics until they intersect the horizontal line $t = 3$ h. That is, the computations follow the order: select $t_0 = t(x = 0)$, calculate $d(x = 0, t_0)$, compute $V(x = 0, t_0)$, $C(x = 0, t_0)$, plot the

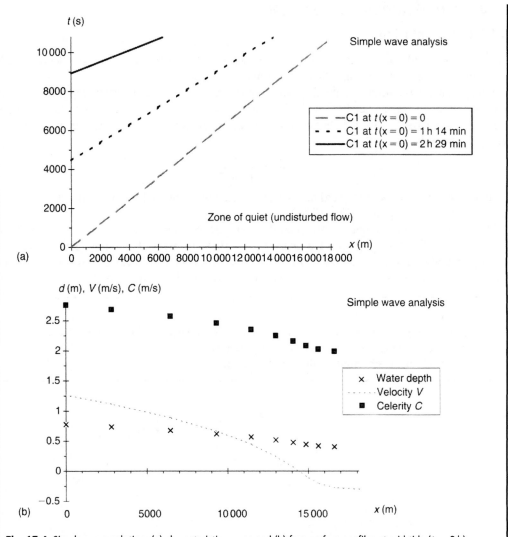

Fig. 17.4 Simple-wave solution: (a) characteristic curves and (b) free-surface profiles at mid-tide ($t = 3$ h).

C1 characteristics and calculate $x(t = 6$ h 12.5 min) along the C1 characteristics. Then compute $V(x, t = 3$ h) and $C(x, t = 3$ h):

$$V(x, t = 3 \text{ h}) = \frac{1}{2}(V(x = 0, t_o) + 2C(x = 0, t_o) + V_o - 2C_o)$$

$$C(x, t = 3 \text{ h}) = \frac{1}{4}(V(x = 0, t_o) + 2C(x = 0, t_o) - V_o + 2C_o)$$

Lastly the flow depth equals $d = C^2/g$.

- In Fig. 17.4b, note that the velocity is positive (i.e. upriver flow) for $x < 14.4$ km at $t = 3$ h.
- Note that the C2 characteristics lines are not straight lines.
- Another relevant example is Henderson (1966: pp. 192–294).

17.4 POSITIVE AND NEGATIVE SURGES

17.4.1 Presentation

A surge (or wave) induced by a rise in water depth is called a *positive surge* (Section 4.3). A *negative surge* is associated with a reduction in water depth.

Considering a rise in water depth at the boundary, the increase in water depth induces an increase in small wave celerity C with time because $C = \sqrt{gd}$ for a rectangular channel. As a result, the slope of the forward characteristics in the (x, t) plane decreases with increasing time along the boundary $(x = 0)$ as:

$$\frac{dt}{dx} = \frac{1}{V + C}$$

It follows that the series of forward characteristics issuing from the boundary condition forms a network of converging lines (Fig. 17.5a).

Similarly, a negative surge is associated with a reduction in water depth, hence a decrease in wave celerity. The resulting forward characteristics issuing from the origin form a series of diverging straight lines. The negative wave is said to be dispersive. Locations of constant water depths move further apart as the wave moves outwards from the point of origin.

Remarks
1. A positive surge is also called bore, positive bore, moving hydraulic jump and positive wave. A positive surge of tidal origin is a tidal bore (Chapter 4).
2. A negative surge is sometimes called negative wave.

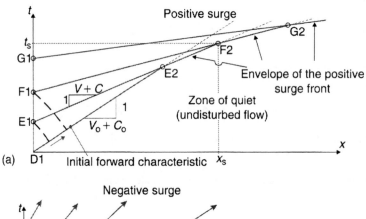

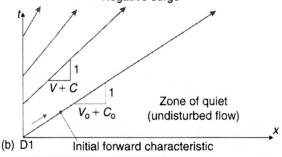

Fig. 17.5 Characteristic trajectories for positive and negative surges: (a) positive surge and (b) negative surge.

17.4.2 Positive surge

Although the positive surge may be analysed using a quasi-steady flow analogy (Chapter 4), its inception and development is studied with the method of characteristics. When the water depth increases with time, the forward characteristics converge and eventually intersect (Fig. 17.5a, points E2, F2 and G2). The intersection of two forward characteristics implies that the water depth has two values at the same time. This anomaly corresponds to a wave front which becomes steeper and steeper until it forms an abrupt front: i.e. the positive surge front (Fig. 17.7, Plate 16). For example, the forward characteristic issuing from E1 intersects the initial forward characteristic at E2 (Fig. 17.5a). For $x > x_{E2}$, the equation of the initial forward characteristic (D1–E2) is no longer valid as indicated by the thin dashed line in Fig. 17.5a.

After the formation of the positive surge, some energy loss takes place across the surge front and the characteristic cannot be projected from one side of the positive surge to the other. The forward characteristic E2–F2 becomes the 'initial characteristic' for $x_{E2} < x < x_{F2}$. The points of intersections (e.g. points E2, F2 and G2) form an envelope defining the zone of quiet (Fig. 17.5a).

Practically, the positive surge forms at the first intersecting point: i.e. the intersection point with the smallest time t (Fig. 17.6). Figure 17.6 illustrates the development of a positive surge from the onset of the surge (i.e. first intersection of forward characteristics) until the surge reaches its final form. After the first intersection of forward characteristics, there is a water depth discontinuity along the forward characteristics forming the envelope of the surge front: e.g. between points E2 and F2, and F2 and G2 in Fig. 17.5. The flow conditions across the surge front satisfy the continuity and momentum principles (Chapter 4).

Remarks

1. The flow properties at the positive surge front may be calculated from the momentum principle based upon a quasi-steady flow analogy (Chapter 4). After formation of the surge front, the celerity of surge front is $U = V + C$ along the relevant forward characteristics.
2. A positive surge has often a breaking front. It may also consist of a train of free-surface undulations if the surge Froude number is less than 1.35–1.4 (e.g. Chanson, 1995a). Figure 17.7a illustrates a non-breaking undular surge in a horizontal rectangular channel.
3. Henderson (1966: pp. 294–304) presented a comprehensive treatment of the surges using the method of characteristics.

Discussion: tidal bores

When a river mouth has a flat, converging shape and when the tidal range exceeds 6–9 m, the river may experience a tidal bore. A tidal bore is basically a wave or a series of waves propagating upstream as the tidal flow turns to rising. It is a positive surge (Section 4.3). As the surge progresses inland, the river flow is reversed behind it (e.g. Lynch, 1982; Chanson, 2001b). For example, in the Qiantang River, fishermen waited for the bore to sail their junks upriver (Tricker, 1965).

Tidal bores induce very-strong turbulent mixing in the estuary and in the river (e.g. Chanson, 2001b). The effect may be felt along considerable distances. With appropriate boundary conditions, a tidal bore may travel far upstream: e.g. the tidal bore on the Pungue River (Mozambique) is still about 0.7 m high about 50 km upstream of the mouth and it may reach 80 km inland.

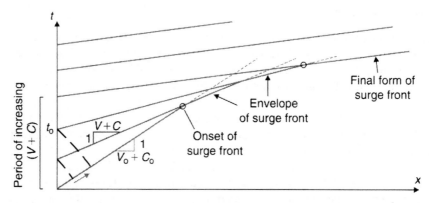

Fig. 17.6 Characteristic trajectories for the development of a positive surge.

(a)

(b)

Fig. 17.7 Photographs of positive surges: (a) Advancing undular surge front in a 0.8 m wide rectangular channel. Surge propagation from left to right. Surge Froude number: 1.12, initial water depth: 0.4 m, initial flow velocity: zero and surge propagation speed: 2.21 m/s. (b) The Hangzhou tidal bore (China), looking downstream at the incoming bore. View from the left bank on 23 September 1999 (courtesy of Dr Dong Zhiyong).

Simple-wave calculations of a positive surge

For a simple wave, the apparition of the positive surge corresponds to the first intersecting point (point E2, Fig. 17.5a). The location and time of the wave front are deduced from the characteristic equation:

$$V_{E1} - 2C_{E1} = V_o - 2C_o \qquad \text{Backward characteristic to E1}$$

and from the trajectory equations:

$$x = (V_o + C_o)t \qquad \text{Initial forward characteristic}$$

$$x = (V_{E1} + C_{E1})(t - t_{E1}) \qquad \text{E1–E2 forward characteristic}$$

where V_o and C_o are the initial flow conditions. The first intersection of the two forward characteristics occurs at:

$$x_{E2} = \frac{(V_o + C_o)(V_o + 3C_{E1} - 2C_o)}{3((C_{E1} - C_o)/t_{E1})}$$

$$t_{E2} = \frac{x_{E2}}{V_o + C_o}$$

The reasoning may be extended to the intersection of two adjacent forward characteristics: e.g. the forward characteristics issuing from E1 and F1 and intersecting at F2, or the forward characteristics issuing from F1 and G1 and intersecting at G2. It yields a more general expression of the surge location:

$$x_s = \frac{(V_o + 3C(x = 0, t_o) - 2C_o)^2}{3(\partial C(x = 0, t_o)/\partial t)} \qquad (17.13a)$$

$$t_s = t_o + \frac{x_s}{V_o + 3C(x = 0, t_o) - 2C_o} \qquad (17.13b)$$

where $C(x = 0, t_o)$ is the specified disturbance celerity at the origin and x_s is the surge location at the time t_s (Fig. 17.5a). If the specified disturbance is the flow rate $Q(x = 0, t_o)$ or the flow velocity $V(x = 0, t_o)$, equation (17.12b) may be substituted into equation (17.13).

Equation (17.13) is the equation of the envelope delimiting the zone of quiet and defining the location of the positive surge front. The location where the surge first develops would correspond to point E2 in the (x, t) plane in Fig. 17.5a.

Application

At the opening of a lock into a navigation canal of negligible slope, water flows into the canal with a linear increase between 0 and 30 s and a linear decrease between 30 and 60 s:

t (s)	0	30	60
Q (m³/s)	0	15	0

In the navigation canal, the water is initially at rest, the initial water depth is 1.8 m and the canal width is 22 m. Calculate the characteristics of the positive surge and its position with time.

Solution

Step 1. The initial conditions are $V_o = 0$ and $C_o = 4.2$ m/s.

Step 2. The trajectory of the initial forward characteristic is $t = x/4.2$ where $x = 0$ at the water lock and x is positive in the downstream direction (i.e. toward the navigation canal).

Step 3. The boundary condition prescribes the flow rate. The velocity and celerity at the boundary satisfy:

$$Q(x = 0, t_o) = V(x = 0, t_o)d(x = 0, t_o)W \qquad \text{Continuity equation}$$

$$V(x = 0, t_o) = V_o + 2(C(x = 0, t_o) - C_o) \qquad \text{C2 characteristic} \qquad (17.11a)$$

with $C(x = 0) = \sqrt{gd(x = 0)}$. It yields:

t_o (s)	0	5	10	15	20	25	30	40	50	60
$C(x = 0, t_o)$(m/s)	1.8	4.23	4.26	4.29	4.32	4.35	4.37	4.32	4.26	4.20
$V(x = 0, t_o)$(m/s)	4.2	0.06	0.12	0.18	0.24	0.29	0.35	0.24	0.12	0.0

Step 4. We draw a series of characteristics issuing from the boundary.

The characteristic curves for $0 < t_o < 30$ s are converging resulting in the formation of positive surge (Fig. 17.8). For $t_o > 30$ s, the reduction in flow rate is associated with the development of a negative surge (behind the positive surge) and a series of diverging characteristics. The position of the bore may be estimated from the first intersection of the converging lines with the initial characteristic or from equation (17.13). The results show a positive surge formation about 970 m downstream of the lock.

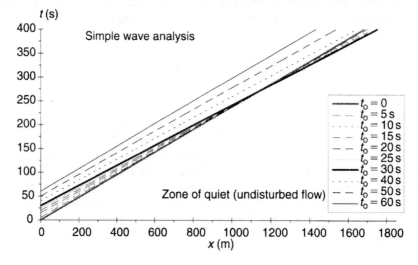

Fig. 17.8 Characteristic curves of the positive surge application.

Remarks

1. The simple-wave approximation basically neglects the bed slope and flow resistance (i.e. $S_o = S_f = 0$). In a frictionless horizontal channel, a rise in water depth must eventually form a surge with a steep front.

2. Barré de Saint-Venant predicted the development of a tidal bore (Barré de Saint-Venant, 1871b: p. 240). He considered both cases of a simple wave and a surge propagating in uniform equilibrium flow (see below).

Positive surge propagating in uniform equilibrium flow

For a wide rectangular channel, the friction slope equals:

$$S_f = \frac{f}{2} \frac{V^2}{gd}$$

At uniform equilibrium, the momentum principle (Chapter 4) states:

$$S_f = \frac{f}{2} \frac{V_o^2}{gd_o} = S_o$$

Hence the term $(S_f - S_o)$ may be rewritten as:

$$S_f - S_o = S_o\left(\frac{Fr^2}{Fr_o^2} - 1\right)$$

where Fr is the Froude number and Fr_o is the uniform equilibrium flow Froude number ($Fr_o = V_o/\sqrt{gd_o}$). Note that the celerity of a small disturbance is $C = \sqrt{gd}$ and the Froude number becomes $Fr = V/C$ (Chapter 3).

Assuming that the initial flow conditions are uniform equilibrium (i.e. $S_o = S_f$), the characteristic system of equations becomes:

$$\frac{D}{Dt}(C(Fr + 2)) = -gS_o\left(\frac{Fr^2}{Fr_o^2} - 1\right) \qquad \text{Forward characteristic} \qquad (17.14a)$$

$$\frac{D}{Dt}(C(Fr - 2)) = -gS_o\left(\frac{Fr^2}{Fr_o^2} - 1\right) \qquad \text{Backward characteristic} \qquad (17.14b)$$

along:

$$\frac{dx}{dt} = C(Fr + 1) \qquad \text{Forward characteristic C1} \qquad (17.15a)$$

$$\frac{dx}{dt} = C(Fr - 1) \qquad \text{Backward characteristic C2} \qquad (17.15b)$$

Discussion

The initial flow conditions being uniform equilibrium (i.e. $S_o = S_f$), a positive disturbance is introduced at one end of the channel (i.e. $x = 0$). The initial forward characteristic propagates in the uniform equilibrium flow ($S_o = S_f$) and it is a straight line with a slope $dt/dx = 1/(V_o + C_o)$ where V_o is the uniform equilibrium flow depth.

Considering a backward characteristic upwards from the initial C1 characteristic, and at the intersection of the C2 characteristics with the initial forward characteristic, the following relationship holds:

$$\frac{D}{Dt}(C(Fr - 2)) = 0$$

Further upwards, the quantity $(S_f - S_o)$ may increase or decrease in response to a small change in Froude number ∂Fr:

$$\frac{\partial(S_f - S_o)}{\partial Fr} = \frac{2S_o}{Fr_o}$$

Note that the sign of both Fr_o and S_o is a function of the initial flow direction. That is, $Fr_o > 0$ and $S_o > 0$ if $V_o > 0$. For $t > 0$, the equations of the forward characteristics are:

$$x = (V_o + C_o)(t - t_o)\frac{3(C(x = 0, t_o) - C_o)}{gS_o((Fr_o - 2)/2C_oFr_o)}\left(\exp\left(gS_o\frac{Fr_o - 2}{2C_oFr_o}(t - t_o)\right) - 1\right) \quad \begin{array}{l}\text{Forward}\\\text{characteristic C1}\end{array}$$

(17.16)

where $t_o = t(x = 0)$ (Henderson, 1966). Equation (17.16) accounts for the effect of flow resistance on the surge formation and propagation. For $Fr_o < 2$, the term in the exponential is negative: i.e. the flow resistance delays (1) the intersection of neighbouring forward characteristics and (2) the positive surge formation. On the other hand, for $Fr_o > 2$, surge formation may occur earlier than in absence of flow resistance (i.e. simple wave). Henderson (1966: pp. 297–304) showed that flow resistance makes positive waves more dispersive for uniform equilibrium flow conditions with Froude number less than 2.

Remarks

1. The above development implies that the disturbance at the boundary is small enough to apply a linear theory and to neglect the terms of second order (Henderson, 1966).
2. The friction slope has the same sign as the flow velocity. If there is a flow reversal ($V < 0$), the friction slope becomes negative to reflect that boundary friction opposes the flow motion. In such a case, the friction slope must be rewritten as:

$$S_f = \frac{f}{2}\frac{V|V|}{gd} = \frac{f}{2}Fr|Fr|$$

where $|V|$ is the magnitude of the velocity.
3. A wave propagating upstream, against the flow, is called an *adverse wave*. A wave propagating downstream is named a *following wave*. Note that a small disturbance cannot travel upstream against a supercritical flow (Chapter 3). That is, an adverse wave can exist only for $Fr_o < 1$.

Application

The Qiantang River discharges into the Hangzhou Wan in East China Sea. Between Laoyancang and Jianshan, the river is 4 km wide, the bed slope corresponds to a 5 m bed elevation drop over the 50 km reach. At the river mouth (Ganpu, located 30 km downstream of Jianshan), the tidal range at Jianshan is 7 m and the tidal period is 12 h 25 min. At low tide, the river flows at uniform equilibrium (2.4 m water depth). Discuss the propagation of the tidal bore.

Assume $S_o = S_f$, a wide rectangular prismatic channel between Laoyancang and Ganpu[2], $S_o = S_f$ at the initial conditions (i.e low tide) and $f = 0.015$. Seawater density and dynamic viscosity are, respectively, 1024 kg/m^3 and 1.22 $\times 10^{-3}$ Pa s.

---

[2] This is an approximation. Between Ganpu and Babao (upstream of Jianshan), the channel width contracts from 20 down to 4 km while the channel bed rises gently. The resulting funnel shape amplifies the tide in the estuary.

Solution

Uniform equilibrium flow calculations are conducted in the Qiantang River at low tide (Chapter 4). It yields normal velocity = 1.12 m/s (positive in the downstream direction), C_o = 4.85 m/s for d_o = 2.4 m, f = 0.015 and S_o = 1×10^{-4}.

Let us select a coordinate system with x = 0 at the river mouth (i.e. at Ganpu) and x positive in the upstream direction. The initial conditions are V_o = −1.12 m/s and C_o = 4.85 m/s. The prescribed boundary condition at the river mouth (Ganpu) is the water depth:

$$d(x = 0,\ t_o) = 3.6 + \frac{7}{2}\left(1 + \cos\left(\frac{2\pi}{T}t - \pi\right)\right)$$

where t_o = $t(x = 0)$ and T is the tide period (T = 44 700 s).

The initial forward characteristic trajectory is a straight line:

$$t = \frac{1}{V_o + C_o}x \qquad \text{Initial forward characteristic}$$

Preliminary calculations are conducted assuming a simple wave for $0 < x < 70$ km (i.e. Laoyancang). The results showed formation of the tidal bore at x = 48 km and t = 12 950 s, corresponding to the intersection of the initial characteristic with the C1 characteristic issuing from $t_o \sim 5400$ s. (Simple-wave calculations would show that the bore does not reach its final form until $x > 300$ km.)

Discussion

The Qiantang bore, also called Hangchow or Hangzhou bore, is one of the world's most powerful tidal bores with the Amazon River bore (*pororoca*) (Fig. 17.7b). As the tide rises into the funnel-shaped Hangzhou Bay, the tidal bore develops and its effects may be felt more than 60 km upstream. Relevant references include Dai and Chaosheng (1987), Chyan and Zhou (1993) and Chen *et al.* (1990).

In the Hangzhou Bay, a large sand bar between Jianshan and Babao divides the tidal flow forming the East Bore and the South Bore propagating East-North-East and North-North-East, respectively. At the end of the sand bar, the intersection of the tidal bores can be very spectacular: e.g. water splashing were seen to reach heights in excess of 10 m!

17.4.3 Negative surge

A negative surge results from a reduction in water depth. It is an invasion of deeper waters by shallower waters. In still water, a negative surge propagates with a celerity $U = \sqrt{gd_o}$ for a rectangular channel, where d_o is the initial water depth. As the water depth is reduced, the celerity C decreases, the inverse slope of the forward characteristics increases and the family of forward characteristics forms a diverging line.

A simple case is the rapid opening of a gate at t = 0 in a channel initially at rest (Fig. 17.9a). The coordinate system is selected with x = 0 at the gate and x positive in the upstream direction. The initial forward characteristic propagates upstream with a celerity $U = +\sqrt{gd_o}$ where d_o is the initial water depth. The location of the leading edge of the negative surge is:

$$x_s = \sqrt{gd_o}\, t$$

Note that the gate opening induces a negative flow velocity.

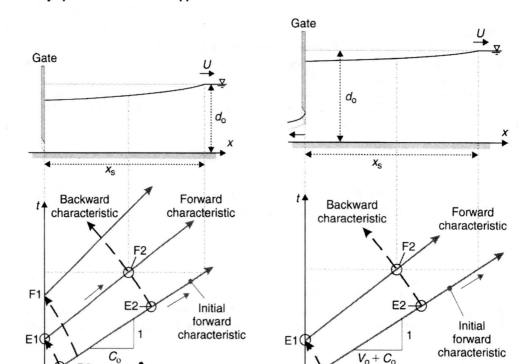

Fig. 17.9 Sketch of negative surges: (a) negative surge in a channel initially at rest and (b) sudden opening of a gate from a partially opened position.

A backward characteristic can be drawn issuing from the initial forward characteristic for $t > 0$ (e.g. point D2) and intersecting the boundary conditions at the gate (e.g. point E1). The backward characteristic satisfies:

$$V(x = 0, t_o) - 2C(x = 0, t_o) = -2C_o \qquad (17.17)$$

At the intersection of a forward characteristic issuing from the gate (e.g. at point E1) with a backwater characteristic issuing from the initial forward characteristic, the following conditions are satisfied:

$$V_{F2} + 2C_{F2} = V(x = 0, t_o) + 2C(x = 0, t_o) \qquad (17.18)$$

$$V_{F2} - 2C_{F2} = -2C_o \qquad (17.19)$$

where the point F2 is sketched in Fig. 17.9a. Equations (17.17)–(17.18), plus the prescribed boundary condition, form a system of four equations with four unknowns V_{F2}, C_{F2}, $V(x = 0, t_o)$ and $C(x = 0, t_o)$. In turn all the flow properties at the point F2 can be calculated.

In the particular case of a simple wave (Section 17.3), the forward characteristics are straight lines because the initial forward characteristic is a straight line (Section 17.3.1). The equation of forward characteristics issuing from the gate (e.g. from point E1) is:

$$\frac{dx}{dt} = V(x = 0, t_o) + C(x = 0, t_o) = V + C = 3C - 2C_o \qquad \text{Simple wave approximation}$$

The integration gives the water surface profile between the leading edge of the negative wave and the wave front:

$$\frac{x}{t - t_o} = 3\sqrt{gd} - 2\sqrt{gd_o} \qquad \text{For } 0 \leqslant x \leqslant x_s \tag{17.20}$$

assuming a rectangular channel. At a given time $t > t_o$, the free-surface profile (equation (17.20)) is a parabola.

> ### Remarks
> 1. The above application considers the partial opening of a gate from an initially closed position. The gate opening corresponds to an increase of flow rate beneath the gate. It is associated with both the propagation of a negative surge upstream of the gate and the propagation of the positive surge downstream of the gate (Chapter 4).
> 2. In a particular case, experimental observations showed that the celerity of the negative surge was greater than the ideal value $U = -\sqrt{gd_o}$ (see Dam break in a dry channel in Section 17.7.2). The difference was thought to be caused by streamline curvature effects at the leading edge of the negative surge.

Sudden complete opening
A limiting case of the above application is the sudden, complete opening of a gate from an initially closed position. The problem becomes a dam break wave and it is developed in Section 17.7.2.

Sudden partial opening
A more practical application is the sudden opening of a gate from a partially opened position (Fig. 17.9b). Using a coordinate system with $x = 0$ at the gate and x positive in the upstream direction, the initial velocity V_o must be negative. The initial forward characteristic propagates upstream with a celerity:

$$U = V_o + \sqrt{gd_o}$$

where d_o is the initial water depth and V_o is negative. Assuming a simple-wave analysis, the development is nearly identical to the sudden opening of a gate from an initially closed position (equation (17.17)–(17.20)). It yields:

$$V(x = 0, t_o) - 2C(x = 0, t_o) = V_o - 2C_o \tag{17.21}$$

$$V_{F2} + 2C_{F2} = V(x = 0, t_o) + 2C(x = 0, t_o) \tag{17.22}$$

$$V_{F2} - 2C_{F2} = V_o - 2C_o \tag{17.23}$$

where the point F2 is sketched in Fig. 17.9b. Equations (17.21)–(17.23), plus the prescribed boundary condition, form a system of four equations with four unknowns. In turn all the flow properties at the point F2 can be calculated.

Application: negative surge in a forebay

A particular application is the propagation of a negative surge in a forebay canal associated with a sudden increase in discharge into the penstock (Fig. 17.10). For example, the starting discharge of a hydropower plant supplied by a canal of small slope. A forebay is a canal or reservoir from which water is taken to operate a waterwheel or hydropower turbine. A penstock is a gate for regulating the flow. The term is also used for a conduit leading water to a turbine.

The problem may be solved for a simple wave in a horizontal forebay channel with initial flow conditions: $d = d_o$ and $V_o = 0$. Considering a point E1 located at $x = 0$ (i.e. penstock), the backward characteristic issuing from this point intersects the initial forward characteristic and it satisfies:

$$V_{E1} - 2C_{E1} = V_o - 2C_o \qquad (17.11a)$$

It yields:

$$V(x = 0) = 2(C(x = 0) - C_o)$$

where V is negative with the sign convention used in Fig. 17.10. The discharge obtained from a rectangular forebay canal is:

$$Q(x = 0) = 2Bd(x = 0)\left(\sqrt{gd(x = 0)} - \sqrt{gd_o}\right)$$

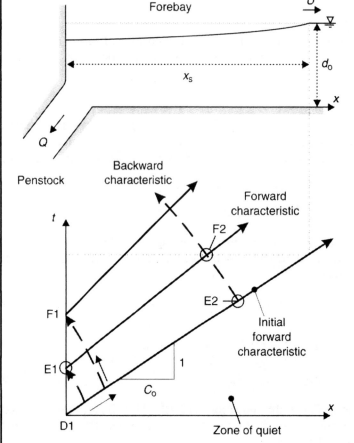

Fig. 17.10 Sketch of a negative surge in a forebay.

where B is the channel width. The discharge is maximum for $d = 4/9d_o$ and the maximum flow rate equals:

$$Q_{max} = \frac{8}{27} d_o \sqrt{gd_o} B \qquad \text{Maximum discharge}$$

Practically, if the water demand exceeds this value, the forebay canal cannot supply the penstock and a deeper canal must be designed.

Remarks

The maximum flow rate for the simple-wave approximation corresponds to $Q^2/(gB^2d^3) = 1$. That is, critical flow conditions. The flow rate equals the discharge at the dam break site for a dam break wave in an initially dry, horizontal channel bed (Section 17.7.2).

The operation of penstocks is associated with both sudden water demand and stoppage. The latter creates a positive surge propagating upstream in the forebay channel.

17.5 THE KINEMATIC WAVE PROBLEM

17.5.1 Presentation

In a kinematic wave model, the differential form of the Saint-Venant equations is written as:

$$\frac{\partial A}{\partial t} + \frac{\partial Q}{\partial x} = 0 \qquad \text{Continuity equation} \qquad (17.24)$$

$$S_f - S_o = 0 \qquad \text{Kinematic wave equation} \qquad (17.25a)$$

That is, the dynamic wave equation is simplified by neglecting the acceleration and inertial terms, and the free surface is assumed parallel to the channel bottom (Section 16.4.1). The kinematic wave equation may be rewritten as:

$$Q = \sqrt{\frac{8g}{f}} A \sqrt{\frac{D_H}{4}} \sqrt{S_o} \qquad (17.25b)$$

where Q is the total discharge in the cross-section, f is the Darcy friction factor, A is the flow cross-sectional area and D_H is the hydraulic diameter. Equation (17.25a) expresses an unique relationship between the flow rate Q and the water depth d, hence the cross-sectional area at a given location x. The differentiation of the flow rate with respect to time may be transformed as:

$$\frac{\partial Q}{\partial t} = \left(\frac{\partial Q}{\partial A}\right)_{x=constant} \times \frac{\partial A}{\partial t}$$

The continuity equation becomes:

$$\frac{\partial Q}{\partial t} + \left(\frac{\partial Q}{\partial A}\right)_{x=constant} \times \frac{\partial Q}{\partial x} = 0 \qquad \text{Continuity equation} \qquad (17.26)$$

It may be rewritten as:

$$\frac{DQ}{Dt} = \frac{\partial Q}{\partial t} + \frac{dx}{dt}\frac{\partial Q}{\partial x} = 0 \qquad (17.27)$$

along

$$\frac{dx}{dt} = \left(\frac{\partial Q}{\partial A}\right)_{x=\text{constant}}$$ (17.28)

That is, the discharge is constant along the characteristic trajectory defined by equation (17.28).

Note

The term

$$\frac{dx}{dt} = \left(\frac{\partial Q}{\partial A}\right)_{x=\text{constant}}$$

is sometimes called the speed of the kinematic wave.

17.5.2 Discussion

The trajectories defined by equation (17.28) are the characteristics of equation (17.26). There is only one family of characteristics, which all propagate in the same direction: i.e. in the flow direction (Fig. 17.11). The solution of the continuity equation in terms of the flow rate $Q(x, t)$, in a river reach ($x_1 \leq x \leq x_2$) and for $t \geq 0$, requires one prescribed initial condition at $t = 0$ and $x_1 \leq x \leq x_2$, and one prescribed upstream condition ($x = x_1$) for $t > 0$. No prescribed downstream condition is necessary.

The kinematic wave model can only describe the downstream propagation of a disturbance. In comparison, the dynamic waves can propagate both upstream and downstream. Practically, the kinematic wave equation is used to describe the translation of flood waves in some simple cases. One example is the monoclinal wave (Section 17.2.2). *But* the kinematic wave routing cannot predict the subsidence of flood wave.

Notes
1. The *subsidence* of the flood wave is the trend to flatten out the flood peak discharge.
2. A kinematic wave model may predict the translation of a flood wave but not its subsidence (i.e. flattening out). The latter may be predicted with the diffusion wave model (Section 17.6).

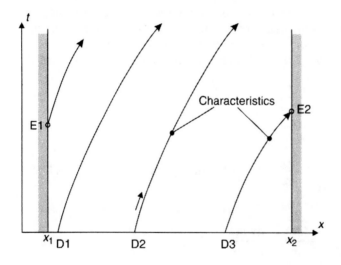

Fig. 17.11 Characteristics of the kinematic wave problem.

17.6 THE DIFFUSION WAVE PROBLEM

17.6.1 Presentation

The diffusion wave equation is a simplification of the dynamic equation assuming that the acceleration and inertial terms are negligible. The differential form of the Saint-Venant equations becomes:

$$B\frac{\partial d}{\partial t} + \frac{\partial Q}{\partial x} = 0 \qquad \text{Continuity equation} \qquad (17.29a)$$

$$\frac{\partial d}{\partial x} + S_f - S_o = 0 \qquad \text{Diffusion wave equation} \qquad (17.29b)$$

The definition of the friction slope gives:

$$S_f = \frac{Q|Q|}{(8g/f)A^2(D_H/4)} \qquad (17.30)$$

where $|Q|$ is the magnitude of the flow rate.

Assuming a constant free-surface width B, the continuity equation (equation (17.29a)) is differentiated with respect to x, and the diffusion wave equation (equation (17.29b)) is differentiated with respect to the time t. It yields:

$$B\frac{\partial^2 d}{\partial x\,\partial t} + \frac{\partial^2 Q}{\partial x^2} = 0 \qquad (17.31a)$$

$$\frac{\partial^2 d}{\partial t\,\partial x} + \frac{\partial}{\partial t}(S_f - S_o) = 0 \qquad (17.31b)$$

The latter equation becomes:

$$\frac{\partial^2 d}{\partial t\,\partial x} + \frac{2|Q|}{(8g/f)A^2(D_H/4)}\frac{\partial Q}{\partial t} - \frac{2Q|Q|}{((8g/f)A^2(D_H/4))^{3/2}}\frac{\partial}{\partial t}\left(A\sqrt{\frac{8g}{f}\frac{D_H}{4}}\right) = 0 \quad (17.31b)$$

Noting that

$$\frac{\partial}{\partial t}\left(A\sqrt{\frac{8g}{f}\frac{D_H}{4}}\right) = \frac{\partial}{\partial d}\left(A\sqrt{\frac{8g}{f}\frac{D_H}{4}}\right)\frac{\partial d}{\partial t} = \frac{\partial}{\partial d}\left(A\sqrt{\frac{8g}{f}\frac{D_H}{4}}\right)\left(-\frac{1}{B}\frac{\partial Q}{\partial x}\right)$$

and by eliminating $(\partial^2 d/\partial x\,\partial t)$ between equations (17.31a) and (17.31b), the diffusive wave equation becomes:

$$-\frac{1}{B}\frac{\partial^2 Q}{\partial x^2} + \frac{2|Q|}{(8g/f)A^2(D_H/4)}\frac{\partial Q}{\partial t} + \frac{2Q|Q|}{B((8g/f)A^2(D_H/4))^{3/2}}\frac{\partial}{\partial d}\left(A\sqrt{\frac{8g}{f}\frac{D_H}{4}}\right)\frac{\partial Q}{\partial x} = 0 \qquad (17.32a)$$

This equation may be rewritten as:

$$\frac{\partial Q}{\partial t} + \frac{Q}{B((8g/f)A^2(D_H/4))^{1/2}}\frac{\partial}{\partial d}\left(A\sqrt{\frac{8g}{f}\frac{D_H}{4}}\right)\frac{\partial Q}{\partial x} = \frac{(8g/f)A^2(D_H/4)}{2B|Q|}\frac{\partial^2 Q}{\partial x^2} \qquad (17.32b)$$

Equation (17.32b) is an advective diffusion equation in terms of the discharge:

$$\frac{\partial Q}{\partial t} + U \frac{\partial Q}{\partial x} = D_t \frac{\partial^2 Q}{\partial x^2}$$

(17.32b)

where U is the speed of the diffusion wave and D_t is the diffusion coefficient:

$$U = \frac{Q}{B((8g/f)A^2(D_H/4))^{1/2}} \frac{\partial}{\partial d}\left(A\sqrt{\frac{8g}{f}\frac{D_H}{4}}\right) \qquad \text{Celerity of diffusion wave}$$

$$D_t = \frac{(8g/f)A^2(D_H/4)}{2B|Q|} \qquad \text{Diffusion coefficient}$$

Considering a limited river reach ($x_1 \leqslant x \leqslant x_2$), the solution of the diffusion wave equation in terms of the flow rate $Q(x, t)$ requires the prescribed initial condition $Q(x, 0)$ for $t = 0$ and $x_1 \leqslant x \leqslant x_2$, and one prescribed condition at each boundary (i.e. $Q(x_1, t)$ and $Q(x_2, t)$) for $t > 0$.

Once the diffusion wave equation (equation (17.32b) is solved in terms of the flow rates, the water depths are calculated from the continuity equation:

$$B\frac{\partial d}{\partial t} + \frac{\partial Q}{\partial x} = 0 \qquad \text{Continuity equation}$$

(17.29a)

Notes

1. The diffusion wave problem is also called *diffusion routing*.
2. The term 'routing' or 'flow routing' refers to the tracking in space and time of a flood wave.
3. The above development was performed assuming a constant free-surface width B.
4. The diffusion coefficient may be rewritten as:

$$D_t = \frac{Q}{2BS_f}$$

5. For a rectangular channel and assuming a constant Darcy friction factor, the following relationship holds:

$$\frac{\partial}{\partial d}\left(A\sqrt{\frac{8g}{f}\frac{D_H}{4}}\right) = \frac{3}{2}B\sqrt{\frac{8g}{f}\frac{A}{P_w}}\left(1 - \frac{2}{3}\frac{A}{BP_w}\right) = \frac{3}{2}\frac{Q}{\sqrt{S_f}d}\left(1 - \frac{2}{3}\frac{A}{BP_w}\right)$$

Hence the speed of the diffusion wave equals:

$$U = \frac{3}{2}V\left(1 - \frac{2}{3}\frac{A}{BP_w}\right)$$

6. For a wide rectangular channel, the friction slope equals:

$$S_f = \frac{Q|Q|}{(8g/f)B^2d^3}$$

Assuming a constant Darcy friction factor f, the following relationship holds:

$$\frac{\partial}{\partial d}\left(A\sqrt{\frac{8g}{f}\frac{D_H}{4}}\right) = \frac{3}{2}B\sqrt{\frac{8g}{f}}d = \frac{3}{2}\frac{Q}{\sqrt{S_f}d}$$

7. For a wide rectangular channel and assuming a constant Darcy friction factor, the speed of the diffusion wave equals:

$$U = \frac{3}{2}V$$

The celerity of the wave is equal to the celerity of the monoclinal wave. But the diffusion wave flattens out with longitudinal distance while the monoclinal wave has a constant shape (Section 17.2.2).

17.6.2 Discussion

For constant diffusion wave celerity U and diffusion coefficient D_t, equation (17.32b) may be solved analytically for a number of basic boundary conditions. Further, since equation (17.32b) is linear, the *theory of superposition* may be used to build-up solutions with more complex problems and boundary conditions. Mathematical solutions of the diffusion and heat equations were addressed in two classical references (Carslaw and Jaeger, 1959; Crank, 1956). Simple analytical solutions of the diffusion wave equation are summarized in Table 17.1 assuming constant diffusion wave speed and diffusion coefficient. The flow situations are sketched in Fig. 17.12.

Note some basic limitations of such analytical solutions. First, the diffusion coefficient D_t is a function of the flow rate. It may be assumed constant only if the change in discharge is small. Secondly, the diffusion wave celerity is a function of the flow velocity and hence of the flow conditions.

Table 17.1 Analytical solutions of the diffusion wave equation assuming constant diffusion wave celerity and diffusion coefficient

Problem (1)	Analytical solution (2)	Initial/boundary conditions (3)
Advective diffusion of a sharp front	$Q(x, t) = Q_o + \dfrac{\delta Q}{2}\left(1 - \text{erf}\left(\dfrac{x - Ut}{\sqrt{4D_t t}}\right)\right)$	$Q(x, 0) = Q_o$ for $x > 0$ $Q(x, 0) = Q_o + \delta Q$ for $x < 0$
Initial volume slug introduced at $t = 0$ and $x = 0$ (i.e. sudden volume injection)	$Q(x, t) = Q_o + \dfrac{\delta Q \text{Vol}}{A\sqrt{4\pi D_t t}}\exp\left(-\dfrac{(x - Ut)^2}{4D_t t}\right)$	$Q(x, 0) = Q_o$ for $t = 0$ Sudden injection of a water volume (Vol) at a rate δQ at origin at $t = 0$
Sudden discharge injection in a river at a steady rate	$Q(x, t) = Q_o + \dfrac{\delta Q}{2}$ $\times\left(1 - \text{erf}\left(\dfrac{x - Ut}{\sqrt{4D_t t}}\right) + \exp\left(\dfrac{Ux}{D_t}\right)\left(1 - \text{erf}\left(\dfrac{x + Ut}{\sqrt{4D_t t}}\right)\right)\right)$	$Q(0, t) = Q_o + \delta Q$ for $0 < t < +\infty$ $Q(x, 0) = Q_o$ for $0 < x < +\infty$

Note: erf = Gaussian error function (Appendix A3.4).

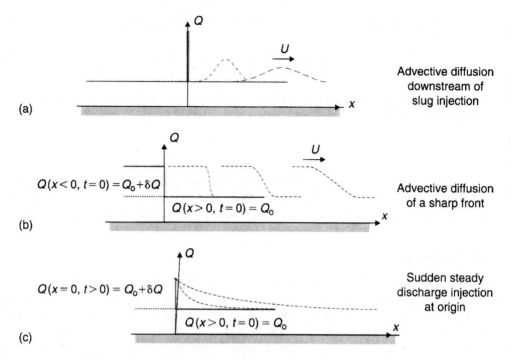

Fig. 17.12 Basic examples of advective diffusion problems.

17.6.3 The Cunge–Muskingum method

A simplification of the diffusion wave model is the Cunge–Muskingum method which was developed as an extension of the empirical Muskingum method.

Empiricism: the Muskingum method
The Muskingum method is based upon the continuity equation for a river reach ($x_1 \leqslant x \leqslant x_2$):

$$\frac{dVol}{dt} = -(Q_2 - Q_1) \qquad \text{Continuity equation}$$

where Vol is the reach volume, $Q_1 = Q(x = x_1)$ is the inflow and $Q_2 = Q(x = x_2)$ is the outflow. The method *assumes* a relationship between the reach volume, and the inflow and outflow rates:

$$Vol = K_M(X_M Q_1 + (1 - X_M)Q_2)$$

where K_M and X_M are empirical routing parameters of the river reach. By differentiating the above equation with respect to time and replacing into the continuity equation, it yields:

$$Q_2 + K_M(1 - X_M)\frac{dQ_2}{dt} = Q_1 - K_M X_M \frac{dQ_1}{dt}$$

assuming constant routing parameters K_M and X_M. If the inflow rate is a known function of time, the equation may be solved analytically or numerically in terms of the outflow Q_2.

Discussion

The Muskingum method may be applied to specific situations to solve the downstream flow conditions as functions of the upstream flow conditions. Note that the downstream flow conditions have no influence on the upstream conditions, and the parameters K_M and X_M must be calibrated and validated for each reach. The routing parameters are calculated from the inflow and outflow hydrographs of past floods in the reach.

Notes

1. The Muskingum method was developed in the 1930s for flood control schemes in the Muskingum River catchment (USA). It is an empirical, intuitive method: 'despite the popularity of the basic method, it cannot be claimed that it is logically complete' Henderson (1966: p. 364); 'although the conventional flood-routing methods [incl. Muskingum] may give good results under some conditions, they have the tendency to be treacherous and unreliable' (Thomas in Miller and Cunge, 1975: p. 246); 'there is often considerable doubt as to how accurate the [flood routing] results are for any application' (Miller and Cunge, 1975: p. 247).

2. The parameters K_M and X_M are completely empirical: i.e. they have no theoretical justification. The routing parameter K_M is homogeneous to a time while the parameter X_M is dimensionless between 0 and 1.

3. X_M must be between 0 and 0.5 for the method to be stable. For $X_M > 0.5$, the flood peak increases with distance.

4. For $X_M = 0$, the Muskingum method becomes a simple reservoir storage problem. For $X_M = 0.5$, the flood motion is a pure translation without attenuation and the Muskingum method approaches the kinematic wave solution.

Cunge–Muskingum method

The above method approaches a diffusion wave problem when:

$$U = \frac{x_2 - x_1}{K_M}$$

$$D_t = \left(\frac{1}{2} - X_M\right) U(x_2 - x_1)$$

where $(x_2 - x_1)$ is the river reach length and K_M and X_M are the Muskingum routing parameters of the river reach. In turn, the routing parameters become functions of the flow rate:

$$X_M = \frac{1}{2}\left(1 - \frac{\left(A\sqrt{(8g/f)(D_H/4)}\right)^3}{(x_2 - x_1)Q|Q|\,\frac{\partial}{\partial d}\left(A\sqrt{\frac{8g}{f}\frac{D_H}{4}}\right)}\right)$$

$$K_M = \frac{x_2 - x_1}{\dfrac{Q}{B((8g/f)A^2(D_H/4))^{1/2}}\,\dfrac{\partial}{\partial d}\left(A\sqrt{\dfrac{8g}{f}\dfrac{D_H}{4}}\right)}$$

This technique is called the Cunge–Muskingum method.

> **Discussion**
> The Cunge–Muskingum method may be applied to specific situations to solve the downstream flow conditions as functions of the upstream flow conditions. Limitations include:
>
> - the inertial terms must be negligible,
> - the downstream flow conditions have no influence on the upstream conditions,
> - the parameters K_M and X_M must be calibrated and validated for the reach.
>
> In the Cunge–Muskingum method, the routing parameters K_M and X_M may be calculated from the physical characteristics of the channel reach, rather than on previous flood records only.
>
> **Notes**
> 1. The extension of the Muskingum method was first presented by Cunge (1969).
> 2. Jean A. Cunge worked in France at Sogreah in Grenoble and he lectured at the Hydraulics and Mechanical Engineering School of Grenoble (ENSHMG).

17.7 DAM BREAK WAVE

17.7.1 Presentation

Failures of several dams during the 19th and 20th centuries attracted a lot of interest on dam break wave flows (Fig. 17.13). Major catastrophes included the failures of the Puentes dam in 1802, Dale Dyke dam in 1864, Habra dam in 1881 and South Fork (Johnstown) dam in 1889. During the 20th century, three major accidents were the St Francis and Malpasset dam failures in 1928 and 1959, respectively, and the overtopping of the Vajont dam in 1963.

Plate 26 and Fig. 17.13 show the remains of the Malpasset dam. Located on the Reyran River upstream of Fréjus, southern France, the dam was completed in 1954, the reservoir was not full until late November 1959. On 2 December 1959 around 9:10 pm, the dam wall collapsed completely. More than 300 people died in the catastrophe. Field observations showed that the surging waters formed a 40 m high wave at 340 m downstream of the dam site and the wave height was still about 7 m about 9 km downstream (Faure and Nahas, 1965). The dam break wave took about 19 min to cover the first 9 km downstream of the dam site. Relatively accurate time records were obtained from the destruction of a series of electrical stations located in the downstream valley. The dam collapse was caused by uplift pressures in faults of the gneiss rock foundation which lead to the complete collapse of the left abutment (Fig. 17.13a).

17.7.2 Dam break wave in a horizontal channel

Dam break in a dry channel

Considering an ideal dam break surging over a dry river bed, the method of characteristics may be applied to completely solve the wave profile as first proposed by Ritter in 1892 (e.g. Henderson, 1966; Montes, 1998). Interestingly, Ritter's work was initiated by the South Fork dam's (Johnstown) catastrophe.

The dam break may be idealized by a vertical wall that is suddenly removed (Fig. 17.14). After removal of the wall, a negative wave propagates upstream and a dam break wave moves downstream. Although there is considerable vertical acceleration during the initial instants of fluid motion, such acceleration is not taken into account by the method of characteristics and the pressure distributions are assumed hydrostatic. For an ideal dam break over a dry horizontal

Fig. 17.13 Photographs of the Malpasset dam break accident: a mistake not to repeat! Reservoir characteristics: dam height: 102.5 m and volume: 50 Mm³. Dam break on 2 December 1959: (a) Remains of the Malpasset dam in 1981. View from upstream looking toward the left abutment. Note the outlet system at the bottom of the valley and the location of rock foundation failure between the bottom outlets and the top left abutment wall. (b) Malpasset dam, view from downstream in 1981. Note the dam right abutment in the background on the top right of the photograph. The concrete blocks in the foreground are more than 4 m high and they were transported by the flood wave.

channel, the basic equations are those of the simple wave (Section 17.3.1):

$$\frac{D}{Dt}\left(V + 2C\right) = 0 \qquad \text{Forward characteristics}$$

$$\frac{D}{Dt}\left(V - 2C\right) = 0 \qquad \text{Backward characteristics}$$

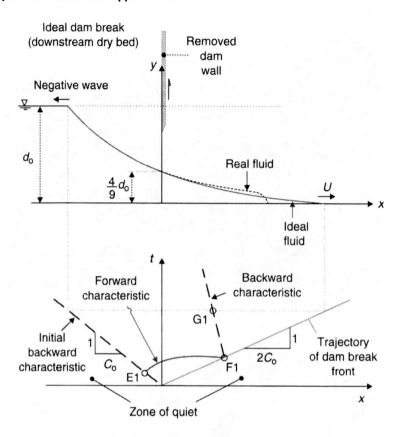

Fig. 17.14 Sketch of dam break wave in a dry horizontal channel.

along:

$$\frac{dx}{dt} = V + C \qquad \text{Forward characteristics C1}$$

$$\frac{dx}{dt} = V - C \qquad \text{Backward characteristics C2}$$

The instantaneous dam break creates a negative wave propagating upstream into a fluid at rest with known water depth d_o. In the (x, t) plane, the initial negative wave characteristics has a slope $dt/dx = -1/C_o$ where $C_o = \sqrt{gd_o}$ assuming a rectangular channel (Fig. 17.14).

A forward characteristics can be drawn issuing from the initial backward characteristics for $t > 0$ and intersecting the trajectory of the leading edge of the dam break wave front (Fig. 17.14, trajectory E1–F1). The forward characteristics satisfy:

$$V + 2C = V_o + 2C_o = 2C_o \qquad (17.33a)$$

At the leading edge of the dam break wave front, the water depth is zero, hence $C = 0$, and the propagation speed of the dam break wave front equals:

$$U = 2C_o = 2\sqrt{gd_o} \qquad \text{Ideal dam break} \qquad (17.34)$$

Considering any backward characteristics issuing from the dam break wave front (Fig. 17.14, trajectory F1–G1), the C2 characteristics is a straight line because the initial backward characteristics is a straight line (Section 17.3.1). The inverse slope of the backward characteristics is a constant:

$$\frac{dx}{dt} = V - C = 2C_o - 3C$$

using equation (17.33a). The integration of the inverse slope gives the water surface profile at the intersection of the C2 characteristics with a horizontal line $t =$ constant (Fig. 17.14, point G1). That is, at a given time, the free-surface profile between the leading edge of the negative wave and the wave front is a parabola:

$$\frac{x}{t} = 2\sqrt{gd_o} - 3\sqrt{gd} \qquad \text{for } -\sqrt{gd_o} \le \frac{x}{t} \le +2\sqrt{gd_o} \tag{17.35a}$$

At the origin ($x = 0$), equation (17.35a) predicts a constant water depth:

$$d(x = 0) = \frac{4}{9}d_o \tag{17.36}$$

Similarly the velocity at the origin is deduced from equation (17.33a):

$$V(x = 0) = \frac{2}{3}\sqrt{gd_o} \tag{17.37}$$

After dam break, the flow depth and velocity at the origin are both constants, and the water discharge at $x = 0$ equals:

$$Q(x = 0) = \frac{8}{27}d_o\sqrt{gd_o}\,B \tag{17.38}$$

where B is the channel width.

Discussion

For a dam break wave down a dry channel, the boundary conditions are *not* the vertical axis in the (x, t) plane. The two boundary conditions are:

1. the upstream edge of the initial negative wave where $d = d_o$,
2. the dam break wave front where $d = 0$.

At the upstream end of the negative surge, the boundary condition is $d = d_o$: i.e. this is the initial backward characteristics in the (x, t) plane (Fig. 17.14). The downstream boundary condition is at the leading edge of the dam break wave front where the water depth is zero.

In Sections 17.3 and 17.4, most applications used a boundary condition at the origin: i.e. it was the vertical axis in the (x, t) plane. For a dam break wave on a horizontal smooth bed, the vertical axis is not a boundary.

Notes

1. Important contributions to the dam break wave problem in a dry horizontal channel include Ritter (1892), Schoklitsch (1917), Ré (1946), Dressler (1952, 1954) and Whitham (1955).

2. The above development is sometimes called Ritter's theory. Calculations were performed assuming a smooth rectangular channel, an infinitely long reservoir and for a quasi-horizontal free surface. That is, bottom friction is zero and the pressure distribution is hydrostatic. Experimental results (e.g. Schoklitsch, 1917; Faure and Nahas, 1961; Lauber, 1997) showed that the assumptions of hydrostatic pressure distributions and zero friction are reasonable, but for the initial instants and at the leading tip of the dam break wave.

3. Bottom friction significantly affects the propagation of the leading tip. Escande *et al.* (1961) investigated specifically the effects of bottom roughness on dam break wave in a natural valley. They showed that, with a very-rough bottom, the wave celerity could be about 20–30% lower than for a smooth bed. The shape of a real fluid dam break wave front is sketched in Fig. 17.14 and experimental results are presented in Fig. 17.18 (see Effects of flow resistance in Section 17.7.3).

4. The assumption of hydrostatic pressure distributions has been found to be reasonable, but for the initial instants: i.e. for $t > 3\sqrt{d_o/g}$ (Lauber, 1997).

5. Ritter's theory implies that the celerity of the initial negative wave characteristics is $U = -\sqrt{gd_o}$. Experimental observations suggested, however, that the real celerity is about $U = -\sqrt{2}\sqrt{gd_o}$ (Lauber, 1997; Leal *et al.*, 2001). It was proposed that the difference was caused by streamline curvature effects at the leading edge of the negative wave.

6. At the origin ($x = 0$), the flow conditions satisfy:

$$\frac{V(x=0)}{\sqrt{gd(x=0)}} = \frac{(2/3)\sqrt{gd_o}}{\sqrt{g(4/9)d_o}} = 1$$

That is, critical flow conditions take place at the origin (i.e. initial dam site) and the flow rate is a constant:

$$Q(x=0) = \sqrt{gd(x=0)^3} = \frac{8}{27}d_o\sqrt{gd_o}\,B$$

where B is the channel width. Importantly, the result is valid only within the assumptions of the Saint-Venant equations. That is, the water free surface is quasi-horizontal and the pressure distribution is hydrostatic at the origin.

7. The above development was conducted for a semi-infinite reservoir. At a given location $x > 0$, equation (17.35a) predicts an increasing water depth with increasing time:

$$d = \frac{4}{9}d_o\left(1 - \frac{3}{2}\frac{x}{\sqrt{gd_o}\,t}\right)^2 \tag{17.35b}$$

At a distance x from the dam site, the water depth d tends to $d = 4/9d_o$ for $t = +\infty$.

8. Henderson (1966) analysed the dam break wave problem by considering the sudden horizontal displacement of a vertical plate behind which a known water depth is initially at rest. His challenging approach yields identical results.

Dam break in a horizontal channel initially filled with water

The propagation of a dam break wave over still water with an initial depth $d_1 > 0$ is possibly a more practical application: e.g. sudden flood release downstream of a dam in a river. It is a different situation from an initially dry channel bed because the dam break wave is lead by a positive surge (Fig. 17.15).

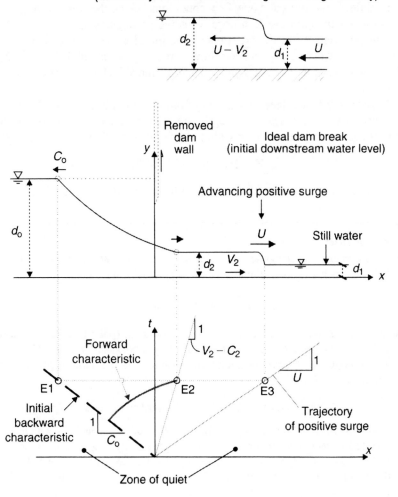

Fig. 17.15 Sketch of dam break wave in a horizontal channel initially filled with water for $d_1 < 0.1383d_0$.

The basic flow equations are the characteristic system of equations (Dam break in a dry channel in Section 17.7.2), and the continuity and momentum equations across the positive surge front. That is:

$$\frac{D}{Dt}(V + 2C) = 0 \quad \text{along } \frac{dx}{dt} = V + C \text{ (forward characteristics)}$$

$$\frac{D}{Dt}(V - 2C) = 0 \quad \text{along } \frac{dx}{dt} = V - C \text{ (backward characteristics)}$$

$$d_1 U = d_2(U - V_2) \quad \text{Continuity equation} \quad (17.39a)$$

$$d_2(U - V_2)^2 - d_1 U^2 = \frac{1}{2}gd_1^2 - \frac{1}{2}gd_2^2 \quad \text{Momentum equation} \quad (17.40a)$$

where U is the positive surge celerity for an observer fixed on the channel bank, and the subscripts 1 and 2 refer, respectively, to the flow conditions upstream and downstream of the positive surge front: i.e. initial flow conditions and new flow conditions behind the surge front (Fig. 17.15).

Immediately after the dam break, a negative surge propagates upstream into a fluid at rest with known water depth d_o. In the (x, t) plane, the initial negative wave characteristics has a slope $dt/dx = -1/C_o$ where $C_o = \sqrt{gd_o}$ assuming a rectangular channel (Fig. 17.15). The initial backward characteristics is a straight line, hence all the C2 characteristics are straight lines (Section 17.3.1).

For $t > 0$, the forward characteristics issuing from the initial backward characteristics cannot intersect the downstream water level ($d = d_1$) because it would involve a discontinuity in velocity. The velocity is zero in still water ($V_1 = 0$) but, on the forward characteristics, it satisfies:

$$V + 2C = V_o + 2C_o = 2C_o \tag{17.33a}$$

Such a discontinuity can only take place as a positive surge which is sketched in Fig. 17.15.

Detailed solution

Considering a horizontal, rectangular channel (Fig. 17.15), the water surface profile between the leading edge of the negative wave (point E1) and the point E2 is a parabola:

$$\frac{x}{t} = 2\sqrt{gd_o} - 3\sqrt{gd} \qquad x_{E1} \leq x \leq x_{E2} \tag{17.35a}$$

Between the point E2 and the leading edge of the positive surge (point E3), the free-surface profile is horizontal. The flow depth d_2 and velocity V_2 satisfy equations (17.39a) and (17.40a), as well as the condition along the C_1 forward characteristics issuing from the initial negative characteristics and reaching point E2:

$$V_2 + 2\sqrt{gd_2} = 2\sqrt{gd_o} \qquad \text{Forward characteristics} \tag{17.33a}$$

Equations (17.33a), (17.39a) and (17.40a) form a system of three equations with three unknowns V_2, d_2 and U. The system of equations may be solved graphically (Fig. 17.16). Figure 17.16 shows $U/\sqrt{gd_1}$, d_2/d_1 and $V_2/\sqrt{gd_2}$ (right axis), and ($V_2 - C_2/\sqrt{gd_1}$ (left axis) as functions of the ratio d_o/d_1.

Once the positive surge forms (Fig. 17.15), the locations of the points E2 and E3 satisfy, respectively:

$$x_{E2} = (V_2 - C_2)t$$

$$x_{E3} = Ut$$

where $C_2 = \sqrt{gd_2}$ for a rectangular channel. Figure 17.16 shows that $U > (V_2 - C_2)$ for $t > 0$ and $0 < d_1/d_o < 1$. That is, the surge front (point E3) advances faster than the point E2.

Montes (1998) showed that the surge celerity satisfies:

$$\sqrt{\frac{d_o}{d_1}} = \frac{1}{2}\frac{U}{\sqrt{gd_1}}\left(1 - \frac{1}{X}\right) + \sqrt{X} \tag{17.41}$$

where

$$X = \frac{1}{2}\left(\sqrt{1 + 8\frac{U^2}{gd_1}} - 1\right)$$

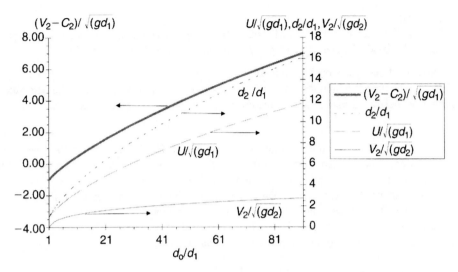

Fig. 17.16 Flow conditions for a dam break wave in a horizontal channel initially filled with water.

The flow conditions behind the surge front are then deduced from the quasi-steady flow analogy (Section 4.3.2).

Notes

1. The above equations were developed for initially still water ($V_1 = 0$). Experimental observations compared well with the above development and they showed that bottom friction is negligible.
2. Figure 17.16 is valid for $0 < d_1/d_o < 1$.
3. The flow depth downstream of (behind) the positive surge is deduced from the continuity and momentum equations. d_2 is independent of the time t and distance x. Practically, the flow depth d_2 may be correlated by:

$$\frac{d_2}{d_o} = 0.9319671 \left(\frac{d_1}{d_o}\right)^{0.371396}$$

Good agreement was observed between large-size experimental data and the above correlation (e.g. Chanson *et al.*, 2000).

4. Equation (17.41) may be empirically correlated by:

$$\frac{U}{\sqrt{gd_1}} = \frac{0.63545 + 0.3286(d_1/d_o)^{0.65167}}{0.00251 + (d_1/d_o)^{0.65167}}$$

5. Figure 17.15 illustrates the solution of the Saint-Venant equations, it shows some discontinuities in terms of free-surface slope and curvatures at the points E1 and E2. Such discontinuities are not observed experimentally and this highlights some limitation of the Saint-Venant equations.
6. The above development may be easily extended for initial flow conditions with a non-zero, constant initial velocity V_1. Equations (17.33a), (17.39a) and (17.40a) become:

$$V_2 + 2\sqrt{gd_2} = 2\sqrt{gd_o} \qquad \text{Forward characteristic} \qquad (17.33b)$$

$$d_1(U - V_1) = d_2(U - V_2) \qquad \text{Continuity equation} \qquad (17.39b)$$

$$d_2(U - V_2)^2 - d_1(U - V_1)^2 = \frac{1}{2}gd_1^2 - \frac{1}{2}gd_2^2 \qquad \text{Momentum equation} \qquad (17.40b)$$

Application

The 40 m high Zayzoun dam failed on Tuesday 4 June 2002. The dam impounded a 35 m depth of water and failed suddenly. The depth of water in the downstream channel was 0.5 m.

1. Estimate the free-surface profile 7 min after the failure.
2. Calculate the time at which the wave will reach a point 10 km downstream of the dam and the surge front height. *Assume an infinitely long reservoir and a horizontal, smooth channel.*

Solution

The downstream channel was initially filled with water at rest. The flow situation is sketched in Fig. 17.15. The x coordinate is zero ($x = 0$) at the dam site and positive in the direction of the downstream channel. The time origin is taken at the dam collapse.

First, the characteristic system of equation, and the continuity and momentum principles at the wave front must be solved graphically using Fig. 17.16 or theoretically.

Using equation (17.41), the surge front celerity is: $U/\sqrt{gd_1} = 10$ and $U = 22.3$ m/s.

At the wave front, the continuity and momentum equations yield:

$$\frac{d_2}{d_1} = \frac{1}{2}\left(\sqrt{1 + 8\frac{U^2}{gd_1}} - 1\right) = 13.7 \qquad \text{Hence } d_2 = 6.9\,\text{m}$$

Equation (17.33a) may be rewritten as:

$$\frac{V_2}{\sqrt{gd_2}} = 2\left(\sqrt{\frac{d_o}{d_2}} - 1\right) = 2.52 \qquad \text{Hence } V_2 = 20.6 \text{ m/s}$$

Note that these results are independent of time.

At $t = 7$ min, the location of the points E1, E2 and E3 sketched in Fig. 17.15 is:

$$x_{E1} = -\sqrt{gd_o}\,t = -7.8\,\text{km}$$

$$x_{E2} = (V_2 - \sqrt{gd_2})t = +5.2 \text{ km}$$

$$x_{E3} = Ut = +9.4\,\text{km}$$

Between the points E1 and E2, the free-surface profile is a parabola:

$$\frac{x}{t} = 2\sqrt{gd_o} - 3\sqrt{gd} \qquad x_{E1} \leqslant x \leqslant x_{E2} \qquad\qquad (17.35\text{a})$$

The free-surface profile at $t = 7$ min is:

x (m)	−100 000	−7783	−5318	−2084	+1894	5124	5225	9353	9353	12 000
d (m)	35	35	28	20	12	7	6.86	6.86	0.5	0.5
Remark		Point E1					Point E2	Point E3		

At a point located 10 km downstream of the dam site, the wave front arrives at $t = 449$ s (7 min 29 s). The height of the surge front is $\Delta d = d_2 - d_1 = 6.4$ m.

Remarks

- Located in Syria, the Zeyzoun dam (or Zayaiun dam) was completed in 1996. The embankment dam cracked on 4 June 2002, releasing about 71 Mm³ of water. A 3.3 m high wall of water rushed though the downstream villages submerging over 80 km². The final breach was 80 m wide. At least 22 people were killed in the catastrophe.

- As $d_1/d_o < 0.138$, the water depth and flow rate at the dam site equal:

$$d(x = 0) = \frac{4}{9}d_o = 15.5 \text{ m}$$

$$q(x = 0) = \frac{8}{27}\sqrt{gd_o^3} = 192 \text{ m}^2/\text{s}$$

- For a sudden dam break wave over an initially dry channel, the wave front celerity would be 37 m/s (equation (17.34), for an ideal dam break).

Discussion

Figure 17.15 sketches a situation where the point E2 is located downstream of the initial dam wall. At the origin, the water depth and velocity are, respectively (Dam break in a dry channel in Section 17.7.2):

$$d(x = 0) = \frac{4}{9}d_o \tag{17.36}$$

$$V(x = 0) = \frac{2}{3}\sqrt{gd_o} \tag{17.37}$$

Another situation is sketched in Fig. 17.17 where the point E2 is located upstream of the dam wall. Between these two situations, the limiting case is a fixed point E2 at the origin. This occurs for:

$$\frac{d_1}{d_o} = 0.1383$$

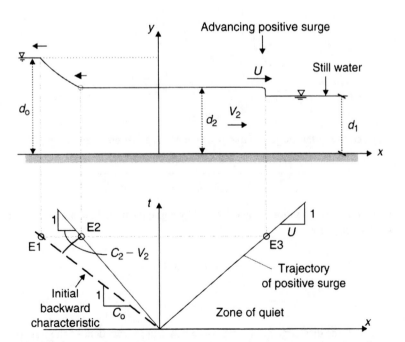

Fig. 17.17 Sketch of a dam break wave in horizontal channel initially filled with water for $0.1383 < d_1/d_o < 1$.

For $d_1/d_o < 0.1383$, the flow situation is sketched in Fig. 17.15. For $d_1/d_o > 0.1383$, the point E2 is always located upstream of the origin and the flow pattern is illustrated in Fig. 17.17.

Application

This demonstrates that the intersection of the forward characteristics issuing from the initial characteristics with the horizontal free surface behind the positive surge (i.e. point E2, Fig. 17.15) is at the origin ($x = 0$) if and only if:

$$\frac{d_1}{d_o} = 0.1383$$

Solution

In the limiting case for which the point E2 is at the origin, the water depth and velocity at the origin are, respectively:

$$d(x = 0) = \frac{4}{9}d_o = d_{E2} = d_2$$

$$V(x = 0) = \frac{2}{3}\sqrt{gd_o} = V_{E2} = V_2$$

where d_2 and the V_2 are the flow velocity behind (downstream of) the positive surge.

First, the flow conditions at the origin satisfy $V_2/\sqrt{gd_2} = 1$ Second, the continuity and momentum equations across the positive surge give:

$$\frac{d_2}{d_1} = \frac{1}{2}\left(\sqrt{1 + 8\frac{U^2}{gd_1}} - 1\right) = \frac{4}{9}\frac{d_o}{d_1}$$

$$\frac{U - V_2}{\sqrt{gd_2}} = \frac{2^{3/2}\left(U/\sqrt{gd_1}\right)}{\left(\sqrt{1 + 8(U^2/gd_1)} - 1\right)^{3/2}} = \frac{27}{8}\left(\frac{d_1}{d_o}\right)^{3/2} = \frac{U}{\sqrt{gd_1}}\sqrt{\frac{d_1}{d_2} - 1}$$

where U is the celerity of the surge (Chapter 4). One of the equations may be simplified into:

$$\frac{U}{\sqrt{gd_1}} = \frac{1}{(3/2)\sqrt{(d_1/d_o)} - (27/8)(d_1/d_o)^{3/2}}$$

This system of non-linear equations may be solved by a graphical method and by test and trial. The point E2 is at the origin for:

$$\frac{d_1}{d_2} = 0.1382701411$$

Note

For $d_1/d_o > 0.1383$, the discharge at the origin is a constant independent of the time. But the flow rate (at $x = 0$) becomes a function of the flow depth d_2 and it is less than $8/27\sqrt{g}Bd_o^{3/2}$.

17.7.3 Discussion

Effects of flow resistance

For a dam break wave in a dry horizontal channel, observations showed that Ritter's theory (Dam break in a dry channel in Section 17.7.2) is valid, but for the leading tip of the wave front. Experimental data indicated that the wave front has a rounded shape and its celerity U is less than $2C_o$ (e.g. Schoklitsch, 1917; Dressler, 1954; Faure and Nahas, 1961). Figure 17.18 compares the simple-wave theory (zero friction) with measured dam break wave profiles and it illustrates the round shape of the leading edge.

Dressler (1952) and Whitham (1955) proposed analytical solutions of the dam break wave that include the effects of bottom friction, assuming constant friction coefficient. Some results are summarized in Table 17.2 while their respective theories are discussed below. Escande *et al.* (1961)

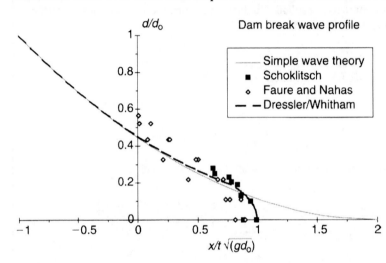

Fig. 17.18 Comparison of the dam break wave profile in an initially dry horizontal channel with and without bottom friction.

Table 17.2 Analytical solutions of dam break wave with and without bottom friction

Parameter	Simple wave (Ritter's solution)	Dressler (1952)	Whitham (1955)	Remarks
(1)	(2)	(3)	(4)	(5)
Wave front celerity U	$U = 2\sqrt{gd_o}$	—	$\dfrac{U}{\sqrt{gd_o}} = 2$ $-3.452\sqrt{\dfrac{f}{8}}t_s\sqrt{\dfrac{g}{d_o}}$	For $t = t_s$ small
$\dfrac{Q(x=0)}{W}$	$\dfrac{8}{27}\sqrt{gd_o^3}$	$\dfrac{8}{27}\sqrt{gd_o^3}$ $\times\left(1 - 0.239\dfrac{f}{8}\sqrt{\dfrac{g}{d_o}}t\right)$	—	
Location of critical flow conditions	$x = 0$	$\dfrac{x}{d_o} = 0.395\dfrac{f}{8}\sqrt{\dfrac{g}{d_o}}t$	—	

Note: Theoretical results obtained for an initially dry horizontal channel.

investigated specifically the effects of bottom roughness on dam break wave in a natural valley. They showed that, with a very-rough bottom, the wave celerity could be about 20–30% lower than for a smooth bed. Faure and Nahas (1965) conducted both physical and numerical modelling of the catastrophe of the Malpasset dam (Fig. 17.13). In their study, field observations were best reproduced with a Gauckler–Manning coefficient $n_{\text{Manning}} = 0.025$–$0.033 \text{ s/m}^{1/3}$.

Discussion: dam break wave developments by Dressler and Whitham

The theoretical solutions of Dressler and Whitham give very-close results and they are in reasonable agreement with experimental data. Both methods yield results close to the simple-wave solution but next to the leading tip.

Dressler (1952) used a perturbation method. His first order correction for the flow resistance gives the velocity and celerity at any position:

$$\frac{V}{\sqrt{gd_o}} = \frac{2}{3}\left(1 + \frac{x}{t\sqrt{gd_o}}\right) + F_1\frac{f}{8}\sqrt{\frac{g}{d_o}}t$$

$$\frac{C}{\sqrt{gd_o}} = \frac{1}{3}\left(2 - \frac{x}{t\sqrt{gd_o}}\right) + F_2\frac{f}{8}\sqrt{\frac{g}{d_o}}t$$

where f is the Darcy friction factor. The functions F_1 and F_2 are, respectively:

$$F_1 = -\frac{108}{7\left(2 - x/t\sqrt{gd_o}\right)^2} + \frac{12}{2 - x/t\sqrt{gd_o}} - \frac{8}{3} + \frac{8\sqrt{3}}{189}\left(2 - \frac{x}{t\sqrt{gd_o}}\right)^{3/2}$$

$$F_2 = \frac{6}{5\left(2 - x/t\sqrt{gd_o}\right)} - \frac{2}{3} + \frac{4\sqrt{3}}{135}\left(2 - \frac{x}{t\sqrt{gd_o}}\right)^{3/2}$$

Dressler compared successfully his results with the data of Schoklitch (1917).

At $t = t_s$, the location x_s of the dam break wave front satisfies:

$$\frac{f}{8}\sqrt{\frac{g}{d_o}}\,t_s = \left(6\left(\frac{54}{7\left(2 - x_s/t_s\sqrt{gd_o}\right)^3} - \frac{3}{\left(2 - x_s/t_s\sqrt{gd_o}\right)^2} + \frac{\sqrt{3}}{63}\left(2 - \frac{x_s}{t_s\sqrt{gd_o}}\right)^{1/2}\right)\right)^{-1}$$

Furthermore, results are summarized in Table 17.2.

Whitham (1955) analysed the wave front as a form of boundary layer using an adaptation of the Pohlhausen method. For a horizontal dry channel, his estimate of the wave front celerity is best correlated by:

$$\frac{U}{\sqrt{gd_o}} = \frac{2}{1 + 2.90724\left((f/8)\sqrt{gt^2/d_o}\right)^{0.4255}}$$

where f is the Darcy friction factor. Whitham commented that his work was applicable only for $U/\sqrt{gd_o} > 2/3$. He further showed that the wave front shape would follow:

$$\frac{d}{d_o} = \sqrt{\frac{f}{4}\frac{x_s - x}{d_o}}\left(2 - 3.452\left(\frac{f}{8}t\sqrt{\frac{g}{d_o}}\right)^{1/3}\right) \qquad \text{Wave front shape } (x < x_s)$$

for $\sqrt{gd_o}\,t(f/8)$ small.

Notes

1. Faure and Nahas (1965) conducted detailed physical modelling of the Malpasset dam failure using a Froude similitude. They used both undistorted and distorted scale models, as well as a numerical model. For the distorted model, the geometric scaling ratios were $X_r = 1/1600$ and $Z_r = 1/400$. Field data were best reproduced in the undistorted physical model ($L_r = 1/400$). Both numerical results and distorted model data were less accurate.

2. In an earlier development, Whitham (1955) assumed the shape of the wave front to follow:

$$d\alpha\sqrt{\frac{f}{4g}}U\sqrt{x_s - x}$$

where x_s is the dam break wave font location, but he discarded the result as inaccurate (?).

Dam break wave down a sloping channel

Considering the dam break wave down a sloping channel, the kinematic wave equation[3] may be solved analytically (Hunt, 1982). Hunt's analysis gives:

$$\frac{V_H t}{L} = \frac{1 - (d_s/H_{dam})^2}{(d_s/H_{dam})^{3/2}} \qquad (17.42)$$

$$\frac{x_s}{L} = \frac{3/2}{d_s/H_{dam}} - \frac{1}{2}\frac{d_s}{H_{dam}} - 1 \qquad (17.43a)$$

$$\frac{U}{V_H} = -\frac{3}{4}\frac{V_H t}{L} + \sqrt{\frac{x_s + L}{L} + \left(\frac{3}{4}\frac{V_H t}{L}\right)^2} \qquad (17.44)$$

where t is the time with $t = 0$ at dam break, d_s is the dam break wave front thickness, x_s is the dam break wave front position measured from the dam site, H_{dam} is the reservoir height at dam site, L is the reservoir length and S_o is the bed slope, $S_o = H_{dam}/L$ (Fig. 17.19). The velocity V_H is the uniform equilibrium flow velocity for a water depth H_{dam}:

$$V_H = \sqrt{\frac{8g}{f}H_{dam}S_o} \qquad (17.45)$$

where f is the Darcy friction factor which is assumed constant. Equations (17.42), (17.43a) and (17.44) are valid for $S_o = S_f$ where S_f is the friction slope and when the free

---

[3] That is, the kinematic wave approximation of the Saint-Venant equations (section 5).

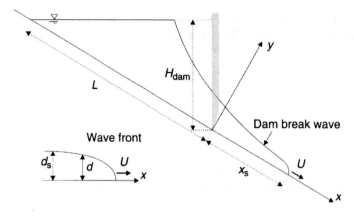

Fig. 17.19 Definition sketch of a dam break wave down a sloping channel.

surface is parallel to the bottom of the sloping channel. Equation (17.43a) may be transformed as:

$$\frac{d_s}{H_{dam}} = \frac{3}{2}\frac{L}{x_s - L} \qquad \text{for } x_s/L > 4 \qquad (17.43\text{b})$$

Hunt (1984, 1988) developed an analytical expression of the shock front shape:

$$\frac{x - x_s}{L} = \frac{d_s}{H_{dam}}\left(\frac{d}{d_s} + \ln\left(1 - \frac{d}{d_s}\right) + \frac{1}{2}\right) \qquad (17.46)$$

where d is the depth (or thickness) measured normal to the bottom and d_s and x_s are functions of time which may be calculated using equations (17.42) and (17.43a), respectively.

Discussion

The kinematic wave approximation gives the relationship between the velocity and the water depth as:

$$V = V_H\sqrt{\frac{d}{H_{dam}}}$$

Once the flood wave has travelled approximately four lengths of the reservoir downstream of the dam site, the free-surface profile of the dam break wave follows:

$$\frac{x + L}{L} = \frac{d}{H_{dam}} + \frac{3}{2}\frac{V_H t}{L}\sqrt{\frac{d}{H_{dam}}} \qquad \text{for } x \leqslant x_s \text{ and } x_s/L > 4$$

Notes

1. The elegant development of Hunt (1982, 1984) was verified by several series of experiments (e.g. Hunt, 1984; Nsom et al., 2000). It is valid however after the flood wave has covered approximately four reservoir lengths downstream of the dam site.
2. Hunt called equations (17.42)–(17.44) the *outer solution* of the dam break wave while equation (17.46) was called the *inner solution*. Note that Hunt's analysis accounts for bottom friction assuming a constant Darcy friction factor.
3. Bruce Hunt is a reader at the University of Canterbury, New Zealand.

Fig. 17.20 Example of stepped waterway susceptible to be subjected to rapid flood waves: Reinstatement of a stepped cascade following mining in a confined steep valley at Bileola (Qld, Australia) during construction in 2002 (courtesy of Dr J. Macintosh). $B = 100\,\text{m}$, $S_o = 0.015$, design flow: $390\,\text{m}^3/\text{s}$ and step height: $2\,\text{m}$.

Further dam break wave conditions

A particular type of dam break wave is the sudden release of water down a stepped chute (Fig. 16.1b). Applications include the sudden release of water down a stepped spillway and flood runoff in stepped storm waterways during tropical storms. Plate 27 and Fig. 17.20 present stepped waterways that may be subjected to such sudden flood waves.

The writer studied the release of a known discharge on an initially dry stepped waterway ($\theta = 3.4°$) with two step geometries (Chanson, 2001, 2003). Visual observations showed two dominant features of the wave propagation: i.e. strong aeration of the wave leading edge and a nappe flow propagation. Furthermore, wave front propagation data were successfully compared with Hunt's (1982, 1984) theory. Best fit was achieved assuming a Darcy–Weisbach friction factor of $f = 0.05$ independently of flow rates and step heights. The same study showed a strongly aerated flow region at the leading edge characterized by large interfacial areas (Chanson, 2003).

Additional dam break wave conditions were experimentally studied. Chanson *et al.* (2000) investigated the propagation of a dam break wave downstream of a free-falling nappe impact. They observed strong splashing and mixing at the nappe impact, and their data showed consistently a greater wave front celerity than for the classical dam break analysis for $x/d_o < 30$, where d_o is a function of the initial discharge.

Khan *et al.* (2000) studied the effects of floating debris. The results showed an accumulation of debris near the wave front and a reduction of the front celerity both with and without initial water levels.

Nsom *et al.* (2000) investigated dam break waves downstream of a finite reservoir in horizontal and sloping channels with very-viscous fluids. The dam break wave propagation was first dominated by inertial forces and then by viscous processes. In the viscous regime, the wave front location followed:

$$x_s \propto \sqrt{\frac{\cos\theta\, t}{\mu}}$$

where θ is the bed slope and μ is the fluid viscosity.

> **Note**
>
> In debris flow surges, large debris and big rocks are often observed 'rolling' and 'floating' at the wave front (e.g. Ancey, 2000).

17.8 EXERCISES

List the key assumptions of the Saint-Venant equations.

What is the celerity of a small disturbance in (1) a rectangular channel, (2) a 90° V-shaped channel and (3) a channel of irregular cross-section? (4) Application: a flow cross-section of a flood plain has the following properties: hydraulic diameter = 5.14 m, maximum water depth = 2.9 m, wetted perimeter = 35 m and free-surface width = 30 m. Calculate the celerity of a small disturbance.
  Solution: (3) $C = \sqrt{gd/2}$. (4) $C = 3.8$ m/s.

A 0.2 m high small wave propagates downstream in a horizontal channel with initial flow conditions $V = +0.1$ m/s and $d = 2.2$ m. Calculate the propagation speed of the small wave.
  Solution: $U = +5.1$ m/s and $C = 4.64$ m/s.

Uniform equilibrium flow conditions are achieved in a long rectangular channel ($B = 12.8$ m, concrete lined and $S_o = 0.0005$). The observed water depth is 1.75 m. Calculate the celerity of a small monoclinal wave propagating downstream. *Perform your calculations using the Darcy friction factor.*
  Solution: $U = +3.0$ m/s (for $k_s = 1$ mm).

The flow rate in a rectangular canal ($B = 3.4$ m, concrete lined and $S_o = 0.0007$) is 3.1 m³/s and uniform equilibrium flow conditions are achieved. The discharge suddenly increases to 5.9 m³/s. Calculate the celerity of the monoclinal wave. How long will it take for the monoclinal wave to travel 20 km?
  Solution: $d_2 = 1.23$ m, $U = +1.81$ m/s and $t = 11\,100$ s (3 h 5 min).

What is the basic definition of a simple wave? May the simple-wave theory be applied to (1) a sloping, frictionless channel, (2) a horizontal, rough canal, (3) a positive surge in a horizontal, smooth channel with constant water depth and (4) a smooth, horizontal canal with an initially accelerating flow?
  Solution: (1) No. (2) No. (3) Yes. (4) No.

What is the 'zone of quiet'?

Considering a long, horizontal rectangular channel ($B = 4.2$ m), a gate operation, at one end of the canal, induces a sudden withdrawal of water resulting in a negative velocity. At the gate, the boundary conditions for $t > 0$ are $V(x = 0, t) = -0.2$ m/s. Calculate the extent of the gate operation influence in the canal at $t = 1$ h. The initial conditions in the canal are $V = 0$ and $d = 1.4$ m.
  Solution: The problem may be analysed as a simple wave. The initial flow conditions are $V_o = 0$ and $C_o = 3.7$ m/s. Let us select a coordinate system with $x = 0$ at the gate and x positive in the upstream direction. In the (x, t) plane, the equation of the initial forward characteristics (issuing from $x = 0$ and $t = 0$) is given as:

$$\frac{dt}{dx} = \frac{1}{V_o + C_o} = 0.27 \text{ s/m}$$

At $t = 1$ h, the extent of the influence of the gate operation is 13.3 km.

Water flows in an irrigation canal at steady state ($V = 0.9$ m/s and $d = 1.65$ m). The flume is assumed smooth and horizontal. The flow is controlled by a downstream gate. At $t = 0$, the gate is very-slowly raised and the water depth upstream of the gate decreases at a rate of 5 cm/min until the water depth becomes 0.85 m: (1) Plot the free-surface profile at $t = 10$ min. (2) Calculate the discharge per unit width at the gate at $t = 10$ min.

Solution: The simple-wave problem corresponds to a negative surge. In the absence of further information, the flume is assumed wide rectangular.

Let us select a coordinate system with $x = 0$ at the gate and x positive in the upstream direction. The initial flow conditions are $V_0 = -0.9$ and $C_0 = 4.0$ m/s. In the (x, t) plane, the equation of the initial forward characteristics (issuing from $x = 0$ and $t = 0$) is given as:

$$\frac{dt}{dx} = \frac{1}{V_0 + C_0} = 0.32 \text{ s/m}$$

At $t = 10$ min, the maximum extent of the disturbance is $x = 1870$ m. That is, the zone of quiet is defined as $x > 1.87$ km.

At the gate $(x = 0)$, the boundary condition is $d(x = 0, t_0 \leqslant 0) = 1.65$ m, $d(x = 0, t_0) = 1.65 - 8.33 \times 10^{-4}t_0$, for $0 < t_0 < 960$ s and $d(x = 0, t_0 \geqslant 960$ s$) = 0.85$ m. The second flow property is calculated using the backward characteristics issuing from the initial forward characteristics and intersecting the boundary at $t = t_0$:

$$V(x = 0, t_0) = V_0 + 2(C(x = 0, t_0) - C_0) \quad \text{Backward characteristics}$$

where $C(x, t_0) = \sqrt{gd(x = 0, t_0)}$.

At $t = 10$ min, the flow property between $x = 0$ and 1.87 km are calculated from:

$$V(x, t = 600) + 2C(x, t = 600) = V(x = 0, t_0) + 2C(x = 0, t_0) \quad \text{Forward characteristics}$$

$$V(x, t = 600) - 2C(x, t = 600) = V_0 - 2C_0 \quad \text{Backward characteristics}$$

where the equation of the forward characteristics is:

$$t = t_0 + \frac{x}{V(x = 0, t_0) + C(x = 0, t_0)} \quad \text{Forward characteristics}$$

These three equations are three unknowns: $V(x, t = 600)$, $C(x, t = 600)$ and $t_0 = t(x = 0)$ for the C1 characteristics. The results of the calculation at $t = 12$ min are presented in Table E.17.1 and Fig. E.17.1.

The flow rate at the gate is -2.56 m²/s at $t = 600$ s. The negative sign shows that the flow direction is in the negative x-direction.

A 200 km long rectangular channel ($B = 3.2$ m) has a reservoir at the upstream end and a gate at the downstream end. Initially the flow conditions in the canal are uniform at $V = 0.35$ m/s and $d = 1.05$ m. The water surface level in the reservoir begins to rise at a rate of 0.2 m/h for 6 h. Calculate the flow conditions in the canal at $t = 2$ h. *Assume* $S_0 = S_f = 0$.

Table E.17.1 Negative surge calculations at $t = 10\,\text{min}$

$t_0(x = 0)$ (C1) (1)	$d(x = 0)$ (2)	$C(x = 0)$ (3)	$V(x = 0)$ (C2) (4)	$Fr(x = 0)$ (5)	x ($t = 10\,\text{min}$) (6)	$V(x)$ ($t = 10\,\text{min}$) (7)	$C(x)$ ($t = 10\,\text{min}$) (8)	$d(x)$ ($t = 10\,\text{min}$) (9)
0	1.65	4.02	−0.90	−0.22	1873	−0.90	4.02	1.65
60	1.6	3.96	−1.02	−0.26	1586	−1.02	3.96	1.60
120	1.55	3.90	−1.15	−0.29	1320	−1.15	3.90	1.55
180	1.5	3.83	−1.27	−0.33	1075	−1.27	3.83	1.50
240	1.45	3.77	−1.40	−0.37	852	−1.40	3.77	1.45
300	1.4	3.70	−1.53	−0.41	651	−1.53	3.70	1.40
360	1.35	3.64	−1.67	−0.46	473	−1.67	3.64	1.35
420	1.3	3.57	−1.80	−0.51	318	−1.80	3.57	1.30
480	1.25	3.50	−1.94	−0.55	187	−1.94	3.50	1.25
540	1.2	3.43	−2.08	−0.61	80.7	−2.08	3.43	1.20
600	1.15	3.36	−2.23	−0.66	0	−2.23	3.36	1.15

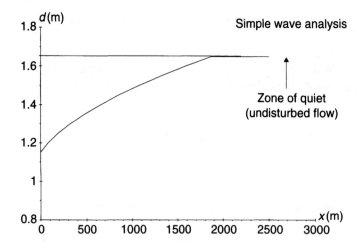

Fig. E.17.1 Negative surge application: free-surface profile at $t = 10\,\text{min}$.

Solution:

$t_0(x = 0)$ (C1) (1)	$d(x = 0)$ (2)	$C(x = 0)$ (3)	$V(x = 0)$ (C2) (4)	$Fr(x = 0)$ (5)	x ($t = 2\,\text{h}$) (6)	$V(x)$ ($t = 2\,\text{h}$) (7)	$C(x)$ ($t = 2\,\text{h}$) (8)	$d(x)$ ($t = 2\,\text{h}$) (9)
0	1.05	3.21	0.35	0.11	25 616	0.35	3.21	1.05
900	1.10	3.28	0.50	0.15	23 841	0.50	3.28	1.10
1800	1.15	3.36	0.65	0.19	21 630	0.65	3.36	1.15
2700	1.20	3.43	0.79	0.23	19 000	0.79	3.43	1.20
3600	1.25	3.50	0.93	0.27	15 964	0.93	3.50	1.25
4200	1.28	3.55	1.03	0.29	13 720	1.03	3.55	1.28
4800	1.32	3.59	1.12	0.31	11 306	1.12	3.59	1.32
5400	1.35	3.64	1.21	0.33	8723	1.21	3.64	1.35
6000	1.38	3.68	1.30	0.35	5976	1.30	3.68	1.38
6600	1.42	3.73	1.39	0.37	3067	1.39	3.73	1.42
7200	1.45	3.77	1.47	0.39	0	1.47	3.77	1.45

Note that the flow situation corresponds to a positive surge formation. However the wave does not have time to develop and steepen enough in 2 h to form a discontinuity (i.e. surge front) ahead. In other terms, the forward characteristics issuing from the reservoir do not intersect for $t \leqslant 2\,\text{h}$.

Water flows in a horizontal, smooth rectangular channel. The initial flow conditions are $d = 2.1$ m and $V = +0.3$ m/s. The flow rate is stopped by sudden gate closure at the downstream end of the canal. Using the quasi-steady flow analogy, calculate the new water depth and the speed of the fully developed surge front.

Solution: $U = 4.46$ m/s and $d_2 = 2.24$ m (gentle undular surge).

Let us consider the same problem as above (i.e. a horizontal, smooth rectangular channel, $d = 2.1$ m and $V = 0.3$ m/s) but the downstream gate is closed slowly at a rate corresponding to a linear decrease in flow rate from 0.63 m^2/s down to zero in 15 min: (1) Predict the surge front development. (2) Calculate the free-surface profile at $t = 1$ h after the start of gate closure.

Solution: At $t = 1$ h, the positive surge has not yet time to form (Fig. E.17.2). That is, the forward characteristics do not intersect yet.

x ($t = 1$ h) (m)	V ($t = 1$ h) (m/s)	C ($t = 1$ h) (m/s)	d ($t = 1$ h) (m)
17 000	−0.30	4.54	2.10
15 251.5	−0.30	4.54	2.10
15 038.2	−0.27	4.55	2.11
14 941.1	−0.21	4.58	2.14
14 875.8	−0.15	4.61	2.17
14 763.4	−0.09	4.64	2.20
14 559.9	−0.04	4.67	2.22
14 262.4	−0.02	4.68	2.23
13 896.2	−0.01	4.68	2.24
13 492.8	0.00	4.69	2.24
8 435.73	0.00	4.69	2.24
1 405.96	0.00	4.69	2.24

Water flows in a horizontal, smooth rectangular irrigation canal. The initial flow conditions are $d = 1.1$ m and $V = +0.35$ m/s. Between $t = 0$ and $t = 10$ min, the downstream gate is slowly rised at a rate implying a decrease in water depth of 0.05 m/min. For $t > 10$ min, the gate position is maintained constant. Calculate and plot the free-surface profile in the canal at $t = 30$ min. *Use a simple-wave approximation.*
Solution:

x ($t = 30$ min)	$V(x)$ ($t = 30$ min)	$C(x)$ ($t = 30$ min)	$d(x)$ ($t = 30$ min)
6000	−0.35	3.28	1.1
5280	−0.35	3.28	1.1
6000	−0.35	3.28	1.1
5280	−0.35	3.28	1.1
4710	−0.50	3.21	1.05
4158	−0.66	3.13	1
3624	−0.81	3.05	0.95
3109	−0.98	2.97	0.9
2613	−1.14	2.89	0.85
2136	−1.32	2.8	0.8
1679	−1.49	2.71	0.75
1242	−1.68	2.62	0.7
825	−1.87	2.52	0.65
430	−2.07	2.42	0.6
408	−2.07	2.43	0.6

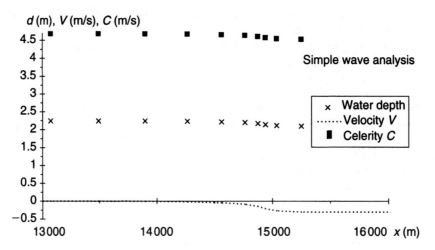

Fig. E.17.2 Positive surge application: free-surface profile at $t = 1$ h.

A 5 m wide forebay canal supplies a penstock feeding a Pelton turbine. The initial conditions in the channel are $V = 0$ and $d = 2.5$ m: (1) The turbine starts suddenly operating with 6 m³/s. Predict the water depth at the downstream end of the forebay canal. (2) What is the maximum discharge that the forebay channel can supply? *Use a simple-wave theory.*

   Solution: (1) $d(x = 0) = 2.24$ m. (2) $Q = -18.3$ m³/s.

The Virdoule River flows in southern France. It is known for its flash floods associated with extreme rainfall events, locally called 'épisode cévenol'. These are related to large masses of air reaching the Cévennes mountain range. The Virdoule River is about 50 m wide. Initially the water depth is 0.5 m and the flow rate is 25 m³/s. During an extreme rainfall event, the discharge at Quissac increases from 25 to 1200 m³/s in 6 h. (Assume a linear increase of flow rate with time.) The flow rate remains at 1200 m³/s for an hour and then decreases down to 400 m³/s in the following 12 h.

(a) Compute the celerity of the monoclinal wave for the flash flood (i.e. increase in flow rate from 25 to 1200 m³/s). *Assume uniform equilibrium flows and $S_o = 0.001$.*

(b) Approximate the flash flood with a dam break wave such that the flow rate at the origin (i.e. Quissac) is 1200 m³/s after dam break.

(c) Calculate the water depth in the village of Sommières, located 21 km downstream of Quissac, for the first 24 h. *Use a simple-wave theory.*

(d) Compare the results between the three methods at Sommières.

   Remarks: Such an extreme hydrological event took place between Sunday 8 September and Monday 9 September 2002 in southern France. More than 37 people died. At Sommières, the water depth of the Virdoule River reached up to 7 m. Interestingly, the old house in the ancient town of Sommières had no ground floor because of known floods of the Virdoule River.

   Solution:

(a) For a monoclinal wave, uniform equilibrium flow calculations are assumed upstream and downstream of the flood wave. The celerity of the monoclinal wave is:

$$U = \frac{Q_2 - Q_1}{A_2 - A_1}$$

(17.10a)

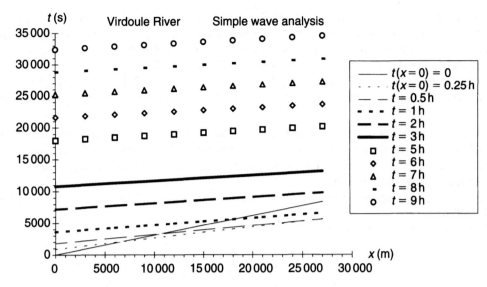

Fig. E.17.3 Forward characteristics curves for Virdoule River flash flood.

Assuming an identical Darcy friction factor for both initial and new flow conditions, the new flow conditions yield $d_2 = 7.1$ m at uniform equilibrium and $U = 3.54$ m/s. The monoclinal wave will reach Sommières 1.6 h after passing Quissac.

(b) For a dam break wave, the flow rate at the dam site ($x = 0$) is 1200 m³/s:

$$Q(x = 0) = \frac{8}{27} d_o \sqrt{g d_o} W \tag{17.38}$$

It yields $d_o = 8.74$ m. Note that $d_1/d_o = 0.057$ and it is reasonable to assume critical flow conditions at $x = 0$ (i.e. equation (17.38)). Downstream of the dam, the initial flow conditions are $d_1 = 0.5$ m and $V_1 = +1$ m/s. Complete calculations show that $U = 10.66$ m/s, $d_2 = 2.46$ m and $V_2 - C_2 = 3.79$ m/s.

The positive surge leading the dam break wave will reach Sommières 33 min after the dam break at Quissac and the wave front height will be 1.96 m. The water depth at Sommières will remain $d = d_2$ until $t = 1.53$ h. Afterwards, the water depth will gradually increase. (Ultimately it would reach 3.9 m assuming an infinitely long reservoir.)

(c) For a simple-wave analysis, the boundary conditions correspond to the formation of a positive surge between $t = 0$ and 6 h, and the formation of a negative surge between $t = 7$ and 19 h. The forward characteristics are plotted in Fig. E.17.3. A positive surge will form rapidly downstream of Quissac and the surge front will reach the township of Sommières at about $t = 1.3$ h.

(d) Discussion: between the three calculations, the last one (i.e. simple-wave calculations) would be the most accurate. Practically, an important result is the arrival time of the flash flood at Sommières. If the flood is detected at Quissac, the travel time between Quissac and Sommières is basically the warning time. The simple calculation predict 1.3 h. The monoclinal wave calculations predict 1.6 h and the dam break wave theory gives 33 min. Interestingly, all the results are of the same order of magnitude.

Write the kinematic wave equation for a wide rectangular channel in terms of the flow rate, bed slope and water depth. What is the speed of a kinematic wave? Does the kinematic wave routing predict subsidence?

A wide channel has a bed slope $S_o = 0.0003$ and the channel bed has an equivalent roughness height of 25 mm. The initial flow depth is 2.3 m and uniform equilibrium flow conditions are achieved. The water depth is abruptly increased to 2.4 m at the upstream end of the channel. Calculate the speed of the diffusion wave and the diffusion coefficient.

  Solution: Initial uniform equilibrium flow conditions are $d = 2.3$ m, $V = +1.42$ m/s and $f = 0.026$. Diffusion wave: $U = 2.1$ m/s and $D_t = 5.4 \times 10^3$ m²/s (for a wide rectangular channel).

A 8 m wide rectangular canal (concrete lining) operates at uniform equilibrium flow conditions for a flow rate of 18 m³/s resulting in a 1.8 m water depth. At the upstream end, the discharge is suddenly increased to 18.8 m³/s. Calculate the flow rate in the canal 1 h later at a location $x = 15$ km. *Use diffusion routing.*

  Solution: Initially, uniform equilibrium flow conditions are $V = +1.25$ m/s, $S_o = S_f = 0.00028$ and $f = 0.017$.

  The celerity of the diffusion wave is:

$$U = \frac{Q}{B((8g/f)A^2(D_H/4))^{1/2}} \frac{\partial}{\partial d}\left(A\sqrt{\frac{8g}{f}\frac{D_H}{4}}\right) = \frac{\sqrt{S_f}}{B}\frac{\partial}{\partial d}\left(A\sqrt{\frac{8g}{f}\frac{D_H}{4}}\right)$$

$$= \frac{\sqrt{S_f}}{B}\frac{3}{2}B\sqrt{\frac{8g}{f}\frac{A}{P_W}}\left(1 - \frac{2}{3}\frac{A}{BP_w}\right) = \frac{3}{2}V\left(1 - \frac{2}{3}\frac{A}{BP_w}\right) = 1.68 \text{ m/s}$$

The diffusion coefficient is:

$$D_t = \frac{(8g/f)A^2(D_H/4)}{2B|Q|} = 4.02 \times 10^3 \text{ m}^2/\text{s}$$

The analytical solution of the diffusion wave equation yields:

$$Q(x, t) = Q_o + \frac{\delta Q}{2}\left(1 - \text{erf}\frac{x - Ut}{\sqrt{4D_t t}} + \exp\left(\frac{Ux}{D_t}\right)\left(1 - \text{erf}\frac{x + Ut}{\sqrt{4D_t t}}\right)\right)$$

assuming constant diffusion wave celerity and diffusion coefficient, where $Q_o = 18$ m³/s and $\delta Q = 0.8$ m³/s. For $t = 1$ h and $x = 15$ km, $Q = 18.04$ m³/s.

A 15 m high dam fails suddenly. The dam reservoir had a 13.5 m depth of water and the downstream channel was dry: (1) Calculate the wave front celerity, and the water depth at the origin. (2) Calculate the free-surface profile 2 min after failure. *Assume an infinitely long reservoir and use a simple-wave analysis* ($S_o = S_f = 0$).

  Solution: $U = 23$ m/s and $d(x = 0) = 6$ m.

A vertical sluice shut a trapezoidal channel (3 m bottom width, 1V:3H side slopes). The water depth was 4.2 m upstream of the gate and zero downstream (i.e. dry channel). The gate is suddenly removed. Calculate the negative celerity. Assuming an ideal dam break wave, compute the wave front celerity and the free-surface profile 1 min after gate removal.

  Solution: The assumption of hydrostatic pressure distribution is valid for $t > \sqrt{d_o/g} \sim 2$ s. Hence the Saint-Venant equations may be applied for $t = 60$ s. Note the non-rectangular channel cross-section. For a trapezoidal channel, the celerity of a small disturbance is:

$$C\sqrt{g\frac{A}{B}} = \sqrt{g\frac{d(W + d\cot\delta)}{W + 2d\cot\delta}} \quad \text{(Section 3.4.2)}$$

where W is the bottom width and δ is the angle with the horizontal (i.e. $\cot\delta = 3$).

The method of characteristics predicts that the celerity of the negative wave is $-C_o = -4.7$ m/s. The celerity of the wave front is $U = +2C_o = +9.6$ m/s. Considering a backward characteristics issuing from the dam break wave front, the inverse slope of the C2 characteristics is a constant:

$$\frac{dx}{dt} = V - C = 2C_o - 3C$$

The integration gives the free-surface profile equation at a given time t:

$$\frac{x}{t} = 2\sqrt{g\frac{d_o(W + d_o\cot\delta)}{W + 2d_o\cot\delta}} - 3\sqrt{g\frac{d(W + d\cot\delta)}{W + 2d\cot\delta}}$$

At $t = 60$ s, the free-surface profile between the leading edge of the wave front and the negative wave most upstream location is:

d (m)	4.2	3	2	1.725	1	0.5	0
x (m)	-282	-160	-38.4	0	119	231	564

A 5 m high spillway gate fails suddenly. The water depth upstream of the gate was 4.5 m depth and the downstream concrete channel was dry and horizontal: (1) Calculate the wave front location and velocity at $t = 3$ min. (2) Compute the discharge per unit width at the gate at $t = 3$ min. *Use Dressler's theory assuming $f = 0.01$ for new concrete lining.* (3) Calculate the wave front celerity at $t = 3$ min using Whitham's theory.

Solution: $x_s = 330$ m, $U = V(x = x_s) = 4.05$ m/s and $q(x = 0) = 0.21$ m²/s (Dressler's theory) and $U = 4.7$ m/s (Whitham's theory).

A horizontal, rectangular canal is shut by a vertical sluice. There is no flow motion on either side of the gate. The water depth is 3.2 m upstream of the gate and 1.2 m downstream. The gate is suddenly lifted: (1) Calculate the wave front celerity, and the surge front height. (2) Compute the water depth at the gate. Is it a function of time?

Solution: (1) $d_1/d_o = 0.375$, $U = 5.25$ m/s and $d(x = 0) = 2.07$ m. (2) $d_2 - d_1 = 0.87$ m.

A 35 m high dam fails suddenly. The initial reservoir height was 31 m above the downstream channel invert and the downstream channel was filled with 1.8 m of water initially at rest: (1) Calculate the wave front celerity and the surge front height. (2) Calculate the wave front location 2 min after failure. (3) Predict the water depth 10 min after gate opening at two locations: $x = 2$ km and $x = 4$ km. *Assume an infinitely long reservoir and use a simple-wave analysis ($S_o = S_f = 0$).*

Solution: (1) $d_1/d_o = 0.06$, $U = 18.1$ m/s and $d_2 - d_1 = 8.34$ m. (2) $x_s = 2.2$ km ($t = 2$ min). (3) $d(x = 2$ km, $t = 10$ min) $= 11.3$ m and $d(x = 4$ km, $t = 10$ min) $= 10.1$ m

A narrow valley is closed by a tall arch dam. The average bed slope is 0.08 and the valley cross-section is about rectangular (25 m width). At full reservoir level, the water depth immediately upstream of the dam is 42 m. In a dam break wave situation (a) predict the wave front propagation up to 45 km downstream of the dam and (b) calculate the wave front celerity and depth as it reaches a location 25 km downstream of the dam site. *Use Hunt's theory assuming $f = 0.06$.*

Solution: (b) $d_s = 1.3$ m and $U = 11.6$ m/s at $x_s = 25$ km (that is more than four reservoir lengths downstream of the dam site).

A 8 m wide rectangular canal is controlled at the downstream end by a regulation gate. The invert is concrete lined and the bed slope is 0.4 m/km. The initial flow conditions are $V = 0.15$ m/s and $d = 1.3$ m everywhere in the canal. The gate is slowly opened and the water depth upstream of the gate decreases at a rate of 10 cm/h for 6 h. *Neglect bed slope and flow resistance. Assume $t = 0$ at the start of gate opening*: (1) Calculate the unsteady flow properties in the canal. (1a) At $t = 1$ h, how far upstream will the water level have changed? (1b) At $t = 3$ h, what is the velocity magnitude at the gate? (2) Plot the free-surface profile at $t = 3$ h.

A rectangular water reservoir is held by a sluice gate upstream of a wide horizontal canal (Fig. E.17.4). The initial flow conditions are $L = 5$ m (reservoir length), $H = 0.5$ m, $d_{tw} = 0.2$ m and zero velocity upstream and downstream of the gate. The gate is *suddenly* opened (completely) at $t = 0$. *Assume $S_o = S_f = 0$.*

(a) At what time does the water level at the upstream boundary of the reservoir start to change? (b) Before the negative wave reaches the reservoir upstream boundary, what is the shape of the free-surface profile in the canal and reservoir based upon the Saint-Venant equations? In Fig. E.17.5, which is the correct free-surface profile? (c) At $t = 1$ s, what is the wave front celerity downstream of the sluice gate? (That is, on the right of the gate in Fig. E.17.4.) (d) At $t = 1$ s, what is the water depth immediately behind the wave front?

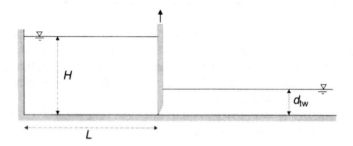

Fig. E.17.4 Definition sketch of a rectangular water reservoir at $t = 0$.

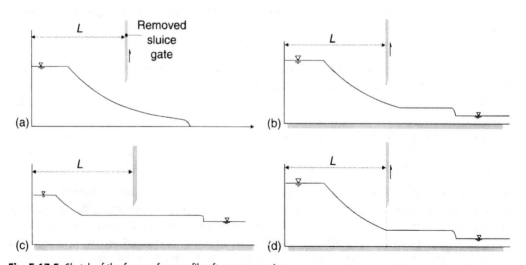

Fig. E.17.5 Sketch of the free-surface profile after gate opening.

Part 3 Revision exercises

REVISION EXERCISE NO. 1

An undershot sluice gate is to be tested in laboratory to determine the discharge characteristics for various opening heights. The maximum prototype discharge will be $220 \, \text{m}^3/\text{s}$ and the gate is installed in a rectangular channel of 20 m wide. Note that the channel bed will be horizontal and concrete lined (i.e. smooth). A 50:1 scale model of the gate is to be built. What similitude should be used? (Justify your selection in words.)

Determine (a) the maximum model discharge required and (b) the minimum prototype discharge for which negligible-scale effects occur in the model.

For one particular gate opening, the *laboratory* flow conditions are upstream flow depth of 0.21 m and downstream flow depth of 0.01 m. (c) Compute the model discharge. State the basic principle(s) involved. (d) Compute the model force acting on the sluice gate. (State the basic principle(s) involved.) (e) What will be the corresponding prototype discharge and force on gate? (f) What will be the scale for the force ratio?

Gate operation may result in unsteady flow situations. If a prototype gate operation has the following characteristics: gate opening duration = 3 min and gate opening height = 0.5 m, what (g) gate opening duration and (h) gate opening height should be used in the model tests?

REVISION EXERCISE NO. 2

An artificial concrete channel model is to be built. Laboratory facilities limit the scale ratio to 50:1 and maximum model discharge is 50 l/s. The maximum full-scale discharge is $150 \, \text{m}^3/\text{s}$, the cross-section of the channel is approximately rectangular (50 m bottom width) and the bed slope is 0.14 m/km.

Note: The roughness height of the prototype is estimated as 3 mm while the smoothest model surface feasible has a Darcy friction factor of about $f = 0.03$.

Discharges ranging between the maximum flow rate and 10% of the maximum flow rate are to be reproduced in the model.

1. *Undistorted* model (assume normal flow conditions)
Compute the model discharge at maximum full-scale discharge, the model flow depth, the Darcy coefficient of the model flow and the Darcy coefficient of the prototype channel?

2. A *distorted* model is to be built
Determine the acceptable maximum (largest) value of the vertical-scale ratio Z_r. Determine the acceptable minimum (smallest) value of the vertical-scale ratio Z_r. For the acceptable minimum value of the vertical-scale ratio Z_r, calculate the corresponding model maximum discharge.

REVISION EXERCISE NO. 3

A senior coastal engineer wants to study sediment motion in the swash zone. For 0.5 m high breaking waves, the resulting swash is somehow similar to a dam break wave running over retreating waters. (1) Assuming an initial reservoir water depth of 0.5 m, an initial water depth $d_1 = 0.07$ m and an initial flow velocity $V_1 = -0.4$ m/s, calculate the surge front celerity and height. *Assume a simple wave* ($S_o = S_f = 0$). (2) Calculate the bed shear stress immediately behind the surge front. The beach is made of fine sand ($d_{50} = 0.3$ mm and $d_{90} = 0.8$ mm). *Assume* $k_s = 2d_{90}$ (*Section 12.4.2, Table 12.2*). *For sea water,* $\rho = 1024$ kg/m^3 *and* $\mu = 1.22 \times 10^{-3}$ Pa s. (3) Predict the occurrence of bed load motion and sediment suspension. (4) During a storm event, breaking waves near the shore may be 3–5 m high. For a 2 m high breaking wave, calculate the surge front height and bed shear stress behind the surge front assuming an initial reservoir water depth of 2 m, an initial water depth $d_1 = 0.15$ m and an initial flow velocity $V_1 = -1$ m/s.

Solution

Let us select a positive x-direction toward the shore. The dam break wave ($d_o = 0.5$ m) propagates in a channel initially filled with water ($d_1 = 0.045$ m) with an opposing flow velocity ($V_1 = -0.4$ m/s). The x-coordinate is zero ($x = 0$) at wave breaking (i.e. pseudo-dam site) and the time origin is taken at the start of wave breaking.

The characteristic system of equation, and the continuity and momentum principles at the wave front must be solved theoretically. The free-surface profile is horizontal between the leading edge of the positive surge (point E_3) and the intersection with the C_1 forward characteristics issuing from the initial negative characteristics. The flow depth d_2 and velocity V_2 behind the surge front satisfy the continuity and momentum equations as well as the condition along the C_1 forward characteristics:

$$d_1(U - V_1) = d_2(U - V_2) \qquad \text{Continuity equation (17.39)}$$

$$d_2(U - V_2)^2 - d_1(U - V_1)^2 = \frac{1}{2}gd_1^2 - \frac{1}{2}gd_2^2 \qquad \text{Momentum equation (17.40)}$$

$$V_2 + 2\sqrt{gd_2} = 2\sqrt{gd_o} \qquad \text{Forward characteristics (17.33)}$$

Equations (17.33), (17.39) and (17.40) form a system of three non-linear equations with three unknowns V_2, d_2 and U.

An iterative calculation shows that the surge front celerity is $U/\sqrt{gd_1} = 3.2$ and $U = 2.12$ m/s. At the wave front, the continuity and momentum equations yield:

$$\frac{d_2}{d_1} = \frac{1}{2}\left(\sqrt{1 + 8\frac{(U - V_1)^2}{gd_1}} - 1\right) = 4.6 \qquad \text{Hence } d_2 = 0.21 \text{ m}$$

Equation (17.33) may be rewritten as:

$$\frac{V_2}{\sqrt{gd_2}} = 2\left(\sqrt{\frac{d_o}{d_2}} - 1\right) = 1.1 \qquad \text{Hence } V_2 = 1.58 \text{ m/s}$$

Behind the surge front, the boundary shear stress equals:

$$\tau_o = \frac{f}{8}\rho V_2^2 = 7.7 \text{ Pa}$$

The Shields parameter τ_* equals 1.55 which is almost one order of magnitude greater than the critical Shields parameter for bed load motion $(\tau_*)_c \sim 0.035$ (Section 10.3). For a 0.3 mm sand particle, the settling velocity is 0.034 m/s. The ratio V_*/w_o equals 2.5 implying sediment suspension (Section 9.2).

For a 2 m high breaking wave during a storm event, the surge front height equals: $d_2 - d_1 = 0.69$ m. The boundary shear stress behind the surge front equal $\tau_o = 21$ Pa.

Remarks

The above development has a number of limitations. The reservoir is assumed infinite, although a breaking wave has a finite volume, the beach slope is assumed frictionless and horizontal.

Note that the calculations of U, V_2 and d_2 are independent of time.

Appendices to Part 3

A3.1 PHYSICAL MODELLING OF MOVABLE BOUNDARY HYDRAULICS

A3.1.1 Introduction

Physical modelling of free-surface flows is most often performed using a Froude similitude. In the presence of sediment motion, additional parameters must be introduced.

In the following sections, we will develop further similitude criterion for movable bed in open channel flows. These derive from the definition of the boundary shear stress:

$$\tau_o = \frac{f}{8} \rho V^2 \tag{A3.1}$$

where f is the Darcy friction factor and V is the flow velocity.

A3.1.2 Bed-load motion

Occurrence of bed-load motion

Experimental investigations and dimensional analysis showed that the occurrence of bed-load motion may be predicted from the Shields diagram:

$$(\tau_*)_c = \frac{(\tau_o)_c}{\rho g (s-1) d_s} = \mathscr{F}_1\left(\rho \frac{d_s V_*}{\mu}\right) \qquad \text{Occurrence of bed-load motion} \tag{A3.2}$$

where d_s is the sediment size, s is the relative density of sediment particle (i.e. $s = \rho_s/\rho$) and V_* is the shear velocity (i.e. $V_* = \sqrt{\tau_o/\rho}$).

Perfect model–prototype similarity of occurrence of bed-load motion will occur only and only if:

$$((\tau_*)_c)_m = ((\tau_*)_c)_p \tag{A3.3}$$

and

$$\left(\rho \frac{d_s V_*}{\mu}\right)_m = \left(\rho \frac{d_s V_*}{\mu}\right)_p \tag{A3.4}$$

where the subscripts m and p refer to the model and prototype characteristics, respectively.

Expressing the bed shear stress as a function of the Darcy friction factor (equation (A3.1)) and for a Froude similitude,[1] the equality of Shields parameter in model and prototype (equation (A3.3)) implies that the scale ratios for the sediment size $(d_s)_r$ and for the relative density $(s-1)_r$ must satisfy:

$$(d_s)_r (s-1)_r = L_r \qquad \text{Un-distorted model} \tag{A3.5a}$$

[1] Assuming an identical gravity acceleration in model and prototype.

and for a distorted model:

$$(d_s)_r (s-1)_r = \frac{Z_r^{\,2}}{X_r} \qquad \text{Distorted model} \qquad \text{(A3.5b)}$$

Equation (A3.5) implies that the scale ratio of the sediment grain size may be different from the geometric length scale ratio by changing the relative density of the model grains.

Note

Let us keep in mind that a Froude similitude implies that the model flow resistance will be similar to that in the prototype:

$$f_r = 1$$

$$f_r = \frac{Z_r}{X_r} \qquad \text{Distorted model (wide channel and flat slope)}$$

If the sediment size scale ratio $(d_s)_r$ differs from the geometric length scale ratio L_r, the scale ratio of bed friction resistance might be affected.

Discussion

Note that the equality of shear Reynolds number (i.e. equation (A3.4)) in model and prototype yields:

$$(d_s)_r = \frac{\mu_r}{\rho_r} \frac{1}{\sqrt{L_r}} \qquad\qquad \text{(A3.6a)}$$

$$(d_s)_r = \frac{\mu_r}{\rho_r} \frac{\sqrt{X_r}}{Z_r} \qquad \text{Distorted model} \qquad \text{(A3.6b)}$$

Equation (A3.6) requires model grain sizes larger than prototype sizes (for $L_r > 1$). This is not acceptable in most cases and equations (A3.4) and (A3.6) must be ignored in practice.

Bed-load sediment transport rate

The bed-load transport rate per unit width is related to the bed-load layer characteristics (see Section 10.3.2) as:

$$q_s = C_s \delta_s V_s \qquad\qquad \text{(A3.7)}$$

where V_s is the average sediment velocity in the bed-load layer and C_s is the mean sediment concentration in the bed-load layer of thickness δ_s.

Using existing correlations (e.g. Nielsen, 1992), the similarity condition of bed-load motion (i.e. equations (A3.3) and (A3.5)) yields:

$$(q_s)_r = \sqrt{L_r} \qquad\qquad \text{(A3.8a)}$$

$$(q_s)_r = \frac{Z_r}{\sqrt{X_r}} \qquad \text{Distorted model} \qquad \text{(A3.8b)}$$

Summary
Similarity of bed-load motion may be achieved if equations (A3.3) and (A3.4) are respected. In practice, only equation (A3.3) can be achieved. If equation (A3.3) (or equation (A3.5)) is satisfied, the similarity of bed-load motion inception and bed-load transport rate may be achieved.

A3.1.3 Suspension

Occurrence of suspension
The occurrence of suspended load is a function of the balance between the particle fall velocity and the upward turbulent flow motion. Experimental investigations showed that suspension occurs for:

$$\frac{V_*}{w_o} > 0.2\text{–}2 \tag{A3.9}$$

where V_* is the shear velocity and w_o is the settling velocity. For spherical particles the terminal fall velocity equals:

$$w_o = -\sqrt{\frac{4gd_s}{3C_d}(s-1)} \qquad \text{Spherical particle} \tag{A3.10}$$

where the drag coefficient C_d is a function of the ratio $w_o d_s/\nu$.

For turbulent settling motion (i.e. $w_o d_s/\nu > 1000$), the drag coefficient is a constant. In addition, the scale ratio of the sediment suspension parameter is:

$$\left(\frac{V_*}{w_o}\right)_r = \frac{\sqrt{L_r}}{\sqrt{(d_s)_r(s-1)_s}} \qquad \text{Turbulent particle settling} \tag{A3.11a}$$

If the condition for the similarity in bed-load motion (i.e. equations (A3.3) and (A3.5)) is satisfied, it yields:

$$\left(\frac{V_*}{w_o}\right)_r = 1 \qquad \text{Turbulent particle settling} \tag{A3.11b}$$

In other words, if the similarity in bed-load motion is achieved, the similarity of turbulent particle settling is also satisfied (provided that the particle settling motion is turbulent in both model and prototype).

Suspended-sediment transport rate
The distribution of suspended matter across the flow depth may be estimated as:

$$c_s = (C_s)_{y=y_s}\left(\frac{(d/y)-1}{(d/y_s)-1}\right)^{w_o/(KV_*)} \tag{A3.12}$$

where $(C_s)_{y=y_s}$ is the reference sediment concentration at the reference location $y = y_s$ and d is the flow depth.

The similarity of suspended-sediment concentration is achieved if the ratio w_o/V_* is identical in model and prototype and if the integration parameters $(C_s)_{y=y_s}$ and y_s are appropriately scaled.

It is presently impossible to reproduce sediment suspension distributions and suspended-load transport rates in small-scale models which reproduce prototype sediment concentration distributions and transport rates.

A3.1.4 Time scale in sediment transport

The *hydraulic time scale* characterizes the duration of an individual event (e.g. a flood event). For a Froude similitude, the scale ratio is:

$$t_r = \sqrt{L_r} \tag{A3.13a}$$

$$t_r = \frac{X_r}{\sqrt{Z_r}} \qquad \text{Distorted model} \tag{A3.13b}$$

In sediment transport another time scale is the *sedimentation time*. In movable-bed channels, the continuity equation for sediment is:

$$\frac{\partial q_s}{\partial x} = -(1 - Po)\frac{\partial z_o}{\partial t}\cos\theta \tag{A3.14}$$

where q_s is the total sediment transport rate, Po is the sediment bed porosity, z_o is the bed elevation and θ is the longitudinal bed slope. For a flat channel, this gives the scale ratio of the sedimentation time t_s:

$$(t_s)_r = (1 - Po)_r\frac{L_r^2}{(q_s)_r} \tag{A3.15a}$$

$$(t_s)_r = (1 - Po)_r\frac{X_r Z_r}{(q_s)_r} \qquad \text{Distorted model} \tag{A3.15b}$$

If most sediment transport occurs by bed-load motion, the above result may be combined with equation (A3.8):

$$(t_s)_r = (1 - Po)_r L_r^{3/2} \qquad \text{Bed-load transport} \tag{A3.16a}$$

$$(t_s)_r = (1 - Po)_r X_r^{3/2} \qquad \text{Bed-load transport – Distorted model} \tag{A3.16b}$$

Comparing equations (A3.13) and (A3.15) (or equation (A3.16)), it appears that the sedimentation time ratio differs from the hydraulic time ratio in most cases.

*   *   *

A3.2 EXTENSION OF THE BACKWATER EQUATION

A3.2.1 Introduction

Presentation

For a rectangular channel, the continuity equation states:

$$Q = VBd \tag{A3.17}$$

where Q is the total discharge, V is the average flow velocity, B is the channel width and d is the flow depth (measured normal to the flow direction).

At any position along a streamline, the (local) total head H is defined as:

$$H = \frac{P}{\rho g} + z + \frac{V^2}{2g}$$ (A3.18)

where P is the pressure, z is the elevation (positive upwards) and V is the velocity along the streamline.

Assuming a hydrostatic pressure gradient, the average total head, as used in the energy equation, equals:

$$H = d \cos \theta + z_o + \frac{\alpha}{2g} \left(\frac{Q}{A} \right)^2$$ (A3.19a)

where d is the flow depth measured normal to the channel bottom, θ is the channel slope, z_o is the bed elevation, α is the kinetic energy correction coefficient (i.e. Coriolis coefficient), Q is the total discharge and A is the cross-sectional area. For a rectangular channel it yields:

$$H = d \cos \theta + z_o + \frac{\alpha}{2g} \left(\frac{Q}{Bd} \right)^2 \qquad \text{Rectangular channel}$$ (A3.19b)

Energy equation

Considering a non-uniform and steady flow, the backwater calculations are performed assuming that the flow is gradually varied and that, at a given section, the flow resistance is the same as for an uniform flow for the same depth and discharge, regardless of trends of the depth. In the s-direction, the energy equation becomes:

$$\frac{\partial H}{\partial s} = -S_f$$ (A3.20)

where S_f is the friction slope, H is the mean total head and s is the co-ordinate along the flow direction. Using the mean specific energy this equation is transformed as:

$$\frac{\partial E}{\partial s} = S_o - S_f$$ (A3.21)

where S_o is the bed slope (i.e. $S_o = \sin \theta$).

Equation (A3.20) is the basic *backwater equation*, first developed in 1828 by J.B. Bélanger.

Note

For uniform or non-uniform flows, the bed slope S_o and friction slope S_f are defined, respectively, as:

$$S_f = f \frac{1}{D_H} \frac{V^2}{2g} = -\frac{\partial H}{\partial s}$$

$$S_o = \sin \theta = -\frac{\partial z_o}{\partial s}$$

In gradually varied flow, the friction slope and Darcy coefficient are calculated as in uniform equilibrium flow (see Chapter 4) but using the local non-uniform flow depth.

A3.2.2 Extension of the backwater equations

Flat channel of constant width

For a flat channel of rectangular cross-section and constant width, the mean specific energy may be rewritten as a function of the Froude number and flow depth (see Chapter 3):

$$E = d\left(1 + \alpha\frac{1}{2}Fr^2\right) \tag{A3.22}$$

where Fr is the Froude number defined as: $Fr = q/\sqrt{gd^3}$, and q is the discharge per unit width.
   The energy equation (A3.21) leads to:

$$\frac{\partial d}{\partial s}\left(1 - \alpha Fr^2\right) = S_o - S_f \tag{A3.23}$$

Note
For a flat rectangular channel, the Froude number is defined as:

$$Fr = \frac{V}{\sqrt{gd}} = \frac{q}{\sqrt{gd^3}} = \frac{Q}{\sqrt{gB^2 d^3}}$$

Flat channel of non-constant width

Considering a *rectangular* flat channel, the differentiation of equation (A3.19b) with respect to the curvilinear co-ordinate s leads to:

$$\frac{\partial H}{\partial s} = \frac{\partial d}{\partial s} + \frac{\partial z_o}{\partial s} - \frac{\alpha}{g}\frac{Q^2}{B^2 d^2}\left(\frac{1}{B}\frac{\partial B}{\partial s} + \frac{1}{d}\frac{\partial d}{\partial s}\right) \tag{A3.24}$$

Introducing the Froude number $Fr = q/\sqrt{gd^3}$, the energy equation yields:

$$\frac{\partial d}{\partial s}\left(1 - \alpha Fr^2\right) = S_o - S_f + \alpha Fr^2\frac{d}{B}\frac{\partial B}{\partial s} \tag{A3.25}$$

Note
The reader must understand that the above development implies that:

- the variation of the Coriolis coefficient α with respect to s is neglected in first approximation;
- the total discharge Q is constant;
- it is reasonable to assume ($\cos\theta \sim 1$) for a flat channel.

Channel of constant width and non-constant slope

Considering a *rectangular* channel of constant width, the differentiation of equation (A3.19) in the s-direction is:

$$\frac{\partial H}{\partial s} = \frac{\partial d}{\partial s}\cos\theta - d\sin\theta\frac{\partial\theta}{\partial s} + \frac{\partial z_o}{\partial s} - \frac{\alpha}{g}\frac{Q^2}{B^2 d^3}\frac{\partial d}{\partial s} \tag{A3.26}$$

The energy equation (A3.20) may be rewritten as:

$$\frac{\partial d}{\partial s}\left(\cos\theta - \alpha Fr^2\right) = S_o - S_f + d\sin\theta\frac{\partial\theta}{\partial s} \tag{A3.27}$$

where $Fr = q/\sqrt{gd^3}$.

Discussion

Considering an inclined channel (slope θ) and assuming a hydrostatic pressure distribution, the mean specific energy equals:

$$E = d\cos\theta + \frac{1}{2g}\frac{Q^2}{A^2}$$

For a rectangular channel, the specific energy is minimum for:

$$\frac{\partial E}{\partial d} = \cos\theta - \frac{Q^2}{gB^2 d^2} = 0$$

In other words, at critical flow conditions, the Froude number equals:

$$Fr = \frac{q}{\sqrt{gd^3}} = \cos\theta$$

Channel of non-constant width and non-constant slope

For a *rectangular* channel of non-constant width and slope, the differentiation of equation (A3.19) with respect to s leads to:

$$\frac{\partial H}{\partial s} = \frac{\partial d}{\partial s}\cos\theta - d\sin\theta\frac{\partial\theta}{\partial s} + \frac{\partial z_o}{\partial s} - \frac{\alpha}{g}\frac{Q^2}{B^2 d^3}\left(\frac{d}{B}\frac{\partial B}{\partial s} + \frac{\partial d}{\partial s}\right) \tag{A3.28}$$

The energy equation yields to:

$$\frac{\partial d}{\partial s}\left(\cos\theta - \alpha Fr^2\right) = S_o - S_f + d\sin\theta\frac{\partial\theta}{\partial s} + \alpha Fr^2\frac{d}{B}\frac{\partial B}{\partial s} \tag{A3.29}$$

where $Fr = q/\sqrt{gd^3}$.

General case

In the general case of a channel of non-constant width and slope, the differentiation of equation (A3.19) with respect to s leads to:

$$\frac{\partial H}{\partial s} = \frac{\partial d}{\partial s}\cos\theta - d\sin\theta\frac{\partial\theta}{\partial s} + \frac{\partial z_o}{\partial s} - \alpha Fr^2\frac{1}{B}\frac{\partial A}{\partial s} \tag{A3.30}$$

where A is the cross-sectional area, B is the free-surface width and $Fr = Q/\sqrt{gA^3/B}$.

> **Notes**
> 1. Both the cross-sectional area A and the free-surface width B are functions of the flow depth d. Note the relationship: $\partial A / \partial d = B$.
> 2. z_o and d must be taken at the same position: i.e. usually at the deepest point in the channel.
> 3. The definition of the Froude number satisfies minimum specific energy for a flat channel. For a steep channel, critical flow conditions are satisfied for:
>
> $$\frac{\partial E}{\partial d} = \cos \theta - \frac{Q^2 B}{gA^3} = 0$$

*   *   *

A3.3 COMPUTER CALCULATIONS OF BACKWATER PROFILES

A3.3.1 Introduction

Several computer models have been developed to compute backwater profiles. The simplest are one-dimensional models based on the backwater equation (i.e. equation (15.1)). Most of these use a step method/depth calculated from distance: e.g. HydroChan and HEC. More complicated hydraulic models are based on the Saint-Venant equations (see Chapter 16) and they are two dimensional or three dimensional.

The use of computer model requires a sound understanding of the basic equations and a good knowledge of the limitations of the models. It is *uppermost important* that computer program users have a thorough understanding of the hydraulic mechanisms and of the key physical processes that take place in the system (i.e. river and catchment). The writer knows too many 'computer users' who have little idea of the model limitations.

Practically, hydraulic engineers must consider both physical and numerical modelling. Physical modelling assists in understanding the basic flow mechanisms as well as visualizing the flow patterns. In addition, physical model data may be used to calibrate and to verify a numerical model. Once validated, the computational model is a valuable tool to predict a wide range of flow conditions.

A3.3.2 Practical class

Before using a computer program for hydraulic calculations, the users must know the basic principles of hydraulics. In the particular case of backwater profiles, they must further be familiar with the basic equations and they should have seen composite longitudinal profiles (i.e. laboratory models and prototypes).

As part of the hydraulic course at the University of Queensland, the author developed a pedagogic tool to introduce students to backwater calculations. It is a long channel laboratory study which includes a physical model of a composite backwater, a spreadsheet integration of the backwater equations and numerical computations. During the afternoon, students are first introduced to the backwater equation before performing measurements in a 20 m long tilting flume (Fig. A3.1). The model has glass sidewalls and the students can see the basic flow patterns, including the flow around the gates. Then the same flow conditions are input in a spreadsheet, and comparisons between the physical model and computational results are conducted. Later the comparative results are discussed in front of the physical model (Fig. A3.1).

In addition the students are shown an audio-visual technical documentary (NHK, 1989) developed as a pedagogic tool to teach backwater effects in estuaries. Indeed the tidal bore is a

Fig. A3.1 Long channel experiment. Flow from right to left, students standing next to a sluice gate and hydraulic jump located about 5 m upstream of the gate.

visual example of the backwater effects associated with the (tidal) rise of the downstream water level inducing the upstream propagation of the bore.

- Physical model: Physical model characteristics: $Q_{des} = 20\,l/s$, $B = 0.25\,m$, $L_{chan} = 20.215\,m$, $S_o = 0.008$.
- Computer program: Excel spreadsheet, HydroChan Version 1.1.
- Audio-visual documentary: NHK Japan Broadcasting Corp (1989). 'Pororoca: the backward flow of the Amazon.' Videocassette VHS colour, NHK, Japan, 29 min.

A3.3.3 Use of HydroChan

A simple tool for backwater calculations is the program HydroChan, designed as a Windows-based software. The program is based on simple hydraulic calculations and the outputs are easy to use. It is further a shareware program which may be obtained from: HydroTools Software: dwilliams@compusmart.ab.ca.

Inputs
The inputs of the program are the channel cross-section, bed slope, roughness parameters and flow conditions. The channel cross-section consists of a main channel and two side flood plains. The flow resistance may be computed using:

- the Gauckler–Manning formula,
- the Darcy equation (assuming a constant value of f), or the Darcy equation assuming fully rough turbulent flow (the input being the equivalent roughness height).

Outputs
The output includes the critical flow conditions, uniform equilibrium flow conditions and the backwater profile. Basically the main result is the free-surface profile (e.g. Fig. A3.2). A graphical solution compares the computed free-surface profile, critical depth and normal depth in the barrel.

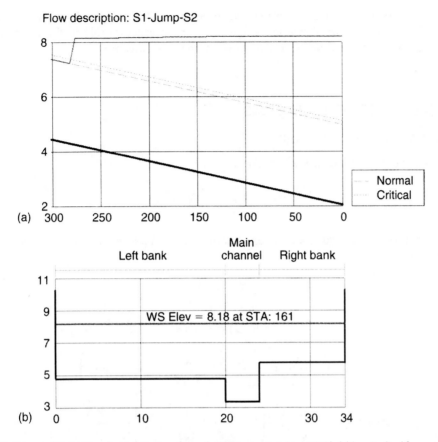

Fig. A3.2 Output of HydroChan (calculations performed with Version 1.1, May 2001): (a) longitudinal free-surface profile and (b) transverse cross-section at the downstream end.

Discussion

Flow resistance calculations in open channels must be performed in term of the *Darcy friction factor*. The calculations depend *critically* upon the flow resistance calculations and assumed equivalent roughness height.

The software HydroChan was designed for prismatic cross-sections and fixed bed.

A3.3.4 Application

A simple application is a prismatic channel with two lateral flood plains. The calculation input are as follows.

HydroChan parameter	Left flood plain	Main channel	Right flood plain
Cross-sectional shape:	Rectangular	Rectangular	Rectangular
k_s (m):	0.1	0.01	0.02
B (m):	20	4	10
z_o (m RL):	3.48	2.048	4.48

For a flow rate of $200\,\text{m}^3/\text{s}$ and a tailwater level of $8.2\,\text{m}$ RL, the computed results are presented in Fig. A3.2. Figure A3.2a shows the longitudinal free-surface profile in the main channel. Figure A3.2b presents the cross-sectional profile at the downstream end.

*   *   *

A3.4 GAUSSIAN ERROR FUNCTIONS

A3.4.1 Gaussian error function

The Gaussian error function erf is defined as:

$$\text{erf}(u) = \frac{2}{\sqrt{\pi}} \int_0^u \exp(-\tau^2)\,d\tau \tag{A3.31}$$

Values of the Gaussian error function are summarized in Table A3.1. Basic properties of the function are:

$$\text{erf}(0) = 0 \tag{A3.32}$$

$$\text{erf}(+\infty) = 1 \tag{A3.33}$$

$$\text{erf}(-u) = -\text{erf}(u) \tag{A3.34}$$

$$\text{erf}(u) = \frac{1}{\sqrt{\pi}}\left(u - \frac{u^3}{3 \times 1!} + \frac{u^5}{5 \times 2!} - \frac{u^7}{7 \times 3!} + \cdots\right) \tag{A3.35}$$

$$\text{erf}(u) \approx 1 - \frac{\exp(-u^2)}{\sqrt{\pi}\,u}\left(1 - \frac{1}{2u^2} + \frac{1 \times 3}{(2x^2)^2} - \frac{1 \times 3 \times 5}{(2x^2)^3} + \cdots\right) \tag{A3.36}$$

where $n! = 1 \times 2 \times 3 \cdots \times n$.

Notes

In first approximation, the function $\text{erf}(u)$ may be correlated by:

$$\text{erf}(u) \approx u(1.375511 - 0.61044u + 0.088439u^2) \qquad 0 \leqslant u < 2$$

$$\text{erf}(u) \approx \tan h(1.198787u) \qquad -\infty < u < +\infty$$

In many applications, the above correlations are *not accurate enough*, and equation (A3.35) and Table A3.1 should be used.

For small values of u, the error function is about:

$$\text{erf}(u) \approx \frac{u}{\sqrt{\pi}} \qquad \text{Small values of } u$$

For large values of u, the erf function is about:

$$\text{erf}(u) \approx 1 - \frac{\exp(-u^2)}{u\sqrt{\pi}} \qquad \text{Large values of } u$$

Plate 17 Photographs of alluvial streams and bed load material: Turkey River, east of Osterdock, IA (USA) in August 1966 (courtesy of Dr Lou Maher). Looking downstream along a meandering section.

Plate 18 Peinan river, Taitung, Taiwan in December 1998. Looking downstream. Note the amount of bed load material.

Plate 19 Dry reservoir of Koorawatha weir, railway dam completed in 1911. View from the right bank in December 1997; 9.1 m high thin arch wall, reservoir capacity: 4×10^4 m^3. Reservoir siltation by bed load predominantly.

Plate 20 Eroded badlands in South Dakota in April 1966 (courtesy of Dr Lou Maher). Note the grass terrace on the right.

Plate 21 Extreme reservoir siltation caused by hyper-concentrated flows: Le Saignon dam (France), June 1998. Dam height: 14.5 m (completed in 1961), reservoir capacity: $1.4 \times 10^5 \text{m}^3$, catchment area: 3.5km^2. Fully silted reservoir after 2 years of operation. View from the right dam abutment: the dam wall being on the right edge of the photograph, and the silted intake tower in the centre.

Plate 22 Anti-dunes in a sunny day, looking downstream (Takatoyo beach, 24 April 1999).

Plate 23 The intricacy of open channel hydraulics: (a) Water crossing at Bridge No. 2, National Road No. 6 (Cambodia): dry channel on March 2002 (courtesy of Peter Ward). (b) The same water crossing on September 2002: flood discharge of about 2000 m^3/s (courtesy of Peter Ward).

Plate 24 Mississippi River in flood on 27 April 1965 (courtesy of Dr Lou Maher). Partially submerged houses north of Guttenburg IA, view to West.

Plate 25 Hydraulic jump roller in a steep stream in München, English garden on June 1997 (courtesy of Mr Dale Young). Flow from left to right.

(a)

(b)

Plate 26 Remains of the Malpasset dam: a mistake not to repeat! Reservoir characteristics: dam height: 102.5 m, volume: 50 Mm³. Dam break on 2 December 1959. (a) Photograph taken in December 1981, view from upstream looking at the right abutment. (b) View from upstream looking at the right abutment on December 1981.

Plate 27 An example of stepped waterways susceptible to be subjected to rapid flood waves. Stepped river training on the Oyana–gawa (Japan) on 1 November 2002. The Oyana River is a tributary of the Fuji gawa.

Plate 28 Photograph of culvert operation in Brisbane (Australia) discharging about 60–80 m³/s in December 2001. Outlet operation, flow from right to left. The road is overtopped few times per year as a result of improper design.

Plate 29 Neil Turner weir near Mitchell QLD (Australia) in July 2001 (courtesy of Chris Proctor). Completed in 1984 on the Maranoa river, dam height: 5.8 m, storage capacity: 2 Mm³ and stepped weir overflow equipped with five steps.

Plate 30 Opuha dam spillway, New Zealand (courtesy of Tonkin and Taylor). Dam height: 47 m (completed in 1999), reservoir capacity: $85 \times 10^6 \, \text{m}^3$, stepped chute ($\theta = 27°$) and hydropower capacity: 7.6 MW. Small overflow down the stepped chute with stilling basin in foreground and embankment dam in background right.

Plate 31 Morning-Glory spillway intake (Chaffey dam NSW, 1979) (June 1997). Design spillway capacity: 800 m^3/s.

Plate 32 MEL spillway inlet at Lake Kurwongbah on 12 September 1999 (structure MEL-SP-1).

Table A3.1 Values of the error function erf

u	erf(u)	u	erf(u)
0	0	1	0.8427
0.1	0.1129	1.2	0.9103
0.2	0.2227	1.4	0.9523
0.3	0.3286	1.6	0.9763
0.4	0.4284	1.8	0.9891
0.5	0.5205	2	0.9953
0.6	0.6309	2.5	0.9996
0.7	0.6778	3	0.99998
0.8	0.7421	$+\infty$	1
0.9	0.7969		

A3.4.2 Complementary error function

The complementary Gaussian error function erfc is defined as:

$$\text{erfc}(u) = 1 - \text{erf}(u) = \frac{2}{\sqrt{\pi}} \int_{u}^{+\infty} \exp(-\tau^2)\,\mathrm{d}\tau \tag{A3.37}$$

Basic properties of the complementary function include:

$$\text{erfc}(0) = 1 \tag{A3.38}$$

$$\text{erfc}(+\infty) = 0 \tag{A3.39}$$

Part 4 Design of Hydraulic Structures

Introduction to the design of hydraulic structures

18

18.1 INTRODUCTION

The *storage of water* is essential for providing Man with drinking water and irrigation water reserves. Storage along a natural stream is possible only if the hydrology of the catchment area is suitable. Hydrological studies provide information on the storage capability and as well as on the maximum (peak) flow in the system. Often the hydrology of a stream does not provide enough supply all year round, and an artificial water storage system (e.g. the reservoir behind a dam) must be developed.

Once a source of fresh water is available (e.g. water storage and rainfall runoff), water must be carried to the location of its use. The *conveyance of water* takes place in channels and pipes. In an open channel, an important difference is the propulsive force acting in the direction of the flow: the fluid is driven by the weight of the flowing water resolved down a slope. The course of open channels derives from the local topography and it may include drops (and cascades) and tunnels underneath embankments (e.g. roads).

18.2 STRUCTURE OF PART 4

Part 4 of the book presents an introduction to the hydraulic design of structures: weirs and small dams (Chapter19), drops and cascades (Chapter 20) and culverts (Chapter 21). The former deals with the storage of water while the latter chapters deal with the conveyance of water.

Weirs and dams are used to store water. A main hydraulic feature is the spillway system, used to safely release flood waters above the structure.

Drops and cascades may be used as an alternative to spillway structures. They are used also in waterways built in steep relief to dissipate the kinetic energy of the flow. Culverts are designed to pass water underneath embankments (e.g. railroad and highway).

18.3 PROFESSIONAL DESIGN APPROACH

This lecture material presents the application of the basic hydraulic principles to real design situations (Plate 30). The design approach is based on a system approach. A hydraulic structure must be analysed as part of the surrounding catchment and the hydrology plays an important role. Structural and hydraulic constraints interact, and the design of a hydraulic structure is a complex exercise altogether.

For example, the design of a culvert requires a hydrological study of a stream to estimate the maximum (design) discharge and to predict the risks of exceptional (emergency) floods. The dimensions of the culvert are based on hydraulic and structural considerations, as well as geotechnical matter as the culvert height and width affect the size and cost of the embankment. Furthermore, the impact of the culvert on the environment must be taken into account: e.g. potential flooding of the upstream plain and impact of road overtopping (Plate 28).

Another example is the construction of a weir across a river (Plates 8 and 29). First, the stream hydrology and the catchment characteristics must be studied. If the catchment can provide enough water all the year around (i.e. mean annual characteristics), the maximum peak inflow must be predicted (i.e. extreme events). The design of the weir is based upon structural, geotechnical and hydraulic considerations. Political matters might also affect the weir site location and the decision to build the dam. A consequent cost of the structure is the spillway, designed to safely pass the maximum peak flood. In addition, the impact of the weir on the upstream and downstream valleys must be considered: e.g. sediment trap, fish migration, downstream water quality (e.g. dissolved oxygen content), modifications of the water table and associated impacts (e.g. salinity).

The design process must be a *system approach*. First, the system must be identified. What are the design objectives? What are the constraints? What is the range of options? What is the 'best choice'? Its detailed analysis must be conducted. The engineers should ask: is this solution really satisfactory?

Design of weirs and spillways   19

19.1 INTRODUCTION

19.1.1 Definitions: dams and weirs

Dams and weirs are hydraulic structures built across a stream to facilitate the storage of water.

A *dam* is defined as a large structure built across a valley to store water in the upstream reservoir. All flows up to the probable maximum flood must be confined to the designed spillway. The upstream water level should not overtop the dam wall. Dam overtopping may indeed lead to dam erosion and possibly destruction.

A conventional *weir* is a structure designed to rise the upstream water level: e.g. for feeding a diversion channel. Small flow rates are confined to a spillway channel. Larger flows are allowed to pass over the top of the full length of the weir. At the downstream end of the weir, the kinetic energy of the flow is dissipated in a dissipator structure (Figs 19.1a and 19.2a and Plate 30).

Another type of weirs is the *minimum energy loss (MEL) weir* (Figs 19.1b and 19.2b). MEL weirs are designed to minimize the total head loss of the overflow and hence to induce (ideally) zero afflux. MEL weirs are used in flat areas and near estuaries (see Appendix A4.2).

Practically, the differences between a small dam and a conventional weir are small, and the terms 'weir' or 'small dam' are often interchanged.

19.1.2 Overflow spillway

During large rainfall events, a large amount of water flows into the reservoir, and the reservoir level may rise above the dam crest. A *spillway* is a structure designed to 'spill' flood waters under controlled (i.e. *safe*) conditions. Flood waters can be discharged beneath the dam (e.g. culvert and bottom outlet), through the dam (e.g. rockfill dam) or above the dam (i.e. overflow spillway).

Most small dams are equipped with an overflow structure (called spillway) (e.g. Fig. 19.3). An overflow spillway includes typically three sections: a crest, a chute and an energy dissipator at the downstream end. The *crest* is designed to maximize the discharge capacity of the spillway. The *chute* is designed to pass (i.e. to carry) the flood waters above (or away from) the dam, and the *energy dissipator* is designed to dissipate (i.e. 'break down') the kinetic energy of the flow at the downstream end of the chute (Figs 19.1a and 19.2).

A related type of spillway is the drop structure. As its hydraulic characteristics differ significantly from that of standard overflow weirs, it will be presented in another chapter.

Notes
1. Other types of spillways include the Morning-Glory spillway (or bellmouth spillway) (see Plate 31, the Chaffey dam). It is a vertical discharge shaft, more particularly the circular hole form of a drop inlet spillway, leading to a conduit underneath the dam (or abutment). The shape of the intake is similar to a Morning-Glory flower. It is sometimes called a Tulip intake. The Morning-Glory spillway is not recommended for discharges usually greater than $80\,\text{m}^3/\text{s}$.

2. Examples of energy dissipators include stilling basin, dissipation basin, flip bucket followed by downstream pool and plunge pool (Fig. 19.3).
3. At an MEL weir, the amount of energy dissipation is always small (if the weir is properly designed) and no stilling basin is usually required. The weir spillway is curved in plan, to concentrate the energy dissipation near the channel centreline and to avoid bank erosion (Fig. 19.2b and Appendix A4.2).
4. MEL weirs may be combined with culvert design, especially near the coastline to prevent salt intrusion into freshwater waterways without upstream flooding effect. An example is the MEL weir built as the inlet of the Redcliffe MEL structure (Appendix A4.3).

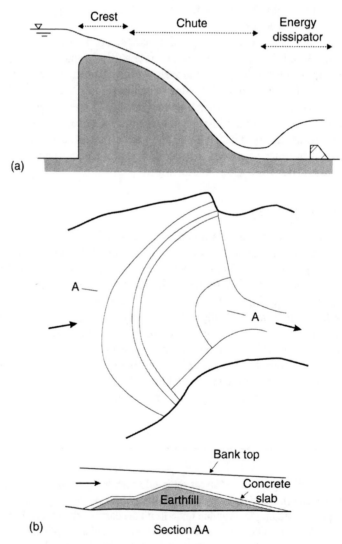

Fig. 19.1 Sketch of weirs: (a) conventional weir and (b) MEL weir.

19.1.3 Discussion

Although a spillway is designed for specific conditions (i.e. design conditions: Q_{des} and H_{des}), it must operate safely and efficiently for a range of flow conditions.

Design engineers typically select the optimum spillway shape for the design flow conditions. They must then verify the safe operation of the spillway for a range of operating flow conditions (e.g. from $0.1 Q_{des}$ to Q_{des}) and for the emergency situations (i.e. $Q > Q_{des}$).

In the following sections, we present first the crest calculations, then the chute calculations followed by the energy dissipator calculations. Later the complete design procedure is described.

(a)

(b)

Fig. 19.2 Examples of spillway operation: (a) Diversion weir at Dalby QLD, Australia on 8 November 1997. Ogee crest followed by smooth chute and energy dissipator (note fishway next to right bank). (b) Overflow above an MEL weir: Chinchilla weir at low overflow on 8 November 1997. Design flow conditions: 850 m^3/s, weir height: 14 m and reservoir capacity: 9.78×10^6 m^3.

(a)

(b)

(c)

Fig. 19.3 Examples of spillway design: (a) Overflow spillway with downstream stilling basin (Salado 10 Auxiliary spillway) (courtesy of USDA natural resources conservation service). (b) Overflow spillway with downstream flip bucket (Reece dam TAS, 1986) (courtesy of Hydro-Electric Commission Tasmania). Design spillway capacity: 4740 m^3/s, overflow event: 365 m^3/s. (c) Stepped spillway chute (Loyalty Road, Australia 1996) (courtesy of Mr Patrick James). Design spillway capacity: 1040 m^3/s, step height: 0.9 m, chute slope: 51° and chute width: 30 m.

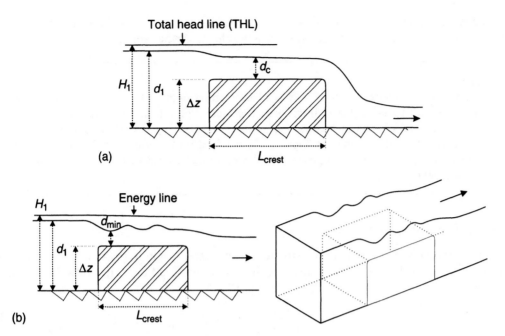

Fig. 19.4 Flow pattern above a broad-crested weir: (a) broad-crested weir flow and (b) undular weir flow.

19.2 CREST DESIGN

19.2.1 Introduction

The crest of an overflow spillway is usually designed to maximize the discharge capacity of the structure: i.e. to pass safely the design discharge at the lowest cost.

In open channels and for a given specific energy, maximum flow rate is achieved for critical flow conditions[1] (Bélanger, 1828). For an ideal fluid overflowing a weir (rectangular cross-section) and assuming hydrostatic pressure distribution, the maximum discharge per unit width may be deduced from the continuity and Bernoulli equations:

$$q = \sqrt{g} \left(\frac{2}{3}(H_1 - \Delta z) \right)^{3/2} \qquad \text{Ideal fluid flow} \qquad (19.1)$$

where g is the gravity acceleration, H_1 is the upstream total head and Δz is the weir height (e.g. Fig. 19.4). In practice the observed discharge differs from equation (19.1) because the pressure distribution on the crest may not be hydrostatic. Furthermore, the weir geometry, roughness and inflow conditions affect the discharge characteristics (e.g. Miller, 1994). The real flow rate is expressed as:

$$q = C_D \sqrt{g} \left(\frac{2}{3}(H_1 - \Delta z) \right)^{3/2} \qquad (19.2)$$

where C_D is the discharge coefficient.

---

[1] Flow conditions for which the specific energy (of the mean flow) is minimum are called critical flow conditions.

The most common types of overflow weirs are the broad-crested weir, the sharp-crested weir and the ogee-crest weir. Their respective discharge characteristics are described below.

Notes

1. Jean-Baptiste Bélanger (1789–1874) was a French professor at the Ecole Nationale Supérieure des Ponts et Chaussées (Paris). In his book (Bélanger, 1828), he first presented the basics of hydraulic jump calculations, backwater calculations and discharge characteristics of weirs.
2. C_D is a dimensionless coefficient. Typically $C_D = 1$ for a broad-crested weir. When $C_D > 1$, the discharge capacity of a weir is greater than that of a broad-crested weir for identical upstream head above crest ($H_1 - \Delta z$).

19.2.2 Broad-crested weir

A broad-crested weir is a flat-crested structure with a crest length large compared to the flow thickness (Fig. 19.4). The ratio of crest length to upstream head over crest must be typically greater than 1.5–3 (e.g. Chow, 1973; Henderson, 1966):

$$\frac{L_{crest}}{H_1 - \Delta z} > 1.5\text{--}3 \tag{19.3}$$

When the crest is 'broad' enough for the flow streamlines to be parallel to the crest, the pressure distribution above the crest is hydrostatic and the critical flow depth is observed on the weir crest. Broad-crested weirs are sometimes used as critical depth meters (i.e. to measure stream discharges). The hydraulic characteristics of broad-crested weirs were studied during the 19th and 20th centuries. Hager and Schwalt (1994) recently presented an authoritative work on the topic.

The discharge above the weir equals:

$$q = \frac{2}{3}\sqrt{\frac{2}{3}g(H_1 - \Delta z)^3} \qquad \text{Ideal fluid flow calculations} \tag{19.1a}$$

where H_1 is the upstream total head and Δz is the weir height above channel bed (Fig. 19.4). Equation (19.1a) may be rewritten conveniently as:

$$q = 1.704(H_1 - \Delta z)^{3/2} \qquad \text{Ideal fluid flow calculations} \tag{19.1b}$$

Notes

1. In a horizontal rectangular channel and assuming hydrostatic pressure distribution, the critical flow depth equals:

$$d_c = \frac{2}{3}E \qquad \text{Horizontal rectangular channel}$$

where E is the specific energy. The critical depth and discharge per unit width are related by:

$$d_c = \sqrt[3]{\frac{q^2}{g}} \qquad \text{Rectangular channel}$$

$$q = \sqrt{gd_c^{\,3}} \qquad \text{Rectangular channel}$$

---

2. At the crest of a broad-crested weir, the continuity and Bernoulli equations yield:

$$H_1 - \Delta z = \frac{3}{2} d_c$$

Note that equation (19.1) derives from the continuity and Bernoulli equations, hence from the above equation.

Discussion

(A) *Undular weir flow*
For low discharges (i.e. $(d_1 - \Delta z)/\Delta z \ll 1$), several researchers observed free-surface undulations above the crest of broad-crested weir (Fig. 19.4b). Model studies suggest that undular weir flow occurs for:

$$\frac{q}{\sqrt{g d_{min}^3}} < 1.5 \qquad \text{Undular weir flow}$$

where d_{min} is the minimum flow depth upstream of the first wave crest (Fig. 19.4). Another criterion is:

$$\frac{H_1 - \Delta z}{L_{crest}} < 0.1 \qquad \text{Undular weir flow}$$

where L_{crest} is the crest length in the flow direction. The second equation is a practical criterion based on the ratio head on crest to weir length.

In practice, design engineers should avoid flow conditions leading to undular weir flow. In the presence of free-surface undulations above the crest, the weir cannot be used as a discharge meter, and waves may propagate in the downstream channel.

(B) *Discharge coefficients*
Experimental measurements indicate that the discharge versus total head relationship departs slightly from equation (19.1) depending upon the weir geometry and flow conditions. Equation (19.1) is usually rearranged as:

$$q = C_D \frac{2}{3} \sqrt{\frac{2}{3} g (H_1 - \Delta z)^3}$$

where the discharge coefficient C_D is a function of the weir height, crest length, crest width, upstream corner shape and upstream total head (Table 19.1).

19.2.3 Sharp-crested weir

A sharp-crested weir is characterized by a thin sharp-edged crest (Fig. 19.5). In the absence of sidewall contraction, the flow is basically two dimensional and the flow field can be solved by analytical and graphical methods: i.e. ideal fluid flow theory (e.g. Vallentine, 1969; p. 79).

For an aerated nappe, the discharge per unit width is usually expressed as:

$$q = \frac{2}{3} C \sqrt{2 g (d_1 - \Delta z)^3} \tag{19.4}$$

where d_1 is the upstream water depth, Δz is the crest height above channel bed (Fig. 19.5) and C is a dimensionless discharge coefficient. Numerous correlations were proposed for C

Table 19.1 Discharge coefficient for broad-crested weirs

Reference (1)	Discharge coefficient C_D (2)	Range (3)	Remarks (4)
Sharp-corner weir Hager and Schwalt (1994)	$0.85\dfrac{9}{7}\left(1 - \dfrac{2/9}{1 + \left((H_1 - \Delta z)/L_{\text{crest}}\right)^4}\right)$	$0.1 < \dfrac{H_1 - \Delta z}{L_{\text{crest}}} < 1.5$	Deduced from laboratory experiments
Rounded-corner weir Bos (1976)	$\left(1 - 0.01\dfrac{L_{\text{crest}} - r}{W}\right)\left(1 - 0.01\dfrac{L_{\text{crest}} - r}{d_1 - \Delta z}\right)$	$\dfrac{d_1 - Dz}{L_{\text{crest}}} > 0.05$ $d_1 - \Delta z > 0.06\,\text{m}$ $\dfrac{H_1 - \Delta z}{\Delta z} < 1.5$ $\Delta z > 0.15\,\text{m}$	Based upon laboratory and field tests
Ackers *et al.* (1978)	0.95^{a}	$0.15 < \dfrac{H_1 - \Delta z}{\Delta z} < 0.6$	

Notes: [a]Re-analysis of experimental data presented by Ackers *et al.* (1978); *r*: curvature radius of upstream corner.

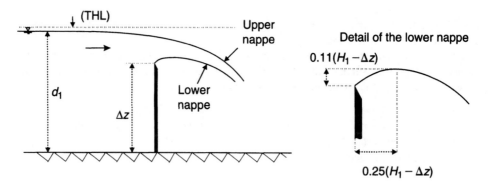

Fig. 19.5 Sharp-crested weir.

(Tables 19.2 and 19.3). In practice, the following expression is recommended:

$$C = 0.611 + 0.08\frac{d_1 - \Delta z}{\Delta z} \tag{19.5}$$

Notes
1. Sharp-crested weirs are very accurate discharge meters. They are commonly used for small flow rates.
2. For a vertical sharp-crested weir, the lower nappe is deflected upwards immediately downstream of the sharp edge. The maximum elevation of the lower nappe location occurs at about $0.11(H_1 - \Delta z)$ above the crest level in the vertical direction and at about $0.25(H_1 - \Delta z)$ from the crest in the horizontal direction (e.g. Miller, 1994).

Table 19.2 Discharge coefficient for sharp-crested weirs (full-width weir in rectangular channel)

Reference (1)	Discharge coefficient C (2)	Range (3)	Remarks (4)
von Mises (1917)	$\dfrac{\pi}{\pi + 2}$	$\dfrac{d_1 - \Delta z}{\Delta z}$ very large	Ideal fluid flow calculations of orifice flow
Henderson (1966)	$0.611 + 0.08\dfrac{d_1 - \Delta z}{\Delta z}$	$0 \leq \dfrac{d_1 - \Delta z}{\Delta z} < 5$	Experimental work by Rehbock (1929)
	1.135	$\dfrac{d_1 - \Delta z}{\Delta z} = 10$	
	$1.06\left(1 + \dfrac{\Delta z}{d_1 - \Delta z}\right)^{3/2}$	$20 < \dfrac{d_1 - \Delta z}{\Delta z}$	
Bos (1976)	$0.602 + 0.075\dfrac{d_1 - \Delta z}{\Delta z}$	$d_1 - \Delta z > 0.03\,\mathrm{m}$ $\dfrac{d_1 - \Delta z}{\Delta z} < 2$ $\Delta z > 0.40\,\mathrm{m}$	Based on experiments performed at Georgia Institute of Technology
Chanson (1999)	1.0607	$\Delta z = 0$	Ideal flow at overfall

Table 19.3 Discharge correlations for sharp-crested weirs (full width in rectangular channel)

Reference (1)	Discharge per unit width q (m²/s) (2)	Comments (3)
Ackers et al. (1978)	$0.564\left(1 + 0.150\dfrac{d_1 - \Delta z}{\Delta z}\right)\sqrt{g}\,(d_1 - \Delta z + 0.0001)^{3/2}$	Range of applications $d_1 - \Delta z > 0.02\,\mathrm{m}$ $\Delta z > 0.15$ $(d_1 - \Delta z)/\Delta z < 2.2$
Herschy (1995)	$1.85(d_1 - \Delta z)^{3/2}$	Approximate correlation ($\pm 3\%$): $(d_1 - \Delta z)/\Delta z < 0.5$

3. At very low flow rates, $(d_1 - \Delta z)/\Delta z$ is very small and equation (19.4) tends to:

$$q \sim 1.803(d_1 - \Delta z)^{3/2} \quad \text{Very small discharge}$$

4. Nappe aeration is extremely important. If the nappe is not properly ventilated, the discharge characteristics of the weir are substantially affected, and the weir might not operate safely. Sometimes, the crest can be contracted at the sidewalls to facilitate nappe ventilation (e.g. Henderson, 1966; pp. 177–178).

19.2.4 Ogee-crest weir

The basic shape of an ogee crest is that of the lower nappe trajectory of a sharp-crested weir flow for the design flow conditions (discharge Q_{des} and upstream head H_{des}) (Figs 19.2a and 19.5–19.8).

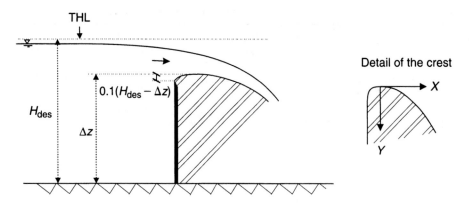

Fig. 19.6 Sketch of a nappe-shaped ogee crest.

(a)

Fig. 19.7 Spillway chute of Hinze dam (Gold Coast QLD, 1976). Dam height: 44 m and design spillway capacity: 1700 m³/s. (a) Ogee-crest chute in September 1997 (view from downstream, on right bank).

Fig. 19.7 (b) Concrete veins at downstream end of chute, directing the supercritical chute flow into the energy dissipator (looking upstream on September 2002). (c) Spillway in operation in the early 1990s for $Q \sim 300$–$400\,\mathrm{m^3/s}$ (courtesy of Gold Coast City Council). View from the left bank, with turning veins in foreground (underwater). Flow from right to left.

Nappe-shaped overflow weir

The characteristics of a nappe-shaped overflow can be deduced from the equivalent 'sharp-crested weir' design. For the design head H_{des}, equation (19.4) leads to:

$$q_{\mathrm{des}} = \frac{2}{3} C \sqrt{g \left(\frac{H_{\mathrm{des}} - \Delta z}{0.89} \right)^3} \tag{19.6}$$

where Δz is the crest elevation (Fig. 19.6). For a high weir (i.e. $(H_{\mathrm{des}} - \Delta z)/\Delta z \ll 1$), the discharge above the crest could be deduced from equations (19.5) and (19.6):

$$q_{\mathrm{des}} = 2.148(H_{\mathrm{des}} - \Delta z)^{3/2} \qquad \text{Ideal flow conditions} \tag{19.7}$$

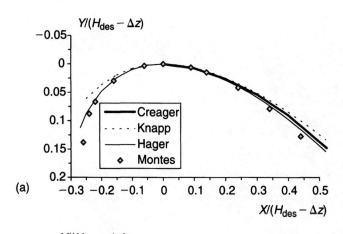

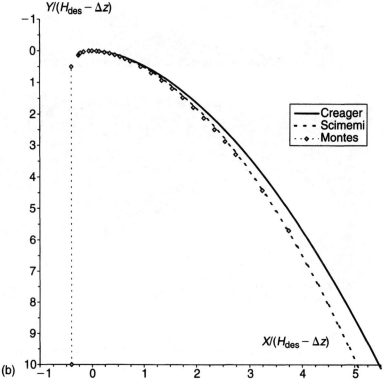

Fig.19.8 Profiles of ogee crest: (a) crest details and (b) Creager and Scimemi profiles.

Discharge characteristics of an ogee crest

In practice, the discharge–head relationship differs from ideal flow conditions (equation (19.7)). For the design flow conditions, the flow rate per unit width is often presented as:

$$q_{des} = C_{des}(H_{des} - \Delta z)^{3/2} \qquad \{\text{Design flow conditions: } q = q_{des}\} \qquad (19.8)$$

where the design discharge coefficient C_{des} is primarily a function of the ogee-crest shape. It can also be deduced from model tests.

Table 19.4 Discharge coefficient of ogee crest (vertical-faced ogee crest)

$\Delta z/(H_{des} - \Delta z)$ (1)	C_{des} (m$^{1/2}$/s) (2)	$(C_D)_{des}$ (3)	Comments (4)
0	1.7	1.0	Broad-crested weir
0.04	1.77	1.04	
0.14	1.92	1.13	
0.36	2.07	1.21	
0.50	2.10	1.23	
1.0	2.14	1.257	
1.5	2.16	1.268	
3.0	2.18	1.28	
>3.0	2.19	1.28	Large weir height

Reference: US Bureau of Reclamation (1987).
Notes: $C_{des} = q_{des}/(H_{des} - \Delta z)^{3/2}$; $(C_D)_{des} = q_{des}/(\sqrt{g}\ (2/3)^{3/2}(H_{des} - \Delta z)^{3/2}$.

For vertical-shaped ogee crest, typical values of the discharge coefficient are reported in Table 19.4 and Fig. 19.9a. In Table 19.4, the broad-crested weir case corresponds to $\Delta z/(H_{des} - \Delta z) = 0$ and it is found $(C_D)_{des} = 1$ (equation (19.1)). For a high weir (i.e. $\Delta z/(H_{des} - \Delta z) > 3$), the discharge coefficient C_{des} tends to 2.19 m$^{1/2}$/s. Such a value differs slightly from equation (19.7) but it is more reliable because it is based upon experimental results.

For a given crest profile, the overflow conditions may differ from the design flow conditions. The discharge versus upstream head relationship then becomes:

$$q = C(H_1 - \Delta z)^{3/2} \quad \{\text{Non-design flow conditions: } q \neq q_{des}\} \quad (19.9)$$

in which the discharge coefficient C differs from the design discharge coefficient C_{des}.

Generally, the relative discharge coefficient C/C_{des} is a function of the relative total head $(H_1 - \Delta z)/(H_{des} - \Delta z)$ and of the ogee-crest shape (e.g. Fig. 19.9b).

Discussion

For $H_1 = H_{des}$ the pressure on the crest invert is atmospheric (because the shape of the invert is based on the lower nappe trajectory of the sharp-crested weir overflow).

When the upstream head H_1 is larger than the design head H_{des}, the pressures on the crest are less than atmospheric and the discharge coefficient C (equation (19.9)) is larger than the design discharge coefficient C_{des} (typically 2.19 m$^{1/2}$/s). For $H_1 < H_{des}$, the pressures on the crest are larger than atmospheric and the discharge coefficient is smaller. At the limit, the discharge coefficient tends to the value of 1.704 m$^{1/2}$/s corresponding to the broad-crested weir case (equation (19.1b)) (Table 19.5).

Standard crest shapes

With nappe-shaped overflow weirs, the pressure on the crest invert should be atmospheric at design head. In practice, small deviations occur because of bottom friction and developing boundary layer. Design engineers must select the shape of the ogee crest such that sub-atmospheric pressures are avoided on the crest invert: i.e. to prevent separation and cavitation-related problems. Several ogee-crest profiles were developed (Table 19.6).

The most usual profiles are the WES profile and the Creager profile. The Creager design is a mathematical extension of the original data of Bazin in 1886–1888 (Creager, 1917). The WES-standard ogee shape is based upon detailed observations of the lower nappe of sharp-crested weir flows (Scimemi, 1930) (Figs 19.5 and 19.8).

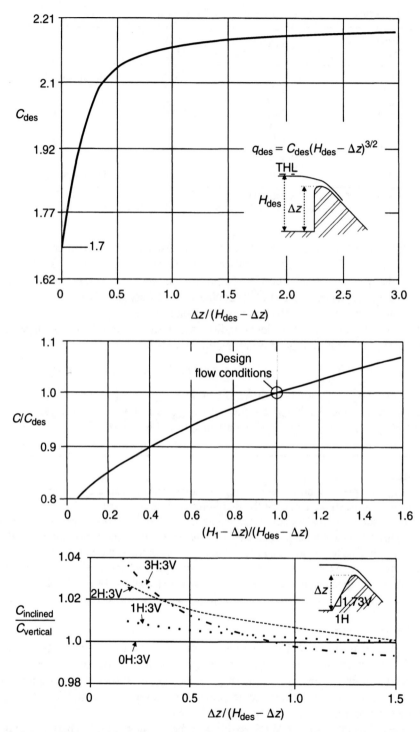

Fig. 19.9 Discharge coefficient of a USBR-profile ogee crest (data: US Bureau of Reclamation, 1987).

Table 19.5 Pressures on an ogee crest invert for design and non-design flow conditions

Upstream head (1)	Pressure on crest (2)	Discharge coefficient (3)
$H_1 = H_{des}$	Quasi-atmospheric	$C = C_{des}$
$H_1 > H_{des}$	Less than atmospheric	$C > C_{des}$
$H_1 < H_{des}$	Larger than atmospheric	$C < C_{des}$
$H1 \ll H_{des}$	Larger than atmospheric	$C \approx 1.704 \, \mathrm{m}^{1/2}/\mathrm{s}$

Table 19.6 Examples of spillway profiles (vertical-faced ogee crest)

Profile (1)	Equations (2)	Comments (3)
Creager (1917) profile	$Y = 0.47 \dfrac{X^{1.80}}{(H_{des} - \Delta z)^{0.80}}$	For $X \geqslant 0$; derived from Bazin's (1888–1898) experiments
Scimemi (1930) profile	$Y = 0.50 \dfrac{X^{1.85}}{(H_{des} - \Delta z)^{0.85}}$	For $X \geqslant 0$; also called WES profile
Knapp (1960)	$\dfrac{Y}{H_{des} - \Delta z} = \dfrac{X}{H_{des} - \Delta z}$ $- \ln\left(1 + \dfrac{X}{0.689(H_{des} - \Delta z)}\right)$	Continuous spilway profile for crest region only (as given by Montes, 1992a)
Hager (1991)	$\dfrac{Y}{H_{des} - \Delta z} = 0.1360$ $+ 0.482625\left(\dfrac{X}{H_{des} - \Delta z} + 0.2818\right)$ $\times \ln\left(1.3055\left(\dfrac{X}{H_{des}\Delta z} + 0.2818\right)\right)$	Continuous spilway profile with continuous curvature radius: $-0.498 < \dfrac{X}{H_{des} - \Delta z} < 0.484$
Montes (1992a)	$\dfrac{R_1}{H_{des} - \Delta z} = 0.05 + 1.47 \dfrac{s}{H_{des} - \Delta z}$	Continuous spillway profile with continuous curvature radius R. Lower asymptote: i.e. for small values of $s/(H_{des} - \Delta z)$
	$\dfrac{R}{H_{des} - \Delta z} = \dfrac{R_1}{H_{des} - \Delta z}\left(1 + \left(\dfrac{R_u}{R_1}\right)^{2.625}\right)^{1/2.625}$	Smooth variation between the asymptotes
	$\dfrac{R_u}{H_{des} - \Delta z} = 1.68\left(\dfrac{s}{H_{des} - \Delta z}\right)^{1.625}$	Upper asymptote: i.e. for large values of $s/(H_{des} - \Delta z)$

Notes: X, Y: horizontal and vertical co-ordinates with dam crest as origin, Y measured positive downwards (Fig. 19.8); R: radius of curvature of the crest; s: curvilinear co-ordinate along the crest shape.

Montes (1992a) stressed out that the ogee-crest profile must be continuous and smooth, and sudden variation of the crest curvature must be avoided to prevent unwanted aeration or cavitation. Ideally, the crest profile should start tangentially to the upstream apron, with a smooth and continuous variation of the radius of curvature.

19.3 CHUTE DESIGN

19.3.1 Presentation

Once the water flows past the crest, the fluid is accelerated by gravity along the chute. At the upstream end of the chute, a turbulent boundary layer is generated by bottom friction and develops in the flow direction. When the outer edge of the boundary layer reaches the free surface, the flow becomes fully developed (Fig. 19.10).

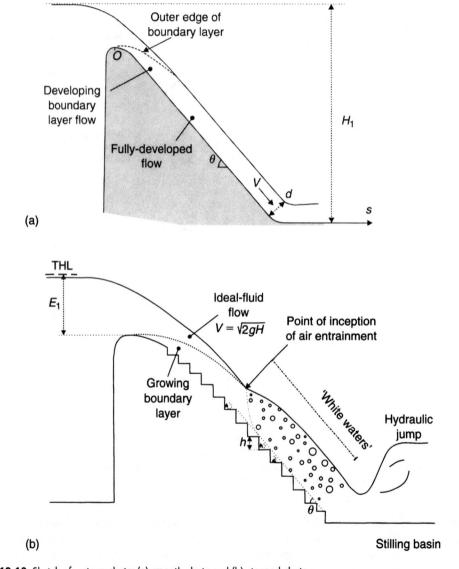

Fig. 19.10 Sketch of a steep chute: (a) smooth chute and (b) stepped chute.

In the developing flow region, the boundary layer thickness δ increases with distance along the chute. Empirical correlations may be used to estimate the boundary layer growth:

$$\frac{\delta}{s} = 0.0212(\sin\theta)^{0.11}\left(\frac{s}{k_s}\right)^{-0.10} \qquad \text{Smooth concrete chute } (\theta > 30°) \qquad (19.10)$$

$$\frac{\delta}{s} = 0.06106(\sin\theta)^{0.133}\left(\frac{s}{h\cos\theta}\right)^{-0.17} \qquad \text{Stepped chute (skimming flow)} \quad (19.11)$$

where s is the distance measured from the crest origin, k_s is the equivalent roughness height, θ is the chute slope and h is the step height. Equations (19.10) and (19.11) are semi-empirical formulations which fit well model and prototype data (Wood *et al.* 1983; Chanson, 1995b, 2001). They apply to steep concrete chutes (i.e. $\theta > 30°$).

In the fully developed flow region, the flow is gradually varied until it reaches equilibrium (i.e. normal flow conditions). The gradually varied flow properties can be deduced by integration of the backwater equation. The normal flow conditions are deduced from the momentum principle (Appendix A4.1).

Note

On steep chutes, free-surface aeration[2] may take place downstream of the intersection of the outer edge of the developing boundary layer with the free surface. Air entrainment can be clearly identified by the 'white water' appearance of the free-surface flow. Wood (1991) and Chanson (1997) presented comprehensive studies of free-surface aeration on smooth chutes while Chanson (1995b, 2001) reviewed the effects of free-surface aeration on stepped channels.

The effects of free-surface aeration include flow bulking, some drag reduction effect and air–water gas transfer. The topic is still under active research (e.g. Chanson, 1997).

19.3.2 Application

For steep chutes (typically $\theta \sim 45$–$55°$), both the flow acceleration and boundary layer development affect the flow properties significantly. The complete flow calculations can be tedious and most backwater calculations are not suitable[3]. Complete calculations of developing flow and uniform equilibrium flow (see Appendix A4.1) may be combined to provide a general trend which may be used for a preliminary design (Fig. 19.11). Ideally, the maximum velocity at the downstream chute end is:

$$V_{max} = \sqrt{2g(H_1 - d\cos\theta)} \qquad \text{Ideal fluid flow} \qquad (19.12)$$

where H_1 is the upstream total head and d is the downstream flow depth: i.e. $d = q/V_{max}$ (Fig. 19.10). In practice the downstream flow velocity V is smaller than the theoretical velocity V_{max} because of friction losses.

---

[2] Air entrainment in open channels is also called free-surface aeration, self-aeration, insufflation or white waters.
[3] Because backwater calculations are valid only for fully developed flows. Furthermore, most softwares assume hydrostatic pressure distributions and neglect the effects of free-surface aeration.

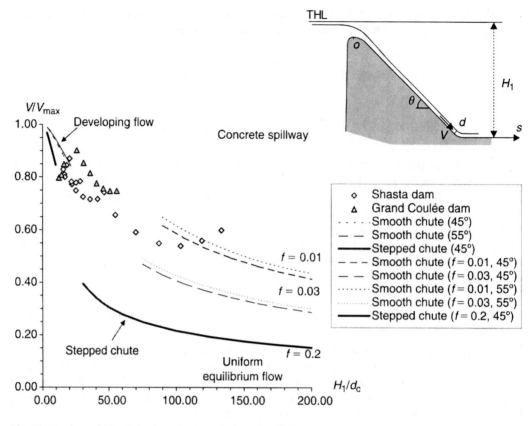

Fig. 19.11 Flow velocity at the downstream end of a steep chute.

In Fig. 19.11, the mean flow velocity at the end of the chute is plotted as V/V_{max} versus H_1/d_c where V_{max} is the theoretical velocity (equation (19.12)), H_1 is the upstream total head (above spillway toe) and d_c is the critical depth. Both developing flow calculations (equations (19.10) and (19.11)) and uniform equilibrium flow calculations are shown. Fitting curves are plotted to connect these lines (Fig. 19.11).

The semi-empirical curves (Fig. 19.11) are compared with experimental results obtained on prototype spillways (Grand Coulée dam and Shasta dam). The curves are valid for smooth and stepped spillways (concrete chutes), with slopes ranging from 45° to 55° (i.e. 1V:1H to 0.7V:1H).

19.3.3 Discussion

It is worth comparing the performances of stepped and smooth chutes. The larger mean bottom shear stress, observed with stepped chute flows, implies larger hydrodynamic loads on the steps than on a smooth invert. Stepped chutes required reinforced stepped profile compared to a smooth chute for identical inflow conditions. On the other hand, Fig. 19.11 also shows that larger energy dissipation takes place along a stepped spillway compared with a smooth chute. Hence, the size of the downstream stilling basin can be reduced with a stepped chute.

> **Note**
> Stepped chute flows are subjected to strong free-surface aeration (Chanson, 1995b, 2001). As a result, flow bulking and gas transfer are enhanced compared to a smooth channel.
>
> **Discussion**
> The energy dissipation characteristics of stepped channels were well known to ancient engineers. During the Renaissance period, Leonardo da Vinci realized that the flow, '*the more rapid it is, the more it wears away its channel*'; if a waterfall '*is made in deep and wide steps, after the manner of stairs, the waters (…) can no longer descend with a blow of too great a force*'. He illustrated his conclusion with a staircase waterfall 'down which the water falls so as not to wear away anything' (Richter, 1939).
>
> During the 19th century, stepped (or staircase) weirs and channels were quite common: e.g. in USA, nearly one third of the masonry dams built during the 19th century were equipped with a stepped spillway. A well-known 19th century textbook stated: '*The byewash*[4] *will generally have to be made with a very steep mean gradient, and to avoid the excessive scour which could result if an uniform*[5] *channel were constructed, it is in most cases advisable to carry the byewash down by a series of steps, by which the velocity will be reduced*' (Humber, 1876: p. 133).

19.4 STILLING BASINS AND ENERGY DISSIPATORS

19.4.1 Presentation

Energy dissipators are designed to dissipate the excess in kinetic energy at the end of the chute before it re-enters the natural stream. Energy dissipation on dam spillways is achieved usually by (1) a standard stilling basin downstream of a steep spillway in which a hydraulic jump is created to dissipate a large amount of flow energy and to convert the flow from supercritical to subcritical conditions, (2) a high velocity water jet taking off from a flip bucket and impinging into a downstream plunge pool or (3) a plunging jet pool in which the spillway flow impinges and the kinetic energy is dissipated in turbulent recirculation (Fig. 19.12). The construction of steps on the spillway chute may also assist in energy dissipation.

The stilling basin is the common type of dissipators for weirs and small dams. Most energy is dissipated in a hydraulic jump assisted by appurtenances (e.g. step and baffle blocks) to increase the turbulence.

> **Notes**
> 1. Other forms of energy dissipator include the drop structure and the impact-type stilling basin. The drop structure is detailed in another section because the flow pattern differs substantially from chute spillways. With impact-type dissipators, dissipation takes place by impact of the inflow on a vertical baffle (e.g. US Bureau of reclamation, 1987: p. 463).
> 2. On stepped chutes, the channel roughness (i.e. steps) contributes to the energy dissipation. In practice, a stilling basin is often added at the downstream end, but its size is smaller than that required for a smooth chute with identical flow conditions (see Section 19.3.2).

---

[4] Channel to carry waste waters: i.e. spillway.
[5] In the meaning of an uniform smooth channel bed (i.e. not stepped).

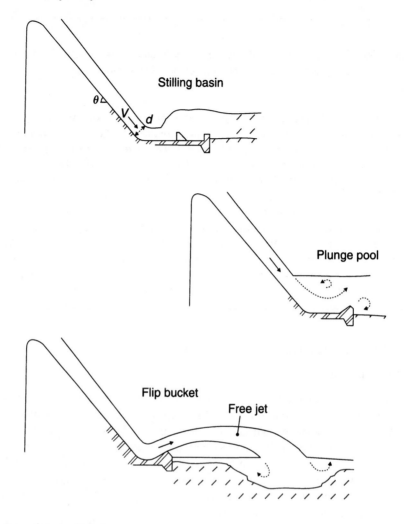

Fig. 19.12 Types of energy dissipators.

19.4.2 Energy dissipation at hydraulic jumps

Introduction

A hydraulic jump is the rapid transition from a supercritical to subcritical flow. It is an extremely turbulent process, characterized by large-scale turbulence, surface waves and spray, energy dissipation and air entrainment (Fig. 19.13). The large-scale turbulence region is usually called the 'roller'.

The downstream flow properties and energy loss in a hydraulic jump can be deduced from the momentum principle as a function of the upstream Froude number Fr and upstream flow depth d. For a horizontal flat rectangular channel, the downstream flow depth equals:

$$\frac{d_{conj}}{d} = \frac{1}{2}\left(\sqrt{1 + 8Fr^2} - 1\right) \tag{19.13}$$

(a)

(b)

Fig. 19.13 Hydraulic jump stilling basin downstream of sluice of gates, Large Colmitage, Cambodia (courtesy of Peter Ward): (a) energy dissipation blocks in foreground and gates in background, looking upstream and (b) operation during the flood season, looking upstream.

where $Fr = V/\sqrt{gd}$. The energy equation gives the head loss:

$$\frac{\Delta H}{d} = \frac{\left(\sqrt{1 + 8Fr^2} - 3\right)^3}{16\left(\sqrt{1 + 8Fr^2} - 1\right)} \tag{19.14}$$

Notes

1. The upstream and downstream depth d and d_{conj} are referred to as conjugate or sequent depths.
2. The upstream Froude number must be greater than unity: $Fr > 1$.
3. Jean-Baptiste Bélanger (1789–1874) was the first to suggest the application of the momentum principle to the hydraulic jump flow (Bélanger, 1828). The momentum equation applied across a hydraulic jump is called the Bélanger equation. Equation (19.13) is sometimes called (improperly) the Bélanger equation.

Application

In a horizontal rectangular stilling basin with baffle blocks, the inflow conditions are $d = 0.95$ m and $V = 16.8$ m/s. The observed downstream flow depth is 6.1 m. The channel is 12.5 m wide. Calculate the total force exerted on the baffle blocks. In which direction does the force applied by the flow onto the blocks act?

Solution

First, we select a control volume for which the upstream and downstream cross-sections are located far enough from the roller for the velocity to be essentially horizontal and uniform. The application of the momentum equation to the control volume yields:

$$\rho q(V_{tw} - V) = \frac{1}{2}\rho g(d^2 - d_{tw}^2) - \frac{F_B}{B}$$

where d_{tw} and V_{tw} are the tailwater depth and velocity, respectively, F_B is the force from the blocks on the control volume and B is the channel width. The total force is $F_B = +605$ kN. That is, the force applied by the fluid on the blocks acts in the downstream direction.

Remark

Equation (19.13) is not applicable. It is valid only for flat horizontal rectangular channels. In the present application, the baffle blocks significantly modify the jump properties.

Types of hydraulic jump

Hydraulic jump flows may exhibit different flow patterns depending upon the upstream flow conditions. They are usually classified as functions of the upstream Froude number Fr (Table 19.7). In practice, it is recommended to design energy dissipators with a steady jump type.

Table 19.8 summarizes the basic flow properties of hydraulic jump in rectangular horizontal channels.

Notes

1. The classification of hydraulic jumps (Table 19.7) must be considered as *rough* guidelines. It applies to hydraulic jumps in rectangular horizontal channels.
2. Recent investigations (e.g. Chanson and Montes, 1995; Montes and Chanson, 1998) showed that undular hydraulic jumps (Fawer jumps) might take place for upstream Froude numbers up to 4 depending upon the inflow conditions. The topic is still actively studied.
3. A hydraulic jump is a very unsteady flow. Experimental measurements of bottom pressure fluctuations indicated that the mean pressure is quasi-hydrostatic below the jump but large pressure fluctuations are observed (e.g. Hager, 1992b).

Table 19.7 Classification of hydraulic jump in rectangular horizontal channels (Chow, 1973)

Fr (1)	Definition (2)	Remarks (3)
1	Critical flow	No hydraulic jump
1–1.7	Undular jump (Fawer jump)	Free-surface undulations developing downstream of jump over considerable distances; *negligible* energy losses
1.7–2.5	Weak jump	Low energy loss
2.5–4.5	Oscillating jump	Wavy free surface, production of large waves of irregular period, unstable oscillating jump, each irregular oscillation produces a large wave which may travel far downstream, damaging and eroding the banks; *to be avoided if possible*
4.5–9	Steady jump	45–70 % of energy dissipation, steady jump, insensitive to downstream conditions (i.e. tailwater depth); *best economical design*
>9	Strong jump	Rough jump, up to 85% of energy dissipation, *risk of channel bed erosion*; to be avoided

Table 19.8 Dimensionless characteristics of hydraulic jump in horizontal rectangular channels

Fr (1)	d_{conj}/d (2)	$\Delta H/H$ (3)	L_r/d (4)
3	3.77	0.26	11.8
3.5	4.47	0.33	15.7
4.5	5.88	0.44	23.4
5.5	7.29	0.53	30.9
6.5	8.71	0.59	38.2
7.5	10.12	0.64	45.3
9	12.24	0.70	55.5
12	16.48	0.77	73.9
15	20.72	0.82	89.6

Length of the roller

The roller length of the jump may be estimated as (Hager *et al.* 1990):

$$\frac{L_r}{d} = 160 \tanh\left(\frac{Fr}{20}\right) - 12 \qquad 2 < Fr < 16 \tag{19.15}$$

where tanh is the hyperbolic tangent function and L_r is the length of the roller. Equation (19.15) is valid for rectangular horizontal channels ($d/B < 0.1$) and it may be used to predict the length of horizontal dissipation basins.

Practically, the length of a hydraulic jump stilling basin must be greater than the roller length for all flow conditions.

19.4.3 Stilling basins

Basically, the hydraulic design of a stilling basin must ensure a safe dissipation of the flow kinetic energy, to maximize the rate of energy dissipation and to minimize the size (and cost) of the structure. In practice, energy dissipation by hydraulic jump in a stilling basin is assisted with elements (e.g. baffle blocks and sill) placed on the stilling basin apron.

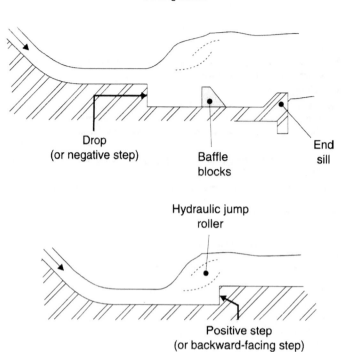

Stilling basin

Drop
(or negative step)

Baffle
blocks

End
sill

Hydraulic jump
roller

Positive step
(or backward-facing step)

Fig. 19.14 Sketch of stilling basins.

Basic shapes

The basic features of stilling basins include drop, backward-facing step (or sill), baffle block(s) and sudden expansion (Fig. 19.14). Hager (1992b) reviewed the advantages of each type. A summary is given below.

Drops and backward-facing steps are simple elements used to stabilize the hydraulic jump. Drops (also called negative steps) are advised when the downstream tailwater level may vary significantly (Fig. 19.14a). Backward-facing steps (also called positive steps) are usually located near the toe of the jump (Fig. 19.14b).

Baffle blocks (or dentated sills) can be placed in one or several rows.[6] The blocks force the flow above them (as a sill) and in between them. Baffle blocks must be designed with standard shapes. They are not recommended when the inflow velocity is larger than 20–30 m/s because of the risks of cavitation damage.

Sudden expansion (in the stilling basin) is another technique to enhance turbulent energy dissipation and to reduce the basin length. Physical modelling is, however, strongly recommended.

Standard stilling basins

Several standardized designs of stilling basins were developed in the 1950s and 1960s (Table 19.9 and Fig. 19.15). These basins were tested in models and prototypes over a considerable range of operating flow conditions. The prototype performances are well known, and they can be selected and designed without further model studies.

---

[6] The single-row arrangement is comparatively more efficient than the multiple-rows geometry.

Table 19.9 Standard types of hydraulic jump energy dissipators

Name (1)	Application (2)	Flow conditions (3)	Tailwater depth d_{tw}[a] (4)	Remarks (5)
USBR Type II	Large structures	$Fr > 4.5$ $q < 46.5\,\text{m}^2/\text{s}$ $H_1 < 61\,\text{m}$ Basin length $\sim 4.4d_{conj}$	$1.05d_{conj}$	Two rows of blocks; the last row is combined with an inclined end sill (i.e. dentated sill); block height $= d$
USBR Type III	Small structures	$Fr > 4.5$ $q < 18.6\,\text{m}^2/\text{s}$ $V < 15\text{–}18.3\,\text{m/s}$ Basin length $\sim 2.8d_{conj}$	$1.0d_{conj}$	Two rows of blocks and an end sill; block height $= d$
USBR Type IV	For oscillating jumps	$2.5 < Fr < 4.5$ Basin length $\sim 6d_{conj}$	$1.1d_{conj}$	One row of blocks and an end sill; block height $= 2d$ Wave suppressors may be added at downstream end
SAF	Small structures	$1.7 < Fr < 17$ Basin length $= 4.5d_{conj}Fr^{-0.76}$	$1.0d_{conj}$	Two rows of baffle blocks and an end sill; block height $= d$
USACE		Basin length $> 4d_{conj}$	$1.0d_{conj}$	Two rows of baffle blocks and end sill

References: Chow (1973), Hager (1992b), Henderson (1966), US Bureau of Reclamation (1987).
Notes: [a]Recommended tailwater depth for optimum stilling basin operation; d: upstream flow depth (inflow depth); d_{conj}: conjugate flow depth (equation (19.13)); d_{tw}: tailwater depth; Fr: inflow Froude number.

In practice, the following types are highly recommended:

- the USBR Type II basin for large structures and $Fr > 4.5$;
- the USBR Type III basin and the SAF basin for small structures;
- the USBR Type IV basin for oscillating jump flow conditions (Fig. 19.16).

Note

The requirements of the USBR for its dissipators are more stringent than other organizations. As a result, the USBR Type III basin is sometimes too conservative and the SAF basin may be preferred for small structures.

19.4.4 Discussion

In practice, design engineers must ensure that a stilling basin can operate safely for a wide range of flow conditions. Damage (scouring, cavitation) to the basin and to the downstream natural bed may occur in several cases:

- the apron is too short and/or too shallow for an optimum jump location (i.e. on the apron),
- poor shapes of the blocks, sill and drop resulting in cavitation damage,
- flow conditions larger than design flow conditions,
- unusual overflow during construction periods,
- poor construction of the apron and blocks,
- seepage underneath the apron, inadequate drainage and uplift pressure built-up,
- wrong conception of the stilling basin.

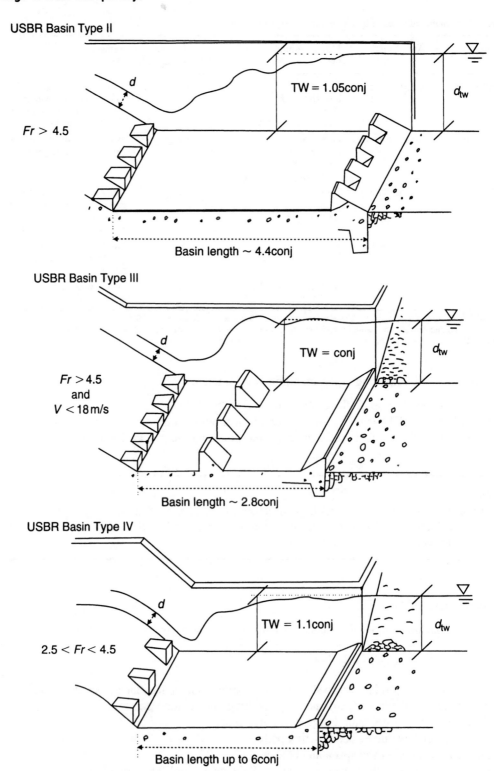

USBR Basin Type II

d

$Fr > 4.5$

TW = 1.05conj

d_{tw}

Basin length ~ 4.4conj

USBR Basin Type III

d

$Fr > 4.5$
and
$V < 18\,m/s$

TW = conj

d_{tw}

Basin length ~ 2.8conj

USBR Basin Type IV

d

$2.5 < Fr < 4.5$

TW = 1.1conj

d_{tw}

Basin length up to 6conj

Fig. 19.15 Sketch of USBR Basins.

Fig. 19.16 Prototype stilling basin: USBR Basin Type IV at Bjelke-Petersen dam, Australia on 7 November 1997. Flow direction from left to right. Design conditions: 3660 m³/s and dam height: 43 m.

19.5 DESIGN PROCEDURE

19.5.1 Introduction

The construction of a small dam across a stream will modify both the upstream and possibly downstream flow conditions. The dam crest elevation must be selected accurately to provide the required storage of water or upstream water level rise. Furthermore, the spillway and stilling basin must operate safely for a wide range of flow rates and tailwater flow conditions.

19.5.2 Dam spillway with hydraulic jump energy dissipator

Considering an overflow spillway with hydraulic jump energy dissipation at the toe, the basic steps in the design procedure are as follows:

Step 1. Select the crest elevation z_{crest} (bed topography and storage level).

Step 2. Choose the crest width B (site geometry, hydrology). The crest width may be smaller than the weir length (across the stream).

Step 3. Determine the design discharge Q_{des} from risk analysis and flood routing. Required informations include catchment area, average basin slope, degree of impermeability, vegetation cover, rainfall intensity, duration and inflow hydrograph. The peak spillway discharge is deduced from the combined analysis of storage capacity, and inflow and outflow hydrographs.

Step 4. Calculate the upstream head above spillway crest $(H_{des} - z_{crest})$ for the design flow rate Q_{des}:

$$\frac{Q_{des}}{B} = C_{des}(H_{des} - z_{crest})^{3/2}$$

Note that the discharge coefficient C_{des} varies with the head above crest $(H_{des} - z_{crest})$.

Step 5. Choose the chute toe elevation (i.e. apron level): $z_{apron} = z_{crest} - \Delta z$. The apron level may differ from the natural bed level (i.e. tailwater bed level).

Step 6. For the design flow conditions, calculate the flow properties d and V at the end of the chute toe (Fig. 19.10) using:

$$H_1 = H_{des} - z_{apron} \qquad \text{Upstream total head}$$

$$V_{max} = \sqrt{2g(H_1 - (q_{des}/V_{max})\cos\theta} \qquad \text{Ideal fluid flow velocity}$$

Step 7. Calculate the conjugate depth for the hydraulic jump:

$$\frac{d_{conj}}{d} = \frac{1}{2}\left(\sqrt{1 + 8Fr^2} - 1\right) \qquad \text{where } Fr = V/\sqrt{gd}$$

Step 8. Calculate the roller length L_r. The apron length must be greater than the jump length.

Step 9. Compare the *jump height rating level* (JHRL) and the natural downstream water level (i.e. the *tailwater rating level*, TWRL). If the jump height does not match the natural water level, the apron elevation, the crest width, the design discharge or the crest elevation must be altered (i.e. go back to Steps 5, 3, 2 or 1 respectively).

Practically, if the JHRL does not match the natural water level (TWRL), the hydraulic jump will not take place on the apron.

Figure 19.17 presents the basic definitions. Figure 19.18 illustrates two cases for which the stilling basin is designed to equal TWRL and JHRL at design flow conditions. In each sketch (Fig. 19.18), the real free-surface line is shown in solid line while the JHRL is shown in dashed line.

Figures 19.19–19.21 show prototype dissipators in operation for different flow rates, highlighting the tailwater effects on the hydraulic jump.

Notes

1. The JHRL is the free-surface elevation downstream of the stilling basin.
2. The TWRL is the natural free-surface elevation in the downstream flood plain. The downstream channel often flows as a subcritical flow, controlled by the downstream flow conditions (i.e. discharge and downstream flood plain geometry).

Discussion: calculations of the JHRL

(A) For a horizontal apron, the JHRL is deduced simply from the apron elevation and the conjugate depth:

$$\text{JHRL} = z_{apron} + d_{conj}$$

(B) For an apron with an end sill or end drop, the JHRL is calculated using the Bernoulli equation:

$$d_{conj} + z_{apron} + \frac{q^2}{2g(d_{conj})^2} = (\text{JHRL} - z_{tw}) + z_{tw} + \frac{q^2}{2g(\text{JHRL} - z_{tw})^2}$$

where z_{tw} is the downstream natural bed elevation. $(z_{tw} - z_{apron})$ is the drop/sill height.

Note that the above calculation assumes that the complete jump is located upstream of the drop/sill, and that no energy loss takes place at the sill/drop.

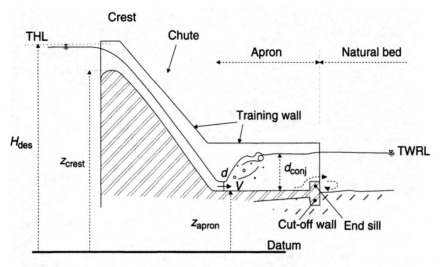

Fig. 19.17 Design of an overflow spillway with hydraulic jump stilling basin.

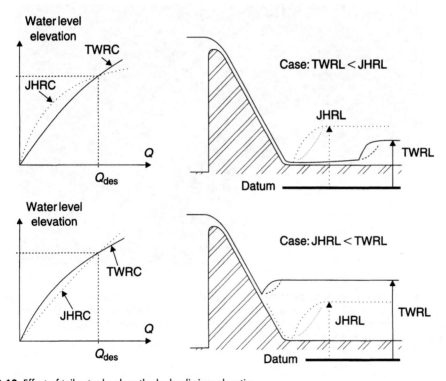

Fig. 19.18 Effect of tailwater level on the hydraulic jump location.

Matching the JHRC and the TWRC

During the design stages of hydraulic jump energy dissipator, engineers are required to compute the JHRL for all flow rates. The resulting curve, called the *jump height rating curve* (JHRC), must be compared with the variations of natural downstream water level with discharges (i.e. TWRC or *tailwater rating curve*).

Fig. 19.19 Effect of tailwater levels on the Waraba Creek weir QLD, Australia (photographs from the collection of late Professor G.R. McKay, Australia). Smooth chute followed by a stilling basin with baffle blocks: (a) Operation at very low overflow Q_1. Flow from the right to the left. Note baffle blocks in the stilling basin. (b) Operation at low overflow Q_2 ($>Q_1$). Note the fully developed hydraulic jump.

First, let us remember that the flow downstream of the stilling basin jump is subcritical, and hence it is controlled by the downstream flow conditions: i.e. by the tailwater flow conditions. The location of the jump is determined by the upstream and downstream flow conditions. The upstream depth is the supercritical depth at the chute toe and the downstream depth is determined by the tailwater flow conditions. The upstream and downstream depths, called conjugate depths, must further satisfy the momentum equation: e.g. equation (19.13) for a horizontal apron in the absence of baffles.

(a)

(b)

Fig. 19.20 Effect of tailwater levels on the Waraba Creek weir QLD, Australia (photographs from the collection of late Professor G.R. McKay, Australia). (a) Operation at medium flow rate Q_3 ($Q_2 < Q_3 < Q_4$). Flow from the left to the right. Note the rising tailwater (compared to Fig. 19.19b). (b) Operation at large discharge Q_4 ($>Q_3$). View from downstream, flow from top right to bottom left. Note the plunge pool formed in the stilling basin caused by the rising tailwater level.

Discussion

In practice, the TWRC is set by the downstream flood plain characteristics for a range of flow rates. It may also be specified. Designers must select stilling basin dimensions such that the JHRC matches the TWRC: e.g. if the curves do not match, a sloping apron section may be used (e.g. Fig. 19.22). The height of the sloping apron, required to match the JHRC and the TWRC for all flows, is obtained by plotting the two curves, and measuring the greatest vertical separation of the curves. Note that if the maximum separation of the two curves occurs at very low flow rates, it is advised to set the height of the sloping apron to the separation of the curves at a particular discharge (e.g. $0.15Q_{des}$).

When the jump height level is higher than the tailwater level, the jump may not take place on the apron but downstream of the apron. The flow above the apron becomes a jet flow and insufficient energy dissipation takes place.

In such a case (i.e. JHRL > TWRL), the apron must be lowered or the tailwater level must be artificially raised. The apron level may be lowered (below the natural bed level) using an inclined upward apron design or a sill at the end of the apron (Fig. 19.22). The tailwater level may be raised also by providing a downstream weir at the end of the basin (and checking that the weir is never drowned).

When the jump height level is lower than the natural tailwater level, the jump may be drowned and a backwater effect takes place. The jump is 'pushed' upstream onto the chute and it may become similar to a plunging jet flow. In this case (JHRL < TWRL), a downward slope or a higher apron level followed by a drop may be incorporated in the apron design (Fig. 19.22).

(a)

(b)

Fig. 19.21 Effect of tailwater levels on the Silverleaf weir QLD, Australia (courtesy of Mr J. Mitchell). Timber crib stepped weir completed in 1953 (5.1 m high structure): (a) operation at very low flow of the stepped weir and (b) operation at large overflow. Note the high tailwater level.

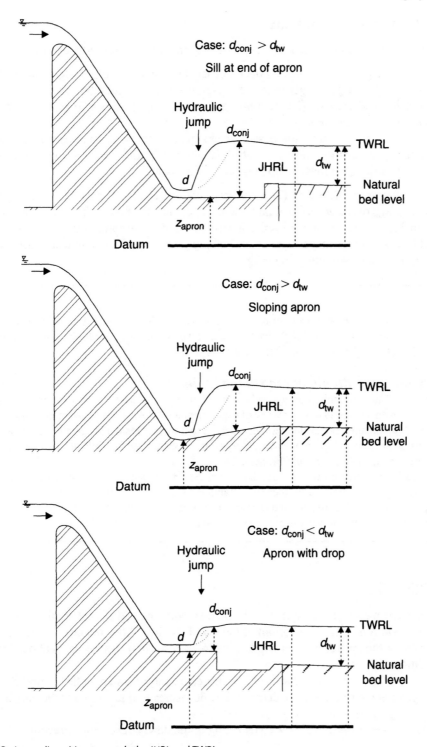

Fig. 19.22 Apron disposition to match the JHRL and TWRL.

19.5.3 Practical considerations

For hydraulic jump energy dissipators, it is extremely important to consider the following points:

- The energy dissipator is designed for the reference flow conditions (i.e. design flow conditions).
- For overflow discharges larger than the design discharge, it may be acceptable to tolerate some erosion and damage (i.e. scouring and cavitation). *However, it is essential that the safety of the dam is ensured.*
- For discharges smaller than the design discharge, perfect performances are expected: i.e. (a) the energy dissipation must be controlled completely and it must occur in the designed dissipator and (b) there must be no maintenance problem.

These objectives are achieved by (1) a correct design of energy dissipation basin dimensions, (2) a correct design of sidewalls (i.e. training walls) to confine the flow to the dissipator and (3) provision of an end sill and cut-off wall at the downstream end of the apron (Fig. 19.17). The training wall must be made high enough to allow for bulking caused by air entrainment in the chute flow and for surging in the hydraulic jump basin.

Note
Practically, it is uppermost important to remember that a hydraulic jump is associated with large bottom pressure fluctuation and high risks of scour and damage. As a result, the jump must be contained within the stilling basin for all flow conditions.

Discussion: air entrainment in stilling basin
The effects of air entrainment on hydraulic jump flow were discussed by Wood (1991). Chanson (1997) presented a comprehensive review of experimental data. Air is entrained at the jump toe and advected downstream in a developing shear layer. Experimental results suggested that the air bubble diffusion process and the momentum transfer in the shear flow are little affected by gravity in first approximation (Chanson and Brattberg, 2000).

Practically, the entrained air increases the bulk of the flow, which is a design parameter that determines the height of sidewalls. Further air entrainment contributes to the air–water gas transfer of atmospheric gases such as oxygen and nitrogen. This process must be taken into account for the prediction of the downstream water quality (e.g. dissolved oxygen content).

19.6 EXERCISES

A broad-crested weir is installed in a horizontal, smooth channel of rectangular cross-section. The channel width is 55 m. The bottom of the weir is 5.20 m above the channel bed. At design flow conditions, the upstream water depth is 6.226 m. Perform hydraulic calculations for design flow conditions.

Compute the flow rate (in the absence of downstream control) assuming that critical flow conditions take place at the weir crest, the depth of flow downstream of the weir and the horizontal sliding force acting on the weir.

Solution: $Q = 98 \, \text{m}^3/\text{s}$ and $F = 9.4 \, \text{MN}$ acting in the downstream flow direction.

Considering a 10 m wide, horizontal, rectangular stilling basin with a baffle row (Fig. E.19.1), the inflow conditions are $d = 0.1 \, \text{m}$ and $V = 10 \, \text{m/s}$: (1) For one flow situation, the observed tailwater depth is 1.2 m. Calculate the magnitude of total force exerted on the baffle row. In which direction does the force applied by the flow onto the blocks act? (2) Variations of the tailwater level are expected. Calculate the force exerted by the flow onto the baffle row for tailwater depths ranging from 0.85 to 2.2 m. *The inflow conditions remain unchanged.* Plot your results on a graph. *The force exerted by the flow on the baffle row is positive in downstream flow direction.*

Solution: (1) $F = 22$ kN acting in the downstream flow direction. (2) The solution is shown in Fig. E.19.2.

An energy dissipator (hydraulic jump type) is to be designed downstream of an undershoot sluice gate. The invert elevation of the transition channel between the sluice gate and the dissipator is set at 70 m RL. The channel bed will be horizontal and concrete lined. The gate and energy dissipator will be located in a 20 m wide rectangular channel. The design inflow conditions of the prototype dissipator are discharge: 220 m³/s and inflow depth: 1.1 m (i.e. flow depth downstream of gate). Four dissipator designs will be investigated:

(i) Dissipation by hydraulic jump in the horizontal rectangular channel (70 m RL elevation) (no blocks).

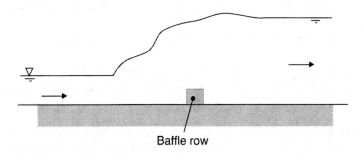

Fig. E.19.1 Sketch of a stilling basin with a baffle row.

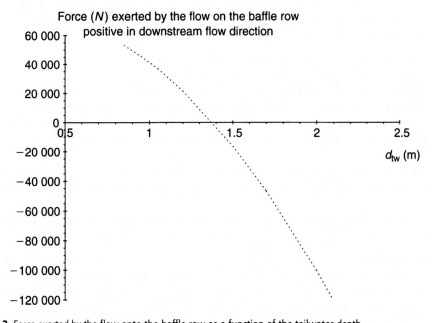

Fig. E.19.2 Force exerted by the flow onto the baffle row as a function of the tailwater depth.

(ii) Dissipation by hydraulic jump in the horizontal rectangular channel (70 m RL elevation) assisted with a single row of nine baffle blocks.

(iii) Dissipation in a standard stilling basin (e.g. USBR and SAF) set at 70 m RL elevation.

(iv) Dissipation in a specially designed stilling basin. The apron elevation will be set to match the tailwater conditions.

For the design (i): (a) compute the JHRL at design flow conditions.

For the design (ii): (b) compute the force on each baffle block when the tailwater level is set at 73.1 m RL at design inflow conditions. (c) In what direction is the hydrodynamic force acting on the blocks: i.e. upstream or downstream?

For the design (iii): (d) what standard stilling basin design would you select: USBR Type II, USBR Type III, USBR Type IV and SAF? (If you have the choice between two (or more) designs, discuss the economical advantages of each option.)

(e) Define in words and explain with sketch(es) the terms TWRL and JHRL. (Illustrate your answer with appropriate sketch(es) if necessary.)

Design a hydraulic jump stilling basin (design (iv)) to match the jump height and tailwater levels. The downstream bed level is at elevation 70 m RL. At design discharge, the tailwater depth equals 3.1 m.

(f) Calculate the apron level in the stilling basin to match the jump height to tailwater at design flow conditions.

(g) Alternatively, if the apron level remains at the natural bed level (70 m RL), determine the height of a broad-crested weir necessary at the downstream end of the basin to artificially raise the tailwater level to match the JHRL for the design flow rate.

An overflow spillway is to be designed with an uncontrolled ogee crest followed by a stepped chute and a hydraulic jump dissipator. The width of the crest, chute and dissipation basin will be 127 m. The crest level will be at 336.3 m RL and the design head above crest level will be 3.1m. The chute slope will be set at 51° and the step height will be 0.5 m. The elevation of the chute toe will be set at 318.3 m RL. The stepped chute will be followed (without transition section) by a horizontal channel which ends with a broad-crested weir, designed to record flow rates as well as to raise the tailwater level:

(a) Calculate the maximum discharge capacity of the spillway.

(b) Calculate the flow velocity at the toe of the chute.

(c) Calculate the residual power at the end of the chute (give the SI units). Comment.

(d) Compute the JHRL at design flow conditions (for a hydraulic jump dissipator).

(e) Determine the height of the broad-crested weir necessary at the downstream end of the dissipation basin to artificially raise the tailwater level to match the JHRL for the design flow rate.

(f) Compute the horizontal force acting on the broad-crested weir at design inflow conditions. In what direction will the hydrodynamic force be acting on the weir: i.e. upstream or downstream?

(g) If a standard stilling basin (e.g. USBR and SAF) is to be designed, what standard stilling basin design would you select: USBR Type II, USBR Type III, USBR Type IV or SAF?

Notes: In calculating the crest discharge capacity, assume that the discharge capacity of the ogee crest is 28% larger than that of a broad crest (for the same upstream head above crest). In computing the velocity at the spillway toe, allow for energy losses by using results presented in the book. The residual power equals $\rho g Q H_{res}$ where Q is the total discharge and H_{res} is the residual total head at chute toe taking the chute toe elevation as datum.

An overflow spillway is to be designed with an un-gated broad crest followed by a smooth chute and a hydraulic jump dissipator. The width of the crest, chute and dissipation basin will be 55 m. The crest level will be at 96.3 m RL and the design head above crest level will be 2.4 m. The chute slope will be set at 45° and the elevation of the chute toe will be set at 78.3 m RL. The stepped chute will be followed (without transition section) by a horizontal channel which ends with a broad-crested weir, designed to record flow rates as well as to raise the tailwater level.

(a) Calculate the maximum discharge capacity of the spillway. (b) Calculate the flow velocity at the toe of the chute. (c) Calculate the Froude number of the flow at the end of the chute. (Comment.) (d) Compute the conjugate flow depth at design flow conditions (for a hydraulic jump dissipator).

The natural tailwater level (TWRL) at design flow conditions is 81.52 m RL: (e) Determine the apron elevation to match the JHRL and the TWRL at design flow conditions.

(f) If a standard stilling basin (e.g. USBR and SAF) is to be designed, what standard stilling basin design would you select: USBR Type II, USBR Type III, USBR Type IV or SAF?

(*Note*: In computing the velocity at the prototype spillway toe, allow for energy losses.)

A weir is to have an overflow spillway with ogee-type crest and a hydraulic jump energy dissipator. Considerations of storage requirements and risk analysis applied to the 'design flood event' have set the elevation of the spillway crest 671 m RL and the width of the spillway crest at $B = 76$ m. The maximum flow over the spillway when the design flood is routed through the storage for these conditions is 1220 m³/s. (*Note*: the peak *inflow* into the reservoir is 3300 m³/s.) The spillway crest shape has been chosen so that the discharge coefficient at the maximum flow is 2.15 (SI units). For the purpose of the assignment, the discharge coefficient may be assumed to decrease linearly with discharge down to a value of 1.82 at very small discharges. The chute slope is 1V:0.8H (i.e. about 51.3°). The average bed level downstream of the spillway is 629.9 m RL and the TWRC downstream of the dam is defined as follows:

Discharge (m³/s)	TWRL (m)	Discharge (m³/s)	TWRL (m)
0	629.9	400	634.45
25	631.5	500	634.9
50	631.95	750	635.75
100	632.6	950	636.35
150	633.0	1220	637.1
250	633.65	1700	638.2

Design three options for the hydraulic jump stilling basin to dissipate the energy at the foot of the spillway as follows: (A) apron level lowered to match the JHRC to the TWRC; (B) apron level set at 632.6 m RL and tailwater level raised artificially with a broad-crested weir at the downstream end of the basin; (C) apron level set at 629.9 m RL (i.e. average bed level) *but* spillway width B is changed to match the JHRC to the TWRC.

In each case use a sloping apron section if this will improve the efficiency and/or economics of the basin. In computing the velocity at the foot of the spillway, allow for energy losses. The design calculations are to be completed and submitted in two stages as specified below: Stage 1 (calculations are to be done for 'maximum' flow *only*) and Stage 2 (off-design calculations).

Stage 1: (a) Calculate apron level to match the JHRC to natural TWRC at 1220 m³/s. (b) Calculate height of broad-crested weir to raise local TWRC to JHRC at 1220 m³/s for an apron level at 632.6 m RL. (c) Calculate B to match JHRC to natural TWRC at 1220 m³/s for an apron level at 629.9 m RL.

Stage 2: For both cases (A) and (B): (a) Calculate the JHRC for all flows up to and including $1220\,\text{m}^3/\text{s}$ in sufficient detail to plot the curve. (b) Use the results from Stage 1 to check the correctness of your calculation for $1220\,\text{m}^3/\text{s}$. (c) Plot the JHRC from (a) and the TWRC (natural or local, as required) on the same graph and determine the height of sloping apron required to match JHRC and TWRC for all flows. (This is obtained from the greatest vertical separation of the two curves as plotted. If the maximum separation occurs at a flow less than $200\,\text{m}^3/\text{s}$, set the height of the sloping apron to the separation of the curves at $200\,\text{m}^3/\text{s}$.) (d) Draw to scale a dimensioned sketch of the dissipator.

Summary sheet (Stage 1)

(a)	Apron level		m RL
(b)	Height of broad-crested weir above apron		m
	Minimum crest length		m
(c)	Spillway crest width		m
(d)	If the crest width is changed from 76 m (design (C)), and for a peak inflow into the reservoir of $3300\,\text{m}^3/\text{s}$, will the 'maximum' spillway overflow change from $1220\,\text{m}^3/\text{s}$?	Yes/No......	
	Give reason for answer		

Summary sheet (Stage 2)

Design A	Apron level		m RL
	Height of sloping apron		m
	Length of horizontal apron		m
Design B	Height of broad-crested weir above apron		m
	Height of sloping apron		m
	Length of horizontal apron		m

For each case (A) and (B), supply plotted curves of JHRC and TWRC (natural or raised as required). Supply dimensioned drawing of dissipators to scale.

A weir is to have an overflow spillway with ogee-type crest and a hydraulic jump energy dissipator. Considerations of storage requirements and risk analysis applied to the 'design flood event' have set the elevation of the spillway crest 128 m RL and the width of the spillway crest at $B = 81\,\text{m}$. The maximum flow over the spillway when the design flood is routed through the storage for these conditions is $1300\,\text{m}^3/\text{s}$. (*Note*: the peak *inflow* into the reservoir is $3500\,\text{m}^3/\text{s}$.) The spillway crest shape has been chosen so that the discharge coefficient at the maximum flow is 2.15 (SI units). For the purpose of the assignment, the discharge coefficient may be assumed to decrease linearly with discharge down to a value of 1.82 at very small discharges. The chute slope is 1V:0.8H (i.e. about 51.3°). (*Note*: both stepped and smooth chute profiles will be considered.)

The average bed level downstream of the spillway is 86.9 m RL and the TWRC downstream of the dam is defined as follows:

Discharge (m^3/s)	TWRL (m)	Discharge (m^3/s)	TWRL (m)
0	86.9	400	91.45
25	88.5	520	91.9
50	88.95	800	92.75
100	89.6	1000	93.35
150	90.0	1300	94.1
250	90.65	1800	95.2

Design three options for the hydraulic jump stilling basin to dissipate the energy at the foot of the spillway as follows: (A) smooth concrete chute with apron level lowered to match the JHRC to the TWRC; (B) stepped concrete chute (step height $h = 0.15$ m) with apron level lowered (or raised) to match the JHRC to the TWRC; (C) smooth concrete chute with apron level set at 86.9 m RL (i.e. average bed level) *but* spillway width B is changed to match the JHRC to the TWRC.

In each case use a sloping apron section if this will improve the efficiency and/or economics of the basin. In computing the velocity at the foot of the spillway, allow for energy losses. The design calculations are to be completed and submitted in two stages: Stage 1 for design flow conditions (1300 m^3/s) and Stage 2 for non-design flow conditions.

Stage 1: (a) Calculate apron level to match the JHRC to natural TWRC at 1300 m^3/s. (b) Calculate apron level to match the JHRC to natural TWRC at 1300 m^3/s (with a stepped chute). (c) Calculate B to match JHRC to natural TWRC at 1300 m^3/s for an apron level at 86.9 m RL.

Stage 2: For both cases (A) and (B): (a) Calculate the JHRC for all flows up to and including 1300 m^3/s in sufficient detail to plot the curve. (b) Use the results from Stage 1 to check the correctness of your calculation for 1300 m^3/s. (c) Plot the JHRC from (a) and the TWRC (natural or local, as required) on the same graph and determine the height of sloping apron required to match JHRC and TWRC for all flows (up to design flow conditions). (This is obtained from the greatest vertical separation of the two curves as plotted. If the maximum separation occurs at a flow less than 200 m^3/s, set the height of the sloping apron to the separation of the curves at 200 m^3/s.) (d) Draw to scale a dimensioned sketch of the dissipator. (e) Discuss the advantages and inconvenience of each design case, and indicate your recommended design (with proper justifications).

Summary sheet (Stage 1)

(a)	Apron level		m RL
(b)	Apron level		m RL
(c)	Spillway crest width		m
(d)	If the crest width is changed from 81 m, will the 'maximum' flow rate change from 1300 m^3/s	Yes/No......	
	Give reason for answer		

Summary sheet (Stage 2)

Design A	Apron level		m RL
	Height of sloping apron		m
	Length of horizontal apron		m
Design B	Apron level		m RL
	Height of sloping apron		m
	Length of horizontal apron		m

For each case (A) and (B), supply curves of JHRC and TWRC (natural or raised as required). Supply dimensioned drawings of dissipators to scale.

Design of drop structures and stepped cascades

20

20.1 INTRODUCTION

For low-head structures, the design of a standard weir (Chapter 19) may be prohibitive and uneconomical, and a drop structure may be preferred (Figs 20.1–20.3). The main features of drop structures are the free overfall, the nappe impact and the downstream hydraulic jump. Energy dissipation takes place at the nappe impact and in the hydraulic jump.

A related form of drop structures is a stepped weir with large steps on which the flow bounces down from one step to the next one, as a succession of drop structures (Figs 20.4 and 20.5). This type of flow is called a nappe flow regime (or jet flow regime).

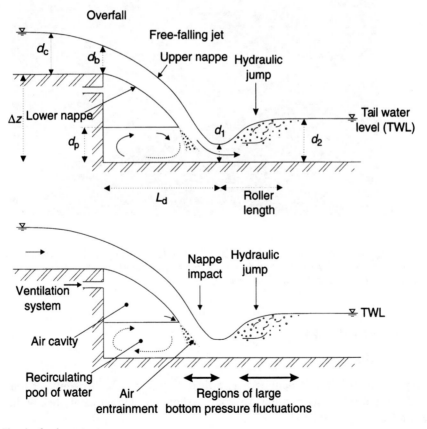

Fig. 20.1 Sketch of a drop structure.

(a)

(b)

Fig. 20.2 Photographs of drop structures: (a) Large drop structure ($\Delta z > 5\,\text{m}$) on Nishizawa-keikoku River near Mitomi (Japan) on November 1998. Flow direction from right to left. (b) Small drop structure on the Kamo-gawa River, Kyoto (Japan) on April 1999. Note the step-pool fishway in the foreground.

20.2 DROP STRUCTURES

20.2.1 Introduction

The simplest case of drop structures is a vertical drop in a wide horizontal channel (Figs 20.1 and 20.2). In the following sections, we shall assume that the air cavity below the free-falling nappe is adequately ventilated.

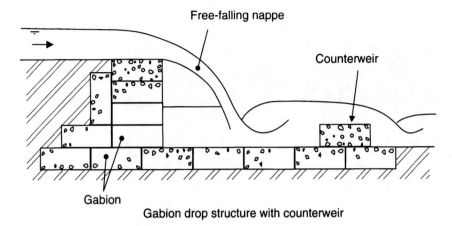

Fig. 20.3 Sketch of a gabion drop structure with counterweir.

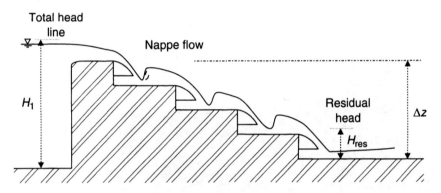

Fig. 20.4 Nappe flow regime above a stepped structure.

Note

A drop structure is also called a vertical weir.

20.2.2 Free overfall

At the overfall, the flow becomes suddenly accelerated. If the upstream flow is subcritical, critical depth is expected to take place at the transition. However, a rapid pressure redistribution is also observed at the brink of the overfall and critical flow depth ($d_c = \sqrt[3]{q^2/g}$) is observed, in practice, at a distance of about three to four times d_c upstream of the step edge (e.g. Henderson, 1966: p. 192).

Analytical developments and experimental measurements showed that, for a horizontal overfall (wide channel), the flow depth at the brink equals:

$$d_b \approx 0.7d_c \tag{20.1}$$

(a)

(b)

(c)

Fig. **20.5** Examples of stepped cascades: (a) Robina stepped weir No. 5 (Gold Coast QLD, 1996) on 3 February 2003 during a small overflow. Embankment weir with precast concrete steps, $Q_{des} = 27\,m^3/s$, $h = 0.5\,m$ and $\theta = 27°$. (b) Stepped diversion weir: Joe Sippel weir (Murgon QLD, 1984) on 7 November 1997. Rockfill embankment with concrete slab steps, $H = 6.5\,m$ and pooled step design. (c) Diversion stepped weir at Chechung, PingTung county (Taiwan) on December 1998. The weir is located less than 5 km upstream of the river mouth.

(d)

(e)

Fig. 20.5 (d) Timber crib weir: Murgon weir (Murgon QLD, 1935) on 8 November 1997. (e) Stepped waterway in a new residential area, Okazaki (Japan) on 10 November 2001.

Notes

1. In a rectangular channel, the critical flow depth equals:

$$d_c = \sqrt[3]{\frac{q^2}{g}}$$

where q is the discharge per unit width and g is the gravity constant. This result is obtained for a horizontal channel with hydrostatic pressure distribution. At an overfall brink the pressure distribution is not hydrostatic and, as a result, $d_b \neq d_c$.

2. For a horizontal rectangular channel, a more accurate estimate of the brink depth is (Rouse, 1936):

$$d_b = 0.715 d_c$$

3. Recent reviews of the overfall characteristics include Montes (1992b) and Chanson (1995b: pp. 230–236). The former reviewed the overfall properties near the edge of the step. The latter describes the nappe trajectory equations and the application of the momentum principle at nappe impact.

20.2.3 Drop impact

At the drop impact, a recirculating pool of water forms behind the overfalling jet (Fig. 20.1). This mass of water is important as it provides a pressure force parallel to the floor which is required to change the jet momentum direction from an angle to the bottom, to parallel to the horizontal bottom.

If the downstream water level (i.e. tailwater level) corresponds to a subcritical flow, a hydraulic jump takes place immediately downstream of the nappe impact (Figs 20.1 and 20.2).

The drop length, pool height and the flow depths d_1 and d_2 (Fig. 20.1) may be estimated from empirical correlations (Rand, 1955):

$$\frac{L_d}{\Delta z} = 4.30 \left(\frac{d_c}{\Delta z} \right)^{0.81} \tag{20.2}$$

$$\frac{d_p}{\Delta z} = \left(\frac{d_c}{\Delta z} \right)^{0.66} \tag{20.3}$$

$$\frac{d_1}{\Delta z} = 0.54 \left(\frac{d_c}{\Delta z} \right)^{1.275} \tag{20.4}$$

$$\frac{d_2}{\Delta z} = 1.66 \left(\frac{d_c}{\Delta z} \right)^{0.81} \tag{20.5}$$

where Δz is the drop height and $d_c = \sqrt[3]{q^2/g}$. Equations (20.2)–(20.5) were successfully verified with numerous laboratory data.

20.2.4 Design criterion

In practice, the design of the drop structure includes:

1. Select the crest elevation z_{crest} (bed topography and storage level).
2. Determine the design discharge Q_{des} from risk analysis and flood routing.
3. Choose a crest width B. The crest width may be smaller than the crest length.
4. Compute the critical depth for design flow conditions:

$$(d_c)_{des} = \sqrt[3]{\frac{Q_{des}^2}{gB^2}}$$

5. Choose the apron level (i.e. $z_{crest} - \Delta z$).

6. For the design flow conditions, compute the drop length and drop impact characteristics (equations (20.2)–(20.5)).

> **Note**
> If the tailwater flow conditions are set, the design procedure must be iterated until the free-surface elevation downstream of the hydraulic jump (i.e. $d_2 + z_{crest} - \Delta z$) matches the tailwater rating level (TWRL).

7. Check the stability of the vertical weir (e.g. Agostini *et al.*, 1987) and the risk of scouring at nappe impact.

> **Notes**
> 1. In practice drop structures are efficient energy dissipators. They are typically used for drops up to 7–8 m and discharges up to about $10 \, m^2/s$ (US Bureau of Reclamation, 1987; Agostini *et al.*, 1987). Note further that vertical drops can operate for a wide range of tailwater depths (usually TWRL $< z_{crest}$).
> 2. Large mean bottom pressures and bottom pressure fluctuations are experienced at the drop impact. The apron must be reinforced at the impingement location of the nappe and underneath the downstream hydraulic jump (Fig. 20.1).
> 3. A counterweir can be built downstream of the drop structure to reduce the effect of erosion (Fig. 20.3). The counterweir creates a natural stilling pool between the two structures which reduces the scouring force below the falling nappe impact.
> 4. An adequate nappe aeration is *essential* for a proper operation of drop structures. Bakhmeteff and Fedoroff (1943) presented experimental data showing the effects of nappe ventilation (and the lack of ventilation) on the flow patterns.

20.2.5 Discussion

The above calculations (equations (20.1)–(20.5)) were developed assuming that the flow upstream of the brink of the overfall is *subcritical*, hence critical immediately upstream of the brink.

If the upstream flow is supercritical (e.g. drop located downstream of an underflow gate), the hydraulic characteristics of supercritical nappe flows are determined by the nappe trajectory, the jet impact on the step and the flow resistance on the step downstream of the nappe impact.

> **Note**
> Several researchers (Rouse, 1943; Rajaratnam and Muralidhar, 1968; Hager, 1983; Marchi, 1993) gave details of the brink flow characteristics and of the jet shape for supercritical flows. For supercritical overfalls, the application of the momentum equation at the base of the overfall, using the same method as White (1943), leads to the result:
>
> $$\frac{d_1}{d_c} = \frac{2Fr^{-2/3}}{1 + (2/Fr^2) + \sqrt{1 + (2/Fr^2)\left(1 + (\Delta z/d_c)Fr^{2/3}\right)}}$$
>
> where Fr is the Froude number of the supercritical flow upstream of the overfall brink.

20.3 NAPPE FLOW ON STEPPED CASCADES

20.3.1 Presentation

When the vertical drop height exceeds 7–8 m or if the site topography is not suitable for a single drop, a succession of drops (i.e. a cascade) can be envisaged (e.g. Plate 6). With a stepped cascade (Figs 20.4 and 20.5), low overflows result in a succession of free-falling nappes. This flow situation is defined as a *nappe flow regime*. Stepped channels with nappe flows may be analysed as a succession of drop structures.

20.3.2 Basic flow properties

Along a stepped chute with nappe flow, critical flow conditions take place next to the end of each step, followed by a free-falling nappe and jet impact on the downstream step. At each step, the jet impact is followed by a hydraulic jump, a subcritical flow region and critical flow next to the step edge.

The basic flow characteristics at each step can be deduced from equations (20.1) to (20.5) (Chanson, 1995b).

Note

At large flow rates, the stepped channel flow becomes a skimming flow regime (i.e. extremely rough turbulent flow) (Plate 9). The re-analysis of model studies (Chanson, 2001) suggests that the transition from nappe to skimming flow is a function of the step height and channel slope. The skimming flow regime occurs for:

$$\frac{d_c}{h} > 1.2 - 0.325\frac{h}{l} \qquad \text{Skimming flow regime}$$

where d_c is the critical depth ($d_c = \sqrt[3]{q^2/g}$), h is the step height and l is the step length. For skimming flow conditions, chute calculations are developed as detailed in Chapter 19 and Appendix A4.1.

Nappe flow regime is observed for:

$$\frac{d_c}{h} < 0.89 - 0.4\frac{h}{l} \qquad \text{Nappe flow regime}$$

There is a range of intermediate flow rates (i.e. $0.89 - 0.4\,h/l < d_c/h \ll 1.2 - 0.325\,h/l$) characterized by a chaotic flow motion associated with intense splashing and called the transition flow regime (Chanson, 2001).

20.3.3 Energy dissipation

In a nappe flow situation, the head loss at any intermediary step equals the step height. The energy dissipation occurs by jet breakup and jet mixing, and with the formation of a hydraulic jump on the step. The total head loss along the chute ΔH equals the difference between the maximum head available H_1 and the residual head at the downstream end of the channel H_{res} (Fig. 20.4):

$$\frac{\Delta H}{H_1} = 1 - \frac{H_{res}}{H_1} \tag{20.6}$$

The residual energy is usually dissipated at the toe of the chute by a hydraulic jump in the dissipation basin. Combining equations (20.4) and (20.6), the total energy loss can be calculated as:

$$\frac{\Delta H}{H_1} = 1 - \left(\frac{0.54(d_c/\Delta z)^{0.275} + (3.43/2)(d_c/\Delta z)^{-0.55}}{(3/2) + (\Delta z/d_c)} \right) \qquad (20.7)$$

where Δz is the dam crest elevation above downstream toe.

Discussion

Equation (20.7) was developed for nappe flow regime with fully developed hydraulic jump. Chanson (1995b, 2001) showed that other types of nappe flow may occur: e.g. nappe flow without hydraulic jump. For such nappe flows, equation (20.7) could overestimate the rate of energy dissipation.

20.4 EXERCISES

A vertical drop structure is to be designed at the downstream end of a mild-slope waterway. The width of the crest and drop will be 15 m. The crest level will be at 23.5 m RL and the design head above crest level will be 1.9 m. The elevation of the downstream channel bed will be set at 17.2 m RL. The impact drop will be followed (without transition section) by a mild-slope channel. The tailwater level at design flow conditions will be set at 19.55 m RL.

(a) Calculate the design flow rate of the drop structure. Assuming the apron level to be at the same elevation as the downstream bed level, calculate (b) supercritical flow depth immediately downstream of the nappe impact, (c) drop length, (d) residual power at the end of the chute (give SI units and comment), (e) jump height rating level (JHRL) at design flow. (f) Determine the apron level to match the tailwater level and JHRL at design flow rate.

A stepped cascade (nappe flow regime) is to be designed for $60 \, \text{m}^3/\text{s}$. The cascade is to have a broad crest followed by five steps (1.5 m high and 2.5 m long). The crest and cascade width will be 51 m. The crest level will be at 74.0 m RL.

At design flow condition (a) calculate the head above crest invert, (b) check that the flow is nappe flow, (c) calculate the residual head and the residual power to be dissipated in a downstream stilling structure. (d) If the spillway flow reaches $210 \, \text{m}^3/\text{s}$, what will happen? (Discuss and comment.) (e) For the same cascade slope and width, what is the smallest step height to have nappe flow (at design flow conditions)?

Culvert design

21

21.1 INTRODUCTION

A culvert is a covered channel of relatively short length designed to pass water through an embankment (e.g. highway, railroad and dam). It is a hydraulic structure and it may carry flood waters, drainage flows, natural streams below earthfill and rockfill structures. From a hydraulic aspect, a dominant feature of a culvert is whether it runs full or not.

The design can vary from a simple geometry (i.e. box culvert) to a hydraulically smooth shape (i.e. minimum energy loss (MEL) culvert[1]) (Fig. 21.1). In this section, we will first review the design of standard culverts, then we will discuss the design of MEL culverts.

Note

A culvert below an embankment dam is usually a 'long' structure which operates full (i.e. as pipe flow). In this section, we shall focus on 'short' culverts operating primarily with free-surface flows.

21.2 BASIC FEATURES OF A CULVERT

21.2.1 Definitions

A culvert consists of three parts: the *intake* (also called inlet or fan), the *barrel* (or throat) and the *diffuser* (also called outlet or expansion fan) (Fig. 21.2). The cross-sectional shape of the barrel may be circular (i.e. pipe), rectangular (i.e box culvert) or multi-cell (e.g. multi-cell box culvert) (Fig. 21.2). The bottom of the barrel is called the *invert* while the barrel roof is called the *soffit* or *obvert*. The training walls of the inlet and outlet are called *wing walls*.

21.2.2 Ideal-flow calculations

A culvert is designed to pass a specific flow rate with the associated natural flood level. Its hydraulic performances are the design discharge, the upstream total head and the maximum (acceptable) head loss ΔH. The design discharge and flood level are deduced from the hydrological investigation of the site in relation to the purpose of the culvert. Head losses must be minimized to reduce upstream backwater effects (i.e. upstream flooding).

The hydraulic design of a culvert is basically the selection of an optimum compromise between discharge capacity and head loss or afflux, and of course construction costs. Hence (short) culverts are designed for free-surface flow with critical (flow) conditions in the barrel. Simplified hydraulic calculations are based upon the assumptions of smooth intake and diffuser, and no energy loss: i.e. the total head is the same upstream and downstream. Assuming a given upstream total head H_1,

---

[1] The design of an MEL culvert is associated with the concept of constant total head. The inlet and outlet must be streamlined in such a way that significant form losses are avoided. For an introduction on MEL culverts, see Apelt (1994). For a complete review of MEL waterways, see Apelt (1983).

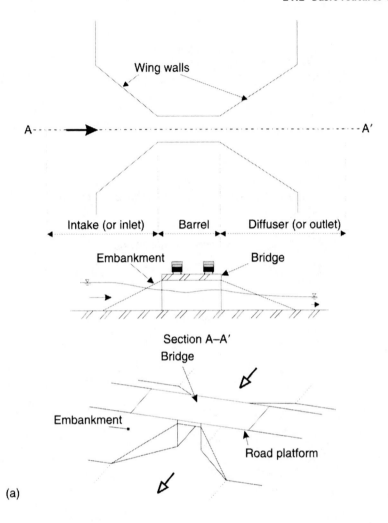

Fig. 21.1 Sketch of a culvert: (a) box culvert.

for a horizontal structure of rectangular cross-section and neglecting energy loss, the maximum discharge per unit width is achieved for critical flow conditions[2] in the barrel:

$$q_{max} = \sqrt{g \left(\frac{2}{3}(H_1 - z_{inlet}) \right)^{3/2}} \qquad (21.1a)$$

where z_{inlet} is the inlet bed elevation.

The minimum barrel width to achieve critical flow conditions is then:

$$B_{min} = \frac{Q_{max}}{\sqrt{g}} \left(\frac{2}{3}(H_1 - z_{inlet}) \right)^{-3/2} \qquad (21.2a)$$

---

[2] In open channel flows, the flow conditions such as the specific energy is minimum are called the critical flow conditions.

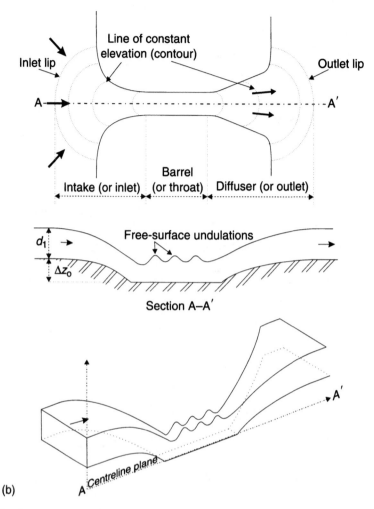

Fig. 21.1 (b) MEL culvert.

The critical depth in the barrel equals:

$$d_c = \frac{2}{3}(H_1 - z_{inlet}) \tag{21.3a}$$

The barrel invert may be lowered to increase the discharge capacity or to reduce the barrel width (e.g. Fig. 21.1b). The above equations become:

$$q_{max} = \sqrt{g}\left(\frac{2}{3}(H_1 - z_{inlet} + \Delta z_o)\right)^{3/2} \tag{21.1b}$$

$$B_{min} = \frac{Q_{max}}{\sqrt{g}}\left(\frac{2}{3}(H_1 - z_{inlet} + \Delta z_o)\right)^{-3/2} \tag{21.2b}$$

$$d_c = \frac{2}{3}(H_1 - z_{inlet} + \Delta z_o) \tag{21.3b}$$

where Δz_o is the bed elevation difference between the inlet invert and the barrel bottom.

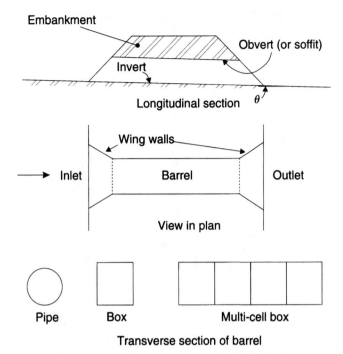

Fig. 21.2 Basic definitions.

> **Notes**
> 1. Equation (21.2) gives the minimum barrel width to obtain near-critical flow without 'choking' effects. If $B < B_{min}$, choking effects will take place in the barrel (e.g. Henderson, 1966: pp. 47–49).
> 2. The barrel width can be reduced by lowering the barrel bottom elevation, but designers must however choose an adequate barrel width to avoid the risks of culvert obstruction by debris (e.g. rocks, trees and cars).

21.2.3 Design considerations

First the function of the culvert must be chosen and the design flow conditions (e.g. Q_{des}, flood level) must be selected.

The primary design constraints in the design of a culvert are:

1. the *cost must be* (always) *minimum*;
2. the afflux[3] must be small and preferably minimum;
3. eventually the embankment height may be given or may be part of the design;
4. a scour protection may be considered, particularly if a hydraulic jump might take place near the culvert outlet.

21.2.4 Flow through a culvert

Most culverts are designed to operate as open channel systems, with critical flow conditions occurring in the barrel in order to maximize the discharge per unit width and to reduce the barrel cross-section (and hence its cost).

---

[3] The afflux is the rise of water level above normal free-surface level on the upstream side of the culvert.

Table 21.1 Flow rate in the barrel as a function of the barrel Froude number (box culvert)

Fr (1)	E/d_c (2)	q/q_{max} (3)
0.3	2.33	0.52
0.5	1.79	0.77
0.7	1.58	0.926
0.8	1.53	0.969
0.9	1.51	0.993
0.95	1.50	0.998
0.99	1.50	1.000
1	1.50	1.000
1.01	1.50	1.000
1.1	1.51	0.994
1.2	1.52	0.977
1.5	1.62	0.89

Notes: q_{max}: maximum discharge per unit width computed as equation (21.1a); E: specific energy in the barrel; Fr: Froude number in the barrel.

The flow upstream and downstream of the culvert is typically subcritical. As the flow approaches the culvert, the channel constriction (i.e. intake section) induces an increase in Froude number. For the design discharge, the flow becomes near critical in the barrel. In practice, perfect-critical flow conditions in the barrel are difficult to establish: they are characterized by 'choking' effects and free-surface instabilities. Usually, the Froude number in the barrel is about 0.7–0.9[4], and the discharge per unit width is nearly maximum as shown in Table 21.1.

Comment

In mountain areas, the stream slope may be steep, and the upstream and downstream flow conditions would typically be supercritical. The culvert is designed to achieve critical conditions in the barrel.

21.2.5 Undular flow in the barrel

In the barrel, the near-critical flow[5] at design discharge is characterized by the establishment of stationary free-surface undulations (e.g. Fig. 21.1b). For the designers, the characteristics of the free-surface undulations are important for the sizing of the culvert height. If the waves leap on the roof, the flow might cease to behave as an open channel flow and become a pipe flow.

Henderson (1966) recommended that the ratio of upstream specific energy to barrel height should be less than 1.2 for the establishment of free-surface flow in the barrel (Table 21.2). Such a ratio gives a minimum clearance above the free-surface level in the barrel of about 20%.

---

[4] If the upstream and downstream flow conditions are supercritical, it is preferable to design the barrel for Froude numbers of about 1.3–1.5.

[5] Near-critical flows are defined as flow situations characterized by the occurrence of critical or nearly critical flow conditions over a 'reasonably long' distance and period time.

Table 21.2 Flow conditions for free-surface inlet flow (standard culvert)

Reference (1)	Conditions (2)	Remarks (3)
Henderson (1966)	$(H_1 - z_{inlet}) \leqslant 1.2D$	Box culvert
	$(H_1 - z_{inlet}) \leqslant 1.25D$	Circular culvert
Hee (1969)	$(H_1 - z_{inlet}) \leqslant 1.2D$	
Chow (1973)	$(H_1 - z_{inlet}) \leqslant 1.2-1.5D$	
US Bureau of Reclamation (1987)	$(H_1 - z_{inlet}) \leqslant 1.2D$	

Notes: D: barrel height; H_1: total head at inlet; z_{inlet}: inlet bed elevation.

Note

Chanson (1995a) observed experimentally that a 20% free-space clearance (between the mean free-surface level and the roof) is a minimum value when the barrel flow conditions are undular. He further showed that the free-surface undulation characteristics are very close to undular weir flow. Both the broad-crested weir and the culvert are designed specifically for near-critical flow above the crest and in the barrel, respectively. The similarity between different types of near-critical flows might be used by designers to gain some order of magnitude of the free-surface undulation characteristics (in absence of additional experimental data).

21.3 DESIGN OF STANDARD CULVERTS

21.3.1 Presentation

A standard culvert is designed to pass waters at a minimum cost without much consideration of the head loss. The culvert construction must be simple: e.g. circular pipes and precast concrete boxes (Fig. 21.3). Figure 21.3 a1–a3 shows a culvert made of a combination of circular pipes and concrete boxes, including inoperation (Fig. 21.3 (a3)) while Fig. 21.3b presents a concrete box culvert.

21.3.2 Operation – flow patterns

For standard culverts, the culvert flow may exhibit various flow patterns depending upon the discharge (hence the critical depth in barrel d_c), the upstream head above the inlet invert $(H_1 - z_{inlet})$, the uniform equilibrium flow depth in the barrel d_o, the barrel invert slope θ, the tailwater depth d_{tw} and the culvert height D.

Hee (1969) regrouped the flow patterns into two classes and eight categories altogether (Fig. 21.4 and Table 21.3):

- Class I for *free-surface inlet flow* conditions,
- Class II for *submerged entrance*.

Free-surface inlet flows (Class I) take place typically for (Table 21.2):

$$\frac{H_1 - z_{inlet}}{D} \leqslant 1.2 \tag{21.4}$$

In each class, the flow patterns can be sub-divided in terms of the control location: i.e. whether the hydraulic control is located at the entrance (i.e. *inlet control*) or at the outlet (i.e. *outlet control*) (Table 21.3).

(a1)

(a2)

(a3)

Fig. 21.3 Examples of standard culverts in Brisbane QLD, Australia: (a) Combination of four concrete pipes and three precast concrete boxes. This culvert is not properly designed and the road is overtopped typically once to twice per year: inlet (photograph taken on September 1996). (a1) Inlet (photograph taken in Sept. 1996). Note the trashrack to trap debris. Note also that the barrel axis is not aligned with the creek. (a2) Outlet (photograph taken on September 1996). (a3) Inlet operation during a rain storm on 31 December 2001. Flow from right to left.

Fig. 21.3 (b) Three concrete boxes. Culvert outlet in August 1999.

(b)

Discussion

With free-surface inlet flow, critical depth is observed at the outlet when S_c is larger than S_o, where S_o is the barrel invert slope and S_c is the barrel critical slope (Fig. 21.4a). Note that S_o and S_c must be calculated in terms of the barrel invert slope and barrel flow conditions.

For submerged entrance cases (Fig. 21.4b), the flow is controlled by the outlet conditions when the barrel is full or drowned (Cases 6 and 7), and by the inlet conditions when free-surface flow is observed in the barrel (Cases 5 and 8).

Case 7 is observed usually for $d_o > D$ where d_o is the barrel normal depth. But it might also occur for $d_o < D$ if a backwater effect (i.e. large tailwater depth) moves the hydraulic jump into the barrel. Case 8 is usually observed for $d_o < d_c$. It may also occur for $d_o > d_c$ because of the 'sluice gate' effect (i.e. vena contracta) induced by the barrel entrance.

For flat (i.e. mild) flood plains, the flow pattern is 'outlet control' if $d > d_c$ in all the waterway, as subcritical flows are controlled from downstream. Culverts flowing full are controlled by the tailwater conditions (i.e. outlet control).

Notes

1. The barrel invert slope S_o is defined as $S_o = \sin\theta$ (Fig. 21.2). The uniform equilibrium flow conditions are related to the bed slope as:

$$S_o = \sin\theta = S_f = f\frac{V_o^{\,2}}{2g(D_H)_o}$$

where V_o and $(D_H)_o$ are the uniform equilibrium flow velocity and hydraulic diameter, respectively, and f is the Darcy friction factor.

2. The channel slope, for which the uniform equilibrium flow is critical, is called the *critical slope* denoted as S_c. The critical slope satisfies:

$$S_c = \sin\theta_c = S_f = f\frac{V_o^{2}}{2g(D_H)_o}$$

where V_o and $(D_H)_o$ must also satisfy $V_o = V_c$ and $(D_H)_o = (D_H)_c$. For a wide rectangular channel, the critical slope satisfies $S_c = \sin\theta_c = f/8$.

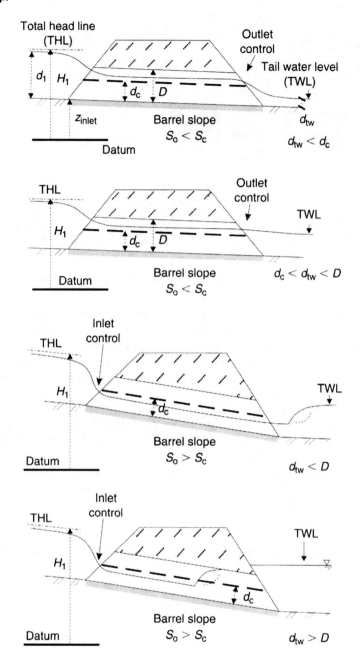

Fig. 21.4 Operation of a standard culvert: (a) free-surface inlet flow conditions.

21.3.3 Discharge characteristics

The discharge capacity of the barrel is primarily related to the flow pattern: free-surface inlet flow, submerged entrance or drowned barrel (Fig. 21.4 and Table 21.3).

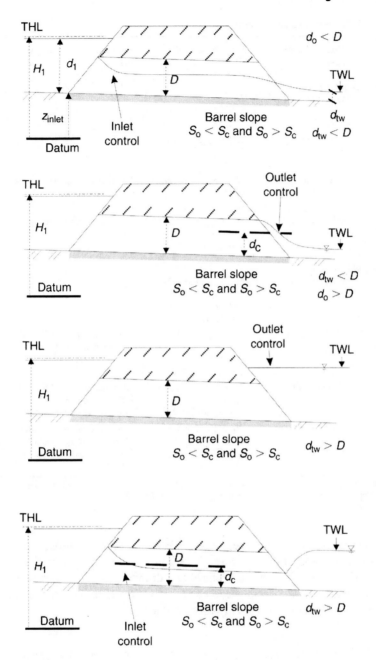

Fig. 21.4 (b) submerged entrance.

When free-surface flow takes place in the barrel, the discharge is fixed only by the entry conditions (Table 21.4), whereas with drowned culverts, the discharge is determined by the culvert resistance. Table 21.4 summarizes several discharge relationships for circular and box culverts.

Table 21.3 Typical conditions of standard culvert operation

Flow pattern (1)	Control location (2)	Flow conditions (3)	Remarks (4)
Class I: Free-surface inlet flow			
Case 1	Outlet control	$(H_1 - z_{inlet}) \leq 1.2D$ $d_{tw} < d_c$ $S_o < S_c$	
Case 2	Outlet control	$(H_1 - z_{inlet}) \leq 1.2D$ $d_c < d_{tw} > D$ $S_o < S_c$	
Case 3	Inlet control	$(H_1 - z_{inlet}) \leq 1.2D$ $d_{tw} < D$ $S_o \geq S_c$	Hydraulic jump takes place at outlet
Case 4	Inlet control	$(H_1 - z_{inlet}) \leq 1.2D$ $d_{tw} < D$ $S_o \geq S_c$	Hydraulic jump takes place in the barrel
Class II: Submerged entrance			
Case 5	Inlet control	$(H_1 - z_{inlet}) > 1.2D$ $d_{tw} < d_c$ $d_o < D$ $S_o < S_c$ or $S_o > S_c$	
Case 6	Outlet control	$(H_1 - z_{inlet}) > 1.2D$ $d_{tw} < d_c$ $d_o > D$ $S_o < S_c$ or $S_o > S_c$	Drowned barrel; critical flow depth is observed at outlet
Case 7	Outlet control	$(H_1 - z_{inlet}) > 1.2D$ $d_{tw} > D$ $S_o < S_c$ or $S_o > S_c$	Drowned barrel; observed usually for $d_o > D$, but might occur for $d_o < D$ if a backwater effect moves the hydraulic jump in barrel
Case 8	Inlet control	$(H_1 - z_{inlet}) > 1.2D$ $d_{tw} > D$ $S_o < S_c$ or $S_o > S_c$	Hydraulic jump takes place at outlet; usually observed for $d_o < d_c$ may occur for $d_o > d_c$ because of vena contracta effect at barrel intake

Notes: D: barrel height; d_c: critical depth in barrel; d_o: uniform equilibrium depth in barrel; d_{tw}: tailwater depth; H_1: total head at inlet invert; S_c: critical slope of barrel; S_o: barrel invert slope; z_{inlet}: inlet invert elevation.
Reference: Hee (1969).

Nomographs are also commonly used (e.g. US Bureau of Reclamation, 1987; Concrete Pipe Association of Australasia, 1991). Figures 21.5 and 21.6 show nomographs to compute the characteristics of box culvert with inlet control and drowned barrel, respectively.

For short box culverts (i.e. free-surface barrel flow) in which the flow is controlled by the inlet conditions, the discharge is typically estimated as:

$$\frac{Q}{B} = C_D \frac{2}{3}\sqrt{\frac{2}{3}g}(H_1 - z_{inlet})^{1.5} \qquad \text{Free-surface inlet flow (Class I)} \qquad (21.5)$$

$$\frac{Q}{B} = CD\sqrt{2g(H_1 - z_{inlet} - CD)} \qquad \text{Submerged entrance and free-surface barrel flow} \qquad (21.6)$$

where B is the barrel width and D is the barrel height (Fig. 21.7). C_D equals 1 for rounded vertical inlet edges and 0.9 for square-edged inlet. C equals 0.6 for square-edged soffit and 0.8 for rounded soffit (Table 21.4).

Table 21.4 Discharge characteristics of standard culverts

Flow pattern (1)	Relationship (2)	Flow conditions (3)	Comments (4)
Free-surface inlet flow			
Circular culvert	$\dfrac{Q}{D} = 0.432\sqrt{g}\,\dfrac{(H_1 - z_{inlet})^{1.9}}{D^{0.4}}$	$0 < \dfrac{H_1 - z_{inlet}}{D} < 0.8$ $0.025 < \sin\theta < 0.361$	Relationship based upon laboratory experiments (Henderson, 1966)
	$\dfrac{Q}{D} = 0.438\sqrt{g}\,(H_1 - z_{inlet})^{1.5}$	$0.8 < \dfrac{H_1 - z_{inlet}}{D} < 1.2$ $0.025 < \sin\theta < 0.361$	Relationship based upon laboratory experiments (Henderson, 1966)
Box culvert	$\dfrac{Q}{B} = C_D\,\dfrac{2}{3}\sqrt{\dfrac{2}{3}g}\,(H_1 - z_{inlet})^{1.5}$ $C_D = 1$ $C_D = 0.9$	$\dfrac{H_1 - z_{inlet}}{D} < 1.2$ Rounded vertical edges (radius > $0.1B$) Vertical edges left square	Relationship based upon laboratory experiments (Henderson, 1966)
Submerged entrance			
Box culvert	$\dfrac{Q}{B} = CD\sqrt{2g(H_1 - z_{inlet} - CD)}$ $C = 0.8$ $C = 0.6$	$\dfrac{H_1 - z_{inlet}}{D} > 1.2$ Rounded soffit edges Square-edged soffit	Free-surface barrel flow; relationship based upon laboratory experiments (Henderson, 1966)
Drowned barrel			
Circular culvert	$\dfrac{Q}{D} = \dfrac{\pi}{4}D\sqrt{2g\,\dfrac{\Delta H}{K'}}$	K': head loss coefficient (primary and secondary losses)	Darcy equation; pipe flow head loss calculations
Box culvert	$\dfrac{Q}{D} = B\sqrt{2g\,\dfrac{\Delta H}{K'}}$	K': head loss coefficient (primary and secondary losses)	Darcy equation; pipe flow head loss calculations

Notes: B: barrel width; D: barrel height; H_1: upstream total head (immediately upstream of inlet); z_{inlet}: inlet invert elevation; ΔH: head loss; θ: barrel invert slope.

Note

Recently, computer programs were introduced to compute the hydraulic characteristics of standard culverts. One is described in Appendix A4.4. Practically, computer programs cannot design culverts: i.e. they do not provide the optimum design. They are, however, useful tools to gain a feel for the operation of a known culvert.

21.3.4 Design procedure

The design process for standard culverts can be divided into two parts (Herr and Bossy, 1965: HEC No. 5). First, a system analysis must be carried out to determine the objectives of the culvert, the design data, the constraints, etc., including the design flow Q_{des} and the design upstream total head H_{des} (basically the design upstream flood height).

In a second stage, the barrel size is selected by a test-and-trial procedure, in which both inlet control and outlet control calculations are performed (Figs 21.8 and 21.9). At the end:

the *optimum size* is the smallest barrel size allowing for *inlet control* operation.

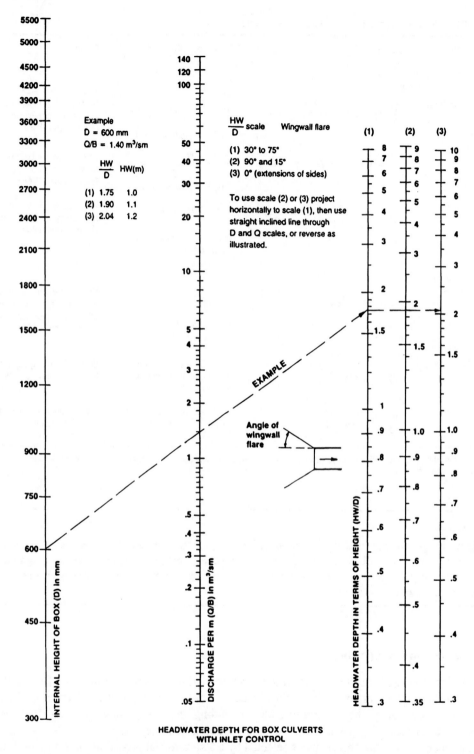

Fig. 21.5 Hydraulic calculations of upstream head above invert bed for box culverts with inlet control (Concrete Pipe Association of Australasia, 1991: p. 39).

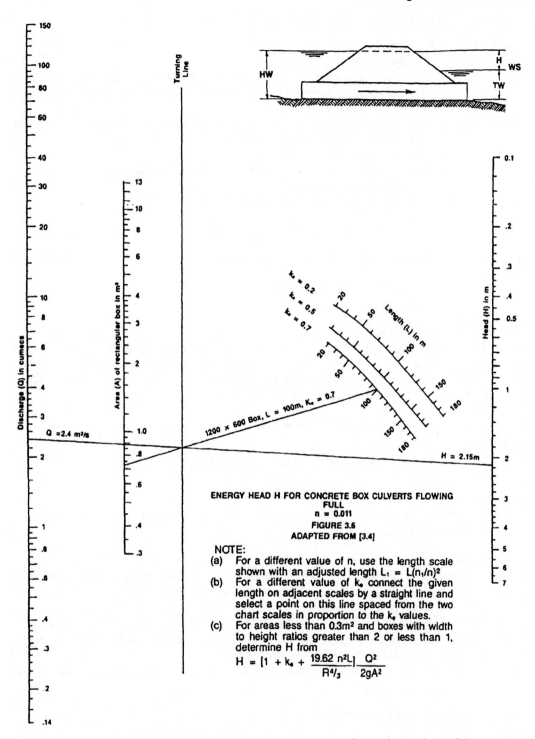

Fig. 21.6 Hydraulic calculations of total head losses for concrete box culverts flowing full (i.e. drowned) (Concrete Pipe Association of Australasia, 1991: p. 41).

Calculations of the barrel size are *iterative*:

Step 1. Choose the barrel dimensions.

Step 2. Assume an *inlet control* (Fig. 21.8a).

   Step 2.1. Calculate the upstream total head $H_1^{(ic)}$ corresponding to the design discharge Q_{des} assuming inlet control. Use the formulae given in Table 21.4 or design charts (e.g. Figs 21.5 or 21.7).

   Step 2.2. Repeat the above procedure (Step 2.1) for different barrel sizes until the upstream head $H_1^{(ic)}$ satisfies the design specifications (i.e. $H_1^{(ic)} = H_{des}$).

Step 3. Assume an *outlet control* (Fig. 21.8b).

   Step 3.1. Use design charts (e.g. Fig. 21.6) to calculate the head loss ΔH from inlet to outlet for the design discharge Q_{des}.

   Step 3.2. Calculate the upstream total head $H_1^{(oc)}$:

$$H_1^{(oc)} = H_{tw} + \Delta H$$

Step 4. Compare the inlet control and outlet control results:

$$H_{des} = H_1^{(ic)} \lessgtr H_1^{(oc)}$$

The larger value controls

When the inlet control design head H_{des} (used in Step 2.2) is larger than $H_1^{(oc)}$, inlet control operation is confirmed and the barrel size is correct. On the other hand, if $H_1^{(oc)}$ is larger than H_{des} (used in Step 2.2), outlet control takes place. Return to Step 3.1 and increase the barrel size until $H_1^{(oc)}$ satisfies the design specification H_{des} (used in Step 2.2).

Notes

1. For a wide flood plain, the tailwater head approximately equals:

$$H_{tw} \approx d_{tw} + z_{outlet}$$

   where z_{outlet} is the outlet invert elevation.

2. Note that $z_{inlet} = z_{outlet} + L_{culv} \tan \theta$, where L_{culv} is the horizontal culvert length and θ is the barrel bed slope (Fig. 21.8).

Discussion

Figure 21.9 illustrates the operation of a box culvert for design and non-design flow conditions. Figure 21.9a shows the design flow conditions with inlet control and free-surface flow in the barrel. Note the presence of a hydraulic jump at the outlet. In Fig. 21.9b, the tailwater depth is larger than the design tailwater depth, leading to the 'drowning' of the barrel. Outlet control takes place and the barrel flow is similar to the flow in a rectangular pipe. Figure 21.9c shows the culvert operation for a discharge lower than design flow, with inlet control and free-surface inlet flow.

21.4 DESIGN OF MEL CULVERTS

21.4.1 Definition

An MEL culvert is a structure designed with the concept of minimum head loss. The flow in the approach channel is contracted through a streamlined inlet into the barrel where the channel

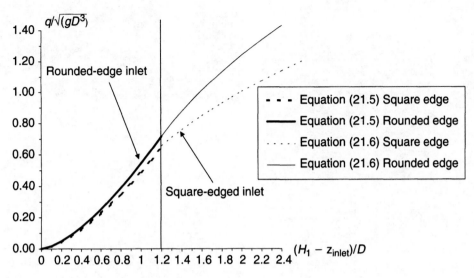

Fig. 21.7 Discharge characteristics of standard-box culverts (inlet control flow conditions).

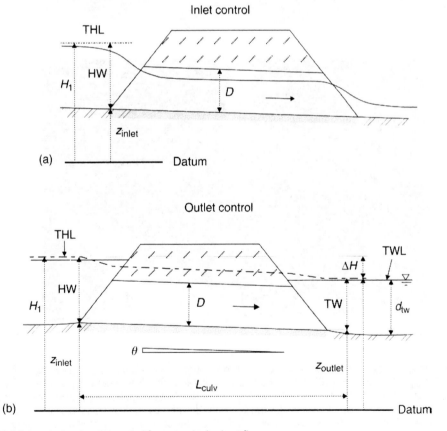

Fig. 21.8 Inlet control and outlet control for a standard culvert flow.

Fig. 21.9 Examples of box culvert operation. Box culvert model. Design flow conditions: $Q_{des} = 10$ l/s, $d_{tw} = 0.038$ m, $S_o = 0.0035$, $B_{min} = 0.15$ m, $D = 0.107$ m, $L_{culv} = 0.5$ m: (a) Design flow conditions: view from downstream (flow from top right to bottom left). $Q = 10$ l/s, $d_{tw} = 0.038$ m and $d_1 = 0.122$ m. Note the hydraulic jump downstream of the culvert. (b) Non-design flow conditions: view from downstream of the drowned barrel (flow from top right to bottom left). $Q = 10$ l/s, $d_{tw} = 0.109$ m, $d_1 = 0.133$ m and outlet control. (c) Non-design flow conditions: view from upstream (flow from bottom right to top left). $Q = 5$ l/s, $d_{tw} = 0.038$ m and $d_1 = 0.082$ m – Inlet control and free-surface inlet flow.

width is minimum, and then it is expanded in a streamlined outlet before being finally released into the downstream natural channel. Both the inlet and outlet must be streamlined to avoid significant form losses. The barrel invert is sometimes lowered to increase the discharge capacity (equation (21.1b)). Figures 21.1b, 21.10–21.12 show typical designs. Appendix A4.3 also presents illustrated examples.

(a1)

(a2)

Fig. 21.10 Photographs of prototype MEL waterways and culverts (see Appendix A4.3): (a) MEL waterway along Norman Creek. Design discharge: ~200 m³/s; throat width: ~4 m. (a1) View from downstream in September 1996. Note the low-flow channel, the outlet (in foreground), the road (Ridge Street) parallel to the waterway (far right) and the freeway bridge (above). (a2) View from downstream on 31 December 2001 around 6:10 am at the end of a rainstorm ($Q \sim$ 60–80 m³/s).

Professor C.J. Apelt presented an authoritative review (Apelt, 1983) and a well-documented audio-visual documentary (Apelt, 1994). The writer highlighted the wide range of design options and illustrated prototypes (Chanson, 1999, 2000b, 2001c). In the following sections, a summary of the design process is given. The same procedure applies to both MEL culverts and MEL waterways (e.g. bridge waterways).

(b1)

(b2)

Fig. 21.10 (b) MEL culvert along Norman Creek, below Ridge Street. Design discharge: 220 m³/s; barrel: seven cells (2 m wide each). (b1) View of the inlet. Looking downstream on Monday 13 May 2002 during a student fieldwork. (b2) Inlet operation on 31 December 2001 around 6:10 am at the end of a rainstorm (Q ~ 60–80 m³/s). Note the hydraulic jump in the inlet that is typical of flow rates less than the design discharge.

Notes

1. MEL culverts are also called constant energy culverts, constant head culverts and minimum energy culverts.
2. MEL waterways are designed with the same principles as MEL culverts. They can be used to reduce the bridge span above the waterway, to increase the discharge capacity of bridge waterways without modification of an existing bridge or to prevent the flooding of an adjacent roadway (e.g. Fig. 21.11).
3. Compared to a standard culvert, an MEL culvert design provides less energy loss of the same discharge Q and throat width B_{min}. Alternatively the throat width can be reduced for the same discharge and head loss.

Discussion: historical development of MEL culverts

The concept of MEL culvert was developed by late Professor Gordon McKay (1913–1989). The first MEL structure was the Redcliffe storm waterway system, also called Humpybong Creek drainage outfall, completed in 1960. It consisted of an MEL weir acting as culvert drop inlet followed by a 137 m long MEL culvert discharging into the Pacific Ocean. The structure is still in use (Appendix A4.3). About 150 structures were built in eastern Australia till date. While a number of small-size structures were built in Victoria, major structures were designed, tested and built in south-east Queensland where little head loss is permissible in the culverts and most MEL culverts were designed for zero afflux (Fig. 21.10). The largest MEL waterway is the Nudgee Road MEL system near Brisbane international airport with a design discharge capacity of 800 m³/s (Fig. 21.11). Built between 1968 and 1970, the channel bed is grass lined and the structure is still in use.

(a)

(b)

Fig. 21.11 Photographs of the Nudgee Road bridge MEL waterway (see Appendix A4.3). Design discharge: 850 m³/s; excavation depth: 0.76 m; natural bed slope: 0.00049 and throat width: 137 m: (a) Inlet view (from left bank) on 14 September 1997. (b) Operation during a flood flow (~400 m³/s) in the 1970s: view of the inlet fan from left bank. Note the low turbulence of the flow (photograph from the collection of late Professor G.R. McKay, Australia).

21.4.2 Basic considerations

The basic concepts of MEL culvert design are *streamlining* and *critical flow conditions throughout all the waterway* (inlet, barrel and outlet).

Streamlining

The intake is designed with a smooth contraction into the barrel while the outlet (or diffuser) is shaped as a smooth expansion back to the natural channel.

The 'smooth' shapes should reduce the head losses (compared with a standard culvert) for the same discharge and barrel width. Practically, small head losses are achieved by streamlining the inlet and outlet forms: i.e. the flow streamlines will follow very smooth curves and no separation is observed.

> ### Discussion: analogy between standard/MEL culverts and orifice/Venturi systems
>
> The shape of the standard and MEL culverts can be, respectively, compared to a sharp-edge orifice and a Venturi meter in a circular pipe.
>
> At an orifice, large head losses take place in the recirculation region immediately downstream of the orifice (i.e. as for a standard culvert). In the Venturi meter, the flow is streamlined and very small head losses are observed. From a top view, the MEL culvert sidewalls usually follow the shape of a Venturi meter (Figs 21.12 and 21.13). As for a Venturi meter, the angle between straight diverging walls and the waterway centreline should be less than about 8° (for straight wing wall outlets).

Critical flow conditions

In an open channel, maximum discharge per unit width for a given specific energy is achieved at critical flow conditions (Section 21.2.2, Table 21.1).

MEL structures are designed to achieve critical flow conditions *in all the engineered waterways*: i.e. in the inlet, at the throat (or barrel) and in the outlet (Fig. 21.12). At the throat, the discharge per unit width q is maximum and it may be increased by lowering the barrel bed (equation (21.1b)).

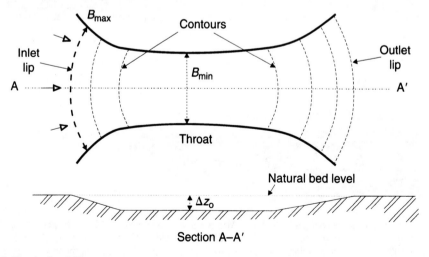

Fig. 21.12 Sketch of an MEL waterway.

Notes
1. MEL structures are usually designed for critical flow conditions from the inlet lip to the outlet lip (i.e. throughout the inlet, barrel and outlet).
2. MEL waterways are not always designed for critical flow conditions. The MEL structure profile may be designed for subcritical (or supercritical) flow conditions: e.g. $Fr \approx 0.2$–0.5.
3. In many cases, it is assumed that the total head gradient through the MEL structure is the same as that which occurred previously in the natural channel: i.e. zero afflux.
4. Lowering the bed level at throat gives an increase in local specific energy E and discharge per unit width q, hence the throat width may be reduced.

21.4.3 A simple design method (by C.J. Apelt)

Professor C.J. Apelt (The University of Queensland) proposed a simple method to calculate the basic characteristics of an MEL culvert. This method gives a preliminary design. Full calculations using the backwater equations are required to accurately predict the free-surface profile.

The simple method is based on the assumption that the flow is critical from the inlet to outlet lips including in the barrel.

Design process
Step 1. Decide the design discharge Q_{des} and the associated total head line (THL). The THL is deduced assuming uniform equilibrium flow conditions in the flood plain.

(a) *Neglect energy losses*

  Step 2.1. Calculate the waterway characteristics in the throat (i.e. barrel) for critical flow conditions. Compute the barrel width B_{min} and/or the barrel invert elevation z_{barrel} (neglecting energy losses).

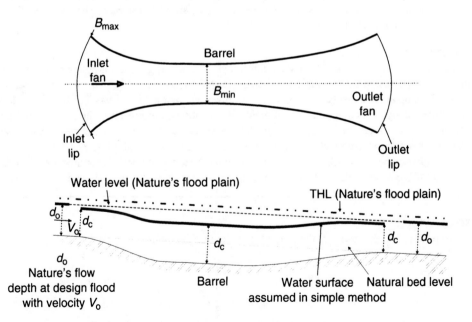

Fig. 21.13 Free-surface profile assumed in the 'simple design method'.

Step 2.2. Calculate the lip width assuming critical flow conditions ($Fr = 1$) and natural bed level (i.e. $\Delta z_o = 0$). The lip width B_{max} is measured along the smooth line normal to the streamlines.

Step 3.1. Decide the shapes of the fans. Select the shape of the fan wing walls (or select the shape of the invert profile).

Step 3.2. Calculate the geometry of the fans to satisfy critical flow conditions every-where. The contours of the fans are each defined by their width B (measured along smooth line normal to streamlines), the corresponding bed elevation (to satisfy $Fr = 1$) and the longitudinal distance from the lip.

In the above steps, either the barrel width B_{min} is selected and the barrel invert drop Δz_o is cal-culated; or Δz_o is chosen and the barrel width is calculated.

(b) *Now include the energy losses*

Step 4. Adjust the bed profile of the waterway to take into account the energy losses. Calculate the total head loss available. Estimate the energy losses (friction loss and form loss) in the waterway (inlet, barrel and outlet). For a long barrel, the barrel slope is selected as the critical slope (of the barrel).

Step 5. Check the 'off-design' performances: i.e. $Q > Q_{des}$ and $Q < Q_{des}$.

Notes

1. In the simple design method, the inlet lip width and the outlet lip width are equal (Fig. 21.13).
2. The barrel width B_{min} is constant all along the barrel.
3. Another way to present the simplified approach is as follows:
   - First, let us assume an ideal-fluid flow (i.e. no friction losses) and a horizontal bed (natural bed and barrel invert). Calculate B_{min}, B_{max}, the three-dimensional shapes of the inlet and of the out-let. If the outlet shape could lead to flow separation at the expansion, increase the outlet length and repeat the outlet design procedure.
   - Secondly, take into account the energy losses (friction and form loss). Calculate the optimum barrel invert slope (see below), and incorporate the natural bed slope and barrel invert slope in the bottom elevation calculations (Fig. 21.13).

Comments

Figure 21.14 presents an example of MEL culvert operation at design and non-design flow conditions. In Fig. 21.14a, design flow conditions are characterized by no afflux ($d_{tw} = d_1$) and the absence of hydraulic jump at the outlet (compared with box culvert operation, Fig. 21.9a). The flow in the barrel is undular and dye injection in the outlet highlights the absence of separation in the diffuser. Figure 21.14b shows a non-design flow ($Q < Q_{des}$). Note the subcritical flow in the barrel, the hydraulic jump in the inlet and the supercritical flow upstream (from the inlet lip) (see also Fig. 21.10d).

A comparison between Figs 21.9 and 21.14 illustrates the different flow behaviour between standard-box and MEL culvert models with the same design flow conditions ($Q_{des} = 10 \text{ l/s}$ and $d_{tw} = 0.038 \text{ m}$).

Discussion

- In Steps 2.1, 2.2, 3.1 and 3.2, the critical flow depth d_c satisfies:

$$q^2 = g d_c^{\ 3}$$

$$d_c = \frac{2}{3} E$$

where E is the specific energy.

(a)

(b)

Fig. 21.14 Examples of MEL culvert operation. MEL culvert model. Design flow conditions: $Q_{des} = 10\,l/s$, $d_{tw} = 0.038\,m$, $S_o = 0.0035$, $B_{min} = 0.10\,m$, $D = 0.17\,m$, $\Delta z_o = 0.124\,m$, no afflux ($d_1 = d_{tw}$): (a) Design flow conditions: view from downstream (flow from top right to bottom left). $Q = 10\,l/s$, $d_{tw} = 0.038\,m$ and $d_1 = 0.038\,m$. Dye injection in the outlet highlight the streamlined flow (i.e. no separation). (b) Non-design flow conditions: view from upstream of the inlet (flow from bottom right to top right). $Q = 5\,l/s$, $d_{tw} = 0.038\,m$ and $d_1 = 0.026\,m$. Note hydraulic jump in the inlet and subcritical flow in barrel.

- B_{max} is deduced from the critical flow conditions at inlet lip (Step 2.2). At the inlet lip, the specific energy is known:

$$E_1 = H_1 - z_{inlet}$$

Hence the flow depth (i.e. critical depth) is known:

$$d_1 = d_c = \frac{2}{3}E_1$$

B_{max} is deduced from the continuity equation:

$$Q_{des} = V_c d_c B_{max}$$

- The profile (and slope) of the inlet and outlet invert is deduced from the complete calculations of (B, z, s) assuming $d = d_c$ at any position (equation (21.2)) and s is the longitudinal coordinate (in the flow direction).
- Upstream of the inlet lip, the flow depth equals the uniform equilibrium flow depth at design flood d_o. At the inlet lip, the flow depth is critical, *and* the 'simple method' assumes that the flow remains critical from the inlet lip down to the outlet lip (Fig. 21.13).
- In practice, the outlet might not operate at critical flow conditions because of the natural flood level (i.e. tailwater level (TWL)) (Fig. 21.15).

  Critical depth at the exit could be achieved by installing a small 'bump' near the exit lip (e.g. Nudgee Road bridge waterway, Queensland). But form losses associated with the 'bump' are much larger than the basic exit losses. In practice, a 'bump' is not recommended.

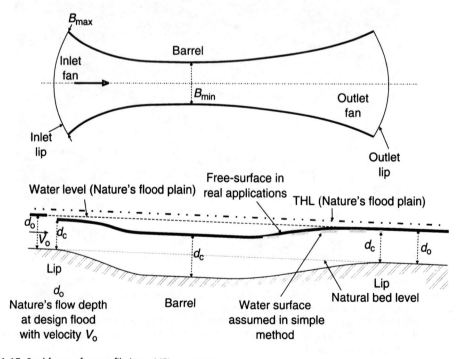

Fig. 21.15 Real free-surface profile in an MEL waterway.

The Nudgee Road waterway was built with a 'bump' as a result of a misunderstanding between the designer and the constructor (see also Appendix A4.3).

- The barrel slope is often designed for the critical slope to account for the friction losses. However the barrel flow is not comparable to a fully developed equilibrium flow for which the momentum, Chézy or Gauckler–Manning equations apply. In practice, Apelt (1983) suggested that most barrel design with critical slope should operate reasonably well. But he mentioned one case where the approximation did not hold: if the slope of the barrel invert is not equal to the critical slope, critical flow conditions might not be achieved in a long barrel. As a result some afflux might occur.
- The simple method calculations must be checked by physical model tests and complete backwater calculations might be conducted.

Notes

1. The lip of the inlet and outlet is the outer edge (upstream for inlet and downstream for outlet) of the inlet and outlet. The 'lip length' (or lip width) is the length of the lip curve measured along the smooth line perpendicular to the streamlines (Figs 21.12 and 21.13).
2. The inlet and outlet may be shaped as a series of curved smooth lines of constant bed elevation (i.e. contour) which are all normal to the flow streamlines. These smooth lines are equi-potential lines (cf. ideal-fluid flow theory). The streamlines and equi-potential lines (i.e. contour lines) form a flow net.
3. In most cases the barrel slope is selected as the critical slope. The critical slope is defined as the slope for which uniform equilibrium flow conditions are critical. It yields:

$$S_c = \sin\theta_c = \frac{f}{8}Fr^2\frac{P_w}{B}$$

or

$$S_c = \sin\theta_c = n_{Manning}^2 Fr^2\frac{g\,P_w^{4/3}}{A^{1/3}B}$$

where Fr, A, B, P_w are, respectively the Froude number, cross-sectional area, width and wetted perimeter of the barrel flow. f and $n_{Manning}$ are the Darcy friction factor and the Gauckler–Manning friction coefficient, respectively.

   Warning: The Gauckler–Manning equation is valid only for wide channels and it is inappropriate for most culvert barrels.
4. The barrel Froude number is usually unity at design flow conditions. In some cases, the barrel flow might be designed for non-unity Froude number (e.g. $Fr = 0.6$–8) to reduce free-surface undulation amplitudes in the barrel and to reduce the outlet energy losses.
5. When the (required) barrel critical slope differs substantially from the natural bed slope, the inlet and outlet shapes must be changed in consequence. This situation requires an advanced design in contrast to the 'simple design method'. The whole waterway must be designed as a 'unity' (i.e. a complete system).

Energy losses

The energy losses in an MEL waterway include the *friction losses* in the inlet, barrel and outlet, and the *form losses* (at the outlet primarily). Figures 21.15 and 21.16 summarize the typical free-surface profile and THL along an MEL waterway. The THL of the natural

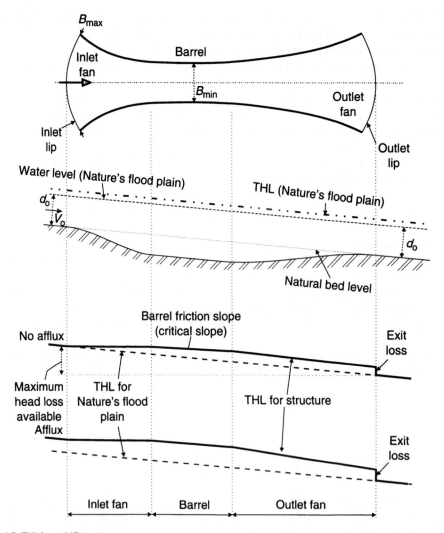

Fig. 21.16 THL in an MEL waterway.

flood plain is indicated in dotted line in Fig. 21.16. A proper design approach must take into account:

- The total head loss available equals the natural bed drop along the culvert length plus the 'acceptable' afflux. The latter is usually zero (e.g. Fig. 21.16).
- The exit losses are always significant. As a result, the outlet flow depth is rarely equal to the critical depth.
- The exit losses are calculated from the exit velocity:

$$\Delta H_{exit} = \frac{V_{exit}^2}{2g} - \frac{V_o^2}{2g} \qquad (21.7)$$

where V_o is the uniform equilibrium flow velocity of the flood plain.

When the outlet does not operate at critical flow conditions, $V_{exit} < V_c$ and the expansion ratio is smaller as $d_{exit} > d_c$. As a result, the exit losses become smaller than that calculated by the simple method (Apelt, 1983).

- The calculations of the exit energy loss are very important. The exit losses must be less than the total energy loss available!
- For the design of MEL culverts the 'simple method' works well when the exit energy loss is less than one third of the total head loss available.

Discussion: exit loss calculations

The exit head loss equals:

$$\Delta H_{exit} = \frac{V_{exit}^2}{2g} - \frac{V_o^2}{2g} \qquad (21.7)$$

where V_{exit} is the mean velocity at the outlet lip and V_o is the normal velocity in the flood plain. Ideally (i.e. 'simple method' design), the exit velocity should equal:

$$V_{exit} = \frac{Q}{B_{max}d_c}$$

where B_{max} is the outlet lip width and d_c is the critical flow depth at the outlet lip.

In practice, the outlet lip depth equals about the normal depth d_o in the flood plain and the 'real' exit velocity is often:

$$V_{exit} = \frac{Q}{B_{max}d_o}$$

21.4.4 Discussion

Challenges and practical considerations

A proper operation of MEL waterways and culverts requires a proper design. Several points are very important and must be stressed out:

- The velocity distribution must be uniform (or as uniform as possible).
- Separation of flow in the inlet and in the outlet must be avoided.
  In one case, an improper inlet design resulted in flow separation at the inlet, and four barrel cells did not operate properly, reducing the discharge capacity of the culvert down by more than one third (at design flow conditions).
  Separation in the outlet is also a common design error. For straight wing outlet walls, the divergence angle between the wall and centreline should be less than about 8° (see Notes below).
- The shapes of the inlet and outlet must be within some reasonable limits.
  For the inlet, Apelt (1983) suggested that the contraction ratio should be limited to a maximum value corresponding to about $\Delta z_o/d < 4$, where Δz_o is the lowering of the bed and d is the approach flow depth.

  Further he indicated that the 'optimum length of inlet fan is the shortest for which the flow approximates the two-dimensional (flow) condition without separation or unacceptable transverse water surface slope'. In practice, this leads to a minimum inlet length of about 0.5 times the maximum inlet width (measured along contour line normal to the flow streamlines): $L_{inlet}/B_{max} > 0.5$.

For the outlet: 'it is more difficult to achieve rapid rates of expansion and large expansion ratios in the outlet fan than it is to achieve rapid rates of contraction and large contraction ratios in the inlet fan' (Apelt, 1983). Head losses at the outlet account for the greater proportion of the losses through the MEL structure.

- Critical flows are characterized by free-surface undulations and flow instabilities.

In practice, free-surface undulations are observed (at design flow conditions) in reasonably long barrel (e.g. Chanson, 1995a). In some cases, the wave amplitude might be as large as $0.5d_c$.

Practically, the waterway may be designed for a barrel Froude number different from unity (e.g. $Fr = 0.6$–0.8) to reduce the wave amplitude and to minimize the obvert elevation.

- The head losses must be accurately predicted.

Notes

1. McKay (1978) recommended strongly to limit MEL design to rectangular cross-sectional waterways. For non-rectangular waterways, the design procedure becomes far too complex and it might not be reliable.

2. *Design of the outlet fan*

   The function of the outlet fan is to decelerate the flow and to expand it laterally before rejoining the natural flood plain (Apelt, 1983). A basic difference between the outlet and inlet fans is the different behaviour of the boundary layers in the two fans. In the inlet fan, the boundary layer is thin and the accelerating convergent flow ensures that the boundary layers remain thin. In contrast, in the outlet fan, the expanding flow is subjected to an adverse (and destabilizing) pressure gradient. The boundary layer growth is more rapid. At the limit, separation might take place and the flow is no longer guided by the fan walls.

   Designers could keep in mind the analogy between MEL culvert and Venturi meter when designing the shape of the outlet fan. Practically, the outlet is often longer than the inlet.

3. *Length of the barrel*

   The selection of the barrel length depends upon the overall design (structural, geomechanics, etc.). Overall, the barrel must be as short as possible. Basically its length would be set at the top width of the embankment plus any extra length needed to fit in with the design of the slopes of the embankment.

   Usually, the barrel length equals the bridge width for low embankments. For high embankment, the barrel length would be larger.

4. *Outlet divergent*

   For a Venturi meter installed in a closed pipe, the optimum divergence angle is about (Levin, 1968; Idelchik, 1969, 1986):

   $$\theta = 0.43 \left(\frac{f}{\alpha_1} \frac{A_{outlet}/A_{inlet} + 1}{A_{outlet}/A_{inlet} - 1} \right)^{4/9} \tag{21.8}$$

   where θ is in radians, f is the Darcy friction factor, A_{outlet} and A_{inlet} are, respectively, the diffuser inlet and outlet cross-sectional areas, and α_1 is an inflow velocity correction coefficient larger than or equal to 1 (Idelchik, 1969, 1986). Equation (21.8) may be applied for straight outlet wing walls of concrete rectangular MEL culverts. Assuming $\Delta z_o = 0$, typical results are shown in Fig. 21.17, suggesting an optimum angle between 5° and 8° for straight divergent walls.

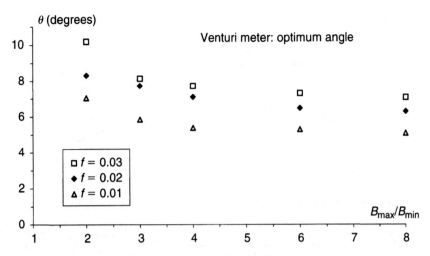

Fig. 21.17 Optimal divergence angle of Venturi meters (equation (21.8)).

In comparison with standard culverts, the design of MEL culverts must include additional factors:

- The flow velocities in the culvert are larger. Wing walls and floors must be adequately protected.
- After a flood, ponding of water is a potential health hazard. It can be avoided by installing drains or anti-ponding pipes (e.g. Fig. 21.10, Appendix A4.3).

Design of multi-cell barrel

For construction reasons, wide rectangular culverts are often built with a number of identical rectangular cells (Figs 21.10 and 21.18). In such cases, the flow cross-section is obstructed by the dividing walls and the *total* wall thickness between adjacent cells must be taken into account into the final design. Indeed, the minimum width of the fan is larger than the 'real' barrel width.

The late Professor Gordon McKay (The University of Queensland) proposed to incorporate into the design a transition section (Fig. 21.18). The transition section is characterized by a progressive change of wall height associated with a change of bed level.

Note
It is recommended that the transition slope be less than 10% (Fig. 21.18).

Operation for non-design flow conditions

Culverts are designed for specific flow conditions (i.e. design flow conditions). However, they operate most of the time at non-design flow rates. In such cases design engineers must have some understanding of the culvert operation.

Figure 21.19 shows a typical sketch of total energy levels (TELs) at the throat of an MEL waterway for non-design discharges. At design flow conditions (i.e. $Q = Q_{des}$), the barrel flow is critical, no afflux takes place and the barrel acts as a (hydraulic) control.

For $Q < Q_{des}$, the barrel flow is typically subcritical, no afflux is observed and the flow is controlled by the normal flow conditions in the flood plain (i.e. downstream flood plain). For discharges larger than the design discharge, the barrel flow becomes critical and acts as a hydraulic control. In addition, the afflux is larger than zero.

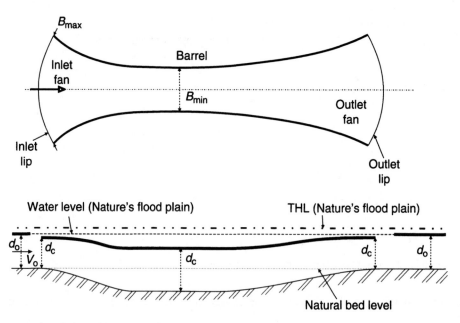

Fig. 21.18 Multi-cell MEL culvert. Design of the transition section.

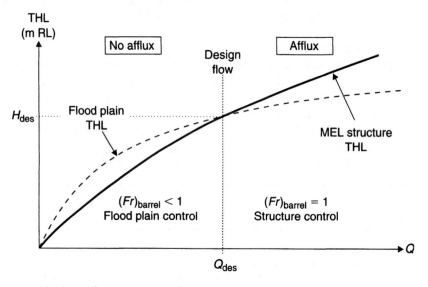

Fig. 21.19 TEL in the throat of an MEL waterway for non-design flow conditions.

21.4.5 Benefits of MEL culverts and waterways

A main characteristic of MEL waterways is the *small head loss*. It results in a small afflux and to a small (or zero) increase of flood level. Furthermore, the *width* of the throat is *minimized*. Moreover, the culvert flow is streamlined and exhibits *low turbulence* (e.g. Fig. 21.11c). As a result, the erosion potential is reduced: e.g. fans can be made of earth with grassed surface (e.g. Fig. 21.11a and c, see also Appendix A4.3).

For zero afflux, the size of an MEL culvert (inlet, barrel, outlet) is much smaller than that of a standard culvert with identical discharge capacity. Further Hee (1969) indicated that, for a very long culvert, the MEL culvert design tends to be more economical.

An additional consideration is the greater factor of safety against flood discharges larger than the design discharge. Model and prototype observations have shown conclusively that MEL culverts can pass safely flood flows significantly larger than the design flow conditions. This is not always the case with standard culverts.

21.5 EXERCISES

A culvert is to be built to pass $25\,m^3/s$ under a road embankment crossing a flood plain. The ground level is 22.000 m RL and the water level corresponding to this flow is expected to be 23.300 m RL. Both levels are at the centreline of the embankment which is 20 m wide at its base. The flood gradient is 0.004. (The culvert will be a multi-cell standard-box culvert made of precast concrete boxes 1.5 m high and 1 m wide. Assume square-edged inlets.)

(a) Design a standard-box culvert (with invert set at ground level) to carry the flood flow without causing any increase in flood level upstream *and operating under inlet control*. (b) If the number of boxes is reduced to five (i.e. 5-cell box culvert), (i) calculate the change in flood level if the design discharge remains unaltered assuming inlet control. (ii) Check whether inlet control is the correct assumption. (Use figure 'flow characteristics of box culverts flowing full (Concrete Pipe Association of Australasia, 1991)', assuming $k_e = 0.5$ for square-edged inlet.)

A culvert is designed to pass $44\,m^3/s$ under a road embankment crossing a flood plain. The ground level is 4.200 m RL and the water level corresponding to this flow is expected to be 5.750 m RL. Both levels are at the centreline of the embankment which is 100 m wide at its base. The flood gradient is 0.005 and the flood plain width is 65 m. The culvert will be a multi-cell box structure built using precast units with inside dimensions 1.8 m wide by 1.8 m high. The total wall thickness between adjacent cells can be taken as 100 mm.

You are required to design a means of carrying the flood flow through the embankment without causing any increase in flood level upstream ('no afflux'). Design the culvert, using the principles of MEL waterways, to minimize the length of the road crossing required. The greatest depth of excavation allowable is 1.1 m below natural surface. Your design will assume that the waterway will be concrete lined. Use the 'simple method' design developed by Professor C.J. Apelt. Calculate the number of cells, the excavation depth and the outlet lip width.

A culvert is to be built to pass $32\,m^3/s$ under a road embankment crossing a flood plain. The ground level is 24.200 m RL and the water level corresponding to this flow is expected to be 25.750 m RL. Both levels are at the centreline of the embankment which is 20 m wide at its base. The flood gradient is 0.005 and the flood plain width is 65 m. The waterway must be designed using the principles of MEL culverts, to minimize the length of the bridge required and to carry the flood flow through the embankment without causing any increase in flood level upstream ('no afflux'). Use the 'simple method' design developed by Professor C.J. Apelt. The greatest depth of excavation allowable is 1.1 m below natural surface. (Assume that the waterway will be concrete lined.):

(a) For the design of MEL culverts, what are the two basic considerations?
(b) What is the inlet lip width?
(c) What is the barrel width?
(d) What is the outlet lip width?
(e) What are the exit losses?
(f) What is the maximum total head loss available? Comment.

A culvert is to be built to pass $87.0\,\mathrm{m^3/s}$ under a road embankment crossing a 350 m wide flood plain. The ground level is 20.000 m RL and the water level corresponding to this flow is expected to be 21.400 m. (Both levels are at the centreline of the embankment which is 20 m wide at its base.) The longitudinal slope of the flood plain is 0.004. The culvert is to be a multi-cell box structure using precast units with inside dimensions 1.8 m wide by 2.0 m high. (Assume square-edged inlets. Use $k_e = 0.5$ for square-edged inlets.)

(I) Design an MEL multi-cell box culvert for this situation. The greatest depth of excavation allowable is 0.900 m below the natural surface. (*Use 'simple' method for design: 'no afflux' design.*) Calculate: (a) inlet lip width, (b) number of cells, (c) required depth of excavation in barrel, (d) critical depth in barrel, (e) specific energy in barrel, (f) freeboard in barrel, (g) exit loss, (h) total head loss available. (Compare the results (g) and (h). Comments.)

(II) Compare the MEL culvert with a standard multi-cell box culvert placed at ground level operating under inlet control for the following cases:

Design 1. The standard culvert has the same number of cells as the MEL one.

(i) Calculate the culvert discharge capacity for the flood level to remain unaltered (i.e. no afflux). (j) Calculate the change in flood level (i.e. afflux) if the discharge remains unaltered (i.e. $87\,\mathrm{m^3/s}$).

Design 2. The standard culvert has sufficient cells to pass the discharge without significant afflux.

(k) Calculate the number of cells.

Design 3. If the standard culvert has the same number of cells as the MEL one and the discharge remains unaltered, determine whether inlet control is the correct assumption.

(l) With outlet control, calculate the upstream head above inlet invert (m). Will the box culvert operate with inlet control or outlet control?

A culvert is to be built to pass $105\,\mathrm{m^3/s}$ under a road embankment crossing a 220 m wide flood plain. The ground level is 15.000 m RL and the water level corresponding to this flow is expected to be 16.600 m. (Both levels are at the centreline of the embankment which is 25 m wide at its base.) The longitudinal slope of the flood plain is 0.004. The culvert is to be a multi-cell box structure using precast units with inside dimensions 1.8 m wide and 2.0 m high.

(a) Design an MEL multi-cell box culvert for this situation. The greatest depth of excavation allowable is 0.900 m below the natural surface. (*Use 'simple' method for design: 'no afflux' design.*)

(b) Compare your MEL culvert with a standard multi-cell box culvert placed at ground level *operating under inlet control* for the following cases: (i) The standard culvert has the same number of cells as the MEL one: calculate (Case a) the change in capacity for the flood level to remain unaltered, and (Case b) the change in flood level if the discharge remains unaltered. (ii) The standard culvert has sufficient cells to pass the discharge without significant afflux. (iii) If the standard culvert has the same number of cells as the MEL one and the discharge remains unaltered, determine whether inlet control is the correct assumption.

Use design charts from Concrete Pipe Association of Australasia (1991), with $k_e = 0.5$ for square-edged inlets.

(c) Considering the MEL culvert design (a), an alternative design includes the design of a broad crest at the inlet lip, with crest elevation set at 15.950 m RL. Design the inlet lip.

The broad crest is introduced to prevent downstream water intrusion into the upstream catchment. (For example, see the Redcliffe MEL culvert design, Appendix A4.3.)

A culvert is to be built to pass $15\,\mathrm{m^3/s}$ under a road embankment crossing a flood plain. The ground level is 16.000 m RL and the water level corresponding to this flow is expected to be 16.900 m RL. Both levels are at the centreline of the embankment which is 20 m wide at its base. The flood gradient is 0.002. (The culvert will be a multi-cell standard-box culvert made of precast concrete boxes 1.0 m high and 0.9 m wide. Assume square-edged inlets.)

(a) Design a standard-box culvert (with invert set at ground level) to carry the flood flow with a maximum afflux of 0.5 m *and operating under inlet control*. (b) If the number of boxes is reduced to five (i.e. 5-cell box culvert): (i) calculate the change in flood level if the design discharge remains unaltered assuming inlet control. (ii) Check whether inlet control is the correct assumption.

Use figures 'hydraulic calculations of upstream head above invert bed for box culverts with inlet control (Concrete Pipe Association of Australasia, 1991: p. 39)' and 'Flow characteristics of box culverts flowing full (Concrete Pipe Association of Australasia, 1991)' assuming $k_e = 0.5$ for square-edged inlet.)

(c) Design an MEL culvert to carry the flood flow with a maximum afflux of 0.5 m and a maximum excavation depth of 0.3 m.

A culvert is to be built to pass 98 m³/s under a road embankment crossing a flood plain. The ground level is 15.000 m RL and the water level corresponding to the design flood is expected to be 16.700 m RL. Both levels are at the centreline of the embankment which is 30 m wide at its base. The flood plain gradient is 3.5 m/km. The culvert will be a multi-cell standard-box culvert made of precast concrete boxes 2.2 m high and 2 m wide. The total wall thickness between adjacent cells may be taken as 0.1 m. (Assume square-edged inlets.)

(a) Design a standard-box culvert (with invert set at ground level) to carry the flood flow without causing any increase in flood level upstream (i.e. no afflux) *and operating under inlet control*. How many cells are required?

(b) If the number of boxes is reduced to five (i.e. 5-cell box culvert): (i) calculate the change in capacity for the flood level to remain unaltered, (ii) calculate the change in flood level if the design discharge remains unaltered assuming inlet control, (iii) check whether inlet control is the correct assumption in (ii) ($k_e = 0.5$ for square-edged inlet).

(c) Design an MEL culvert to carry the flood with four cells using the 'simple method' design developed by Professor C.J. Apelt: (i) calculate the required depth of excavation below the natural surface, (ii) calculate the lip lengths of the inlet and of the outlet, (iii) what is the minimum inlet length that you recommend?

A road embankment is to be carried across a flood plain. You are required to design a means of carrying the flood flow (safely) below the embankment. The design data are:

Design flood flow	98 m³/s
Natural ground level at centreline of embankment	15.000 m RL
Design flood water level (at centreline of embankment)	16.700 m RL
Flood plain slope	3.5 m/km
Width of embankment base	30 m

The culvert will be built as a multi-cell box culvert using precast units with inside dimensions 2 m wide by 2.2 m high. The *total* wall thickness between adjacent cells can be taken as 100 mm. An earlier study of MEL culvert calculations (for this structure) suggested to use a 5-cell structure (i.e. 5-cell MEL culvert). That MEL culvert was designed to carry the flood flow without causing any increase in flood level upstream (i.e. no afflux).

Compare the MEL culvert design with *your design of a standard-box culvert* placed at ground level operating under inlet control for the following cases: (a) The standard culvert has the same number of cells as the MEL one. Calculate (Case 1) the change in capacity for the flood level to remain unaltered, and (Case 2) the change in flood level if the discharge remains unaltered.

(b) The standard culvert has sufficient cells to pass the discharge without significant afflux (Case 3).

Notes: Discharges through box culverts operating under inlet control can be calculated using Fig. 21.7 provided in the book, from Fig. 7.21 on page 264 of Henderson (1966), or from equations (7.31) and (7.32) in Henderson (1966). Assume square-edged inlet.

(c) Repeat calculations of (a) (i.e. Cases 1 and 2) by using the nomograph (Fig. 3.4) of Concrete Pipe Association of Australasia (1991).

(d) Using the nomograph (Fig. 3.6) of Concrete Pipe Association of Australasia (1991), determine whether inlet control is the correct assumption for (Case 2). Use $k_e = 0.5$ for square-edged inlets.

Summary sheet

Flow capacity when water depth in flood plain just upstream is 1.7 m (Case 1):

Question	(a) (m³/s)	(b) (% of 98 m³/s)	(c) (m³/s)
Square-edged inlet			

Depth of water and specific energy on flood plain just upstream (Case 2):

Question	(a) Water depth (m)	(a) Specific energy (m)	(b) Water depth (m)	(b) Specific energy (m)
Square-edged inlet				

(b) Number of cells: (square-edged inlet) (Case 3)
(d) Inlet control (Case 2):   Yes/No
If No, state the new specific energy just upstream: m RL

A culvert is to be built to pass 32 m³/s under a road embankment crossing a flood plain. The ground level is 4.200 m RL and the water level corresponding to this flow is expected to be 5.750 m RL. Both levels are at the centreline of the embankment which is 100 m wide at its base. The flood gradient is 0.005 and the flood plain width is 65 m. Design a means of carrying the flood flow through the embankment without causing any increase in flood level upstream ('no afflux').

(a) Design the waterway, using the principles of MEL culverts, to minimize the length of the bridge required. The greatest depth of excavation allowable is 1.1 m below natural surface. Your design must include the details of inlet and outlet fans. Assume that the waterway will be concrete lined.

Use the 'simple method' design developed by Professor C.J. Apelt. Estimate the exit loss. Comment.

(b) Calculate the width of waterway required to achieve the same objective of no afflux if standard-box culverts are used with *rounded inverts* set at natural ground level and *operating under inlet control*. Select a barrel size which minimizes the construction costs (by minimizing the cross-sectional area of the barrel).

Summary sheet
(a)

Exit loss	
Total head loss available	
Minimum slope required for barrel	
Slope available	
Comment	

Specify in the table below the waterway widths and invert levels at intervals corresponding to depths of excavation equal to 0.25 × (maximum excavation).

Location	Station	Distance from embankment CL (+ve in d/s) (m)	Depth of excavation (m)	Ground level (m RL)	Waterway width (m)
Upstream lip	1				
	2				
	3				
	4				
Barrel (u/s)	5				
Barrel (midway)	6				
Barrel (d/s)	7				
	8				
	9				
	10				
Downstream lip	11				

Provide dimensioned drawing of waterway (plan and longitudinal section).
(b)

Net waterway width in box culverts (rounded inlets)	m
Recommended barrel height	m

Part 4 Revision exercises

REVISION EXERCISE NO. 1

A hydraulic jump stilling basin (equipped with baffle blocks) is to be tested in laboratory to determine the dissipation characteristics for various flow rates. The maximum prototype discharge will be $310\,\mathrm{m^3/s}$ and the rectangular channel is 14 m wide. Note that the channel bed will be horizontal and concrete lined (i.e. smooth). A 42:1 scale model of the stilling basin is to be built. Discharges ranging between the maximum flow rate and 10% of the maximum flow rate are to be reproduced in the model.

(a) Determine the maximum model discharge required. (b) Determine the minimum prototype discharge for which negligible scale effects occurs in the model. (Comment your result.)

For one particular inflow condition, the laboratory flow conditions are upstream flow depth of 0.025 m, upstream flow velocity of 3.1 m/s and downstream flow depth of 0.193 m. (c) Compute the model force exerted on a single baffle block. State the basic principle(s) involved. (d) What is the direction of force in (c): i.e. upstream or downstream? (e) What will be the corresponding prototype force acting on a single block? (f) Compute the prototype head loss. (g) Operation of the basin may result in unsteady wave propagation downstream of the stilling basin. What will be the scale for the time ratio? (h) For a prototype design at maximum design discharge and with a prototype inflow depth of 1.7 m, what standard stilling basin design would you recommend: (i) USBR Type II, (ii) USBR Type III, (iii) USBR Type IV or (iv) SAF? (Justify your answer. If you have the choice between two (or more) designs, discuss the economical advantages of each option.)

REVISION EXERCISE NO. 2 (HYDRAULIC DESIGN OF A NEW GOLD CREEK DAM SPILLWAY)

The Gold Creek dam is an earthfill embankment built between 1882 and 1885. The length of the dam is 187 m and the maximum height of the embankment is at 99.1 m RL. The reservoir storage capacity is about $1.8 \times 10^6\,\mathrm{m^3}$. The catchment area is $10.48\,\mathrm{km^2}$ of protected forest area.

An overflow spillway is located on the left abutment on rock foundation. It is uncontrolled: i.e. there is no gate or control system. The stepped chute (i.e. 'staircase wastewater course') was completed in 1890 (Fig. R.8). The crest and chute width is 55 m. The crest is nearly horizontal and 61 m long in the flow direction. It is followed by 12 identical steps. The spillway crest elevation is at 96.3 m RL. The steps are 1.5 m high and 4 m long. The stepped chute is followed (without transition section) by a smooth downward sloping channel ($\theta = 1.2°$) which ends with a sharp-crested weir, designed to record flow rates. The sloping channel is used as dissipation channel. It is 55 m wide and its shape is approximately rectangular. The sharp-crested weir is rectangular without sidewall contraction (i.e 55 m wide). It is located 25 m downstream of the chute toe and the elevation of the weir crest is set at 79.1 m RL. (For the purpose of the assignment, the spillway crest discharge characteristics may be assumed ideal.)

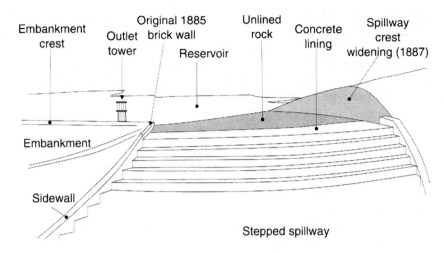

Fig. R.8 Sketch of the 1890 Gold Creek dam spillway, after a 1890s photograph.

The dam owner is investigating an increase of the maximum discharge capacity of the spillway. The new overflow spillway will replace the existing waste waterway structure. Two options are considered:

(i) Design of an overflow spillway with ogee-type crest and a hydraulic jump dissipator. The crest level is at the same level as the existing spillway. The chute slope is set at 45°. The elevation of the chute toe is set at 78.3 m RL.
(ii) Design of an overflow spillway with ogee-type crest and hydraulic jump dissipator. The crest level is lowered by completely erasing one step; i.e. new crest level set at 94.8 m RL. The chute slope is set at 45°. The elevation of the chute toe is set at 78.3 m RL.

In each case, the dam wall, upstream of the spillway crest, has a 3V:1H slope, and the reservoir bottom is set at 77 m RL. In both refurbishment options, the chute is followed by a concrete apron. The tailwater level will be raised artificially with a broad-crested weir located at the downstream end of the concrete apron (i.e. geometry to be determined). In each case, use a sloping apron section if this will improve the efficiency and economics of the basin. In calculating the crest discharge capacity, assume an USBR ogee crest shape and use Fig. 19.9 for the discharge coefficient calculations. In computing the velocity at the spillway toe, allow for energy losses.

The assignment includes three parts.

1. Calculation of the maximum discharge capacity of the existing spillway $Q_{max}^{(1)}$ and estimation of the hydraulic jump location at the downstream end of the spillway. Assume a skimming flow regime on the stepped chute.
2. For the new spillway design (Option (i)), compute the maximum discharge capacity $Q_{max}^{(2)}$ of the new spillway. Compare the result with the maximum discharge capacity of the present spillway.
3. For the new spillway design (Option (ii)), compute the maximum discharge capacity of the spillway $Q_{max}^{(3)}$, design the hydraulic jump stilling basin to dissipate the energy at the foot of the spillway, and select the location, elevation and crest length of the broad-crested weir to ensure that the hydraulic jump stilling basin is always contained between the smooth chute toe and the weir for discharges ranging from $Q_{max}^{(3)}$ down to $0.1 Q_{max}^{(3)}$.

Note: The maximum discharge over the crest is computed usually when the reservoir free-surface elevation reaches the dam crest elevation. For an earth embankment, it is essential that *no* overtopping (at all) occurs. For that reason, in practice, it is safer to allow a safety margin of 0.4 m (to 1 m) for wind wave effects (i.e. dam overtopping caused by wind wave action). For the assignment, use a margin of safety of 0.4 m. The calculations are to be completed and submitted in two stages as specified below:

Stage A: Calculations are to be done for the maximum flow rate *only*.

(a) Calculate the maximum discharge capacity of the present spillway and the flow velocity at the toe of the chute.
(b) Calculate the maximum discharge capacity of the refurbished spillway (Option (i)) and the flow velocity at the toe of the chute.
(c) Calculate the maximum discharge capacity of the refurbished spillway (Option (ii)) and the flow velocity at the toe of the chute.
(d) Compare the residual power at the end of the chute between the three options. In term of energy dissipation, what are the main differences between the three designs? In each case, what type of standard stilling basin would you suggest? (The residual power equals $\rho g Q H_{res}$ where Q is the total discharge and H_{res} is the total head taking 78.3 m RL as datum.)

Stage B
(a) For the present spillway, calculate the hydraulic jump dissipation channel to dissipate the energy at the toe of the spillway for the *maximum flow rate* (assuming skimming flow regime) calculated in Stage A: (1) Calculate the jump length. (2) Calculate the hydraulic jump height rating level (JHRL). (3) Calculate the tailwater rating level (TWRL). (4) Will the hydraulic jump be contained between the chute toe and the sharp-crested weir at maximum flow rate? In the negative, what will happen? Justify and explain clearly your answer in words, and use appropriate relevant sketch(es). (5) Draw to scale the dimensioned sketch of the dissipation channel, showing the spillway toe and the sharp-crested weir. (6) Draw the free-surface profile on the dimensioned sketch.
(b) For the spillway refurbishment (Option B), design the hydraulic jump stilling basin to dissipate the energy at the foot of the smooth chute spillway: (1) Calculate the JHRC for all flows ranging from $Q_{max}^{(3)}$ down to $0.1Q_{max}^{(3)}$. $Q_{max}^{(3)}$ was calculated in Stage A. (2) Plot the JHRC (from (1)). (3) Determine the crest level of the broad-crested weir to match the JHRL and the TWRL for the maximum flow rate $Q_{max}^{(3)}$. (4) Calculate a minimum weir crest length suitable for all flow rates ranging from $Q_{max}^{(3)}$ down to $0.1Q_{max}^{(3)}$. (5) Plot the TWRC (created by the broad-crested weir) on the same graph as the JHRC for all flows ranging from $Q_{max}^{(3)}$ down to $0.1Q_{max}^{(3)}$. (6) Determine the height of sloping apron required to match the JHRC to TWRC for all flows. (This is obtained from the greatest vertical separation of the two curves as plotted in Question (5). If the maximum separation of the two curves occurs at a flow less than $0.1Q_{max}^{(3)}$, set the height of the sloping apron to the separation of the curves at $0.1Q_{max}^{(3)}$.) (7) Draw to scale the dimensioned sketch of the dissipator. (8) Draw the free-surface profile on the dimensioned sketch for $Q_{max}^{(3)}$ and for $0.3Q_{max}^{(3)}$.

Note: Use a sloping apron section if this will improve the efficiency and/or economics of the basin. You may consider an apron elevation different from the chute toe reference elevation (i.e. 78.3 m RL) if this is more suitable.

Summary sheet (Stage A)

	Present spillway	Refurbished spillway (Option (i))	Refurbished spillway (Option (ii))	Units
Maximum discharge capacity				......
Flow velocity at chute toe				......
Froude number at chute toe				......
Residual power at chute toe				......
Recommended type of standard stilling basin[a]				N/A

Note: [a]USBR Types II, III or IV, or SAF.

Summary sheet (Stage B)

(a)

TWRL	m RL
JHRL	m RL
Jump length	m

(b)

JHRL at maximum flow rate	m RL
Broad-crested weir crest elevation	m RL
Broad-crested weir crest length	m
Horizontal length of apron	m
Height of sloping apron	m

Supply the plotted curves of the JHRC and TWRC. Supply dimensioned drawings of the dissipation basin to scale.

Appendix: History of the dam

The Gold Creek dam was built between 1882 and 1885.[1] It is an earthfill embankment with a clay puddle corewall, designed by John Henderson. The fill material is unworked clay laid in 0.23 m layers. The length of the dam is 187 m and the maximum height of the embankment above foundation is 26 m. The reservoir storage capacity is about $1.8 \times 10^6 \, m^3$. The catchment area is 10.48 km^2 of protected forest area. Originally the Gold Creek reservoir supplied water directly to the city of Brisbane. As the Gold Creek reservoir is located close to and at a higher elevation than the Enoggera dam, the reservoir was connected to the Enoggera reservoir in 1928 via a tunnel beneath the ridge separating the Enoggera creek and Gold creek basins. Nowadays the Gold Creek reservoir is no longer in use, the pipeline having been decommissioned in 1991.[2] The reservoir is kept nearly full and it is managed by Brisbane Forest Park. An outlet tower was built between 1883 and 1885 to draw water from the reservoir. The original structure in cast iron failed in 1904, following improper operation while the reservoir was empty. The structure was replaced by the present concrete structure (built in 1905).

An overflow spillway is located on the left abutment on rock foundation. It is uncontrolled: i.e. there is no gate or control system. Since the construction of the dam in 1882–1885, the spillway

---

[1] It is the 14th large dam built in Australia.

[2] The tunnel was first decommissioned in 1977 but the water supply to Enoggera reservoir resumed from 1986 up to 1991.

has been modified three times basically. The original spillway was an unlined-rock overflow. In 1887, the spillway channel was widened to increase its capacity. In January 1887 and early in 1890, large spillway overflows occurred and the unlined-rock spillway was damaged and scoured. A staircase concrete spillway was built over the existing spillway in 1890. The steps are 1.5 m high and 4 m long. In 1975 the crest level was lowered to increase the maximum discharge capacity.

References:
Chanson, H., and Whitmore, R.L. (1996). The stepped spillway of the gold creek dam (built in 1890). *ANCOLD Bulletin*, No. 104, December, 71–80.
Chanson, H., and Whitmore, R.L. (1998). Gold creek dam and its unusual waste waterway (1890–1997): design, operation, maintenance. *Canadian Journal of Civil Engineering*, **25**(4), August, 755–768 and Front Cover.

REVISION EXERCISE NO. 3 (HYDRAULIC DESIGN OF THE NUDGEE ROAD BRIDGE WATERWAY)

The Nudgee Road bridge is a main arterial road in the eastern side of Brisbane (see Appendix A4.3). The bridge crosses the Kedron Brook stream downstream of the Toombul shopping town. At that location, a man-made waterway called the Schultz canal carries the low flows. Today the Schultz canal flows east–north-east and discharges into the deep-water Kedron Brook flood way, flowing parallel to and north of the Brisbane Airport into the Moreton Bay. The natural flood plain (upstream and downstream of the bridge) is 400 m wide and grass lined (assume $k_s = 90$ mm) with a longitudinal slope of 0.49 m/km. At the bridge location, the natural ground level is at 10.0 ft RL.

The assignment will include three parts: the design of the old waterway, the hydraulic design of the present waterway and the upgrading of the waterway.

Part 1: The old waterway (prior to 1968)
The old waterway could pass '7000 cusecs' before overtopping the road bridge:

Design flood flow	7000 ft³/s
Natural ground level at centreline of embankment	10.00 ft RL
Width of embankment base	30 ft
Road elevation	13.00 ft RL
Bridge concrete slab thickness	6 in.
Flood plain slope	0.49 m/km

(a) Calculate the width of the old waterway (i.e. span of the old bridge) assuming that the waterway was a standard box culvert with invert set at natural ground level and operating under inlet control with (i) squared-edged inlets and (ii) rounded inlets.
(b) Use the spreadsheet you developed for the backwater calculations to calculate the flood level (i.e. free-surface elevation) for distances up to 1500 m upstream of the culvert entrance. Plot both backwater profiles and compare with the uniform equilibrium flow profile.
(c) Using the commercial software HydroCulv (see Appendix A4.4), calculate the hydraulic characteristics of the throat flow at design flood flow.

Part 2: The present Nudgee Road waterway (built in 1969)
In 1969, the new Nudgee Road bridge was hydraulically designed based upon a minimum energy loss (MEL) design ('no afflux'). The improvement was required by the rapid development of the Kedron Brook industrial area and the development of the Toombul shopping town upstream. The waterway was designed to pass about 849.5 m³/s (30 000 cusec) which was

the 1:30 year flood. The ground level (at the new bridge location) is 3.048 m RL (10 ft RL) and the water level corresponding to this flow was expected to be 4.7549 m RL (15.6 ft RL). Both levels are at the centreline of the bridge. The waterway throat is 137 m wide, 40 m long and lined with a kikuyu grass (assume $k_s = 0.04$ m). (*Note*: the flood plain gradient is 0.49 m/km.)

Design the MEL waterway culvert for this situation. Include details of the inlet and outlet in your design. (Use the simple design method developed by Professor C.J. Apelt. Estimate exit loss. Comment.)

Part 3: Upgrading the Nudgee Road waterway (1997–1998)
The Nudgee Road waterway must be enlarged to pass 1250 m^3/s. The water level corresponding to this flow is expected to be 5.029 m RL (at bridge centreline, in the absence of waterway and embankment). It is planned to retain the existing road bridge and most of the road embankment by building a 0.275 m high concrete wall along the road (on the top of the bridge and embankment) to prevent overtopping. The waterway inlet, outlet and throat will be concrete lined to prevent scouring which could result from the larger throat velocity. Two MEL designs ('no afflux') are considered: (a) retain the existing bridge and excavate the throat invert and (b) extend the existing bridge and keep the throat invert elevation as in the present waterway (calculation in Part 2).

(a) Design the MEL waterway culverts. (Use the simple design method developed by Professor C.J. Apelt. Estimate exit loss. Comment.)
(b) Calculate the construction cost of both designs.
(c) For the *minimum-cost* MEL waterway design, include details of the inlet and outlet in your design.

Construction costs:

Extension of a two-lane bridge	$10 500 per metre length
Excavation cost	$9 per cubic metre
(Embankment cost	$21 per cubic metre)

Summary sheet (Part 1)

(a)	(i) Net waterway width, square edged	m
	(ii) Net waterway width, round edged	m
(b)	Normal depth in the flood plain	m
	Increase in flood level 800 m upstream	m
(c)	Inlet control? (Yes/No) (*HydroCulv calculations*)	m
	If No, state new specific energy just upstream (*HydroCulv calculations*)	
	Free board at bridge centreline (*HydroCulv calculations*)	

Summary sheet (Part 2)

Invert level at bridge centreline	m RL
d_c in throat	m
Minimum slope required for throat	
Exit loss	m
Total head loss available	m
Comments	

Specify in the table below the waterway widths and invert levels at intervals in the inlet corresponding to depths of excavation equal to 0.25 × (maximum excavation).

Location	Station	Distance from CL of bridge (+ve in d/s) (m)	Depth of excavation (m)	Invert level (m RL)	Waterway width (m)
Upstream lip	1				
	2				
	3				
	4				
Throat (u/s)	5				
Throat (midway)	6				
Throat (d/s)	7				
Downstream lip	8				

Supply dimensioned drawing of waterway (plan and longitudinal section).

Summary sheet (Part 3)

	Design A	Design B	
Inlet lip length			m
Waterway width (at throat)	137.00		m
Invert level at bridge centreline			m RL
d_c in throat			m
E in throat			m
Minimum slope required for throat			
Exit loss			m
Total head loss available			m
Comments			
Excavation volume			m^3
Excavation cost	$	$	
Bridge enlargement cost	$	$	
Total cost	$	$	

Most economical design:
Specify in the table below the waterway widths and invert levels at intervals corresponding to depths of excavation equal to 0.25 × (maximum excavation).

Location	Station	Distance from CL of embankment (+ve in d/s) (m)	Depth of excavation (m)	Invert level (m RL)	Waterway width (m)
Upstream lip	1				
	2				
	3				
	4				
Throat (u/s)	5				
Throat (midway)	6				
Throat (d/s)	7				
	8				
	9				
	10				
Downstream lip	11				

Supply dimensioned drawing of waterway (plan and longitudinal section).

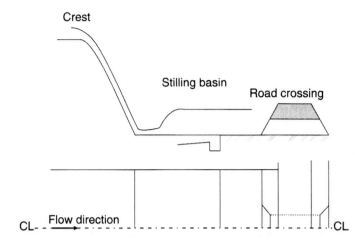

Fig. R.9 Sketch of the weir and downstream river crossing.

REVISION EXERCISE NO. 4

A large weir is to be designed with an overflow section consisting of an un-gated broad crest followed by a smooth concrete chute and a hydraulic jump stilling basin. Considerations of storage requirements and risk analysis applied to the 'design flood event' have set the elevation of the spillway crest 43.25 m RL. The chute slope is 1V:0.8H (i.e. about 51.3°). The toe of the concrete chute and apron level are set at 12.55 m RL, which is the average bed level downstream of the spillway (i.e. 12.55 m RL). The maximum flow over the spillway when the design flood is routed through the storage for these conditions is 4250 m³/s.

The spillway overflow is to be 90 m wide and the stilling basin is followed immediately downstream by a culvert for road crossing (Fig. R.9). (The road is a vital link for the township and it must not be overtopped at design flow conditions.) Two designs will be investigated:

(a) a hydraulic jump stilling basin followed by a standard box culvert,
(b) a hydraulic jump stilling basin equipped with a baffle row (Fig. R.10).

(1) Design (a)
Firstly, the calculations are performed at *design flow conditions* ($Q = 4250$ m³/s) for a hydraulic jump stilling basin (design (a), flat bottom).

(1.1.1) Calculate the free-surface elevation of the reservoir, the flow velocity at the downstream end of the chute, the hydraulic jump rating level (HJRL) and the minimum stilling basin length.

(1.1.2) The spillway flow is to be directed through a standard box culvert underneath the road embankment (Fig. R.9, design (a)). The culvert is to be equipped with *rounded-edged* inlets.

Calculate the number of precast units assuming spillway design flow conditions, the invert set at natural ground level *and culvert operating under inlet control*. Compute the minimum road elevation to prevent overtopping at spillway design flow conditions.

The width of the road embankment base is to be 10 m. The flow downstream of a hydraulic jump is often wavy. For an earth embankment, it is essential that no overtopping occurs. For that reason, in practice, it is safer to allow a safety margin of 0.4 m for wave effects (i.e. embankment overtopping caused by wave action). Assume the bed to be horizontal downstream of the

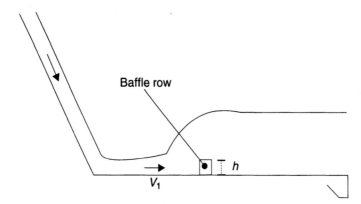

Fig. R.10 Sketch of stilling basin with baffle row.

stilling basin. In computing the velocity at the foot of the spillway, allow for energy losses. The box culvert is to be made of precast units 5.5 m high and 4 m wide. Use Henderson's (1966) formulae for inlet control calculations.

(1.2) Considering the non-design condition (Q = 480 m³/s), calculate the spillway flow conditions: i.e. the free-surface elevation of the reservoir, the flow velocity at the toe of the chute and the HJRL.

(2) Design (b)
Design (b) aims to reduce the tailwater by introducing a baffle row (Fig. R.10). (The baffle height is to be h = 1.1 m.) Calculate the stilling basin properties at maximum flow conditions: i.e. the HJRL and the drag force acting on the baffle row.

Assume that the drag force on a baffle row equals $F_d = C_d \rho V_1^2 / 2 A_f$, where V_1 is the hydraulic jump inflow velocity, the drag coefficient C_d equals 0.7 and A_f is the frontal area of the baffle row ($A_f = Bh$).

Appendices to Part 4

A4.1 SPILLWAY CHUTE FLOW CALCULATIONS

A4.1.1 Introduction

On an uncontrolled chute spillway, the flow is accelerated by the gravity force component in the flow direction. For an ideal-fluid flow, the velocity at the downstream chute end can be deduced from the Bernoulli equation:

$$V_{max} = \sqrt{2g(H_1 - d \cos \theta)} \qquad \text{Ideal fluid flow} \qquad (A4.1)$$

where H_1 is the upstream total head, θ is the channel slope and d is the downstream flow depth (Fig. A4.1).

In practice, however, friction losses occur and the flow velocity at the downstream end of the chute is less than the ideal-fluid velocity (called the maximum flow velocity). At the upstream end of the chute, a bottom boundary layer is generated by bottom friction and develops in the flow direction. When the outer edge of the boundary layer reaches the free surface, the flow becomes fully developed.

First the basic equations are reviewed for both the developing and fully developed flows. Then the complete calculations are developed and generalized.

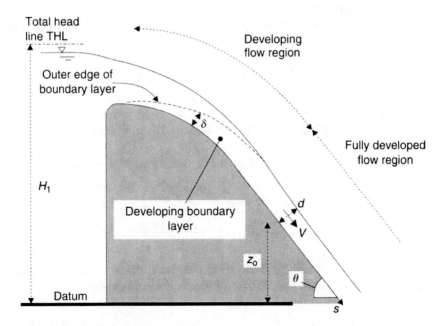

Fig. A4.1 Sketch of a chute spillway flow.

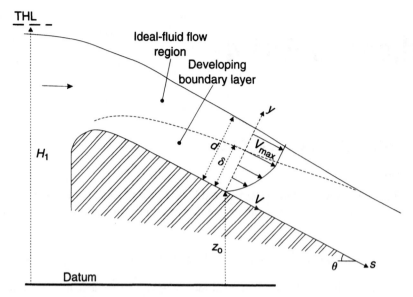

Fig. A4.2 Sketch of the developing flow region.

A4.1.2 Developing flow region

Basic equations

In the developing flow region, the flow consists of a turbulent boundary layer next to the invert and an ideal-fluid flow region above. In the ideal-fluid region, the velocity (called the free-stream velocity) may be deduced from the Bernoulli equation:

$$V_{max} = \sqrt{2g(H_1 - z_o - d\cos\theta)} \qquad \delta < y < d \tag{A4.2}$$

where z_o is the bed elevation and d is the local flow depth.

In the boundary layer, model and prototype data indicate that the velocity distribution closely follows a power law:

$$\frac{V}{V_{max}} = \left(\frac{y}{\delta}\right)^{1/N_{bl}} \qquad 0 < y/\delta < 1 \tag{A4.3}$$

where δ is the boundary layer thickness defined as $\delta = y$ ($v = 0.99V_{max}$) and y is the distance normal to the channel bed (Fig. A4.2). The velocity distribution exponent typically equals $N_{bl} = 6$ for smooth concrete chutes.

Combining equations (A4.2) and (A4.3), the continuity equation gives:

$$q = V_{max}\left(d - \frac{\delta}{N_{bl} + 1}\right) \tag{A4.4}$$

Boundary layer growth

Several researchers investigated the boundary growth on spillway crests (Table A4.1).

For smooth concrete chutes, the following formula is recommended:

$$\frac{\delta}{s} = 0.0212(\sin\theta)^{0.11}\left(\frac{s}{k_s}\right)^{-0.10} \qquad \text{Smooth concrete chute } (\theta > 30°) \tag{A4.5}$$

Table A4.1 Turbulent boundary layer growth on chute spillways

Reference (1)	Boundary layer growth δ/s (2)	Remarks (3)
Spillway flow (smooth chute)		
Bauer (1954)	$0.024\left(\dfrac{s}{k_s}\right)^{-0.13}$	Empirical formula based upon model data
Keller and Rastogi (1977)	Theoretical model with $k - \varepsilon$ turbulence model	WES spillway shape
Cain and Wood (1981)	$a_1\left(\dfrac{s}{k_s}\right)^{-0.10}$	Based upon prototype data
Wood *et al.* (1983)	$0.0212\,(\sin\theta)^{0.11}\left(\dfrac{s}{k_s}\right)^{-0.10}$	Semi-empirical formula; fits prototype data well
Spillway flow (stepped chute)		
Chanson (1995b)	$0.06106\,(\sin\theta)^{0.133}\left(\dfrac{s}{h\cos\theta}\right)^{-0.17}$	Skimming flow only, empirical formula based upon model data and includes models with various crest shapes

Notes: k_s: equivalent roughness height; h: step height.

where s is the distance from the crest measured along the chute invert, θ is the chute slope and k_s is the equivalent roughness height. Equation (A4.5) is a semi-empirical formula which fits well model and prototype data (Wood *et al.*, 1983). It applies to steep concrete chutes (i.e. $\theta > 30°$).

For stepped chutes with skimming flow, the turbulence generated by the steps enhances the boundary layer growth. The following formula can be used in first approximation:

$$\frac{\delta}{s} = 0.06106\,(\sin\theta)^{0.133}\left(\frac{s}{h\cos\theta}\right)^{-0.17} \quad \text{Stepped chute (skimming flow)} \quad (A4.6)$$

where h is the step height. Equation (A4.6) was checked with model and prototype data (Chanson, 1995b, 2001). It applies only to skimming flow on steep chutes (i.e. $\theta > 30°$). Note that it is nearly independent of the crest shape.

Developing flow characteristics
In the developing region, the complete flow characteristics can be deduced from equations (A4.2) to (A4.5) (or equation (A4.6) for stepped chutes). The results give the flow depth, mean flow velocity, free-stream velocity, boundary layer thickness and velocity distribution at any position s measured from the crest.

Results are shown in Fig. A4.3, where the dimensionless mean velocity is plotted as a function of the dimensionless upstream head. Calculations were performed assuming an USBR ogee-shaped crest.

Figure A4.3 presents results for smooth concrete chutes and stepped chutes. The trend indicates that the mean velocity in the developing flow region departs from the theoretical velocity with increasing upstream total head. Indeed, for a given discharge, an increase in upstream total head brings an increase in dam height, in distance s from the crest, in boundary layer thickness (equations (A4.5) and (A4.6)) and hence in the bottom friction effect.

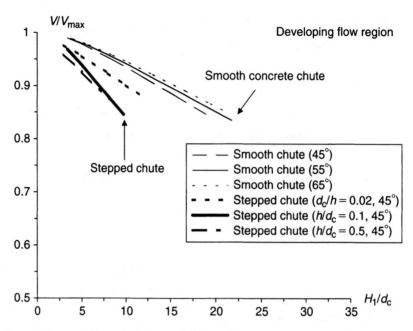

Fig. A4.3 Dimensionless mean flow velocity in the developing flow region. Note: $V = q/d$; Ogee-shaped crest (both smooth and stepped chute calculations).

Note

Figure A4.3 presents stepped chute calculations for the following chute geometry: smooth USBR ogee crest followed by steps, assuming $N_{bl} = 6$. If the steps start further upstream, the bottom roughness would modify the velocity profile and lower values of N_{bl} are advisable.

A4.1.3 Fully developed flow region

Presentation

In the fully developed flow region, the flow is accelerated until the boundary friction counter-balances the gravity force component in the flow direction (i.e. uniform equilibrium flow or normal flow). At equilibrium, the normal conditions can be deduced from the momentum equation.

Uniform equilibrium flow

In uniform equilibrium open channel flows, the momentum equation yields (for a wide channel):

$$V_o = \sqrt{\frac{8g}{f}} \sqrt{d_o \sin \theta} \qquad (A4.7a)$$

where V_o is the uniform equilibrium flow velocity, f is the Darcy friction factor, θ is the channel slope and d_o is the normal flow depth. Equation (A4.7a) is a form of the Chézy equation.

Combining with the continuity equation, it becomes:

$$V_o = \sqrt[3]{\frac{8g}{f} q \sin \theta} \qquad (A4.7b)$$

On a long chute, uniform equilibrium flow conditions are reached before the downstream end of the chute, and the downstream flow velocity V equals:

$$\frac{V}{V_{\max}} = \sqrt{\frac{4 \sin \theta}{f\left(\dfrac{H_1}{d_c} \sqrt[3]{8 \sin \theta / f} - \cos \theta\right)}} \qquad (A4.8)$$

where $V_{\max}$ is the ideal-flow velocity (equation (A4.1)) and d_c is the critical flow depth. Equation (A4.8) is valid for large values of H_1/d_c and for wide rectangular channels.

Figure A4.4 presents typical results for smooth concrete chutes ($f = 0.01, 0.03, 0.005$) and stepped chutes ($f = 0.2$; Chanson, 2001).

Discussion: flow resistance in stepped chute flows

For large flow rates above a stepped chute, the water skims over the pseudo-bottom formed by the step edges with formation of recirculating vortices between the main flow and the step corners. Flow resistance is predominantly form drag. An analytical solution of the dimensionless shear stress between main stream and cavity flow gives:

$$f = \frac{1.128}{\varepsilon}$$

where ($1/\varepsilon$) is the expansion rate of the wake region developing downstream of each step edge (Chanson, 2001).[1] The re-analysis of a large number of model and prototype data demonstrated that the equivalent Darcy friction factor is nearly independent of the step size and Reynolds number: i.e. $f = 0.2$ for steep stepped chutes (Chanson et al., 2002).

Gradually varied flow region

In the fully developed flow region, the flow is gradually varied in the upstream part until becoming uniform equilibrium. In the gradually varied flow region, the flow properties can be deduced from a differential form of the energy equation:

$$\frac{\partial H}{\partial s} = -S_f \qquad (A4.9a)$$

where H is the mean total energy, s is the direction along the channel bed and S_f is the friction defined as:

$$S_f = f \frac{1}{D_H} \frac{V^2}{2g}$$

D_H is the hydraulic diameter and V is the mean flow velocity.

Assuming a hydrostatic pressure distribution, the energy equation becomes:

$$\frac{\partial d}{\partial s}(\cos \theta - \alpha Fr^2) = \sin \theta - S_f \qquad (A4.9b)$$

where d is the flow depth measured normal to the channel bed, α is the kinetic energy correction coefficient[2] and Fr is the Froude number. In the fully developed flow region, the velocity

---

[1] For monophase flow, $\varepsilon = 10$–13 while $\varepsilon = 6$ for air–water flow.
[2] Also called Coriolis coefficient.

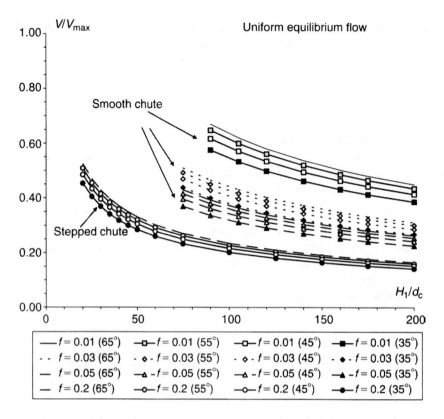

Fig. A4.4 Mean uniform equilibrium flow velocity as a function of the upstream head (equation (A4.8)).

distribution may be approximated by:

$$\frac{V}{V_{max}} = \left(\frac{y}{d}\right)^{1/N}$$ (A4.10)

A value of $N = 6$ is typical for smooth concrete chutes. Lower values are obtained for stepped chute flows (Chanson, 1995b). In any case, for a power law velocity distribution, the kinetic energy coefficient equals:

$$\alpha = \frac{(N + 1)^3}{N^2(N + 3)}$$ (A4.11)

Fully developed flow calculations are started from a position of known flow depth and flow velocity: i.e. typically at the end of the developing region when $\delta/d = 1$. Equation (A4.9) can be integrated numerically to deduce the flow depth and mean velocity at each position along the crest.

A4.1.4 Application and practice

Presentation
For flat-slope chutes, the flow properties in the developing flow region and in the fully developed flow region must be computed using the above development (Sections A4.2 and A4.3).

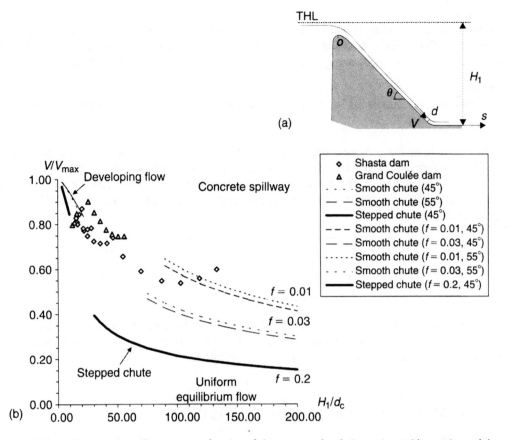

Fig. A4.5 Flow velocity at the spillway toe as a function of the upstream head. Comparison with experimental data (Bradley and Peterka, 1957: paper 1403): (a) definition sketch and (b) results.

For steep chutes (typically $\theta \sim 45\text{--}55°$), both the acceleration and boundary layer development affect the flow properties significantly. In addition, free-surface aeration[3] may take place downstream of the intersection of the outer edge of the developing boundary layer with the free surface. Air entrainment can be clearly identified by the 'white water' appearance of the free-surface flow (e.g. Wood, 1991; Chanson, 1997).

Practical considerations

On a steep channel, the complete calculations of the flow properties can be tedious. In practice, however, the combination of the flow calculations in developing flow (e.g. Fig. A4.3) and in uniform equilibrium flow (e.g. Fig. A4.4) give a general trend which may be used for a preliminary design.

Figure A4.5 combines the results obtained in Figs A4.3 and A4.4 with smooth transition curves. It provides some information on the mean flow velocity at the end of the chute as a function of the theoretical velocity V_{max} (equation (A4.1)), the upstream total head (above spillway toe) and the critical depth. In Fig. A4.5, the general trend is shown for smooth and stepped spillways (concrete chutes), with slopes ranging from 45° to 55° (i.e. 1V:1H to 1V:0.7H). Experimental results obtained on smooth-chute prototype spillways (Grand Coulée dam and Shasta dam) are also shown.

---

[3] Air entrainment in open channels is also called free-surface aeration, self-aeration, insufflation or white waters.

Application

Considering a 45° chute, the dam height above spillway toe is 55 m and the design discharge per unit width is $10 \, \text{m}^2/\text{s}$. Calculate the downstream flow properties for a smooth chute and for a stepped chute (skimming flow regime). Assume a concrete chute with an USBR ogee-shaped crest.

Solution

The critical depth equals 2.17 m. The discharge versus head relationship can be estimated as:

$$q = C(H_1 - \Delta z)^{3/2}$$

with $C = 2.18 \, \text{m}^{1/2}/\text{s}$ at design discharge (the ratio of head above crest to dam height is about 0.03). At design discharge, the ratio of upstream total head to critical depth equals $H_1/d_c = 27$.

The downstream flow velocity can be deduced from Fig. A4.5. For a smooth concrete spillway, a value of $f = 0.03$ is typical. On stepped chutes, a value of $f = 1.0$ is usual. Results are summarized below:

	V/V_{max}	V (m/s)	d (m)
Smooth chute	0.80	26.9	0.37
Stepped chute	0.35	11.8	0.85

Note that such results indicate that larger bottom friction and energy dissipation take place along a stepped spillway compared with a smooth chute. At the end of the chute, $V = 12 \, \text{m/s}$ for a stepped chute design compared to $V = 27 \, \text{m/s}$ for a smooth invert.

*　　　*　　　*

A4.2 EXAMPLES OF MINIMUM ENERGY LOSS WEIRS

A4.2.1 Introduction

The concept of minimum energy loss (MEL) weirs was developed by late Professor G.R. McKay. MEL weirs were designed specifically for situations where the river catchments are characterized by torrential rainfalls (during the wet summer) and by very small bed slope ($S_o \sim 0.001$). The MEL weirs were developed to pass large floods with MEL, hence with minimum upstream flooding. At least four MEL weirs were built in Queensland (Australia): Clermont weir, Chinchilla weir,[4] Lemontree weir and a fourth structure built on the Condamine river (list in chronological order).

A related structure is the MEL spillway inlet designed with the concept of MEL in a fashion somehow similar to the design of an MEL culvert inlet: e.g. at Lake Kurwongbah (Plate 36) and at Swanbank.

Design technique

The purpose of an MEL weir is to minimize afflux and energy dissipation at design flow conditions (i.e. bank full), and to avoid bank erosion at the weir foot. The weir is curved in plan to converge the chute flow and the overflow chute is relatively flat. Hence the downstream hydraulic jump is concentrated near the river centreline away from the banks and usually on (rather than downstream of) the chute toe. The inflow Froude number remains low and the rate of energy dissipation is small compared to a traditional weir.

MEL weirs are typically earthfill structures protected by concrete slabs. Construction costs are minimum. A major inconvenience of an MEL weir design is the risk of overtopping during

---

[4] Classified as a 'large dam' according to the International Commission on Large Dams.

construction (e.g. Chinchilla weir). In addition, an efficient drainage system must be installed underneath the chute slabs.

In the following sections, two practical examples are illustrated.

Hydraulic calculations
Assuming a broad crest and no head loss at the intake (i.e. smooth approach), the discharge capacity of the weir equals:

$$Q = B_{crest}\sqrt{g}\left(\frac{2}{3}(H_1 - \Delta z)\right)^{3/2} \qquad (A4.12)$$

where H_1 is the upstream head, Δz is the weir height above ground level and B_{crest} is the crest width (Fig. A4.6).

Ideally, an MEL weir could be designed to achieve critical flow conditions at any position along the chute and, hence, to prevent the occurrence of a hydraulic jump. Neglecting energy loss along the chute, the downstream width should be:

$$B_2 = B_{crest}\left(\frac{H_1 - \Delta z}{H_1}\right)^{3/2} \qquad \text{'Ideal conditions'} \qquad (A4.13)$$

Practically this is not achievable because critical flow conditions are unstable and the variations of the tailwater flow conditions with discharge are always important. In practice, a weak jump takes place at the chute toe. Note that the jump occurs in an expanding channel and the downstream flow depth (d_{tw}) is fixed by the tailwater conditions.

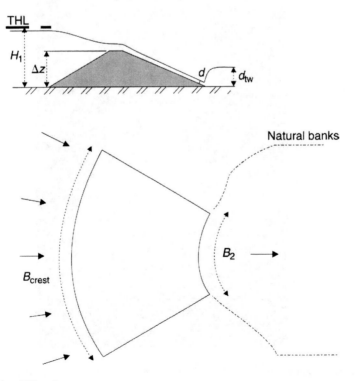

Fig. A4.6 Sketch of an MEL weir.

A4.2.2 Examples of MEL weirs

Ref. No.   MEL-WE-1
Description   MEL weir at Chinchilla (Fig. A4.7)

(a)

(b)

Fig. A4.7 MEL weir at Chinchilla on 8 November 1997 (structure Ref. No. MEL-WE-1): (a) view from the right bank and (b) view from downstream. Note the low turbulence in the stream.

Location	Chinchilla QLD, Australia
River	Condamine river
Purpose	Irrigation water
Design discharge	$850 \, m^3/s$
Weir characteristics	Earth embankment with concrete slabs downstream facing
	Maximum height: 14 m
	Crest length: 410 m
	Maximum reservoir capacity: $9.8 \times 10^6 \, m^3$
Notes	Excellent design. Designed for zero afflux at bank full (Turnbull and McKay, 1974). Completed in 1973. Passed $1130 \, m^3/s$ in 1974 with a measured afflux of less than 100 mm. Listed as a large dam in ICOLD (1984).
	During construction, the unprotected earthfill was overtopped and damaged by a flood.

Ref. No.	MEL-WE-2
Description	MEL weir at Lemontree (Fig. A4.8)
Location	Mirmellan QLD, Australia
River	Condamine river (upstream of Chinchilla weir)
Purpose	Irrigation water
Weir characteristics	Earth embankment with concrete slabs downstream facing
Notes	Excellent design. Completed in the 1980s. Located 245 km upstream of the Chinchilla weir MEL-WE-1.
	A pumping station is located upstream of the weir. The station is equipped with three large pumps to transfer water into a (large) farm storage tank, for future water replenishment of the weir reservoir during dry periods.

Ref. No.	MEL-WE-3
Description	MEL weir at Clermont

Fig. A4.8 MEL weir at Lemontree on 8 November 1997 (structure Ref. No. MEL-WE-2).

Location	Clermont QLD, Australia
River	Sandy Creek
Purpose	Groundwater recharge originally. Today irrigation water.
Design discharge	$850 \, m^3/s$
Weir characteristics	Earth embankment with concrete slabs downstream facing
	Maximum height: 6.1 m
	Crest length: 116 m
Notes	Excellent design. Completed in the 1963. Still in use.
	During construction, the unprotected earthfill was overtopped and damaged twice by flood waters.

Ref. No.	MEL-SP-1
Description	Spillway inlet at Lake Kurwongbah (also called Sideling Creek dam) (Plate 32)
Location	North Pine, Petrie QLD, Australia
River	Pine river
Purpose	Brisbane water supply water
Design discharge	$710 \, m^3/s$ (maximum discharge: $849 \, m^3/s$)
Spillway characteristics	Crest length: 106.7 m
	Channel width: 30.48 m
Dam characteristics	Earth embankment dam (25 m high)
	Reservoir capacity: $15.5 \times 10^6 \, m^3$
Notes	Constructed between 1959 and 1968. The ungated spillway was designed with the concept of MEL in a fashion somehow similar to the design of an MEL culvert inlet (McKay, 1971). The crest inlet fan converges into a 30.48 m wide channel ending with a flip bucket. The efficient crest design allowed an extra 0.4572 m of possible water storage.

*   *   *

A4.3 EXAMPLES OF MINIMUM ENERGY LOSS CULVERTS AND WATERWAYS

A4.3.1 Introduction

In the coastal plains of Queensland (north-east of Australia), torrential rains during the wet season plays a heavy demand on culverts. Furthermore, the natural slope of the flood plains is often very small ($S_o \sim 0.001$) and little fall (or head loss) is permissible in the culverts.

Professors G.R. McKay and C.J. Apelt developed and patented the design procedure of MEL waterways. These structures are designed with the concept of minimum head loss. Professor C.J. Apelt presented an authoritative review of the topic (Apelt, 1983) and a well-documented documentary (Apelt, 1994). The writer highlighted the wide range of design options (Chanson, 1999, 2000).

In the following sections, practical examples are described and illustrated.

A4.3.2 Historical developments of MEL culverts and waterways

The first MEL culvert was the Redcliffe storm waterway system, also called Humpybong Creek drainage outfall, completed in 1960 (structure Ref. No. MEL-C-6). It consisted of an MEL weir acting as culvert drop inlet followed by a 137 m long MEL culvert discharging into the Pacific Ocean. The weir was designed to stop beach sand being washed in and choking the culvert, as well as to prevent salt intrusion in Humpybong Creek without afflux. The culvert discharged flood water underneath a shopping centre parking. The structure is still in use and it passed floods greater than the design flow in several instances without flooding (McKay, 1970).

About 150 structures were built in eastern Australia till date. While a number of small-size structures were built in Victoria, primarily under the influence of Norman Cottman, shire engineer, major structures were designed, tested and built in south-east Queensland, where little head loss is permissible in the culverts and most MEL culverts were designed for zero afflux (next sections). The largest MEL waterway is the Nudgee Road MEL system near Brisbane international airport with a design discharge capacity of 800 m^3/s (structure Ref. No. MEL-W-2). Built between 1968 and 1970, it is believed that the waterway successfully passed floods in excess of the design flow. The channel bed is grass lined and the structure is still in use. Several MEL culverts were built in southern Brisbane during the construction of the south-east Freeway in 1970–1971. The design discharge capacity ranged from 200 to 250 m^3/s. The culverts operate typically several days per year. McKay (1971) indicated further MEL culverts built in northern territory near Alice Springs in 1970. Cottman (1976) described the Newington bridge MEL waterway completed in 1975 (Q_{des} = 142 m^3/s). In 1975 and 1988, the structure successfully passed 122 and 150 m^3/s, respectively, without any damage (Cottman and McKay, 1990).

MEL culvert designs received strong interests in Canada, USA and UK. For example, Lowe (1970), Loveless (1984), Federal Highway Administration (1985: p. 114) and Hamill (1999). Two pertinent studies in Canada (Lowe, 1970) and UK (Loveless, 1984) demonstrated that MEL culverts can successfully pass ice and sediment load without clogging nor silting. These laboratory findings were confirmed by inspections of MEL culverts after major flood events demonstrating the absence of siltation (during site inspections by the writer).

A4.3.3 Examples of MEL waterways

Ref. No.	MEL-W-1
Description	MEL waterway along Norman Creek (Fig. A4.9)
Location	Brisbane, Australia
Purpose	Passage of Norman Creek underneath the south-east Freeway and along (parallel to) Ridge street, Greenslopes
Design discharge	200 m^3/s (to 250 m^3/s without flooding Ridge Street)
Geometry	Throat width: ~10 m
	Throat height: ~4 m
Notes	Excellent design
	In 1999–2000, a busway was built over the waterway on the east of the Freeway. The outlet was slightly modified on the right side to accommodate a busway bridge pier immediately beside the outlet.

Ref. No.	MEL-W-2
Description	MEL waterway under Nudgee Road bridge (built in 1968–1969) (Fig. A4.10)
Location	Brisbane, Australia
Purpose	Crossing of Nudgee Road above Schultz canal connecting Kedron Brook to Serpentine Creek

Fig. A4.9 MEL waterway along Norman Creek (structure Ref. No. MEL-W-1) (September 1996). View from downstream.

(a)

(b)

Fig. A4.10 MEL waterway underneath Nudgee Road bridge (structure Ref. No. MEL-W-2) (14 September 1997): (a) Detail of the throat looking downstream, Nudgee Road bridge in the foreground with the new Gateway freeway bridge over the outlet in the background. Note the cattle wandering in waterway and the absence of lining. (b) Outlet view, looking upstream at the new freeway bridge above the outlet, the Nudgee Road bridge in background and the waterway throat underneath.

Design discharge	$850 \, \text{m}^3/\text{s}$
Design head	Flood level: 4.755 m RL (at bridge centreline)
	Natural ground level: 3.048 m RL (at bridge centreline)
Geometry	Barrel width: $\sim$137 m (bridge length)
	Barrel length: $\sim$18.3 m
	Barrel invert drop: 0.76 m (below natural ground level)
	Culvert length: $\sim$245.3 m (including inlet and outlet, 122 m long each)
	Natural bed slope: 0.00049
Notes	Excellent design. Grassed waterway. Design discharge was the 1:30 year flood observed in 1968. Tested during torrential rains in 1970, 1972 and 1974. No scour observed. Natural flood plain width: 427 m.
	Prior to 1968 the old waterway capacity was $198 \, \text{m}^3/\text{s}$ before overtopping the bridge.

A4.3.4 Examples of MEL culverts

The information is regrouped in two paragraphs. First, efficient designs are described. Each structure has been subjected to large floods and has operated properly. Then poor (inefficient) designs are listed. Experience should be learned from each case.

Efficient designs

Ref. No.	MEL-C-1
Description	MEL culvert underneath the south-east Freeway (built in 1971) (Fig. A4.11)
Location	Brisbane, Australia
Purpose	Passage of Norman Creek underneath the south-east Freeway and Marshall Road, Holland Park West
Design discharge	$170 \, \text{m}^3/\text{s}$ (disaster flow: $250 \, \text{m}^3/\text{s}$)
Geometry	Multi-cell box culvert (2 cells)
	Culvert length: 146 m (including inlet and outlet)
Notes	Excellent original design, but present performances adversely affected by new pier construction in inlet.
	Culvert located in a bend, underneath a major traffic intersection with the freeway above (the intersection).
	In 1999–2000, the construction of a busway on the eastern side of the Freeway lead to a modification of the MEL culvert inlet. The inlet wing walls were modified and a series of circular bridge piers were built in the inflow channel. It is most likely that the performances of the MEL culvert at design flow conditions will be hampered.
Ref. No.	MEL-C-2
Description	MEL culvert underneath the south-east Freeway (built in 1971) (Fig. A4.12)
Location	Brisbane, Australia
Purpose	Passage of Norman Creek underneath the south-east Freeway and parallel to Birdwood street, Holland Park West

(a)

(b)

Fig. A4.11 MEL culvert along Norman Creek, below the south-east Freeway (structure Ref. No. MEL-C-1) (November 1996): (a) Inlet of culvert MEL-C-1 (photograph taken on 5 August 2000). Note the busway bridge piers in the middle of the inlet. (b) View of the outlet. Note the freeway embankment and freeway bridge across a traffic intersection.

Design discharge	$170\,\mathrm{m^3/s}$ (disaster flow: $250\,\mathrm{m^3/s}$)
Geometry	Multi-cell box culvert (4 cells)
	Cell width: $\sim\!2\,\mathrm{m}$
	Cell height: $\sim\!3\,\mathrm{m}$
	Barrel length: $\sim\!110\,\mathrm{m}$
Notes	Excellent design.
	Structure located immediately downstream of the MEL culvert Ref. No. MEL-C-1, both being about 1–2 km upstream of the MEL waterway Ref. No. MEL-W-1.
Ref. No.	MEL-C-3
Description	MEL culvert along Norman Creek (Fig. A4.13)
Location	Brisbane, Australia
Purpose	Passage of Norman Creek underneath Ridge street, Greenslopes

(a)

(b)

Fig. A4.12 MEL culvert along Norman Creek, below the south-east Freeway (structure Ref. No. MEL-C-2): (a) View of the inlet on 13 May 2002. Birdwood Street is on the left. (b) View of the outlet. Note the grassed outlet invert on November 1996.

Design discharge	~220 m^3/s
Geometry	Multi-cell culvert: 7 cells (precast concrete boxes)
	Cell width: ~2 m
	Cell height: ~3.5 m
Notes	Excellent design.
	Culvert located immediately downstream of the MEL waterway Ref. No. MEL-W-1, both structures being downstream of the MEL culvert Ref. No. MEL-C-2.
	Bicycle path located in one cell. Low-flow drain located in another cell.

Fig. A4.13 MEL culvert along Norman Creek, below Ridge Street (structure Ref. No. MEL-C-3) (September 1996). View of the outlet. Note the low-flow channel in the foreground (left).

Fig. A4.14 MEL culvert at Wynnum underneath the Gateway Arterial on 14 September 1997 (structure Ref. No. MEL-C-4): Inlet view.

Ref. No.	MEL-C-4
Description	MEL culvert at Wynnum underneath the Gateway Arterial (Figs A4.14 and A4.15)
Location	Brisbane, Australia
Purpose	Crossing of the Gateway Arterial; 0.3 km south of Wynnum Road
Design discharge	$\sim$220 m^3/s

Fig. A4.15 MEL culvert at Wynnum underneath the Gateway Arterial on 14 September 1997 (structure Ref. No. MEL-C-4): View of the outlet. Note a burned car at the exit of one cell.

Fig. A4.16 MEL culvert underneath the Gateway Arterial on 14 September 1997 (structure Ref. No. MEL-C-5). Inlet view. Note car passing along the Gateway freeway.

Geometry	Multi-cell box culverts (6 cells)
	Cell width: ~3 m
	L_{inlet}: ~34 m
	Cell height: ~3 m
	B_{max}: ~62 m
Notes	Good design. But the low-flow drain does not work properly and water ponding occurs often.
Ref. No.	MEL-C-5
Description	MEL culvert at Wynnum underneath the Gateway Arterial (Fig. A4.16)
Location	Brisbane, Australia
Purpose	Crossing of the Gateway Arterial; 0.6 km north of Wynnum Road

(a)

(b)

(c)

Fig. A4.17 MEL culvert at Redcliffe peninsula on September 1996 (structure Ref. No. MEL-C-6): (a) View of the barrel entrance and inlet from upstream. The inlet is designed as an MEL weir to prevent salt intrusion into the creek as well as to maintain a recreational lake upstream. (b) Inlet view from the road with the upstream lake and fountains. (c) View of the outlet with the sea in the background.

Design discharge	$\sim 100\,\mathrm{m^3/s}$
Geometry	Multi-cell box culverts (11 cells)
	Cell width: $\sim 1.8\,\mathrm{m}$
	Cell height: ~ 1.5 and $1.7\,\mathrm{m}$
	L_{inlet}: $\sim 60\,\mathrm{m}$
	B_{max}: $\sim 90\,\mathrm{m}$
	L_{outlet}: $\sim 75\,\mathrm{m}$
Notes	Good design. But the outlet system is not optimum because the outlet lip is followed by a drop of ground level.

Ref. No.	MEL-C-6
Description	MEL culvert at Redcliffe peninsula (built in 1959) (Fig. A4.17)
Location	Redcliffe QLD, Australia
Purpose	Passage of Humpy Bong Creek underneath a shopping centre parking and prevention of salt intrusion from the sea into the creek (during dry periods)
Design discharge	$25.8\,\mathrm{m^3/s}$
Geometry	One rectangular cell
	Barrel width: $5.48\,\mathrm{m}$
	Barrel height: $\sim 3.5\,\mathrm{m}$
	Barrel length: $137.2\,\mathrm{m}$
	Barrel invert slope: 0.0016
	Barrel invert drop: $1.16\,\mathrm{m}$
	L_{inlet}: $30.5\,\mathrm{m}$
	L_{outlet}: $30.5\,\mathrm{m}$
	Culvert length: $198.1\,\mathrm{m}$ (including inlet and outlet)
Note	Excellent design. Fulfil completely the design expectations. The culvert is regularly used.
	Culvert inlet combined with an MEL weir to maintain a recreational lake upstream of the culvert and to prevent salt water intrusion into the creek (at high tide levels).
	The culvert discharges directly into the sea.

A4.3.5 Poor designs

Ref. No.	MEL-C-X1
Description	MEL culvert along Norman Creek (Fig. A4.18)
Location	Brisbane, Australia
Purpose	Passage of Norman Creek underneath Cornwall street, Stones Corner
Design discharge	$220\,\mathrm{m^3/s}$ (?)
Geometry	Multi-cell culvert: 9 cells (precast concrete boxes)
	Cell width: $\sim 2\,\mathrm{m}$
	Cell height: $\sim 2.8\,\mathrm{m}$
	Barrel length: $16.1\,\mathrm{m}$
	Barrel invert slope: 0.004
	Barrel invert excavation: $\sim 1.2\,\mathrm{m}$ (below natural ground level)

(a)

(b)

Fig. A4.18 MEL culvert along Norman Creek, below Cornwall Street on September 1996 (structure Ref. No. MEL-C-X1): (a) View of the right bank of the inlet, Norman Creek and a bicycle path. Norman creek is on the right of the bicycle path and it connects to the low-flow channel. (b) View of the outlet (from the right bank).

Notes	Poor design. Cornwall Street is often overtopped during torrential rains.
	Culvert located downstream of the MEL culvert Ref. No. MEL-C-3
Discussion	The culvert is not properly aligned with the natural creek. The barrel centreline is oriented almost 90° off the incoming stream (see photograph). Furthermore, the inlet and outlet are both extremely short.
	During flood periods flow separation is observed at the inlet (near the left bank) (Hee, 1978) and four cells become ineffective (i.e. recirculation region). As a result, Cornwall street becomes overtopped. Further heavy siltation is observed after the floods.

Fig. A4.19 MEL culvert along Norman Creek, beneath the freeway on 5 August 2000 (structure Ref. No. MEL-C-X2): (a) Culvert outlet. General view. (b) Culvert outlet. Note the steps at the end of the barrel and upward deflector in foreground.

The real discharge capacity of the culvert is probably about $120\,m^3/s$ (before Cornwall Street overtopping) (Plate 28).

Ref. No.	MEL-C-X2
Description	MEL culvert along Norman Creek (Fig. A4.19)
Location	Brisbane, Australia
Purpose	Passage of Norman Creek under the south-east Freeway in Annerley
Design discharge	$\sim 220\,m^3/s$ (?)
Geometry	Multi-cell culvert: 4 cells (precast concrete boxes)

Notes　　　　　　　　　　　Poor design. The culvert barrel ends with two downward steps ($\Delta z \sim 0.6$ m) about 2 m upstream of the barrel downstream end. The outlet consists of a succession of two upward deflector, one located close to the barrel and another at the outlet lip. It is most likely that, at design flows, the flow will be supercritical at the downstream end of the barrel and that a hydraulic jump might occur in the outlet.

　　　　　　　　　　　　　Culvert located downstream of the MEL culvert Ref. No. MEL-C-2, and immediately upstream of the MEL waterway Ref. No. MEL-W-1.

　　　　　　　　　　　　　Low-flow drain located in one cell.

*　　　*　　　*

A4.4 COMPUTER CALCULATIONS OF STANDARD CULVERT HYDRAULICS

A4.4.1 Introduction

Computer programs have recently been developed to calculate the hydraulic characteristics of standard culverts. Most assume a known culvert geometry and compute the free-surface elevation in the barrel, at the inlet and at the outlet (e.g. HEC-6 and HydroCulv).

Discussion

Before using a computer program for hydraulic calculations, the users must know the basic principles of hydraulics. In the particular case of culvert design, they must further be familiar with the design procedure (Chapter 21) and they should have seen culverts in operation (i.e. laboratory models and prototypes).

　　As part of the hydraulic course at the University of Queensland, the author developed a pedagogic tool to introduce students to culvert design. It is a culvert laboratory study which includes an audio-visual presentation, a physical model of a box culvert and numerical computations using the program HydroCulv. During the afternoon, students are first introduced to culvert design (video-presentation; Apelt, 1994) before performing measurements in a standard box culvert model. The model is in perspex and the students can see the basic flow patterns in the channel, barrel, inlet and outlet (Fig. A4.20). Then the same flow conditions are input in the program HydroCulv, and comparisons between the physical model and computer program results are conducted. Later the comparative results are discussed in front of the physical model (see below).

Audio-visual material

Apelt, C.J. (1994). '*The Minimum Energy Loss Culvert.*' Videocassette VHS, colour, Department of Civil Engineering, University of Queensland, Australia, 18 Minutes.

Physical model

The physical model characteristics are $Q_{des} = 10$ l/s, $B_{max} = 1$ m, $B_{min} = 0.15$ m, $D = 0.11$ m, $L_{culv} = 0.5$ m, $d_{tw} = 0.038$ m, intake and outlet: 45° diffuser, $S_o = 0.0035$.

Computer program

HydroCulv Version 1.1.

Fig. A4.20 Box culvert experiment. Design flow conditions: $Q_{des} = 10\,l/s$, $d_{tw} = 0.038\,m$, $S_o = 0.0035$, $B_{min} = 0.15\,m$, $D = 0.107\,m$ and $L_{culv} = 0.5\,m$. Flow from bottom right to top left. View from upstream at design flow conditions: $Q = 10\,l/s$, $d_{tw} = 0.038\,m$ and $d_1 = 0.122\,m$.

A4.4.2 Use of HydroCulv

A simple tool for culvert calculations is the program HydroCulv, designed as a Windows-based software. The program is based on solid hydraulic calculations and the outputs are easy to use. It is further a shareware program which may be obtained from: HydroTools Software: dwilliams@compusmart.ab.ca

Inputs
The inputs of the program are the geometry of the culvert and the hydraulic parameters. The geometric characteristics include barrel shape, height, width, length and slope. The culvert barrel is assumed prismatic and must have a constant bed slope. The hydraulic parameters include the energy loss coefficients and the boundary conditions (e.g. discharge and tailwater depth).

Outputs
The output may vary depending upon the boundary conditions (i.e. normal flow in flood plain or specified head elevation). Basically, the main result is the free-surface profile in the barrel (e.g. Fig. A4.21). In addition, the results show whether the culvert flow conditions are inlet control or

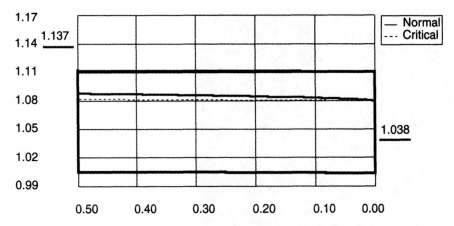

Fig. A4.21 Output of HydroCulv for the flow conditions 1: free-surface profile in the barrel.

outlet control. A graphical solution compares the computed free-surface profile, critical depth and normal depth in the barrel.

Discussion

The calculations depend *critically* upon the energy loss coefficients entered for the calculations. Three types of loss coefficients must be input:

- Roughness
  The calculations may be performed with the Gauckler–Manning coefficient 'n', the Darcy friction factor 'f' or the Keulegan friction coefficient 'k', related to the Chézy coefficient as $C_{\text{Chézy}} = 18 \log_{10}(D_\text{H}/k) + 8.73$ (see Chapter 4).
- Entrance loss coefficient
  The entrance head loss equals, $\Delta H = K_1' V^2/2$, where K_1' is the entrance loss coefficient and V is the barrel velocity. For a pipe intake, $K_1' = 0.5$ and a similar value is reasonable for a culvert intake.
- Exit loss coefficient
  At the culvert outlet, the exit loss equals, $\Delta H = K_2' V^2/2$, where K_2' is the exit loss coefficient and V is the barrel velocity. $K_2' = 1$ for a pipe exit and this value might be used as an order of magnitude for culvert outlet.

4.4.3 Application

Let us consider a simple application of the program HydroCulv. The calculations input are as follows:

HydroCulv parameter	HydroCulv input	HydroCulv units	Remarks
U/S Invert Elevation	1.004768	m RL	
D/S Invert Elevation	1.00301	m RL	
Culvert length	0.5	m	
Roughness 'n'	0.01	SI units	n_{Manning} in s/m$^{1/3}$ for perspex
Entrance loss coefficient	0.5	–	K_1'

(continued)

HydroCulv parameter	HydroCulv input	HydroCulv units	Remarks
Exit loss coefficient	1	–	K_2'
Culvert height	0.107	m	Barrel height D
Culvert width	0.15	m	Barrel width B_{min}
Shape	BOX		Box culvert
Boundary conditions			
Specified head elevation	Q (cms)	TW head elevation (m)	
	0.010	1.038	Conditions 1 (i.e. design flow)
	0.010	1.105	Conditions 2

Comparison physical model–computer results

The same flow conditions (i.e. boundary conditions) were tested in the physical model. For the design flow conditions 1, the physical model has a free-surface barrel flow, submerged entrance and inlet control. For the flow conditions 2, the flow pattern in the physical model is free-surface barrel flow, free-surface inlet flow and inlet control.

Comparative results between physical model and computer program are summarized below:

	Physical model tests	HydroCulv results
Flow conditions 1	Submerged entrance Free-surface barrel flow Inlet control $H_1 = 1.128$ m	Free-surface entrance flow Free-surface barrel flow Inlet control $H_1 = 1.137$ m
Flow conditions 2	Submerged entrance Pipe flow in barrel Outlet control $H_1 = 1.133$ m	Free-surface entrance flow Hydraulic jump in barrel Inlet control $H_1 = 1.336$ m

Basically, close agreement is noted at design flow conditions. But, with non-design flow conditions (e.g. conditions 2), some differences may be observed.

From the author's experience, the above example is typical of the agreement between culvert flow conditions and computer program results. In any situation for which great accuracy is required, design engineers are strongly encouraged to conduct physical model test.

Note

The above tests were performed with the Version 1.1 (32-bits), June 1997.

References

Aberle, J., and Smart, G.M. (2003) The influence of roughness structure on flow resistance on steep slopes. *Journal of Hydraulic Research, IAHR*, **41**(3), 259–269.

Ackers, P., White, W.R., Perkins, J.A., and Harrison, A.J.M. (1978) *Weirs and Flumes for Flow Measurement*. (John Wiley: Chichester, UK) 327 pages.

Agostini, R., Bizzarri, A., Masetti, M., and Papetti, A. (1987) *Flexible Gabion and Reno Mattress Structures in River and Stream Training Works. Section One: Weirs*, 2nd edn (Officine Maccaferri: Bologna, Italy).

Alembert, Jean le Rond d' (1752) *Essai d'une Nouvelle Théorie de la Résistance des Fluides*. (Essay on a New Theory on the Resistance of Fluids.) (David: Paris, France).

Alexander, J., and Fielding, C. (1997) Gravel antidunes in the tropical Burdekin River, Queensland, Australia. *Sedimentology*, **44**, 327–337.

Anderson, A.G. (1942) Distribution of suspended sediment in a natural stream. *Transactions of the American Geophysics Union*, **23**(2), 678–683.

Apelt, C.J. (1983) Hydraulics of minimum energy culverts and bridge waterways. *Australian Civil Engineering Transactions, I.E.Aust.*, **CE25**(2), 89–95.

Apelt, C.J. (1994) *The Minimum Energy Loss Culvert*. Videocassette VHS colour, Department of Civil Engineering, University of Queensland, Australia, 18 minutes.

Ashby, T. (1935) *The Aqueducts of Ancient Rome*. (Clarendon Press: Oxford, UK) I.A. Richmond (ed.), 342 pages.

Bagnold, R.A. (1956) The flow of cohesionless grains in fluids. *Philosophical Transactions of the Royal Society* of *London, Series A*, No. 964, **249**, 235–297.

Bagnold, R.A. (1966) An approach to the sediment transport problem from general physics. *US Geological Survey*, Professional Paper 422-I, Washington DC, USA, 37 pages.

Bakhmeteff, B.A., and Fedoroff, N.V. (1943) Energy loss at the base of a free overfall – discussion. *Transactions, ASCE*, **108**, 1364–1373.

Bakhmeteff, B.A., and Matzke, A.E. (1936) The hydraulic jump in terms of dynamic similarity. *Transactions, ASCE*, **101**, 630–647. Discussion: **101**, 648–680.

Ballance, M.H. (1951) The roman bridges of the via flaminia. *Papers of the British School at Rome*, **19**, 78–117 and Plates xiv–xix.

LeBaron Bowen Jr, R., and Albright, F.P. (1958) *Archaelogical Discoveries in South Arabia*. (The John Hopkins Press: Baltimore, USA).

Barré de Saint-Venant, A.J.C. (1871a) *Théorie du Mouvement Non Permanent des Eaux, avec Application aux Crues de Rivières et à l'Introduction des Marées dans leur Lit*. (Comptes Rendus des séances de l'Académie des Sciences: Paris, France) Séance 17 July 1871, **73**, 147–154 (in French).

Barré de Saint-Venant, A.J.C. (1871b) *Théorie du Mouvement Non Permanent des Eaux, avec Application aux Crues de Rivières et à l'Introduction des Marées dans leur Lit*. (Comptes Rendus des séances de l'Académie des Sciences: Paris, France) **73**(4), 237–240 (in French).

Bauer, W.J. (1954) Turbulent boundary layer on steep slopes. *Transactions, ASCE*, **119**, 1212–1233.

Bazin, H. (1865a) Recherches Expérimentales sur l'Ecoulement de l'Eau dans les Canaux Découverts. ('Experimental Research on Water Flow in Open Channels.) *Mémoires présentés par divers savants à l'Académie des Sciences*, (Paris, France) **19**, 1–494 (in French).

Bazin, H. (1865b) Recherches Expérimentales sur la Propagation des Ondes. (Experimental Research on Wave Propagation.) *Mémoires présentés par divers savants à l'Académie des Sciences*, (Paris, France) **19**, 495–452 (in French).

Bazin, H. (1888–1898) Expériences Nouvelles sur l'Ecoulement par Déversoir. (Recent Experiments on the Flow of Water over Weirs.) *Mémoires et Documents, Annales des Ponts et Chaussées* (Paris, France) 1888: Sér. 6, **16**, 2nd Sem., pp. 393–448; 1890: Sér. 6, **19**, 1st Sem., pp. 9–82; 1891: Sér. 7, **2**, 2nd Sem., pp. 445–520; 1894: Sér. 7, **7**, 1st Sem., pp. 249–357; 1896: Sér. 7, **12**, 2nd Sem., pp. 645–731; 1898: Sér. 7, **15**, 2nd Sem., pp. 151–264 (in French).

Beaver, W.N. (1920) *Unexplored New Guinea. A Record of the Travels, Adventures, and Experiences of a Resident Magistrate amongst the Head-Hunting Savages and Cannibals of the Unexplored Interior of New Guinea.* (Seeley, Service & Co: London, UK) 320 pages.

Bélanger, J.B. (1828) *Essai sur la Solution Numérique de quelques Problèmes Relatifs au Mouvement Permanent des Eaux Courantes.* (Essay on the Numerical Solution of Some Problems relative to Steady Flow of Water.) (Carilian-Goeury: Paris, France) (in French).

Bélanger, J.B. (1849) Notes sur le Cours d'Hydraulique. (Notes on the Hydraulics Subject.) *Mém. Ecole Nat. Ponts et Chaussées* (Paris, France) (in French).

Bélidor, B.F. de (1737–1753) *Architecture Hydraulique. (Hydraulic Architecture.)* 4 Volumes (Charles-Antoine Jombert: Paris, France) (in French).

Belyakov, A.A. (1991) Hydraulic Engineering and The Environment in Antiquity. *Gidrotekhnicheskoe Stroitel'stvo*, No. 8, 46–51 (in Russian) (translated in *Hydrotechnical Construction*, 1992, Plenum, 516–523).

Benet, F., and Cunge, J.A. (1971) Analysis of experiments on secondary undulations caused by surge waves in trapezoidal channels. *Journal of Hydraulics Research, IAHR*, **9**(1), 11–33.

Beyer, W.H. (1982) *CRC Standard Mathematical Tables.* (CRC Press Inc: Boca Raton, Florida, USA).

Bhowmik, N.G. (1996) Impact of the 1993 floods on the Upper Mississippi and Missouri River Basins in the USA. *Water International*, **21**, 158–169.

Blackman, D.R. (1978) The volume of water delivered by the four great aqueducts of Rome. *Papers of the British School at Rome*, **46**, 52–72.

Blackman, D.R. (1979) The length of the four great aqueducts of Rome. *Papers of the British School at Rome*, **47**, 12–18.

Bos, M.G. (1976) *Discharge Measurement Structures.* Publication No. 161, Delft Hydraulic Laboratory, Delft, The Netherlands (also Publication No. 20, ILRI, Wageningen, The Netherlands).

Bos, M.G., Replogle, J.A., and Clemmens, A.J. (1991) *Flow Measuring Flumes for Open Channel Systems.* (ASAE: St. Joseph MI, USA) 321 pages.

Bossut, Abbé C. (1772) Traité Elémentaire d'Hydrodynamique. (Elementary Treaty on Hydrodynamics.) 1st edn (Paris, France) (in French) (2nd edn: 1786, Paris, France; 3rd edn: 1796, Paris, France).

Bossy, G., Fabre, G., Glard, Y., and Joseph, C. (2000) Sur le Fonctionnement d'un Ouvrage de Grande Hydraulique Antique, l'Aqueduc de Nîmes et le Pont du Gard (Languedoc, France). (On the Operation of a Large Ancient Hydraulic structure, the Nîmes Aqueduct and the Point du Gard (Languedoc, France).) *Comptes Rendus de l'Académie des Sciences de Paris, Sciences de la Terre et des Planètes*, **330**,769–775 (in French).

Boussinesq, J.V. (1877) Essai sur la Théorie des Eaux Courantes. ('Essay on the Theory of Water Flow.) *Mémoires présentés par divers savants à l'Académie des Sciences* (Paris, France) **23**, Ser. 3, No. 1, supplément 24, 1–680 (in French).

Boussinesq, J.V. (1896) Théorie de l'Ecoulement Tourbillonnant et Tumultueux des Liquides dans les Lits Rectilignes à Grande Section (Tuyaux de Conduite et Canaux Découverts) quand cet Ecoulement s'est régularisé en un Régime Uniforme, c'est-à-dire, moyennement pareil à travers toutes les Sections Normales du Lit. (Theory of turbulent and tumultuous flow of liquids in prismatic channels of large cross-sections (pipes and open channels) when the flow is uniform, i.e. constant in average at each cross-section along the flow direction.) *Comptes Rendus des séances de l'Académie des Sciences* (Paris, France) **122**, 1290–1295 (in French).

Boys, P.F.D. du (1879) Etude du Régime et de l'Action exercée par les Eaux sur un Lit à Fond de Graviers indéfiniment affouillable. (Study of Flow Regime and Force exerted on a Gravel Bed of infinite Depth.) *Ann. Ponts et Chaussées* (Paris, France) Sér 5, **19**, 141–195 (in French).

Bradley, J.N., and Peterka, A.J. (1957) The hydraulic design of stilling basins. *Journal of Hydraulic Division, ASCE*, **83**, No. HY5, papers 1401, 1402 and 1403. Bradley, J.N., and Peterka, A.J. (1957) The hydraulic design of stilling basins: hydraulic jumps on a horizontal apron (basin I). *Journal of Hydraulic Division, ASCE*, **83**, No. HY5, paper 1401, 1401-1–1401-22. Bradley, J.N., and Peterka, A.J. (1957) The hydraulic design of stilling basins: high dams, earth dams and large canal structures (basin II). *Journal of Hydraulic Division, ASCE*, **83**, No. HY5, paper 1402, 1402-1–1402-14. Bradley, J.N., and Peterka, A.J. (1957) The hydraulic design of stilling basins: short stilling basin for canal structures, small outlet works and small spillways (basin III). *Journal of Hydraulic Division, ASCE*, **83**, No. HY5, paper 1403, 1403-1–1403-22.

Bresse, J.A. (1860) *Cours de Mécanique Appliquée Professé à l'Ecole des Ponts et Chaussées. (Course in Applied Mechanics lectured at the Pont-et-Chaussées Engineering School.)* (Mallet-Bachelier: Paris, France) (in French).

Brown, G.O. (2002) Henry Darcy and the making of a law. *Water Resources Research*, **38**(7), paper 11, 11-1–11-12.

Brown, P.P., and Lawler, D.F. (2003) Sphere drag and settling velocity revisited. *Journal of Environmental Engineering, ASCE*, **129**(3), 222–231.

Buat, P.L.G. du (1779) *Principes d'Hydraulique. (Hydraulic Principles.)* 1st edn (Imprimerie de Monsieur: Paris, France) (in French) (2nd edn: 1786, Paris, France, 2 volumes; 3rd edn: 1816, Paris, France, 3 volumes).

Buckingham, E. (1915) Model experiments and the form of empirical equations. *Transactions, ASME*, **37**, 263–296.

Buckley, A.B. (1923) The influence of silt on the velocity of water flowing in open channels. *Minutes of the Proceedings of the Institution Civil Engineers, 1922–1923*, **216**, Part II, 183–211. Discussion, 212–298.

Buffington, J.M. (1999) The legend of A.F. Shields. *Journal of Hydraulic Engineering, ASCE*, **125**(4) 376–387.

Burdy, J. (1979) Lyon. lugudunum et ses quatre aqueducs. (Lyon. Lugudunum and its Four Aqueducts.) *Dossiers de l'Archéologie*, Séries Les Aqueducs Romains, **38**, October–November, 62–73 (in French).

Burdy, J. (2002) *Les Aqueducs Romains de Lyon.* (Presses Universitaires de Lyon: Lyon, France) 204 pages.

Buschmann, M.H., and Gad-El-Hak, M. (2003) Debate concerning the mean-velocity profile of a turbulent boundary layer. *AIAA Journal*, **41**(4), 565–572.

Cain, P. (1978) Measurements within self-aerated flow on a large spillway. Ph.D. Thesis, Ref. 78–18, Department of Civil Engineering, University of Canterbury, Christchurch, New Zealand.

Cain, P., and Wood, I.R. (1981) Measurements of self-aerated flow on a spillway. *Journal Hydraulic Division, ASCE*, **107**, HY11, 1425–1444.

Carslaw, H.S., and Jaeger, J.C. (1959) *Conduction of Heat in Solids*, 2nd edn (Oxford University Press:London, UK) 510 pages.

Carvill, J. (1981) *Famous Names in Engineering.* (Butterworths: London, UK).

Chang, C.J. (1996) A tale of two reservoirs – greater Taipei's water woes. *Sinorama*, **21**(12), 6–19.

Chanson, H. (1995a) Flow characteristics of undular hydraulic jumps. Comparison with near-critical flows. *Report CH45/95*, Department of Civil Engineering, University of Queensland, Australia, June, 202 pages.

Chanson, H. (1995b) *Hydraulic Design of Stepped Cascades, Channels, Weirs and Spillways.* (Pergamon: Oxford, UK) January, 292 pages.

Chanson, H. (1997) *Air Bubble Entrainment in Free-surface Turbulent Shear Flows.* (Academic Press: London, UK) 401 pages.

Chanson, H. (1998) Extreme reservoir sedimentation in Australia: a review. *International Journal of Sediment Research, UNESCO-IRTCES,* **13**(3), 55–63.

Chanson, H. (1998b) The hydraulics of roman aqueducts: steep chutes, cascades and dropshafts. *Research Report No. CE156,* Department of Civil Engineering, University of Queensland, Australia, 97 pages.

Chanson, H. (1999a) *The Hydraulics of Open Channel Flows: An Introduction.* (Butterworth-Heinemann: Oxford, UK) 512 pages (Reprinted in 2001).

Chanson, H. (1999b) *The Hydraulics of Open Channel Flows: An Introduction.* (Edward Arnold: London, UK) 512 pages.

Chanson, H. (2000a) Boundary shear stress measurements in undular flows: application to standing wave bed forms. *Water Resources Research,* **36**(10), 3063–3076.

Chanson, H. (2000b) Introducing originality and innovation in engineering teaching: the hydraulic design of culverts. *European Journal of Engineering Education,* **25**(4), 377–391.

Chanson, H. (2000c) Hydraulics of roman aqueducts: steep chutes, cascades and dropshafts. *American Journal of Archaeology,* **104**(1), January, 47–72.

Chanson, H. (2000d) A hydraulic study of roman aqueduct and water supply. *Australian Journal of Water Resources, I.E.Aust.,* **4**(2), 111–120. Discussion: **5**(2), 217–220.

Chanson, H. (2001) *The Hydraulics of Stepped Chutes and Spillways.* (Balkema: Lisse, The Netherlands) 418 pages.

Chanson, H. (2001b) Flow field in a tidal bore: a physical model. *Proceedings of the 29th IAHR Congress,* Beijing, China, Theme E, Tsinghua University Press, Beijing, G. Li (ed.), 365–373.

Chanson, H. (2001c) Teaching hydraulic design in an Australian undergraduate civil engineering curriculum. *Journal of Hydraulic Engineering, ASCE,* **127**(12), 1002–1008.

Chanson, H. (2002a) An experimental study of roman dropshaft hydraulics. *Journal of Hydraulic Research, IAHR,* **40**(1), 3–12.

Chanson, H. (2002b) Hydraulics of a large culvert beneath the roman aqueduct of nîmes. *Journal of Irrigation and Drainage Engineering, ASCE,* **128**(5), October, 326–330.

Chanson, H. (2002c) Certains aspects de la conception hydrauliques des aqueducs romains. (Some aspect on the hydraulic design of roman aqueducts.) *Journal La Houille Blanche* (in print).

Chanson, H. (2002d) An experimental study of Roman dropshaft operation: hydraulics, two-phase flow, acoustics. *Report CH50/02,* Department of Civil Engineering, University of Queensland, Brisbane, Australia, 99 pages.

Chanson, H. (2003) Sudden flood release down a stepped cascade. Unsteady air–water flow measurements. Applications to wave run-up, flash flood and dam break wave. *Report CH51/03,* Department of Civil Engineering, University of Queensland, Brisbane, Australia, 142 pages.

Chanson, H., Aoki, S., and Maruyama, M. (2000) Experimental investigations of wave runup downstream of nappe impact. Applications to flood wave resulting from dam overtopping and Tsunami wave runup. *Coastal/Ocean Engineering Report, No. COE00-2,* Department of Architecture and Civil Engineering, Toyohashi University of Technology, Japan, 38 pages.

Chanson, H., and Brattberg, T. (2000) Experimental study of the air–water shear flow in a hydraulic jump. *International Journal of Multiphase Flow,* **26**(4), 583–607.

Chanson, H., and James, P. (1998) Rapid reservoir sedimentation of four historic thin arch dams in Australia. *Journal of Performance of Constructed Facilities, ASCE,* No. 2, May.

Chanson, H., and James, D.P. (2002) Historical development of arch dams: from cut-stone arches to modern concrete designs. *Australian Civil Engineering Transactions, I.E.Aust.,* **CE43**, 39–56, and front cover.

Chanson, H., and Montes, J.S. (1995) Characteristics of undular hydraulic Jumps. Experimental apparatus and flow patterns. *Journal of Hydraulic Engineering, ASCE*, **121**(2), 129–144. Discussion: **123**(2), 161–164.

Chanson, H., and Whitmore, R.L. (1996) The stepped spillway of the gold creek dam (built in 1890). *ANCOLD Bulletin*, No. 104, December, 71–80.

Chanson, H., and Whitmore, R.L. (1998) Gold Creek Dam and its unusual waste waterway (1890–1997): design, operation, maintenance. *Canadian Journal of Civil Engineering*, **25**(4), August, 755–768 and front cover.

Chanson, H., Yasuda, Y., and Ohtsu, I. (2002) Flow resistance in skimming flows and its modelling. *Canadian Journal of Civil Engineering*, **29**(6), 809–819.

Chen, C.L. (1990) Unified theory on power laws for flow resistance. *Journal of Hydraulic Engineering, ASCE*, **117**(3), 371–389.

Chen, Jiyu, Liu, Cangzi, Zhang, Chongle, and Walker, H.J. (1990) Geomorphological development and sedimentation in Qiantang estuary and Hangzhou bay. *Journal of Coastal Research*, **6**(3), 559–572.

Cheng, N.S. (1997) Simplified settling velocity formula for sediment particle. *Journal of Hydraulic Engineering, ASCE*, **123**(2), 149–152.

Chien, N. (1954) Meyer-Peter formula for bed-load transport and Einstein bed-load function. *Research Report No. 7*, University California Institute of Engineering, USA.

Chiew, Y.M., and Parker, G. (1994) Incipient sediment motion on non-horizontal slopes. *Journal of Hydraulic Research, IAHR*, **32**(5), 649–660.

Chow, V.T. (1973) *Open Channel Hydraulics*. (McGraw-Hill International: New Yok, USA).

Chyan, Shuzhong, and Zhou, Chaosheng (1993) *The Qiantang Tidal Bore*. (Hydropower Publ., Beijing, China) The World's Spectacular Sceneries, 152 pages (in Chinese).

Clamagirand, E., Rais, S., Chahed, J., Guefrej, R., and Smaoui, L. (1990). L'aqueduc de carthage. (The carthage aqueduct.) *Journal La Houille Blanche*, No. 6, pp. 423–431.

Colebrook, C.F. (1939) Turbulent flow in pipes with particular reference to the transition region between the smooth and rough pipe laws. *Journal Institute Civil Engineering*, 1938–1939, No. 4, 133–156.

Coleman, N.L. (1970) Flume studies of the sediment transfer coefficient. *Water Res. Res.*, **6**(3), 801–809.

Coles, D. (1956) The law of wake in the turbulent boundary layer. *Journal of Fluid Mechanics*, **1**, 191–226.

Comolet, R. (1976) *Mécanique Expérimentale des Fluides*. (*Experimental Fluid Mechanics*.) (Masson editor: Paris, France) (in French).

Concrete Pipe Association of Australasia (1991) *Hydraulics of Precast Concrete Conduits*, 3rd edn (Jenkin Buxton Printers: Australia) 72 pages.

Conseil Général du Rhône (1987) *Préinventaire des Monuments et Richesses Artistiques. I L'Aqueduc Romain du Mont d'Or. (Pre-inventory of the Monuments and Art Treasures. I The Roman Aqueduct of the Mont d'Or.)* (Bosc Frères Publ.: Lyon, France) Henri Hours (ed.), 104 pages (in French).

Conseil Général du Rhône (1991) *Préinventaire des Monuments et Richesses Artistiques. II L'Aqueduc Romain de l'Yzeron. (Pre-inventory of the Monuments and Art Treasures. II The Roman Aqueduct of the Yzeron.)* (Bosc Frères Publ.: Lyon, France) Henri Hours (ed.), 168 pages and 1 map (in French).

Conseil Général du Rhône (1993) *Préinventaire des Monuments et Richesses Artistiques. III L'Aqueduc Romain de la Brévenne. (Pre-inventory of the Monuments and Art Treasures. III The Roman Aqueduct of the Brévenne.)* (*Bosc Frères Publ.*: Lyon, France) Henri Hours (ed.), 230 pages and 1 map (in French).

Conseil Général du Rhône (1996) *Préinventaire des Monuments et Richesses Artistiques. IV Lyon. L'Aqueduc Romain du Gier. (Pre-inventory of the Monuments and Art Treasures. IV Lyon. The Roman Aqueduct of the Gier.)* (Bosc Frères Publ.: Lyon, France) Jean Burdy (ed.), 407 pages and 1 map (in French).

Cook, O.F. (1916) Staircase farms of the ancients. *National Geographic Magazine*, **29**, 474–534.

Coriolis, G.G. (1836) Sur l'établissement de la formule qui donne la figure des remous et sur la correction qu'on doit introduire pour tenir compte des différences de vitesses dans les divers points d'une même section d'un courant. (On the establishment of the formula giving the backwater curves and on the correction to be introduced to take into account the velocity differences at various points in a cross-section of a stream.) *Annales des Ponts et Chaussées*, 1st Sem., Ser. 1, **11**, 314–335 (in French).

Cottman, N.H. (1976) Fivefold increase obtained in the capacity of a small bridge using a shaped minimum energy subway. *Australian Road Research*, **6**(4), 42–45.

Cottman, N.H., and McKay, G.R. (1990) Bridges and culverts reduced in size and cost by use of critical flow conditions. *Proceedings of the Institution of Civil Engineers*, London, Part 1, **88**, 421–437. Discussion: 1992, **90**, 643–645.

Couette, M. (1890) Etude sur les Frottements des Liquides. (Study on the frictions of liquids.) *Ann. Chim. Phys.*, Paris, France, **21**, 433–510 (in French).

Crank, J. (1956) *The Mathematics of Diffusion*. (Oxford University Press: London, UK).

Creager, W.P. (1917) *Engineering of Masonry Dams*. (John Wiley & Sons: New York, USA).

Creager, W.P., Justin, J.D., and Hinds, J. (1945) *Engineering for Dams*. (John Wiley & Sons: New York, USA) 3 Volumes.

Cunge, J.A. (1969) On the subject of a flood propagation computation method (Muskingum method). *Journal of Hydraulic Research, IAHR*, **7**(2), 205–230.

Dai, Zheng, and Zhou, Chaosheng (1987) The qiantang bore. *International Journal of Sediment Research*, 1, November, 21–26.

Danilevslkii, V.V. (1940) History of hydroengineering in Russia before the nineteenth century. *Gosudarstvennoe Energeticheskoe Izdatel'stvo*, Leningrad, USSR (in Russian) (English Translation: *Israel Program for Scientific Translation*, IPST No. 1896, Jerusalem, Israel, 1968, 190 pages).

Darcy, H.P.G. (1856) *Les Fontaines Publiques de la Ville de Dijon*. (*The Public Fountains of the City of Dijon*) (Victor Dalmont: Paris, France), 647 pages (in French).

Darcy, H.P.G. (1858) Recherches expérimentales relatives aux mouvements de l'eau dans les tuyaux. (Experimental research on the motion of water in pipes.) *Mémoires Présentés à l'Académie des Sciences de l'Institut de France*, **14**, 141 (in French).

Darcy, H.P.G., and Bazin, H. (1865) *Recherches Hydrauliques*. (*Hydraulic Research*.) (Imprimerie Impériales: Paris, France) Parties 1ère et 2ème (in French).[1]

Degremont (1979) *Water Treatment Handbook*, 5th edn (Halsted Press Book, John Wiley & Sons: New York, USA).

Dooge, J.C.I. (1991) The manning formula in context. In *Channel Flow Resistance: Centennial of Manning's Formula*, B.C. Yen (ed.) (Water Resources Publishers: Littleton CO, USA) pp. 136–185.

Dressler, R.F. (1952) Hydraulic resistance effect upon the dam-break functions. *Journal of Research, National Bureau of Standards*, **49**(3), 217–225.

Dressler, R. (1954) Comparison of theories and experiments for the hydraulic dam-break wave. *Proceedings of the International Association of Scientific Hydrology Assemblée Générale*, Rome, Italy, **3**(88), 319–328.

Dupuit, A.J.E. (1848) *Etudes Théoriques et Pratiques sur le Mouvement des Eaux Courantes*. (*Theoretical and Practical Studies on Flow of Water*.) (Dunod: Paris, France) (in French).

Einstein, A. (1906) Eine Neue Bestimmung der Moleküldimensionen. *Annals of Physics*, **19**, 289 (in German).

Einstein, A. (1911) Eine Neue Bestimmung der Moleküldimensionen. *Annals of Physics*, **34**, 591 (in German).

[1]Work prepared and published posthumously by Bazin (1865a,b).

Einstein, H.A. (1942) Formulas for the transportation of bed-load. *Transactions, ASCE,* **107**, 561–573.

Einstein, H.A. (1950) The bed-load function for sediment transportation in open channel flows. US *Department of Agriculture Technical Bulletin No. 1026* (Soil Conservation Service, Washington DC, USA).

Elder, J.W. (1959) The dispersion of marked fluid in turbulent shear flow. *Journal of Fluid Mechanics,* **5**(4), 544–560.

Engelund, F. (1966) Hydraulic resistance of alluvial streams. *Journal of Hydraulics Division, ASCE,* **92**, No. HY2, 315–326.

Engelund, F., and Hansen, E. (1967) *A Monograph on Sediment Transport in Alluvial Streams.* (Teknisk Forlag: Copenhagen, Denmark).

Engelund, F., and Hansen, E. (1972) *A Monograph on Sediment Transport in Alluvial Streams,* 3rd edn (Teknisk Forlag: Copenhagen, Denmark) 62 pages.

Escande, L., Nougaro, J., Castex, L., and Barthet, H. (1961). Influence de quelques paramètres sur une onde de crue subite à l'aval d'un barrage. (The influence of certain parameters on a sudden flood wave downstream from a dam.) *Journal La Houille Blanche,* No. 5, 565–575 (in French).

Eurenius, J. (1980) Ancient dams of Saudi Arabia. *International Water Power and Dam Construction,* **32**(3), March, 21–22.

Evans, A.H. (1928) *The Palace of Minos: A Comparative Account of the Successive Stages of the Early Cretan Civilization as Illustrated by the Discoveries at Knossos.* (Macmillan: London, UK), **II**, Part 1, 390 pages and 19 plates.

Fabre, G., Fiches, J.L., and Paillet, J.L. (1991) Interdisciplinary research on the aqueduct of nîmes and the pont du gard. *Journal of Roman Archaeology,* **4**, 63–88.

Fabre, G., Fiches, J.L., and Paillet, J.L. (2000). *L'Aqueduc de Nîmes et le Pont du Gard. Archéologie, Géosystème, Histoire.* (The Nîmes Aqueduct and the Pont du Gard. Archaeology, Geosystem, History.) *CNRS Editions,* CRA Monographies Hors Série, Paris, France, 483 pages and 16 plates (in French).

Fabre, G., Fiches, J.L., Leveau, P., and Paillet, J.L. (1992) The Pont du Gard. Water and the Roman Town. *Presses du CNRS,* Caisse Nationale des Monuments Historiques et des Sites, Collection Patrimoine au Présent, Paris, France, 127 pages.

Falvey, H.T. (1980) *Air–Water Flow in Hydraulic Structures.* (USBR Engineering Monograph, No. 41, Denver, Colorado, USA).

Farrington, I.S. (1980) The archaeology of irrigation canals with special reference to Peru. *World Archaeology,* **11**(3), 287–305.

Farrington, I.S., and Park, C.C. (1978) Hydraulic engineering and irrigation agriculture in the Moche Valley, Peru: c. A.D. 1250–1532. *Journal of Archaeological Science,* **5**, 255–268.

Faure, J., and Nahas, N. (1961) Etude numérique et expérimentale d'Intumescences à forte courbure du front. (A numerical and experimental study of steep-fronted solitary waves.) *Journal La Houille Blanche,* No. 5, 576–586. Discussion: No. 5, 587 (in French).

Faure, J., and Nahas, N. (1965) Comparaison entre observations réelles, calcul, etudes sur modèles distordu ou non, de la propagation d'une onde de submersion. (Comparison between field observations, calculations, distorted and undistorted model studies of a dam break wave.) *Proceedings of the 11th IAHR Biennial Congress,* Leningrad, Russia, **III**, Paper 3.5, 1–7 (in French).

Fawer, C. (1937) Etude de Quelques Ecoulements Permanents à Filets Courbes. (Study of some Steady Flows with Curved Streamlines.) Thesis, Lausanne, Switzerland, Imprimerie La Concorde, 127 pages (in French).

Federal Highway Administration (1985). Hydraulic design of highway culverts. *Hydraulic Design Ser. No. 5, Report No. FHWA-IP-85-15,* US Federal Highway Administration, 253 pages.

Fernandez Luque, R., and van Beek, R. (1976) Erosion and transport of bed-load sediment. *Journal of Hydraulic Research, IAHR,* **14**(2), 127–144.

Fevrier, P.A. (1979) L'armée romaine et la construction des aqueducs. (The roman army and the construction of aqueducts.) *Dossiers de l'Archéologie*, Séries Les Aqueducs Romains, **38**, October/November, 88–93 (in French).

Fick, A.E. (1855) On liquid diffusion. *Philosophical Magazine*, 4(10), 30–39.

Fischer, H.B., List, E.J., Koh, R.C.Y., Imberger, J., and Brooks, N.H. (1979) *Mixing in Inland and Coastal Waters*. (Academic Press: New York, USA).

Fourier, J.B.J. (1822) *Théorie Analytique de la Chaleur. (Analytical Theory of Heat.)* (Didot: Paris, France) (in French).

Franc, J.P., Avellan, F., Belahadji, B., Billard, J.Y., Briancon-Marjollet, L., Frechou, D., Fruman, D.H., Karimi, A., Kueny, J.L., and Michel, J.M. (1995) La Cavitation. Mécanismes Physiques et Aspects Industriels. (The Cavitation. Physical Mechanisms and Industrial Aspects.) *Presses Universitaires de Grenoble*, Collection Grenoble Sciences, France, 581 pages (in French).

Galay, V. (1987) Erosion and sedimentation in the Nepal Himalayas. An assessment of river processes. *CIDA, ICIMOD, IDRC & Kefford Press*, Singapore (also *Report No. 4/3/010587/1/1 Seq. 259*, Government of Nepal).

Garbrecht, G. (1987a) *Hydraulics and Hydraulic Research: A Historical Review*. (Balkema: Rotterdam, The Netherlands).

Garbrecht, G. (1987b) Hydrologic and hydraulic concepts in antiquity. In *Hydraulics and Hydraulic Research: A Historical Review*., (Balkema: Rotterdam, Netherlands), pp. 1–22.

Garbrecht, G. (1996) Historical water storage for irrigation in the Fayum Depression (Egypt). *Irrigation and Drainage Systems*, **10**(1), 47–76.

Ganchikov, V.G., and Munavvarov, Z.I. (1991) The Marib Dam (history and the present time). *Gidrotekhnicheskoe Stroitel' stvo*, No. 4, 50–55 (in Russian) (translated in *Hydrotechnical Construction*, 1991, Plenum pp. 242–248).

Gauckler, P.G. (1867) *Etudes Théoriques et Pratiques sur l'Ecoulement et le Mouvement des Eaux. (Theoretical and Practical Studies of the Flow and Motion of Waters.)* (Comptes Rendues de l'Académie des Sciences, Paris, France), Tome 64, pp. 818–822 (in French).

Gibbs, R.J., Matthews, M.D., and Link, D.A. (1971) The relationship between sphere size and settling velocity. *Journal of Sedimentary Petrology*, **41**(1), 7–18.

Gilbert, G.K. (1914) *The Transport of Debris by Running Water*. (Professional Paper No. 86, US Geological Survey: Washington DC, USA).

Graf, W.H. (1971) *Hydraulics of Sediment Transport*. (McGraw-Hill: New York, USA).

Grewe, K. (1986) Atlas der Römischen Wasserleitungen nach Köln. (Atlas of the Roman Hydraulic Works Near Köln.) (Rheinland Verlag, Köln, Germany), 289 pages (in German).

Grewe, K. (1992) Plannung und Trassierung römisher Wasserleitungen. (Planning and Surveying of Roman Water Supplies.) Verlag Chmielorz GmbH, Wiesbaden, Germany, Schriftenreihe der Frontinus Gesellschaft, Suppl. II, 108 pages (in German).

Guy, H.P., Simons, D.B., and Richardson, E.V. (1966) *Summary of Alluvial Channel Data from Flume Experiments*. (Professional Paper No. 462-I, US Geological Survey: Washington DC, USA).

Hager, W.H. (1983) Hydraulics of Plane Free Overfall. *Journal of Hydraulic Engineering, ASCE, VI.* **109**(12), 1683–1697.

Hager, W. (1991) Experiments on standard spillway flow. *Proceedings of the Institution of the Civil Engineerings, London, Part 2*, **91**, pp. 399–416.

Hager, W.H. (1992a) Spillways, shockwaves and air entrainment – review and recommendations. *ICOLD Bulletin*, No. 81, January, 117 pages.

Hager, W.H. (1992b) *Energy Dissipators and Hydraulic Jump* Water Science and Technology Library, Vol. 8 (Kluwer Academic: Dordrecht, The Netherlands), 288 pages.

Hager, W.H., Bremen, R., and Kawagoshi, N. (1990) Classical hydraulic jump: length of roller *Journal of Hydraulic Research, IAHR*, **28**(5), 591–608.

Hager, W.H., and Schwalt, M. (1994) Broad-crested weir. *Journal of Irrigation and Drainage Engineering, ASCE,* **120**(1), 13–26. Discussion: **12**(2), 222–226.

Hamill, L. (1999) *Bridge Hydraulics*. (E&FN Spon: London, UK), 367 pages.

Hathaway, G.A. (1958) Dams – their effect on some ancient civilizations. *Civil Engineering, ASCE*, **28**(1), January, 58–63.

Hauck, G.W., and Novak, R.A. (1987) Interaction of flow and incrustation in the Roman aqueduct of Nîmes. *Journal of Hydraulic Engineering, ASCE*, **113**(2), 141–157.

HEE, M. (1969) Hydraulics of culvert design including constant energy concept. *Proceedings of the 20th Conference of Local Authority Engineers*, Department of Local Government, Queensland, Australia, Paper 9, pp. 1–27.

Hee, M. (1978) Selected case histories. *Proceedings of the Workshop on Minimum Energy Design of Culvert and Bridge Waterways*, Australian Road Research Board, Melbourne, Australia, Session 4, Paper 1, pp. 1–11.

Hellström, B. (1941) Några Iakttagelser Över Vittring Erosion Och Slambildning i Malaya Och Australien. *Geografiska Annaler*, Stockholm, Sweden, No. 1–2, pp. 102–124 (in Swedish).

Helmholtz, H.L.F. (1868) Über discontinuirliche Flüssigkeits-Bewegungen. *Monatsberichte der königlich preussichen Akademie der Wissenschaft zu Berlin*, pp. 215–228 (in German).

Henderson, F.M. (1966) *Open Channel Flow*. (MacMillan Company, New York, USA).

Herr, L.A., and Bossy, H.G. (1965) Hydraulic charts for the selection of highway culverts. *Hydraulic Engineering Circular*, US Department of Transportation, Federal Highway Administration, HEC No. 5, December.

Herschy, R. (1995) General purpose flow measurement equations for flumes and thin plate weirs. *Flow Measurement and Instrumentation*, **6**(4), 283–293.

Hinze, J.O. (1975) *Turbulence*, 2nd edn (McGraw-Hill: New York, USA).

Hodge, A.T. (1992) *Roman Aqueducts and Water Supply*. (Duckworth: London, UK), 504 pages.

Horn, A., Joo, M., and Poplawski, W. (1999) Sediment transport rates in highly episodic river systems: a preliminary comparison between empirical formulae and direct flood measurements. *Proceedings of the Water 99 Joint Congress, 25th Hydrology and Water Research Symposium* and *2nd International Conference on Water Research and Environment Research*, Brisbane, Australia, Vol. 1, pp. 238–243.

Howe, J.W. (1949) Flow measurement. *Proceedings of the 4th Hydraulic Conference*, Iowa Institute of Hydraulic Research, H. Rouse (ed.) (John Wyley & Sons) June, pp. 177–229.

Humber, W. (1876) *Comprehensive Treatise on the Water Supply of Cities and Towns with Numerous Specifications of Existing Waterworks*. (Crosby Lockwood: London, UK).

Hunt, B. (1982) Asymptotic solution for dam-break problems. *Journal of Hydraulic Division, Proceedings, ASCE*, **108**, No. HY1, 115–126.

Hunt, B. (1984) Perturbation solution for dam break floods. *Journal of Hydraulic Engineering, ASCE*, **110**(8),1058–1071.

Hunt, B. (1988) An asymptotic solution for dam break floods in sloping channels. In *Civil Engineering Practice*. P.N. Cheremisinoff, N.P. Cheremisinoff and S.L. Cheng, Lancaster Pen. (eds) (Technomic: USA, Lancaster, USA), Vol. 2, Section 1, Chapter 1, pp. 3–11.

Hydropower and Dams (1997) Mini hydro scheme for Egyptian oasis. *International Journal of Hydropower and Dams*, **4**(4), 12.

Idel'cik, I.E. (1969) *Mémento des Pertes de Charge*. (Handbook of Hydraulic Resistance.) Eyrolles (ed.), Collection de la direction des études et recherches d'Electricité de France, Paris, France.[2]

Idelchik, I.E. (1986) *Handbook of Hydraulic resistance*. 2nd revised and augumented edition (Hemisphere: New York, USA).[3]

International Commission on Large Dams (ICOLD) (1984) World Register of Dams – Registre Mondial des barrages – *ICOLD*. (ICOLD: Paris, France), 753 pages.

---

[2]Idel'cik or Idelchik refer to the same Russian author.
[3]*Ibid.*

Ippen, A.T., and Harleman, R.F. (1956) Verification of theory for oblique standing waves. *Transactions, ASCE*, **121**, 678–694.

Isaacson, E., Stoker, J.J., and Troesch, A. (1954) Numerical solution of flood prediction and river regulation problems. Report II. Numerical solution of flood problems in simplified models of the Ohio River and the junction of the Ohio and Mississippi Rivers. *Report No. IMM-NYU-205*, Institute of Mathematical Science, New York University, pp. 1–46.

Isaacson, E., Stoker, J.J., and Troesch, A. (1956) Numerical solution of flood prediction and river regulation problems. Report III. Results of the numerical prediction of the 1945 and 1948 floods on the Ohio River and of the 1947 flood through the junction of the Ohio and Mississippi Rivers, and of the floods of the 1950 and 1948 through Kentucky Reservoir. *Report No. IMM-NYU-235*, Institute of Mathematical Science, New York University, pp. 1–70.

ISO (1979) *Units of Measurements – ISO Standards Handbook 2*. (International Organization for Standardization ISO: Switzerland).

Jevons, W.S. (1858) On clouds; their various forms, and producing causes. *Sydney Magazine of Science and Art*, **1**(8), 163–176.

Jimenez, J.A., and Madsen, O.S. (2003) A simple formula to estimate settling velocity of natural sediments. *Journal of Waterway, Port, Coastal and Ocean Engineering, ASCE*, **129**(2), 70–78.

Julien, P.Y. (1995) Erosion and sedimentation. (Cambridge University Press: Cambridge, UK), 280 pages.

Julien, P.Y., and Raslan, Y. (1998) Upper-regime plane bed. *Journal of Hydraulic Engineering, ASCE*, **124**(11), 1986–1096.

Kamphuis, J.W. (1974) Determination of sand roughness for fixed beds. *Journal of Hydraulic Research, IAHR*, **12**(2), 193–203.

Kazemipour, A.K., and Apelt, C.J. (1983) Effects of irregularity of form on energy losses in open channel flow. *Australian Civil Engineering Transactions, I.E.Aust.*, **CE25**, 294–299.

Keller, R.J., and Rastogi, A.K. (1977) Design chart for predicting critical point on spillways. *Journal of Hydraulic Division, ASCE*, **103**(HY12), 1417–1429.

Kelvin, Lord (1871) The influence of wind and waves in water supposed frictionless. *London, Edinburgh and Dublin Philosophical Magazine and Journal of Science, Series 4*, **42**, 368–374.

Kennedy, J.F. (1963) The Mechanics of dunes and antidunes in erodible-bed channels. *Journal of Fluid Mechanics*, **16**(4), 521–544 (& 2 plates).

Kennedy, J.F. (1995) The albert shields story. *Journal of Hydraulic Engineering, ASCE*, **12**(11), 766–772.

Keulegan, G.H. (1938) Laws of turbulent flow in open channels. *Journal of Research, National Bureau of Standards*, **21**, December., Paper RP1151, pp. 707–741.

Khan, A.A., Steffler, P.M., and Gerard, R. (2000) Dam-break surges with floating debris *Journal of Hydraulic Engineering, ASCE*, **126**(5), 375–379.

Knapp, F.H. (1960) Ausfluss, Überfall and Durchfluss im Wasserbau. Verlag G. Braun, Karlsruhe, Germany (in German).

Korn, G.A., and Korn, T.M. (1961) *Mathematical Handbook for Scientist and Engineers*. (McGraw-Hill: New York, USA).

Kosok, P. (1940) The role of irrigation in ancient peru. *Proceedings of the 8th American Scientific Congress*, Vol. 2, Washington DC, USA, pp. 168–178.

Lagrange, J.L. (1781) Mémoire sur la Théorie du Mouvement des Fluides. ('Memoir on the Theory of Fluid Motion.') in *Oeuvres de Lagrange*, Gauthier-Villars, Paris, France (printed in 1882) (in French).

Lane, E.W., and Kalinske, A.A. (1941) Engineering calculations of suspended sediment. *Transactions of the American Geophysics Union*, **20**.

Lanning, E.P. (1967) *Peru before the Incas*. (Prentice-Hall: Englewood Cliffs NJ, USA).

Larsonneur, C. (1989) La Baie du Mont-Saint-Michel: un Modèle de Sédimentation en Zone Tempérée. (The Bay of Mont-Saint-Michel: A Sedimentation Model in Temperate Environment.) *Bulletin Inst. Géol. Bassin d'Aquitaine, Bordeaux*, **46**, 5–73 & 4 plates (in French).

Lauber, G. (1997) Experimente zur Talsperrenbruchwelle im glatten geneigten Rechteckkanal. ('Dam Break Wave Experiments in Rectangular Channels.') *Ph.D. thesis*, VAW-ETH, Zürich, Switzerland (in German). (also *Mitteilungen der Versuchsanstalt fur Wasserbau, Hydrologie und Glaziologie*, ETH-Zurich, Switzerland, No. 152).

Laursen, E.M. (1958) The total sediment load of streams. *Journal of Hydraulic Division, ASCE*, **84**(HY1), Paper 1530, 1–36.

Lesieur, M. (1994) La Turbulence. (The Turbulence.) *Presses Universitaires de Grenoble*, Collection Grenoble Sciences, France, 262 pages (in French).

Leveau, P. (1979) La Construction des Aqueducs. (The Construction of Aqueduct.) *Dossiers de l'Archéologie*, Séries Les Aqueducs Romains, **38**, October–November, pp. 8–19 (in French).

Leveau, P. (1991) Research on Roman Aqueducts in the past Ten Years. *Future Currents in Aqueduct Studies*, Leeds, UK, T. HODGE (ed.), pp. 149–162.

Leveau, P. (1996) The barbegal water mill in its environment: archaeology and the economic and social history of antiquity. *JRA*, Vol. **9**, pp. 137–153.

Levin, L. (1968) Formulaire des Conduites Forcées, Oléoducs et Conduits d'Aération. ('Handbook of Pipes, Pipelines and Ventilation Shafts.') *Dunod*, Paris, France (in French).

Li, D., and Hager, W.H. (1991). Correction coefficients for uniform channel flow. *Canadian Journal of Civil Engineering*, **18**, 156–158.

Liggett, J.A. (1975) Basic equations of unsteady flow. in *Unsteady Flow in Open Channels*. Vol. 1, K. Mahmood and V. Yevdjevich (eds) (WRP Publ.: Fort Collins, USA), pp. 29–62.

Liggett, J.A. (1993) Critical depth, velocity profiles and averaging. *Journal of Irrigation and Drain Engineering, ASCE*, **119**(2), 416–422.

Liggett, J.A. (1994) *Fluid Mechanics*. (McGraw-Hill: New York, USA).

Loveless, J.H. (1984) A comparison of the performance of standard and novel culvert designs including the effects of sedimentation. *Proceedings of the 1st International Conference on Hydraulic Design in Water Resources Engineering: Channels and Channel Control Structures*, K.V.H. Smith (ed.) (Springer Verlag: Southampton, UK), **1**, pp. 183–193.

Lowe, S.A. (1970) Comparison of energy culverts to standard three cell box culverts. *Masters thesis*, University of Manitoba, Canada.

Lynch, D.K. (1982) Tidal bores. *Scientific American*, **247**(4), October, 134–143.

McKay, G.R. (1970) Pavement drainage. *Proceedings of the 5th Australian Road Research Board Conference*, **5**, Part 4, pp. 305–326.

McKay, G.R. (1971) Design of minimum energy culverts. *Research Report*, Department of Civil Engineering, University of Queensland, Brisbane, Australia, 29 pages & 7 Plates.

McKay, G.R. (1978) Design principles of Minimum energy waterways. *Proceedings of the Workshop on Minimum Energy Design of Culvert and Bridge Waterways*, Australian Road Research Board, Melbourne, Australia, Session 1, pp. 1–39.

McMath, R.E. (1883) Silt movement by the mississippi. *Van Nostrand's Engineering Magazine*, pp. 32–39.

Malandain, J.J. (1988) La Seine au Temps du Mascaret. ('The Seine River at the Time of the Mascaret.') *Le Chasse-Marée*, No. 34, pp. 30–45 (in French).

Manning, R. (1890) On the flow of water in open channels and pipes. *Institution of Civil Engineers of Ireland*.

Marchi, E. (1993) On the Free-Overfall. *Journal of Hydraulic Research*, IAHR, **31**(6), 777–790.

Mariotte, E. (1686) Traité du Mouvement des Eaux et des Autres Corps Fluides. ('Treaty on the Motion of Waters and other Fluids.') Paris, France (in French) (translated by J.T. Desaguliers, *Senex and Taylor*, London, UK, 1718).

Mason, J.A. (1957) The Ancient Civilizations of Peru. (Penguin books: Harmondsworth, UK).

Massau, J. (1889) Appendice au Mémoire sur l'Intégration Graphique. *Annales de l'Association des Ingénieurs sortis des Ecoles Spéciales de Gand*, Belgique, Vol. 12, pp. 135–444 (in French).

Massau, J. (1900) Mémoire sur l'Intégration Graphique des Equations aux Dérivées Partielles. *Annales de l'Association des Ingénieurs sortis des Ecoles Spéciales de Gand*, Belgique, Vol. 23, pp. 95–214 (in French).

Maynord, S.T. (1991) Flow resistance of riprap. *Journal of Hydraulic Engineering, ASCE*, **117**(6), 687–696.

Melville, B.W., and Coleman, S.E. (2000) *Bridge Scour*. (Water Resources Publications: Highlands Ranch, USA), 550 pages.

Meunier, M. (1995) Compte-Rendu de Recherches No. 3 BVRE de DRAIX. (Research report No. 3 Experimental Catchment of Draix) *Etudes CEMAGREF, Equipements pour l'Eau et l'Environnement*, No. 21, 248 pages (in French).

Meyer-Peter, E. (1949) Quelques Problèmes concernant le Charriage des Matières Solides. (Some Problems related to Bed Load Transport.) *Soc. Hydrotechnique de France*, No. 2 (in French).

Meyer-Peter, E. (1951) Transport des matières Solides en Général et problème Spéciaux. *Bull. Génie Civil d'Hydraulique Fluviale*, Tome 5 (in French).

Meyer-Peter, E., Favre, H., and Einstein, A. (1934) Neuere Versuchsresultate über den Geschiebetrieb. *Schweiz. Bauzeitung*, **103**(13), (in German).

Miller, A. (1971) *Meteorology*, 2nd edn (Charles Merrill: Colombus OH, USA), 154 pages.

Miller, D.S. (1994) Discharge characteristics. *IAHR Hydraulic Structures Design Manual No. 8*, Hydraulic Design Considerations (Balkema: Rotterdam, The Netherlands), 249 pages.

Miller, W.A., and Cunge, J.A. (1975) Simplified equations of unsteady flows. In *Unsteady Flow in Open Channels*. Vol. 1, K. Mahmood and V. Yevdjevich (eds) (WRP: Fort Collins, USA), pp. 183–257.

Mises, R. von (1917) Berechnung von Ausfluss und Uberfallzahlen. *Z. ver. Deuts. Ing.*, **61**, 447 (in German).

Molchan-Douthit, M. (1998) Alaska bore tales. *National Oceanic and Atmospheric Administration* (Anchorage, USA), revised, 2 pages.

Montes, J.S. (1992a) Curvature analysis of spillway profiles. *Proceedings of the 11th Australasian Fluid Mechanics Conference AFMC*, **2**, Paper 7E-7, Hobart, Australia, pp. 941–944.

Montes, J.S. (1992b) A potential flow solution for the free overfall. *Proceedings of the Institution Civil Engineers Water Maritime and Energy*, **96**(December), pp. 259–266. Discussion: 1995, **112**(March), pp. 81–87.

Montes, J.S. (1998) *Hydraulics of Open Channel Flow*. (ASCE Press: New-York, USA), 697 pages.

Montes, J.S., and Chanson, H. (1998) Characteristics of undular hydraulic jumps. Results and calculations. *Journal of Hydraulic Engineering, ASCE*, **124**(2), 192–205.

Moody, L.F. (1944) Friction factors for pipe flow. *Transactions, ASME*, **66**, 671–684.

Morelli, C. (1971) *The International Gravity Standardization Net 1971* (*I.G.S.N.71*). (Bureau Central de l'Association Internationale de Géodésie: Paris, France.)

Navier, M. (1823) *Mémoire sur les Lois du Mouvement des Fluides*. (*Memoirs on the Laws of Fluid Motion*.) (Mém. Acad. des Sciences: Paris, France), Vol. 6, pp. 389–416.

Nials, F.L., Deeds, E.E., Moseley, M.E., Pozorski, S.G., Pozorski, S.G., and Feldman, R. (1979a) El Niño: the catastrophic flooding of coastal Peru. Part I. *Field Museum of Natural History Bulletin*, **50**(7), 4–14.

Nials, F.L., Deeds, E.E., Moseley, M.E., Pozorski, S.G., Pozorski, S.G., and Feldman, R. (1979b) El Niño: the catastrophic flooding of coastal Peru. Part II. *Field Museum of Natural History Bulletin*, **50**(8), 4–10.

Nielsen, P. (1992) Coastal bottom boundary layers and sediment transport. *Advanced Series on Ocean Engineering*, **4**, (World Scientific: Singapore).

Nielsen, P. (1993) Turbulence effects on the settling of suspended particles. *Journal of Sedimentary Petrology*, **63**(5), 835–838.

Nikuradse, J. (1932) *Gesetzmässigkeit der turbulenten Strömung in glatten Rohren*. (*Laws of Turbulent Pipe Flow in Smooth Pipes*.) (VDI-Forschungsheft, No. 356) (in German) (translated in NACA TT F-10, 359).

Nikuradse, J. (1933) *Strömungsgesetze in rauhen Rohren.* (*Laws of Turbulent Pipe Flow in Rough Pipes.*) (VDI-Forschungsheft, No. 361) (in German) (translated in NACA Tech. Memo. No. 1292, 1950).

Nsom, B., Debiane, K., and Piau, J.M. (2000) Bed Slope Effect on the dam break problem. *Journal of Hydraulic Research, IAHR,* **38**(6), 459–464.

O'Connor, C (1993) *Roman Bridges.* (Cambridge University Press: Cambridge, UK), 235 pages.

Phillips, W. (1955) Qataban and Sheba – *Exploring Ancient Kingdoms on the Biblical Spice Routes of Arabia.* (Victor Gollancz: London, UK).

Pitlick, J. (1992) Flow resistance under conditions of intense gravel transport. *Water Resources Research,* **28**(3), 891–903.

Poiseuille, J.L.M. (1839) Sur le mouvement des liquides dans le tube de très petit diamètre. (On the movement of liquids in the pipe of very small diameter.) *Comptes Rendues de l'Académie des Sciences de Paris,* **9**, 487 (in French).

Prandtl, L. (1904) Über Flussigkeitsbewegung bei sehr kleiner Reibung. (On fluid motion with very small friction.) *Verh. III Intl. Math. Kongr.,* Heidelberg, Germany (in German) (Also NACA Tech. Memo. No. 452, 1928).

Prandtl, L. (1925) Über die ausgebildete turbulenz. (On fully developed turbulence.) *Z.A.M.M.,* **5**, 136–139 (in German).

Rajaratnam, N. (1967). Hydraulic jumps. *Advances in Hydroscience,* Vol. 4, V.T. Chow (ed.) (Academic Press: New York, USA), pp. 197–280.

Rajaratnam, N., and Muralidhar, D. (1968) Characteristics of the rectangular free overfall. *Journal of Hydraulic Research, IAHR,* **6**(3), 233–258.

Rakob, F. (1974) Das Quellenheigtum in Zaghouan und die Römische Wasserleitung nach Karthago. *Mitt. des Deutschen Archaeologischen Instituts Roemische Abteilung,* Vol. 81, pp. 41–89, Plates 21–76 and maps (in German).

Rakob, F. (1979) L'Aqueduc de Carthage. (The Aqueduct of Carthage.) *Dossiers de l'Archéologie,* Séries Les Aqueducs Romains, Vol. 38, October–November, pp. 34–42 (in French).

Rand, W. (1955) Flow geometry at straight drop spillways. *Proceedings, ASCE,* **81**(791), September, pp. 1–13.

Raudkivi, A.J. (1990) *Loose Boundary Hydraulics,* 3rd edn (Pergamon Press: Oxford, UK).

Raudkivi, A.J., and Callander, R.A. (1976) *Analysis of Groundwater Flow.* (Edward Arnold Publisher: London, UK).

Rayleigh, Lord (1883) Investigation on the character of the equilibrium of an incompressible heavy fluid of variable density. *Proceedings of London Mathematical Society,* Vol. 14, pp. 170–177.

Ré, R. (1946) Etude du Lacher Instantané d'une Retenue d'Eau dans un Canal par la Méthode Graphique. (Study of the sudden water release from a reservoir in a channel by a graphical method.) *Journal La Houille Blanche,* **1**(3), May, 181–187 and 5 plates (in French).

Rehbock, T. (1929) The River Hydraulic Laboratory of the Technical University of Karlsruhe. In *Hydraulic Laboratory Practice* (ASME, New York, USA), 111–242.

Reynolds, O. (1883) An experimental investigation of the circumstances which determine whether the motion of water shall be direct or sinuous, and the laws of resistance in parallel channels. *Philosophical Transactions of the Royal Society London,* **174**, 935–982.

Richter, J.P. (1939) *The Literary Works of Leonardo da Vinci,* 2nd edn (Oxford University Press: London, UK), 2 volumes.

Rijn, L.C. van (1984a) Sediment transport, Part I: Bed load transport. *Journal of Hydraulic Engineering, ASCE,* **110**(10), 1431–1456.

Rijn, L.C. van (1984b) Sediment transport, Part II: Suspended load transport. *Journal of Hydraulic Engineering, ASCE.* **110**(11), 1613–1641.

Rijn, L.C. van (1984c) Sediment transport, Part III: Bed forms and alluvial roughness. *Journal of Hydraulic Engineering, ASCE,* **110**(12), 1733–1754.

Rijn, L.C. van (1993) *Principles of Sediment Transport in Rivers, Estuaries and Coastal Seas.* (Aqua: Amsterdam, The Netherlands).

Ritter, A. (1892) Die Fortpflanzung der Wasserwellen. *Vereine Deutscher Ingenieure Zeitswchrift,* **36**(2), 33, 13 August, 947–954 (in German).

Rouse, H. (1936) Discharge characteristics of the free overfall. *Civil Engineering,* **6**(April), 257.

Rouse, H. (1937) Modern conceptions of the mechanics of turbulence. *Transactions, ASCE,* **102**, 463–543.

Rouse, H. (1938) *Fluid Mechanics for Hydraulic Engineers.* (McGraw-Hill: New York, USA) (also Dover: New York, USA, 1961, 422 pages).

Rouse, H. (1943) Energy loss at the base of a free overfall – Discussion. *Transactions, ASCE,* **108**, 1383–1387.

Rouse, H. (1946) *Elementary Mechanics of Fluids.* (John Wiley & Sons: New York, USA), 376 pages.

Runge, C. (1908) Uber eine Method die partielle Differentialgleichung $\Delta u =$ constant numerisch zu integrieren. *Zeitschrift der Mathematik und Physik,* Vol. 56, pp. 225–232 (in German).

Sarrau (1884) *Cours de Mécanique. (Lecture Notes in Mechanics.)* (Ecole Polytechnique: Paris, France) (in French).

Schetz, J.A. (1993) *Boundary Layer Analysis.* (Prentice Hall: Englewood Cliffs, USA).

Schlichting, H. (1979) *Boundary Layer Theory,* 7th edn (McGraw-Hill: New York, USA).

Schnitter, N.J. (1967) A short history of dam engineering. *Water Power,* **19**(April), 142–148.

Schnitter, N.J. (1994) *A History of Dams: The Useful Pyramids.* (Balkema: Rotterdam, The Netherlands).

Schoklitsch, A. (1914) *Über Schleppkraft un Geschiebebewegung.* (Engelmann: Leipzige, Germany) (in German).

Schoklitsch, A. (1917) Über Dambruchwellen. *Sitzungberichten der Königliche Akademie der Wissenschaften, Vienna,* Vol. 126, Part IIa, pp. 1489–1514.

Schoklitsch, A. (1930). *Handbuch des Wasserbaues. (Handbook of Hydraulic Structures.)* (Springer: Vienna, Austria).

Schoklitsch, A. (1950) *Handbuch des Wasserbaues. (Handbook of Hydraulic Structures.)* (Springer: Vienna, Austria).

Scimemi, E. (1930) *Sulla Forma delle Vene Tracimanti. (The Form of Flow over Weirs.)* (L'Energia Elettrica: Milano), Vol. 7, No. 4, pp. 293–305 (in Italian).

Shields, A. (1936) *Anwendung der Aehnlichkeitsmechanik und der Turbulenz Forschung auf die Geschiebebewegung.* (Mitt. der Preussische Versuchanstalt für Wasserbau und Schiffbau: Berlin, Germany), No. 26.

Smith, N. (1971) *A History of Dams.* (The Chaucer Press: Peter Davies, London, UK).

Smith, N.A.F (1992–1993). The Pont du gard and the aqueduct of nîmes. *Transactions of Newcomen Society,* **64**, 53–76. Discussion: **64**, 76–80.

Spiegel, M.R. (1968) *Mathematical Handbook of Formulas and Tables.* (McGraw-Hill: New York, USA).

Stoker, J.J. (1953) Numerical solution of flood prediction and river regulation problems. Report I. Derivation of basic theory and formulation of numerical method of attack. *Report No. IMM-200,* Institute of Mathematical Science, New York University.

Stoker, J.J. (1957) *Water Waves. The Mathematical Theory with Applications.* (Interscience Publishers: New York, USA), 567 pages.

Stokes, G. (1845) *Transactions of the Cambridge Philosophical Society,* **8**.

Stokes, G. (1851) On the effect of internal friction of fluids on the motion of pendulums. *Transactions of the Cambridge Philosophical Society,* **9**(Pt II), 8–106.

Straub, L.G. (1935) Missouri River Report. *House Document 238,* (US Government Printing Office: Washington DC, USA).

Streeter, V.L., and Wylie, E.B. (1981) *Fluid Mechanics,* 1st SI metric edition (McGraw-Hill: Singapore).

Strickler, A. (1923) *Beiträge zur Frage der Geschwindigkeitsformel und der Rauhligkeitszahlen für Ströme, Kanäle und geschlossene Leitungen.* (*Contributions to the Question of a velocity Formula and Roughness data for Streams, Channels and Closed Pipelines.*) Vol. 16 (Mitt. des Eidgenössischen Amtes für Wasserwirtschaft, Bern, Switzerland) (in German) (Translation T-10, W.M. Keck Laboratory of Hydraulics and Water Resources, California Institute of Technology, USA, 1981)

Sumer, B.M., Kozakiewicz, A., Fredsøe, J., and Deigaard, R. (1996). Velocity and concentration profiles in sheet-flow layer on movable bed. *Journal of Hydraulic Engineering, ASCE,* **122**(10), 549–558.

Sutherland, W. (1893) The viscosity of gases and molecular forces. *Philosophical Magazine,* Ser. 5, 507–531.

Swanson, W.M. (1961) The magnus effect: a summary of investigations to date. *Journal of Basic Engineering, Transactions ASME,* Ser. D, **83**, 461–470.

Takahashi, T. (1991) *Debris Flow.* IAHR Monograph (Balkema: Rotterdam, The Netherlands).

Thompson, P.A. (1972) *Compressible Fluid Dynamics.* (McGraw-Hill:, New York, USA), 665 pages.

Tison, L.J. (1949) Origine des ondes de sable (ripple-marks) et des bancs de sable sous l'action des courants. (Origin of sand waves (ripple marks) and dunes under the action of the flow.) *Proceedings of 3rd Meeting International Association for Hydraulic Structures Research,* Grenoble, France, 5–7 September, Paper II-13, 1–15 (in French).

Treske, A. (1994) Undular bores (favre-waves) in open channels – Experimental Studies. *Journal of Hydraulic Research, IAHR,* **32**(3), 355–370. Discussion: **33**(3), 274–278.

Tricker, R.A.R. (1965) *Bores, Breakers, Waves and Wakes.* (American Elsevier: New York, USA).

Tsubaki, T., Kawasumi, T., and Yasutomi, T. (1953) On the Influence of Sand Ripples upon the Sediment Transport in Open Channels. *Reports of Research Institute for Applied Mechanics,* Kyushu University, Vol. II, No. 8, December, pp. 211–256.

Turnbull, J.D., and McKay, G.R. (1974) The design and construction of chinchilla weir – condamine River Queensland. *Proceedings of the 5th Austalasian Conference on Hydraulics and Fluid Mechanics,* Christchuch, New Zealand, **II**, pp. 1–8.

Ubbelohde-Doering, H. (1967) *On the Royal Highways of the Inca Civilizations of Ancient Peru.* (Thames and Hudson: London, UK).

US Department of the Interior (1987). *Design of Small Dams,* 3rd edn (Bureau of Reclamation: Denver CO, USA).

Valenti, V. (1995a) Aqueduc Romain de Mons à Fréjus. 1. Etude Descriptive et Technique. Son Tracé, son Profil, son Assise, sa Source. *Research Report,* Fréjus, France, 97 pages.

Valenti, V. (1995b) Aqueduc Romain de Mons à Fréjus. 2. Etude Hydraulique. Son Débit. de sa Mise en Service à son Déclin. *Research Report,* Fréjus, France, 123 pages.

Vallentine, H.R. (1969) *Applied Hydrodynamics,* SI edition (Butterworths: London, UK).

Van Deman, E.B. (1934) The building of the Roman aqueducts. *Carnegie Institution of Washington,* Publication No. 423, Washington DC, USA, 440 pages (also *McGrath Publ. Co.,* Washington DC, USA, 440 pages, 1973).

Vanoni, V.A. (1946) Transportation of suspended sediment in water. *Transactions, ASCE,* **111**, 67–133.

Villanueva, A.V. (1993) El Abastecimiento de Agua a la Cordoba Romana. I: El Acueducto de Valdepuentes. (The Water Supply of the Roman Cordoba. I: Aqueduct of Valdepuentes.) *Monografias No. 197,* Universidad de Cordoba, Servicio de Publicaciones, Cordoba, Spain, 172 pages (in Spanish).

Villanueva, A.V. (1996) El Abastecimiento de Agua a la Cordoba Romana. II: Acueductos, Ciclo de Distribución y Urbanismo. (The Water Supply of the Roman Cordoba. II: Aqueduct, Distribution System and Urbanism.) *Monografias No. 251,* Universidad de Cordoba, Servicio de Publicaciones, Cordoba, Spain, 222 pages (in Spanish).

Wan, Zhaohui, and Wang, Zhaoyin (1994) *Hyperconcentrated Flow.* IAHR Monograph, (Balkema: Rotterdam, The Netherlands), 290 pages.

Wang, Zhaoyin (2002) Free surface instability of non-newtonian laminar flows. *Journal of Hydraulics Researches, IAHR*, **40**(4), 449–460.

Wasson, R.J., and Galloway, R.W. (1986) Sediment yield in the barrier range before and after European settlement. *Australian Rangeland Journal*, **2**(2), 79–90.

White, M.P. (1943) Energy loss at the base of a free overfall – discussion. *Transactions, ASCE*, **108**, 1361–1364.

Witham, G.B. (1955) The effects of hydraulic resistance in the dam-break problem. *Proceedings of the Royal Society of London*, Ser. A, **227**, pp. 399–407.

Willcocks, W. (1919) *From the Garden of Eden to the Crossing of the Jordan*. (E&FN SPON: New York, USA).

Wisner, P. (1965) Sur le rôle du critère de froude dans l'etude de l'entraînement de l'air par les courants à grande vitesse. (On the role of the froude criterion for the study of air entrainment in high velocity flows.) *Proceedings of the 11th IAHR Congress*, Leningrad, USSR, paper 1.15 (in French).

Wood, I.R. (1991) Air Entrainment in Free-Surface Flows. *IAHR Hydraulic Structures Design Manual No. 4, Hydraulic Design Considerations* (Balkema: Rotterdam, The Netherlands), 149 pages.

Wood, I.R., Ackers, P., and Loveless, J. (1983) General method for critical point on spillways. *Journal of Hydraulic Engineering, ASCE*, **109**(2), 308–312.

Yalin, M.S. (1964) Geometrical properties of sand waves. *Journal of Hydraulics Division, ASCE*, **90**, No. HY5, 105–119.

Yalin, M.S., and Karahan, E. (1979) Inception of sediment transport. *Journal of Hydraulics Division, ASCE*, **105**, No. HY11, 1433–1443.

Yen, B.C. (1991a) Hydraulic resistance in open channels. In *Channel Flow Resistance: Centennial of Manning's Formula*, B.C. Yen (ed.) (Water Resources: Littleton CO, USA) pp. 1–135.

Yen, B.C. (1991b) *Channel Flow Resistance: Centennial of Manning's Formula*. (Water Resources: Littleton CO, USA), 453 pages.

Yen, B.C. (2002) Open channel flow resistance. *Journal of Hydraulic Engineering, ASCE*, **128**(1), 20–39.

Yevdjevich, V. (1975) Introduction. in unsteady flow in open channels. Vol. 1 K. Mahmood and V. Yevdjevich (ed.) (WRP: Fort Collins, USA), pp. 1–27.[4]

---

[4]Yevdjevich is sometimes spelt Jevdjevich or Yevdyevich.

Additional bibliography

The following includes several materials of pedagogical value. They may assist the reader (student and lecturer) in gaining a good feel for open channel hydraulics and to visualize practical applications of the lecture material.

Bibliography: history of hydraulics

Carvill, J. (1981) *Famous Names in Engineering*. (Butterworths: London, UK).

Galbrecht, G. (1987) *Hydraulics and Hydraulic Research: A Historical Review*. (Balkema: Rotterdam, The Netherlands).

Rouse, H. and Ince, S. (1957) *History of Hydraulics*. (Iowa Institute of Hydraulic Research: Iowa City, USA), 269 pages.

Schnitter, N.J. (1994) *A History of Dams: The Useful Pyramids*. (Balkema: Rotterdam, The Netherlands).

Smith, N. (1971) *A History of Dams*. (The Chaucer Press, Peter Davies, London, UK).

Viollet, P.L. (2000) L'Hydraulique dans les Civilisations Anciennes. 5000 ans d'Histoire. *The Hydraulics of Ancient Civilisations. 5,000 years of History*. (Presses de l'Ecole Nationale des Ponts et Chaussées, Paris, France), 374 pages (in French).

Bibliography: audio-visual material

Apelt, C.J. (1994) *The Minimum Energy Loss Culvert*. Videocassette VHS colour, Department of Civil Engineering, University of Queensland, Australia, 18 minutes.
Comments: Utilized to reduce flooding of stormwater plains drains, the benefits of minimum energy loss culverts, designed by Gordon McKay and Colin Apelt, are illustrated by comparison with the flow capacity of the standard culvert. *It is a very good teaching tool* to introduce the concept of specific energy and the application to culvert design.

Lerner, B. (1994) *After the Flood*. Videocassette VHS colour, SBS – The Cutting Edge, 48 minutes.
Comments: In order to control flooding of the river Brahmaputra, Bangladesh, water engineers propose to change the width and course of the river. Along the Mississippi, USA, similar water engineering is the alleged cause of the Mississippi's flooding in 1993. Archival film helps to illustrate some of the problems to be overcome. Produced by Bettina Lerner. *Very good documentary* dealing with practical applications of open channel hydraulics, sediment transport, catchment hydrology and environmental impact of hydraulic structures.

Geology: A Search for Order (1989) *Pororoca: The Backward Flow of the Amazon*. Videocassette VHS colour, Series Geology: A Search for Order, 30 minutes.
Comments: This 25-minutes documentary shows a team of Japanese researchers investigating the tidal bore of the Amazon river: i.e. the Pororoca. There are *amazing shots of the world's largest tidal bore*. The documentary is a superb illustration of backwater effects in an estuary and the propagation of a tidal bore. (Reference: Geology: A Search for Order Series, A/V Distribution Services, Florida State University: V070311).

Mississippi Floods 1993. Videocassette VHS colour, Australian Channel News, 4 minutes.

Comments: News footage of the Mississippi flood in 1993. Footage from Australian News Channels 7, 9, 10, SBS.

St Anthony's Falls Hydraulic Laboratory (1947) *Some Phenomena of Open Channel Flow*. Videocassette NTSC B&W, SAF Hydraulic Laboratory, Minneapolis MN, USA, 33 minutes.

Comments: In this programme, open channel flow lecture material is demonstrated. It looks at supercritical and subcritical flow, hydraulic jumps, hydraulic drops, specific energy curve, pressure momentum curve, critical depth, travel of surface waves in channels flowing at critical, subcritical and supercritical velocities, uphill flow, abrupt gate closure, movable bed channels and more.

US Bureau of Reclamation (1988?) *Challenge at Glen Canyon Dam*. Videocassette VHS colour, US Department of Interior, Denver, Colorado, USA, 27 minutes.

Comments: This program is divided into two parts. The first part examines flood waters of the Colorado River system. The second part describes the damage caused to the Glen Canyon dam spillways following the excessive amount of water which flowed into Lake Powell due to heavy snow falls late in the season. The program then goes on to examine the method used to repair the damage after the flood has passed. It is a *superb educational movie* for both civil, environmental and hydraulic engineering students. It is quite entertaining.

US National Committee for Fluid Mechanics (1967) *The Hydraulic Surge Wave*. Videocassette VHS B&W, Education Development Center, USA, 4 minutes.

Comments: Film of experiments illustrating the hydraulic surge wave, the hydraulic jump and the analogy between hydraulic jump and surge.

Bibliography: internet references

Gallery of photographs in hydraulic engineering {http://www.uq.edu.au/~e2hchans/photo.html}.

Air entrainment on chute and stepped spillways {http://www.uq.edu.au/~e2hchans/self_aer.html}.

Air entrainment at circular plunging jet: physical and acoustic characteristics {http://www.uq.edu.au/~e2hchans/bubble/}.

Air entrainment in the developing flow region of two-dimensional plunging jets – databank {http://www.uq.edu.au/~e2hchans/data/jfe97.html}.

Embankment overflow stepped spillways: earth dam spillways with precast concrete blocks {http://www.uq.edu.au/~e2hchans/over_st.html}.

Formal water gardens {http://www.uq.edu.au/~e2hchans/wat_gard.html}.

Free-surface undulations in open channels: undular jups, undular surges, standing waves {http://www.uq.edu.au/~e2hchans/undular.html}.

Gold Creek dam and its historical stepped spillway {http://www.uq.edu.au/~e2hchans/gold_crk.html}.

History of Arch dams {http://www.uq.edu.au/~e2hchans/arch_dam.html}.

History of steel dams {http://www.uq.edu.au/~e2hchans/steel_da.html}.

Hydraulic, environmental and ecological assessment of a sub-tropical stream in eastern Australia {http://www.uq.edu.au/~e2hchans/eprapa.html}.

Hydraulics of Roman aqueducts: myths, fables, realities. A hydraulician's perspective {http://www.uq.edu.au/~e2hchans/rom_aq.html}.

Hydraulics of rubber dams {http://www.uq.edu.au/~e2hchans/rubber.html}.

Minimum energy loss (MEL) culverts and bridge waterways {http://www.uq.edu.au/~e2hchans/mel_culv.html}.

Minimum energy loss (MEL) weir design: an overflow earthfill embankment dam {http://www.uq.edu.au/~e2hchans/mel_weir.html}.

Extreme reservoir siltation {http://www.uq.edu.au/~e2hchans/res_silt.html}.

Spillway aeration devices to prevent cavitation damage in high-head chutes {http://www.uq.edu.au/~e2hchans/aer_dev.html}.

Sabo check dams. Mountain protection systems in Japan {http://www.uq.edu.au/~e2hchans/sabo.html}.

Tidal bore of the Seine river (mascaret) {http://www.uq.edu.au/~e2hchans/mascaret.html}.

Timber crib weirs {http://www.uq.edu.au/~e2hchans/tim_weir.html}.

Whirlpools {http://www.uq.edu.au/~e2hchans/whirlpl.html}.

Rivers seen from space {http://www.athenapub.com/rivers1.htm}.

Rivers of the world {http://www.kented.org.uk/ngfl/rivers/Photo%20Gallery/photogalleryfront.htm}.
   Map {http://www.kented.org.uk/ngfl/rivers/Feature%20Articles/riversoftheworld.htm}.

Structurae, International Database and Gallery of Structures {http://www.structurae.de/}.

US Army Corps of Engineering Photographic Image Library {http://images.usace.army.mil/photolib.html}.

NASA Earth observatory {http://earthobservatory.nasa.gov/}.
   NASA Visible Earth {http://visibleearth.nasa.gov/}.

Unesco World Heritage Listing {http://whc.unesco.org/nwhc/pages/home/pages/homepage.htm}.
   Virtual Tours {http://whc.unesco.org/nwhc/pages/sites/main.htm}.

Aerial photographs of American rivers and valleys {http://www.geology.wisc.edu/%7Emaher/air/air00.htm}.

LacusCurtius – Roman Waterworks and Hydraulic Engineering {http://www.ukans.edu/history/index/europe/ancient_rome/E/Gazetteer/Periods/Roman/Topics/Engineering/waterworks/home.html}.

Gorze Aqueduct at Metz (France) {http://www.ac-nancy-metz.fr/ia57/jussy/netsco/English/accueil.htm}.

Highlights in the History of Hydraulics by Hunter Rouse {http://www.lib.uiowa.edu/spec-coll/Bai/hydraul.htm}.

International Association for Hydraulic engineering and Research, IAHR {http://www.iahr.org/}.

ICOLD (international commission on large dams) {http://genepi.louis-jean.com/cigb/index.html}.

Dam Safety Committe of new South Wales (Australia) {http://www.damsafety.nsw.gov.au/}.

British Dam Society {http://www.britishdams.org/}.

US Bureau of Reclamation, Dams and Reservoirs {http://www.usbr.gov/dataweb/dams/index.html}.

US Geological Survey Water Resources {http://water.usgs.gov/}.

Hydrotools (softwares) {http://www.compusmart.ab.ca/dwilliam/hydtools.htm}.

Problems

A study of the Marib dam and its sluice system (BC 115 to AD 575)

Preface

The following problem covers lecture material relevant to Basic Principles (Part I), Sediment Transport (Part II) and Physical Modelling and Numerical Modelling (Part III). The assignment was first used in 1992. Based upon real historical facts, the geometry of the hydraulic structures has been simplified to facilitate the calculations.

P1.1 INTRODUCTION

P1.1.1 Historical background

The site of the Marib dam is located on the Wadi Dhana, upstream of the ancient city of Marib in northern Yemen (Fig. P1.1). Marib was the capital over which Queen of Sheba (or Sabah) ruled around BC 950. The Queen of Sheba is mentioned in the Old Testament for her famous visit to King Solomon in Jerusalem. The Kingdom of Sheba was prosperous and its power was based on trade, by both sea (to and from India and the Persian Gulf) and land routes (the spices roads in Arabia) (Philips, 1955; LeBaron Bowen and Albright, 1958). Marib, capital city of the kingdom, was the focal point of the trade routes, and the Sabaens built the Marib dam to irrigate the land around the city. The dam was considered one of the wonders of the ancient world and played an important role. The Kingdom of Sheba is described as a fertile land. The land was cultivated by controlling flood water in the 'wadis', and had fertile gardens and fields with fruits and spices (Eurenius, 1980).

The Sabaens developed large irrigation systems in south-west Arabia using earth dams to divert flood waters into the land. Such waters contained a large amount of silt, sand and gravel sediments.

The dam was built in order to intercept the floods in the Wadi Dhana. Its purpose was to raise the level of the wadi's flow during periods of run-off, following a fall of rain in the mountains and to divert this water into the canal systems which took the waters to the city (Fig. P1.2).

P1.1.2 Dam construction

The first Marib dam was built around BC 750. The dam was a simple earth structure, 580 m in length and is estimated to have been 4 m high. It ran straight across the Wadi between high rocks on the southern side to a rock shelf on the northern bank. The dam was built slightly downstream of the narrowest point in the Wadi Dhana to allow the space for a natural spillway and sluices between the northern end of the dam and the high rocky cliff to the west (Fig. P1.3).

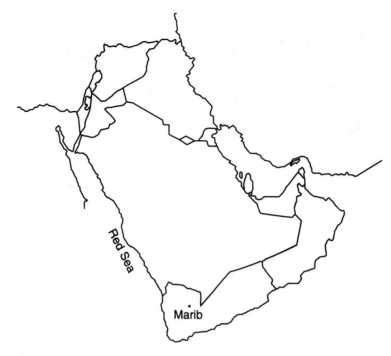

Fig. P1.1 Map of the Arabic peninsula.

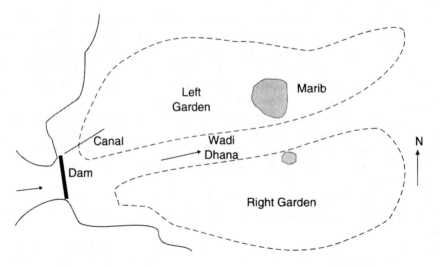

Fig. P1.2 Sketch of the Marib area.

Around BC 500 the dam was heightened. The second structure was a 7 m high earth dam. The cross-section was triangular with both faces sloping at 45° (Fig. P1.3). The water face (i.e. the upstream slope) was covered with stones set in mortar to make the dam watertight and to resist the erosive effects of waves (Smith, 1971).

The final form of the Marib dam was reached after the end of the Kingdom of Sheba. From BC 115 onwards the ruling people in southern Arabia were the Himyarites and the next major reconstruction appears to be a Himyarite work (Smith, 1971). This reconstruction led to a new 14 m dam with elaborate water works at both ends (Fig. P1.3).

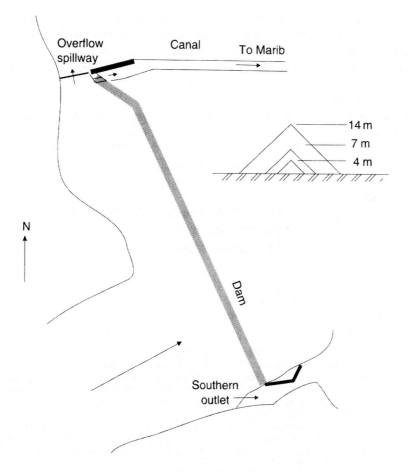

Fig. P1.3 Sketch of the successive dams on the site shown in plan and cross-section.

Southern sluice system

The southern sluices system was known as 'Marbat el-Dimm'. The spillway was an overfall located about 7 m below the top of the dam, and the spillway width was 3.5 m.

Northern sluice system

The northern system of sluices included an overfall spillway and a channel outlet (Fig. P1.3).

The channel outlet was located between the spillway and the earth dam, and it contained one great wall over 140 m long, about 9 m thick and from 5 to 9 m high. This system conducted water from the northern end of the dam through a single 1000 m long and 12 m wide canal to a rectangular structure which divided the water entering into 12 different streams (LeBaron Bowen and Albright, 1958).

At the dam the discharge flowing into the canal was controlled by two gates. The overfall gates were 2.5 m wide, and the maximum gate openings were 1 and 4 m below the dam crest. In the initial section of the canal, the canal walls widened to form a basin of about 23 m wide and 65 m long (essentially the length of the northern wall) which acted as a settling basin for heavier material carried by the Wadi. The 1000 m long canal had earth walls which were covered with a cemented stone lining on the inside (i.e. water side). The cross-sections of the canal and of the settling basin were approximately rectangular.

The spillway was a 40 m wide overfall with a broad unlined rock crest. The crest elevation of the overflow spillway was located 3 m below the height of the dam and the spillway was not gated. As the crest was 3 m below the crest of the dam, it is believed that the reservoir volume corresponding to the top 3 m of the dam was used to damp the flood and flood catchment.

End of the Marib dam

The dam suffered numerous breaching caused by overtopping and the maintenance works were substantial. In AD 575, the dam was overtopped and never repaired. The final destruction of the dam was a milestone in the history of the Arabic peninsula. The fame of the Marib dam was such that its final destruction is recorded in the Koran.

'New' Marib dam

In 1986 a new 38 m high earth dam was built across the Wadi Dhana. The dam site is located 3 km upstream of the ruins of the old Marib dam (Ganchikov and Munavvarov, 1991) . The new dam was designed to store water for irrigating the Marib plains.

P1.1.3 Chronological summary

Date	Comments
BC 750	Construction of the 1st Marib dam. 4 m high earth dam. Length: 580 m. Original northern spillway and canal sluices. Irrigation of the northern bank of the wadi.
BC 500	Second structure: the initial dam was heightened. 7 m high earth dam. Triangular cross-section with both faces sloping at 45°. Water face covered with stones set in mortar. Construction of southern sluices. Irrigation of both the northern and southern side.
After BC 115	Dam reconstruction (Himyarite work) New 14 m high earth dam. New water works at both ends. Northern water works: raise of the spillway floor, construction of five spillway channels, two masonry sluices, masonry tank used as settling pond and a paved channel (1000 m long) to a distribution tank.
AD 325	Final form of the northern outlet works
AD 449	Flood. Dam restoration
AD 450	Flood. Dam breach. Complete renovation
AD 542	Flood. Dam breach
AD 543	Dam restoration
AD 575	Destruction of the dam. The dam was never restored.
AD 1984–1986	Construction of a new 38 m high earth dam

P1.1.4 Bibliography

Books

LeBaron Bowen Jr, R. and Albright, F.P. (1958) *Archaeological Discoveries in South Arabia*. (The John Hopkins Press: Baltimore, USA).

Phillips, W. (1955) *Qataban and Sheba – Exploring Ancient Kingdoms on the Biblical Spice Routes of Arabia*. (Victor Gollancz: London, UK).

Schnitter, N.J. (1994) *A History of Dams: the Useful Pyramids*. (Balkema Publication: Rotterdam, The Netherlands).

Smith, N. (1971) *A History of Dams*. (The Chaucer Press: Peter Davies, London, UK).

Journal articles

Eurenius, J. (1980) Ancient dams of Saudi Arabia. *International Water Power and Dam Construction*, **32**(3), 21–22.

Ganchikov, V.G. and Munavvarov, Z.I. (1991) The Marib Dam (history and the present time). *Gidrotekhnicheskoe Stroitel'stvo*, No. 4, pp. 50–55 (in Russian). (Translated in *Hydrotechnical Construction*, 1991, Plenum, pp. 242–248).

Hathaway, G.A. (1958) Dams – their effect on some ancient civilizations. *Civil Engineering*, **28**(1), 58–63.

Schnitter, N.J. (1967) A short history of dam engineering. *Water Power*, **19**, 142–148.

P1.2 HYDRAULICS PROBLEM

Introductory note

In the problem we shall consider the final stage of the dam and its auxiliary structures (e.g. spillways, channel) with a dam height of 14 m.

P1.2.1 Study of the spillways

In the following questions, let us assume frictionless flows.

(A) Draw a sketch of the dam, both spillways and the outlet, view from the reservoir. Indicate the main dimensions on the sketch.

(B) *Southern spillway*
(B.1) We assume that the water level in the reservoir is equal to the crest of the northern spillway, and we assume that the canal gates are closed (i.e. no discharge in the canal). Using the continuity equation, what is the formula giving the total discharge Q as a function of the spillway width and the flow depth above the crest? Explain in words how this formula is obtained.
(B.2) Application: what is the maximum discharge of the southern spillway for this flow configuration?

(C) *Northern spillway*
(C.1) If the flow depth above the crest is 1 m, what is the total discharge above the spillway? Explain in words your calculations.
(C.2) What is the maximum flood discharge that the northern spillway can absorb before the dam becomes submerged? Explain in detail your calculations.

(D) We assume that the canal gates are closed (i.e. no discharge in the canal). What are the maximum discharge capacities of the southern bank spillway, the northern bank spillway and both spillways, before the dam is overtopped?

P1.2.2 Study of the channel outlet and settling basin

(A) *Basic hydraulic knowledge*
(a) What are the assumptions used to derive the Bernoulli equation? Write the Bernoulli equation.
(b) What is the definition and significance of the Coriolis coefficient? Rewrite the Bernoulli equation using the Coriolis coefficient.
(c) Assuming that the velocity distribution follows a power law:

$$\frac{V}{V_{max}} = \left(\frac{y}{d}\right)^{1/N}$$

Give the relation between the maximum velocity V_{max} and the average flow velocity. Give the expression of the Coriolis coefficient as a function of the exponent of the power law.

(B) *Change in channel width*

Considering a horizontal, rectangular channel with a change of channel width, and assuming a frictionless flow:

(a) Write the continuity and Bernoulli equation for the channel.

(b) Combining the continuity equation and the Bernoulli equation, develop the differentiation of the Bernoulli equation, along a streamline in the *s*-direction. Introducing the Froude number, how do you rewrite the differentiation of the Bernoulli equation in terms of the Froude number? How would the flow depth vary when the channel width varies? Explain your answer in words.

(c) The upstream flow conditions are subcritical. What is the optimum width such that the flow becomes critical? Explain each step of your calculations.

(d) We consider now the flow entering into the settling basin. The upstream flow conditions are: $B = 12\,\text{m}$, $d = 1.7\,\text{m}$, $Q = 25\,\text{m}^3/\text{s}$. What are the flow conditions in the settling basin? Explain in words your calculations. Discuss the aim of the settling basin. Why did the designers adopt a settling basin at this location?

(C) *Settling basin*

We consider the same flow entering the settling basin with the following inflow conditions (upstream of the widening) $B = 12\,\text{m}$, $d = 1.7\,\text{m}$ and $Q = 25\,\text{m}^3/\text{s}$. We shall assume that the flow conditions in the settling basin are those calculated above (by assuming a smooth transition between the upstream flow and the settling basin).

During a flood event, the canal inflow is usually heavily sediment-laden. Taking into account the bed roughness, we will consider the sediment motion of the inflow and of the settling basin flow.

(a) Compute the bed shear stress in the inflow channel and in the settling basin.

(b) Find out the critical particle size for bed-load motion in the inflow channel and in the settling basin. Discuss the findings.

(c) Calculate the critical particle size for bed-load motion in the inflow channel and in the settling basin.

(d) Discuss your results. For example, what is the heaviest bed-load particle size that could enter the 1000 m long canal for the above flow conditions.

P1.2.3 Study of the canal

(A) *Model study*

The King of Sheba wants a study of the 1000 m canal. He orders a (undistorted) scale model study where the length of the model is 40 m.

The prototype flow conditions to be investigated are:

$$\{1\}\ d = 1\,\text{m};\ Q = 12\,\text{m}^3/\text{s} \qquad \{2\}\ d = 3\,\text{m};\ Q = 72\,\text{m}^3/\text{s}$$

where *d* is the flow depth and *Q* is the total discharge.

(a) What type of similitude would you choose? What is the geometric scale of the model? Explain and discuss your answer in words.

(b) For both prototype flow conditions, what is (are) the width(s) of the scale model(s)?

(c) For both prototype flow conditions, what are the model flow depth(s), flow velocity(ies) and total discharge(s)? Explain carefully your answer in words.

(d) Determine the minimum prototype discharge for which negligible scale effects occurs in the model.

(B) *Normal flow conditions*

The 1000 m long canal is a paved channel of 12 m width and the slope is 0.1°.

(a) What values would you choose for the (equivalent sand) roughness height and the Gauckler–Manning coefficient? Discuss and justify your choice.

(b) Considering a uniform equilibrium flow, develop the momentum equation, expressing the bed shear stress as a function of: (1) the Darcy coefficient, and (2) the Gauckler–Manning coefficient. Explain in words your development.

(c) The water discharge in the canal is $25\,\mathrm{m}^3/\mathrm{s}$.

(c1) For the selected roughness height, what are the values of the Darcy coefficient and the normal depth computed from the Darcy coefficient? Explain and discuss your calculations.

(c2) What is the uniform flow depth using the formula of the uniform depth as a function of the Gauckler–Manning coefficient?

(c3) Discuss your results and eventual the discrepancies between the results of the above questions. Compare both the friction coefficients and the flow depth.

(C) *Backwater calculations*

We consider now the gradually varied flows of real fluid in the settling basin and in the $1000\,\mathrm{m}$ canal.

(a) Write the differential form of the energy equation in terms of the flow depth and local Froude number for a channel of variable width and variable slope. Comment on your result.

(b) What are all the assumptions made in order to obtain the above equation?

(c) Explain (in words) how you would integrate the resulting differential equation. Where would you start the calculations?

(d) Table P1.1 provides you the geometric characteristics of the $1000\,\mathrm{m}$ canal and the settling basin at the start of the canal. We consider a flow rate of $Q = 60\,\mathrm{m}^3/\mathrm{s}$.

(d1) Plot the canal profile on graph paper (dimensioned sketched).

(d2) Locate the occurrence of critical flow conditions and hydraulic controls (if any).

(d3) Select the position where you will start your calculations.

(d4) Compute the flow depth, flow velocity, Darcy coefficient, Froude number and friction slope at all the positions defined in Table P1.1.

(d5) Plot the backwater curves (i.e. the flow depth curves) on the graph paper (see Question 9.4.1).

Notes

1. Question (d3) is very important for correct calculations.
2. Flow resistance calculations must be performed using the Darcy friction coefficient.
3. Assume a linear variation of the channel width B, the channel slope θ and roughness height k_s between each position (Table P1.1).

Table P1.1 Canal geometry

Point	Bed elevation (m)	s(m)	Roughness	θ (degree)	B(m)	Remarks
A		0	$k_s = 0.5\,\mathrm{mm}$	0	12	
B		30	$k_s = 0.5\,\mathrm{mm}$	0	23	Start of the settling basin
C		65	$k_s = 0.5\,\mathrm{mm}$	0	23	
D		95	$k_s = 0.5\,\mathrm{mm}$	0	23	End of the settling basin
E		120	Paved channel (*)	0.1	12	Start of the canal
F		900	Paved channel (*)	0.1		
G		901	Paved channel (*)	3.0	12	
H		1060	Paved channel (*)	3.0	12	End of the canal

Note: (*) Students are asked to use the roughness height of paved channel previously obtained in question B and to compute the flow resistance using the Darcy equation.

P1.3 HYDROLOGICAL STUDY: FLOOD ATTENUATION OF THE MARIB RESERVOIR

The objective of this exercise is to estimate the peak flow rate (for the undammed river) of the flood that would have just caused overtopping of the 14 m high, old Marib dam. To do this, the hydraulic performances of the two old spillways must first be understood (Section P.1.2.1).

Information about the surface area of the old Marib reservoir is not easily available, but reasonable scaling down from the 30 km^2 lake that now serves the much larger new Marib dam, suggests that the old lake would have been about 4 km^2.

Assumptions

Assume that this area applies with the lake free surface level the same as the crest of the 40 m wide northern spillway, which is also the assumed initial condition for the approaching flood wave.

As the water level rises to overtop the old dam, assume that the lake area increases linearly to 5.5 km^2.

Assume that the time-of-concentration for this catchment is about 12 hours (*Note*: This value is considerably less than the typical time-to-peak for the Brisbane river which has almost exactly the same catchment area, but the Marib catchment is very sparsely vegetated in relation to the local catchment).

A study of the Moeris reservoir, the Ha-Uar dam and the canal connecting the Nile River and Lake Moeris around BC 2900 to BC 230

P2

Preface

The following problem covers lecture material relevant to Basic Principles (Part I). The assignment was first used in 1993. Based upon real historical facts, the geometry of the hydraulic structures has been simplified to facilitate the calculations.

P2.1 INTRODUCTION

P2.1.1 Presentation

In the history of dams and the story of civilizations, the first dams were built in Egypt and Iraq around BC 3000 where they controlled canals and irrigation works. Often, the civilizations originated in areas where irrigation was a necessity. The history of dams followed closely the rise and fall of civilizations, especially where these depended on the development of the water resources.

A typical example is the Egyptian civilization. For centuries, the prosperity of Egypt relied on the annual flood of the Nile River from July to September and the irrigation systems. One of the most enormous efforts of the Egyptian Kings was the creation of the Lake Moeris in the Fayum depression and the construction of a 16 km long canal connecting the lake to the Nile. The lake was used to regulate the Nile River and as a water reservoir for irrigation purposes.

P2.1.2 The Moeris Lake[1]

The location of the lake was the Fayum (or Fayoum) depression, located 80 km south-west of the city of Cairo (Fig. P2.1). The depression is a vast area of 1700 km^2 and the lowest point in the depression is 45 m below the sea level.

Egyptian engineers connected the Nile River and the Fayum depression to lead flood flows into the depression during high floods. The connection between the river and the depression was a natural cut in the mountains. It was in existence at the time of the King Menes, founder of the 1st Egyptian dynasty (BC 2900). At that time, the Fayum depression contained only a natural lake filled from the Nile during large floods.

[1]There are still some arguments upon whether the Lake Moeris existed or not. There are suggestions that there were in fact two lakes (e.g. Schnitter, 1994, pp. 4–6).

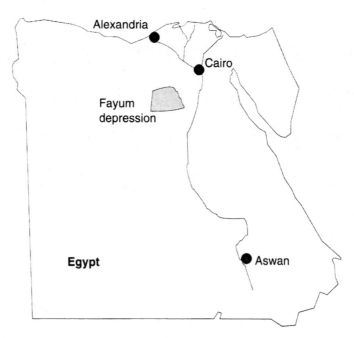

Fig. P2.1 Map of Egypt and the Nile River.

King Amenembat (BC 2300) widened and deepened the canal between the Nile and the Fayum depression. He converted the existing lake into an artificial reservoir (i.e. Lake Moeris) which controlled the highest flood of the Nile. The canal connecting the river to the reservoir was regulated by the Ha-Uar dam. The regulation system consisted of two earthen dams at both ends of the canal, with gates by means of which the architects regulated the rise and fall of the water.

The Lake Moeris had three main purposes:

1. the control of the highest floods of the Nile during the July–September periods,
2. the regulation of the Nile River during the dry season by releasing water from the lake,
3. the irrigation of a large surface area around the lake.

From its size and depth, the Lake Moeris was capable of receiving the overflow of the Nile during its rising, and preventing the flooding of downstream cities like Memphis (see Fig. P2.1). When the river fell, the lake waters discharged again via the canal connecting the river to the lake, and these waters were available for irrigation.

The Ha-Uar dam had also a strategic interest. At the time when Egypt was divided in two kingdoms, the Lower Egypt (i.e. Northern Egypt) and the Upper Egypt (i.e. Southern Egypt), the frontier fortress of Lower Egypt was at the head of the Lake Moeris canal (Hathaway, 1958). The capture of the dams controlling the canal, and the injudicious or malicious use of the reservoir could deprive a great part of the Lower Egypt (i.e. Northern Egypt) of any basin irrigation at all, for such irrigation utilized only the surface waters of the Nile flood. The importance of the fortress commanding the regulator of the canal ceased when the kingdoms were re-united.

The abandonment of the Lake Moeris was primarily caused by the fact that the Lahoun branch of the Nile (i.e. the west branch of the river, Fig. P2.2) dwindled in size and reduced the use of the reservoir. From BC 230, the canal was abandoned and the area inundated by Lake

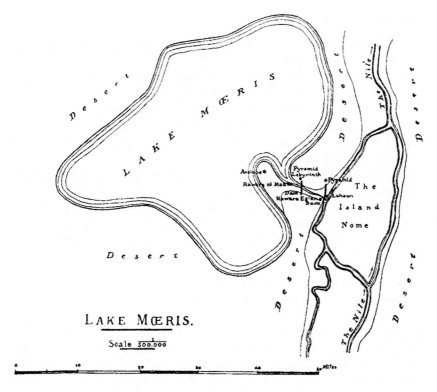

Fig. P2.2 Sketch of the Lake Moeris (after Willcocks, 1919).

Moeris became the province of Fayoum as it is today. Nile shells can still be found in the Fayoum area near the limits of the ancient Lake Moeris (Willcocks, 1919).

Lowest point	45 m below the sea level
Mean free surface level	22.5 m above the sea level
Capacity	$50 \times 10^9 \, m^3$ of water (50 TL)
Surface area of the lake	$1700 \times 10^6 \, m^2$ (17 Ma)
Irrigated land surface area	$1.46 \times 10^9 \, m^2$ (14.6 Ma)

Geometrical characteristics
The Lake Moeris was located in the Fayoum (or Fayum) depression. The geometry of the lake was:
The flood regulation capacity of the lake was $13 \times 10^9 \, m^3$ of water each year, and $3 \times 10^9 \, m^3$ extra for every year it was not used. As an example, if the reservoir was used only every 2 years, it was capable to take from a flood: $13 \times 10^9 + 3 \times 10^9 = 16 \times 10^9 \, m^3$ of water.

Climatic conditions
The evaporation from the lake was about 2.5 m per annum in depth. When the lake was full, the evaporation was around $4.25 \times 10^9 \, m^3$ of water per year.

It must be noted that the artificial regime of the lake had an effect on the climate. The regulation of the Nile by the Moeris reservoir prevented or reduced stagnant and dirty waters downstream, and hence suffocating air in the cities downstream of the Ha-Uar dam (Belyakov, 1991).

P2.1.3 The Ha-Uar dam

The canal connecting the Nile River to Lake Moeris was controlled by two dams at each end of the canal (Fig. P2.2). The two regulators were earthen dams.

As shown in Fig. P2.2, the Nile flowed in two channels opposite the head of the Lake Moeris canal. The upper regulator (i.e. eastern dam) consisted of the existing Lahoun[2] bank (i.e. west bank of the river), a broad spill channel, cut out of the rock to a suitable level for passing ordinary floods and a massive earthen dam across a ravine, which was cut in dangerously high floods. The other end of the canal (i.e the western end) was a much simpler earth dam. During a flood, the dam was cut: the cutting was easy enough.

Fortresses and barracks were located on either sides of the two dams to protect them. Access to the eastern dam and to the eastern end of the canal was difficult: it was written that a fleet was essential to gain possession of the lower great dam (i.e. western dam) (Willcocks, 1919).

The passing of very large floods was possible by cutting the dams. But their reconstructions after the passing of a flood entailed an expense of labour which even an Egyptian Pharaoh considered excessive!

Canal connecting the Nile and the Lake Moeris

The canal connecting the river to the Fayum depression was initially a natural cut through the mountain bordering the Libyan desert (Fig. P2.2). The natural canal was 16 km long and 1.5 km wide (Garbrecht, 1987b). This place is now called the Fayoum Bahr Yusuf Canal.

Around BC 2300, King Amenembat started the construction of an artificial canal along the natural valley. The artificial channel was 16 km long and 5 m deep. Its shape was trapezoidal with a 600 m width at the bottom (Fig. P2.3). The slope of the banks was 1V:10H to allow the use of non-cohesive rockfill and earth materials. The protection of the channel bottom and the bank slopes consisted of cut stones and cement joints. The covering blocks were placed on the bottom and on the slopes, and they were cemented together.

The average slope of the channel bed was 0.01°. The canal was inclined towards the Fayum depression.

Eastern dam

The Eastern dam of the Ha-Uar dam structure blocked the valley connecting the Nile River to the Fayum depression. The axis of the valley was east–west. The dam stretched over 1550 m

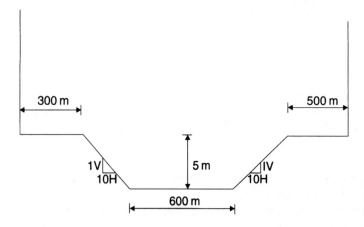

Fig. P2.3 Cross-section of the artificial canal and the natural valley (view from the Nile River).

[2]Also spelled Lahun.

from the south to the north, and it consisted of three large parts:

1. In the north, the Lahoun bank blocked the natural valley over a 550 m length. The elevation of the top of the bank was 10 m above the canal bottom. The bank consisted of non-erodable material.
2. Next to the Lahoun bank, there was a broad spill channel. The channel was an ungated broad-crested weir. The weir was 400 m wide with a rectangular cross-section. The crest elevation was 3 m above the canal bottom.
3. An earthen dam blocked the southern extremity of the natural valley. The top of the dam was 9 m above the channel bottom.

A hydraulic structure was located between the broad-crested weir and the earthen dam. The structure supported a sluice gate used to release waters to the river. The gate was 5 m high and 10 m wide. The bottom of the gate opening was at the same elevation as the canal bottom.

P2.1.4 Historical events

History indicates that Joseph arrived in Egypt in the time of the Hyksos who ruled the Lower Egypt (i.e. Northern Egypt) while Theban dynasties ruled Upper Egypt. It was a time of unending wars between the two kingdoms.

One of the famines of long duration in Egypt occurred during Joseph's time and was described in the Book of Genesis. It was caused by the capture and the breaching of the Ha-Uar dam by the King of Upper Egypt (i.e. Southern Egypt). The famines were ended by the recapture and repair of the dam by the King of Lower Egypt. According to a tradition, Joseph (BC 1730) worked on the restoration of the canal and the dams. Later Jewish slaves were used for the maintenance of the works.

When the Jews fled from Egypt with Moses, stories relate that the Egyptian army was destroyed by the sea when it crossed the 'Red Sea'. In fact, the drowning of the army was probably caused by the breaching of the Ha-Uar dam.

P2.1.5 Important dates

Date	Comments
BC 2900	The connection between the Nile and the Fayum depression was in existence at the time of the King Menes, founder of the 1st Egyptian dynasty. This Egyptian King built the city of Memphis.
BC 2300	King Amenembat (12th dynasty) widened and deepened the canal between the Nile and the Fayum depression. He converted the existing lake into an artificial reservoir which controlled the highest flood of the Nile.
BC 230	Abandonment of Lake Moeris primarily due to the fact that the Lahoun branch of the Nile (see Fig. P2.2) dwindled in size and reduced the use of the reservoir.

P2.1.6 Discussion

Although there are a lot of stories about Lake Moeris, the earliest of these historians wrote his account of the lake in BC 430. There is no trace left of the western dam and only the remains of the eastern structure may be seen (e.g. Garbrecht, 1987; Schnitter, 1994).

In fact there are still doubts and arguments about the existence of Lake Moeris. Schnitter (1967) mentioned the enigmatic Lake Moeris. Smith (1971) and Belyakov (1991) implied that Lake Moeris was a natural lake and not an artificial reservoir. Garbrecht (1987b) suggested that the Fayum depression was transformed by large-scale reclamation works into a fertile province around BC 1700 and was never used as an artificial reservoir. Recent studies (Schnitter, 1994; Garbrecht, 1996) suggested the existence of two lakes since around BC 300–250. Prior to BC 300 the water level in the Fayum depression was about 15–20 m above sea level. It dropped down to about 2 m below sea level (and later to about 36 m below sea level) as a result of land reclamation works associated with high evaporation rate. An artificial reservoir was created around BC 300–250 to irrigate the reclaimed land. The artificial lake was in use up to the 19th century.

In any case, Willcocks (1919) and Hathaway (1958) provided a lot of evidences supporting the existence of Lake Moeris. Furthermore the constructions of the pyramids of large temples and of a large dam (i.e. the Sadd el-Kafara dam) indicate that Egyptian engineers had the expertise and the knowledge to build large-scale civil engineering works. There is no doubt that the Egyptians were able to divert the floods of the Nile River into Lake Moeris. Furthermore they had the technology and the engineering skills to build canals and earthen dams.

Interestingly, a new mini hydro scheme will be installed by the year 2000 (*Hydropower & Dams* 1997). The purpose of the project is the regulation of irrigation water in the depression as well as the generation of hydro-electricity using low-head Kaplan turbines.

P2.1.7 Bibliography

Books

Garbrecht, G. (1987b) Hydrologic and hydraulic concepts in antiquity. In *Hydraulics and Hydraulic Research: A Historical Review*. (Balkema Publication: Rotterdam, The Netherlands) pp. 1–22.

Schnitter, N.J. (1994) *A History of Dams: the Useful Pyramids*. (Balkema Publication: Rotterdam, The Netherlands).

Smith, N. (1971) *A History of Dams*. (The Chaucer Press: Peter Davies, London, UK).

Willcocks, W. (1919) *From the Garden of Eden to the Crossing of the Jordan*. (E & FN Spon Ltd.: New York, USA).

Journal articles

Belyakov, A.A. (1991) Hydraulic engineering and the environment in antiquity. *Gidrotekhnicheskoe Stroitel'stvo*, No. 8, pp. 46–51 (in Russian). (Translated in *Hydrotechnical Construction*, 1992, Plenum Publication, pp. 516–523).

Buckley, A.B. (1923) The influence of silt on the velocity of water flowing in open channels. *Minutes of the Proceedings of the Institution of Civil Engineers*, 1922–1923, **216**, Part II, 183–211. Discussion, 212–298.

Garbrecht, G. (1996) Historical water storage for irrigation in the Fayum depression (Egypt). *Irrigation and Drainage Systems*, **10**(1), 47–76.

Hathaway, G.A. (1958). Dams – their effect on some ancient civilizations. *Civil Engineering*, January, 58–63.

Hydropower & Dams (1997) Mini hydro scheme for Egyptian oasis. *International Journal of Hydropower and Dams*, 4(4), 12.

Schnitter, N.J. (1967) A short history of dam engineering. *Water Power*, April, 142–148.

P2.2 HYDRAULICS PROBLEM

Introductory note

We shall consider the Lake Moeris and the artificial canal connecting the Nile River to the reservoir (Fig. P2.3). On Fig. P2.3, the flood plain on each side of the trapezoidal channel consisted of grass, bush and rocks. An equivalent roughness height of $k_s = 100$ mm could be considered, if necessary.

P2.2.1 Study of the upper regulator

(A) Draw a sketch of the eastern dam and the canal cross-section, view from the west (i.e. view from the canal). Indicate the main dimensions on the sketch. Show the north and south directions.

(B) During a flood, the discharge in the Nile River south of Ha-Uar is $8000 \, \text{m}^3/\text{s}$. At the same time, the chief engineer records a 1.1 m flow depth above the sill of the broad-crested weir.

   In this question, you will assume that the weir crest and the transition between the weir and the channel are smooth and horizontal. For each sub-question, students are asked to explain in words each formula and assumption used. Indicate clearly each fundamental principle(s).

(B.1) What is the flow rate in the Nile River downstream of the Ha-Uar dam?

(B.2) Sketch a cross-section profile and a top view of both the broad-crested weir and the start of the channel. On the cross-section view, plot the water surface profile.

(B.3) What are the values of the specific energy: (a) upstream of the weir, (b) mid-sill, (c) downstream of the weir where the cross-section is rectangular (and 400 m wide) and (d) downstream of the weir at the start of the canal where the cross-section is trapezoidal (see Fig. P2.3)? You will assume that the channel bed elevation is the same upstream and downstream of the weir.

(B.4) Develop the dimensionless relationship between the specific energy and the flow depth (i.e. E/d_c as a function of d/d_c) for a channel of irregular cross-section. Explain clearly in words each step of your development. Deduce the expression of the dimensionless specific energy E/d_c for a rectangular channel.

(B.5) Plot the dimensionless specific energy diagram (i.e. E/d_c versus d/d_c) on a graph paper. Indicate on the graph the points representing the flow conditions: (a) upstream of the weir, (b) mid-sill and (c) downstream of the weir where the cross-section is rectangular (and 400 m wide).

(B.6) Downstream of the weir, is the flow subcritical or supercritical: (a) at the end of the weir where the cross-section is rectangular (400 m width) and (b) at the entrance of the canal where the cross-section is rectangular? Justify your answers in words.

(C) The flow rate in the trapezoidal canal was initially $1500 \, \text{m}^3/\text{s}$. The initial flow depth in the canal was 2 m. A large flood arrives from the Nile River and the flow depth at the start of the canal becomes instantly 2.5 m. A surge develops and travels downstream in the canal towards Lake Moeris.

In this question, the channel will be assumed smooth and horizontal.

(C.1) Draw the appropriate sketch(es) of the travelling surge. Indicate clearly the direction of the initial flow and the direction of the surge.

(C.2) What type of surge is taking place in the channel? Can you make any appropriate assumption(s) to compute the new flow conditions. Justify your answer. If you are not able to do any calculations, go to the next question. If you can do the calculations, continue this question.

(C.3) Define a control volume across the surge front for the trapezoidal canal. Indicate the control volume on your sketch(es) (Question C.1). Show on your sketch(es) (Question C.1) the forces acting on the control volume. Show also your choice for the positive direction of distance and of force.

(C.4) Write the continuity and momentum equation as applied to the control volume shown on your sketch.

(C.5) Compute the surge velocity and the new flow rate.

(C.6) Neglecting flow resistance, how long would it take for the surge to reach the downstream end of the canal?

(C.7) A horseman would need 25 min to cover the distance between the two ends of the channel. Starting from the upstream end at the same time as the surge, will he reach the downstream end before the surge.

P2.2.2 Study of the channel

(A) The canal is discharging $3000\,m^3/s$.
(A.1) For a canal of irregular shape, how are the critical flow conditions defined? Explain your answer clearly. Use appropriate sketch(es) if necessary.
(A.2) For the trapezoidal artificial canal, what is the critical depth for the above discharge? Is the critical depth a function of the channel roughness and/or slope?

(B) *Uniform equilibrium flow*
(B.1) What is the definition of an uniform flow? Give at least two practical examples of uniform flow situations.
(B.2) Draw a sketch of an uniform flow situation. Choose an appropriate control volume. Show on your sketch the forces acting on the control volume. Show also your choice for the positive direction of distance and of force.
(B.3) Write the momentum equation for an uniform flow in a channel of irregular cross-section.
(B.4) How do you deduce all the uniform flow conditions for a channel of irregular cross-section? Explain your answer in words. *No more* than one equation is required.

(C) The canal is discharging $3000\,m^3/s$ (as for Question P.2.2.2(A)).
(C.1) What roughness height would you choose for the trapezoidal channel? Justify your answer in words.
(C.2) What is the uniform flow depth for that discharge? Is the uniform flow depth a function of the channel roughness and/or slope?
(C.3) What value of the Manning coefficient would you choose for the channel? Justify your choice. Compute the uniform flow depth using the Manning formula? Does your result differ from the result obtained in Question C.2? Discuss the comparison (if any) between the results obtained in Question C.2 and Question C.3.
(C.4) Assuming that uniform flow is obtained in the canal, where can you control the flow in the canal (e.g. upstream, downstream)? Justify and discuss your answer.
(C.5) Give at least two examples of hydraulic controls that could be used to regulate the flow in the canal (Question C.4). Discuss each example and explain clearly the difference(s) between each possibility. Sketch each example.

(D) In BC 2251, a very large flood of the Nile River discharged into the canal to Lake Moeris. The level of water in the valley connecting the river and the reservoir was 2.1 m above the bed elevation of the valley (i.e. 7.1 m above the bed elevation of the canal).
Assuming that uniform flow conditions were reached in the long canal and the valley:
(D.1) Deduce the water discharge into the reservoir.
(D.2) Provide all the flow conditions in the artificial canal (i.e. water discharge, velocity of water, flow depth, hydraulic diameter, cross-section area, wetted perimeter, friction factor).
(D.3) Provide all the flow conditions in the flood plain (i.e. water discharge, velocity of water, flow depth, hydraulic diameter, cross-section area, wetted perimeter, friction factor).
(D.4) Is the flow supercritical? Explain your answer.

Students must detail, discuss and justify every step of their method: e.g. calculation of flow resistance, choice of roughness height(s). For the trapezoidal canal, students should use the roughness height selected in Question P.2.2(C1). The roughness height of the flood plain (i.e. on each side of the artificial canal) is given at the beginning of the assignment.

P2.3 HYDROLOGY OF EGYPT'S LAKE MOERIS

Part A

(a) Using an atlas like 'The Times Atlas of the World', find out which 2 months of the year normally yield the greatest rainfalls in the upper Nile catchment, near Khartoum (Sudan), and near Adis Abeba (Ethiopia). In this region, which direction are winds blowing from during those months? (This question prompts consideration of where is the likely evaporative source of moisture for this rainfall.)

(b) In the upper Nile catchment, why do winds tend to blow in that direction during that period? (This question prompts consideration of what is driving the airflow. If the wind direction does not conform to the general circulation described in class, then the driving force must be a very strong phenomenon.)

(c) Explain in a few sentences why the flood absorbtion capacity of Lake Moeris is significantly less if a major flood occurs during the previous year than if it does not, but greater still if preceded by two or more relatively dry years.

(d) What does the atlas indicate for the general order of magnitude of annual evaporation rates in the Nile region? Explain how this relates to (c) above. How can an (empty) depression form below sea level, and remain empty?

(e) Do you think high flows in the Nile would have snow-melt as a significant contributor? Explain. In what way is soil at the bottom of Lake Moeris likely to be *infertile*?

Part B

This question addresses the potential for attenuating flood flows in the Nile River by means of the off-stream storage provided by Lake Moeris. (The previous question dealt with flood attenuation due to direct (on-stream) storage.) Just upstream of the junction between the Nile and the lake's canal, the river flow typically varies as indicated in Fig. P2.4, reaching a peak rate of just over 700 Gl/day (about 8.2 Ml/s) in early September. (Very wet years probably yield flows about 30% greater than this.)

For the questions below, increase the ordinates of Fig. P2.4 to somewhere between the 'typical' wet season and the 'extreme' wet season, by adding about 30%.

Although Part II of this assignment asks you to find normal depth in the lake's canal at a particular flow rate, for this question you will need the whole stage–discharge curve (Q versus H) up to about 10 Ml/s, which may include a portion of over-bank flow.

From Fig. P2.4, note that the flow rate in June is typically steady (at about 50 Gl/day). During this period, assume that the lake's surface level is the same as the river surface, with no flow in the canal. As the wet season brings higher flows from up-river each week, assume that the river level at the junction must rise until the sum of the lake's canal flow and the downstream river flow equals the given total flow rate from upstream each week. For this exercise, assume that the canal has no dams at either end, and perform all requested calculations manually (without computer assistance). (*Note*: The imprecise calculations requested below are intended merely to provide a rough estimate of the attenuating effect of flood diversion to the lake. More accurate hydraulic calculations would be needed, particularly as the lake approaches 'full'.)

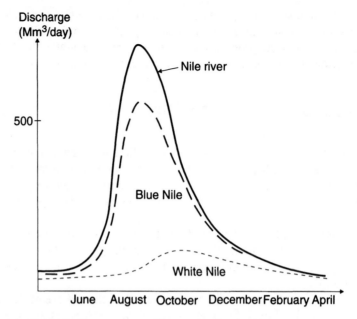

Fig. P2.4 Hydrography of the Nile River near Aswan (data: Garbrecht, 1987b).

(a) Replot the total inflow hydrograph (appropriately rescaling the data from Fig. P2.4) dividing each month into four equal 'weeks' of 7.6 days, and showing the hydrograph in stepped form (as if each 'weekly' flow was constant).

(b) Integrate the total Nile flow volume, from 1 July to when the falling hydrograph passes half its peak value. Deduce what depth of Lake Moeris (assuming constant plan area) would be required to store *all* of this volume.

(c) Plot a stage–discharge curve for the river at (above and below) the canal junction:

$$Q \text{ (Ml/s)} = 0.12H^2$$

(This means, for example, that when $H = 8.2$ m, then $Q = 8.2$ Ml/s $= 8200$ m^3/s approximately, at which time the velocity is about 1.5 m/s, the sectional area is about 5500 m^2, and the river width is about 550 m.)

(d) On the plot of (c), superimpose a hypothetical (but incorrect) stage–discharge curve for the canal:

$$Q = 0.72H^{1.4}$$

(From your Section P2.2 results, plot points to show how incorrect this curve is.)

Set up a tabulation (with each row representing a 'week', starting at the end of June), showing how each 'week's' upstream inflow is divided into two components, one of which represents the diversion to Lake Moeris. A column should show the stage (H) at the canal junction. (The objective of this table is to identify how much attenuation of the upstream river flow is achieved by the diversion.) Superimpose the deduced downstream river flow on the plot of (a). Highlight the 'attenuation'.

(e) Sum the canal flow volumes to the lake each 'week' from 1 July, and deduce the new surface level of the lake at the end of each week. Stop the calculations in (d) as soon as the lake surface level matches the stage level at the canal–river junction (i.e. the canal flow is completely 'drowned'). (From this point, continued falling river levels would result in outflows from Lake Moeris, unless these are intercepted by blocking the canal. (to save the water for later irrigation purposes)

A study of the Moche river irrigation systems (Peru AD 200–1532)

Preface

The following problem covers lecture material relevant to Basic Principles (Part I) and Hydraulic Modelling (Part III). The assignment was first used in 1994. Based upon real historical facts, the geometry of the hydraulic structures has been simplified to facilitate the calculations.

P3.1 INTRODUCTION

P3.1.1 Presentation

South-American agriculture began in the mountain regions of southern Peru and northern Bolivia (Fig. P3.1). The early civilizations grew up in the semi-arid highlands and arid coastal valleys traversed by small rivers. Irrigation was the dominant agricultural technique throughout the prehistory of the north coast of Peru. The establishment of the early civilizations took place in the river valleys (e.g. Moche, Chicama rivers).

The northern coastal valleys of Peru were populated by the Mochicas around AD 200–1000 and later by the Chimus (AD 1000–1466). The Mochicas and the Chimus developed the irrigation of the area particularly on the rivers Moche and Chicama. The Moche valley was the locus of the Mochican civilization and later held the capital of the Chimu empire, Chan-Chan.

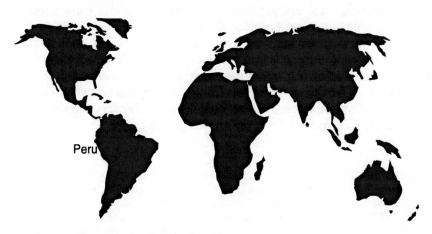

Fig. P3.1 Map of the world and location of the Moche valley.

In AD 1466, the Incas overran the Chimu empire.[1] Although the capital city of the Incas was Cuzco, in the Peruvian highlands, the Inca empire extended at its maximum from Ecuador to Chile. Inca engineers were expert in road and bridge building. Furthermore, they gained some expertise in dam construction and canal building from the Chimus. However, in AD 1535, the power of the Incas was destroyed by the Spaniards. The Spanish conquistadors of Peru, led by Francisco Pizarro,[2] came seeking gold and silver.[3] Their small troupe benefited from a civil war within the Inca empire. After the capture and killing of the Great Inca, the Incas surrendered without fight.

Geography

Peru is located on the Pacific coast of South America. The borders of ancient Peru extended from just south of the Equator to about 20° south latitude. The countryside is dominated by the great mountain mass of the Central Andes.

The Andes, running from north-west to south-east, divide Peru into three long and narrow topographic zones: (a) the desert coast including the western slope of the mountains, (b) the highlands and (c) the tropical forests on the eastern slope of the mountains.

The Peruvian coast is one of the world's driest deserts and life is almost impossible without irrigation: e.g. the average annual rainfall is less than 5 mm per year at the estuary of the Moche river (Nials *et al.*, 1979a,b). The waters along the Peruvian coast are found in small rivers that flow from the nearby Andes across the coastal desert into the Pacific and which are fed by limited seasonal rains in the high mountains.

The Mochican culture

The Mochican civilization (AD 200–1000) was located around the valley of the Moche river. It is believed that the Mochican civilization was dynamic, aggressive and well advanced in irrigation and construction. The Mochicans built large pyramids (e.g. Huaca del Sol, Huaca de la Luna) and some impressive irrigation systems in the Moche and Chicama valleys.

The Mochicans extensively used gold, silver and copper for ornamental purposes. Further, they excelled in the field of ceramics.

The Chimu empire

The Chimu civilization was founded in the Moche valley before it spread over the adjacent valleys (e.g. Chicama, Viru valleys). In the latter stages, the maximum extent of the empire covered around 1000 km of coastline up to the Ecuadorean border (Fig. P3.2).

The Chimu civilization was well developed and its culture was famous in South America. The Chimu engineers were also expert in road building (e.g. inter-valley roads along the Pacific coast), irrigation canal construction (e.g. inter-valley canal) and city development (e.g. Chan-Chan).

The capital city of the Chimu empire was Chan-Chan, located in the lower delta of the Moche valley. Chan-Chan was the largest city ever built in ancient Peru and it covered an area of about 28 km². Torrential rains in 1934 destroyed much of the city ruins.

---

[1] It is thought that the Incas conquered Chan-Chan after cutting its water supply system.

[2] Francisco Pizarro (AD 1475–1541) was born in Trujillo (Spain). With help of his brothers Gonzalo (AD 1502–1548) and Hernando (AD 1508–1578), he conquered Peru. He was killed in Lima by the partisans of his rival Diego de Almagro (AD 1475–1538) and led by de Almagro's son Diego (AD 1518–1542).

[3] The land of Peru was rich in gold, silver, copper, tin and other metals. South-American Indians extensively used silver and gold for ornaments.

Fig. P3.2 Extent of the Chimu and Inca empires.

The development of the Chimu empire and of its capital city Chan-Chan relied heavily upon the development of irrigation systems. The water supply of Chan-Chan was provided by large canals withdrawing waters from the Moche and Chicama rivers (see below).

The Chimu empire was defended in the south by the impressive fortress of Paramonga in the Fortaleza valley. In the north, no frontier defences have been observed and the Incas invaded the Chimu empire in AD 1466 from the north.

The Inca empire

The Incas were a small tribe in the southern highlands of Peru during the early part of the 13th century AD. Their settlements were located around Cuzco. After defeating the other tribal states of southern highlands of Peru (around AD 1438), the Incas emerged as the strongest military power of the region. They conquered, retained and re-organized the territories of the south high-lands and the nearby coastal regions.

The Inca dynasty included 13 emperors (Table P3.1). During the expansion period, Cuzco became the capital city of the Inca empire.

Table P3.1 The Inca dynasty

(1)	Name (2)	Period (3)	Comments (4)
[1]	Manco Capac	AD 1200	
[2]	Sinchi Roca		
[3]	Lloque Yupanqui		
[4]	Mayta Capac		
[5]	Capac Yupanqui		
[6]	Inca Roca		
[7]	Yahuar Huacac		
[8]	Viracocha Inca		
[9]	Pachacuti Inca Yupanqui	1438–1471	Beginning of the Inca empire expansion
[10]	Topa Inca Yupanqui	1471–1493	
[11]	Huayna Capac	1493–1525	
[12]	Huascar	1525–1532	
[13]	Atahuallpa	1532–1533	Killed by the Spanish

Ref: (Mason, 1957)

In the early 1460s, the Inca armies marched north conquering the highlands as far as Quito.[4] Then they invaded the Chimu empire and destroyed the Chimu armies (AD 1466). The Inca empire was extended to its ultimate limits near the end of the 15th century (Fig. P3.2). At the time of the death of Topa Inca Yupanqui (AD 1493), the Inca empire extended from northern Ecuador to central Chile, including the whole of Bolivia and Peru, a coastal distance close to 5000 km.

The Incas made their conquest through a combination of military might and diplomacy. The Incas were recognized as ferocious warriors and great engineers. Settlements were defended by hilltop fortresses (e.g. Sacsahuaman on the hill at the edge of Cuzco). An impressive network of roads and bridges connected the main cities across their empire.

The last Inca emperor, Atahuallpa, was captured and executed by Pizarro in AD 1533.

P3.1.2 The irrigation system of the Moche river valley

The irrigation systems of the Mochicas and Chimus included large and long canals with flat and steep gradients, aqueducts, dams, side weirs. They were built by the Mochica civilization, later extended by the Chimus and the Incas. The two largest canals were the Vichansao canal and the Inter-valley canal (Figs. P3.3 and P3.4).

Vichansao canal

The Vichansao canal was the most important prehistoric canal of the Moche valley. Its construction began around AD 0(?) to AD 250. By AD 1250, the system was over 28 km long. The Chimus subsequently added another 7 km of canal. Later the Incas extended the canal again by another 5–10 km.

The canal was 2 m wide and the height of the sidewalls was 1.5 m. Parts of the canal were made with cobble bed; other parts were made of sand and clay. The canal was fed from an artificial reservoir located 25 km upstream of the Moche river estuary. The river was dammed by a diversion weir across the river bed. The dam was made of stonework masonry and the river flow discharged over the crest. The crest of the dam was horizontal and 20 m long. The dam was 2 m high, its cross-section was trapezoidal with a vertical upstream face. The thickness of the structure was 4 m at the base and 3 m at the crest.

---

[4]Quito is now the capital city of modern Ecuador (altitude: 2540 m).

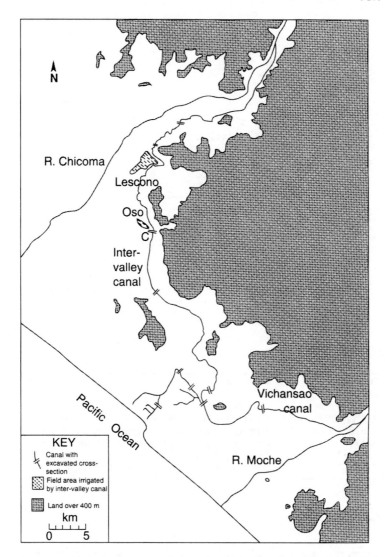

Fig. P3.3 Location of the Vichansao and Inter-valley canals (*source:* Farrington and Park, 1978).

Inter-valley canal

When the Chan-Chan region expanded so that the Vichansao canal capacity became inadequate, the Inter-valley canal was designed to provide additional water supply. The canal was built to withdraw waters from the Chicama valley and it was later connected to the Vichansao canal (Fig. P3.3). The Chicama river is a larger stream than the Moche river and the canal enabled more continuous water supply throughout the year, which could have been used to irrigate a second crop.

The initial canal was 79 km long. At a later stage, another portion (60 km long) was added to connect the inter-valley canal with the Vichansao canal. The canal was 7 m wide and 2 m deep.

Note that, after the Spanish conquest of Peru, the Spaniards continued to use the irrigation system of the Moche and Chicama valleys (Kosok, 1940).

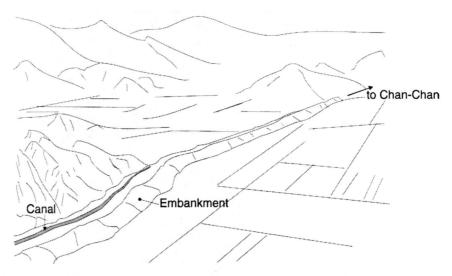

Fig. P3.4 Sketch of the inter-valley canal in the Chicama valley (Acequia de Ascope) (after photographs by Ubbelohde-Doering 1967).

P3.1.3 Important dates

Date (1)	Comments (2)
AD 0(?)–250	First construction works of the Vichansao canal
AD 1250–1462	Extension of the Vichansao canal during the Chimu empire
AD 1466	Annexation of the Chimu empire to the Inca empire
AD 1532	Landing of Pizarro in Ecuador
AD 1533	Execution of the last Inca emperor Atahuallpa

P3.1.4 Bibliography

Books
Lanning, E.P. (1967) *Peru Before the Incas*. (Prentice-Hall: Englewood Cliffs NJ, USA).
Mason, J.A. (1957) *The Ancient Civilizations of Peru*. (Penguin books: Harmondsworth, UK).
Smith, N. (1971) *A History of Dams*. (The Chaucer Press: Peter Davies, London, UK).
Ubbelohde-Doering, H. (1967) *On the Royal Highways of the Inca Civilizations of Ancient Peru*. (Thames and Hudson: London, UK).

Journal articles
Cook, O.F. (1916) Staircase farms of the ancients. *National Geographic Magazine*, **29**, 474–534.
Farrington, I.S. (1980) The archaeology of irrigation canals with special reference to Peru. *World Archaeology*, **11**(3), 287–305.
Farrington, I.S. and Park, C.C. (1978) Hydraulic engineering and irrigation agriculture in the Moche Valley, Peru: c. A.D. 1250–1532. *Journal of Archaeological Science*, **5**, 255–268.
Kosok, P. (1940) The role of irrigation in ancient Peru. *Proceedings of the 8th American Scientific Congress*, Washington DC, USA, **2**, 168–178.
Nials, F.L., Deeds, E.E., Moseley, M.E., Pozorski, S.G. and Feldman, R. (1979a) El Niño: the catastrophic flooding of coastal Peru. Part I. *Field Museum of Natural History Bulletin*, **50**(7), 4–14.

Nials, F.L., Deeds, E.E., Moseley, M.E., Pozorski, S.G. and Feldman, R. (1979b) El Niño: the catastrophic flooding of coastal Peru. Part II. *Field Museum of Natural History Bulletin*, **50**(8), 4–10.

P3.2 HYDRAULICS PROBLEM

P3.2.1 Study of the Vichansao canal and the diversion dam (Part A)

We shall consider the Vichansao canal connecting the Moche river to the Chan-Chan region (Fig. P3.3) and the diversion dam across the Moche river. We shall assume that: (a) the Moche river discharge (upstream of the canal intake) is about $2.1 \, \text{m}^3/\text{s}$, and (b) the canal operates and discharges 30% of the lowest river flow rate.

Note: Later, compare the above river discharge with the results obtained in Part III.3, Question B2.

(A) *Sketch*
(A.1) Draw a sketch in elevation of the Moche river, the diversion dam and the Vichansao canal intake. Indicate the main dimensions on the sketch. Show the north and south directions.
(A.2) Draw also the canal cross-section with the main dimensions.
(A.3) On which river bank is the canal intake: (a) the right bank (i.e. north-west), or (b) the left bank (i.e. south-east)?

(B) *The Vichansao canal*
We consider now the waters flowing into the Vichansao canal.
(B.1) Define the concept of critical flow conditions. Explain your answer in words only. Do not give any equations.
(B.2) For a rectangular channel (channel width B), derive the expression of the critical flow depth. 'Derive' means that you must explain and justify your calculations before giving the expression of the critical flow depth.
(B.3) For the Vichansao canal, give the value (and units) of the critical flow depth.

(C) *The diversion dam*
In operation, the water overflows the entire crest of the diversion dam uniformly. We shall assume that the energy losses can be neglected.
(C.1) What is the flow depth above the crest of the diversion dam?
(C.2) What is the value (and units) of the specific energy: (a) on the diversion weir crest, (b) upstream of the diversion weir and (c) downstream of the diversion weir.
(C.3) What is the free-surface elevation of the upstream reservoir above the dam crest?
(C.4) What is the flow depth downstream of the diversion dam?
(C.5) Sketch a cross-section profile of the diversion dam. On the cross-section view, plot the water surface profile.
For each sub-question, you are asked to explain in words each formula and assumption that you use. Indicate clearly each fundamental principle(s).

P3.2.2 Study of the Inter-valley canal

We shall consider now the Inter-valley canal connecting the Chicama river to the Chan-Chan region (Figs. P3.3 and P3.4).

(A) *Hydraulic model study*
The Chimu emperor wants a hydraulic study of the first 79 km of the canal. He orders a (undistorted) scale model study where the length of the model equals 985 m.

The prototype flow conditions to be investigated are:

$$\{1\}\ d = 0.4\,\text{m};\ Q = 0.7\,\text{m}^3/\text{s}$$
$$\{2\}\ d = 1.5\,\text{m};\ Q = 3\,\text{m}^3/\text{s}$$

where d is the flow depth and Q the total discharge.

(A.1) What type of similitude would you choose? What is (are) the scale(s) of the model(s)? Explain and discuss your answer *with words*.
For both prototype flow conditions (i.e. {1} and {2}), answer the following questions:
(A.2) What is (are) the width(s) of the scale model(s)?
(A.3) What are the model flow depth(s), flow velocity(ies) and total discharge(s)? Explain carefully your answer in words.

(B) *Upstream end of the canal*
The first part of the canal (i.e. first 79 km) is a paved channel (7 m width) and the bottom slope is 0.5°.
(B.1) Draw a sketch of a uniform flow situation. Choose an appropriate control volume. Show on your sketch the forces acting on the control volume. Show also your choice for the positive direction of distance and of force.
(B.2) Write the momentum equation for an uniform flow in the channel.
(B.3) Develop the formula of the uniform flow depth as a function of the Darcy friction factor. Explain in words your development.
(B.4) What values would you choose for: (a) the roughness height and (b) the Manning coefficient? Discuss and justify your choice.
(B.5) The water discharge in the canal is $2\,\text{m}^3/\text{s}$.
(B.5.1) For the selected roughness height (Question B.4), what are the values of the Darcy coefficient and the normal depth computed from the Darcy coefficient? Explain and discuss your calculations.
(B.5.2) What is the uniform flow depth computed using the Manning equation and the Manning coefficient (Question B.4)?
(B.5.3) Discuss your results and the (inevitable) discrepancies between the results of the above questions B.5.1 and B.5.2. Compute the values of the Chézy coefficients for each case. Compare both the Chézy coefficients *and* the flow depths.
(B.5.4) Is the uniform flow subcritical, supercritical or critical? Discuss (in words) the properties of the uniform flow regime.
For the next two Questions (C and D), we shall consider now the downstream end of the inter-valley canal (near the junction with the Vichansao canal). The last 3 km of the canal includes:

(A) a rectangular channel (7 m wide, 2850 m long, 3° bed slope) with concrete bottom and sidewalls,
(B) a 150 m long rectangular channel (7 m wide, 0.1° slope) with gravel bed ($d_{50} = 0.2\,\text{m}$),[5] followed by
(C) a wooden sluice gate (7 m wide) to regulate the canal.

(C) *Downstream end of the canal*
We consider that the canal discharges $1.4\,\text{m}^3/\text{s}$ of water and that the flow depth downstream of the sluice gate is 0.05 m.

---

[5]For a gravel bed, it is reasonable to assume that the equivalent roughness height equals nearly the median gravel size (d_{50}).

(C.1) Compute the flow velocity and Froude number downstream of the sluice gate.

(C.2) Calculate the flow depth, flow velocity and Froude number immediately upstream of the sluice gate. Explain clearly each step of your calculations and highlight each assumption used.

(C.3) Compute the uniform flow depth in: (a) the 2850 m long channel, and (b) the 150 m long channel (immediately upstream of the sluice gate).

(C.4) Compute the free-surface profile in both channels. Explain very clearly *in words* each step of your calculations. Discuss any assumptions used during the calculations.

(C.5) On graph paper, plot the channel bottom, the sluice gate and the free-surface profile along the 3 km long channel. Use an appropriate scale.

(D) *Gate operation*

The canal discharge remains constant (i.e. $1.4 \, \text{m}^3/\text{s}$). The initial flow conditions are the same as defined in Question C and plotted in Question C.5. Suddenly, the wooden sluice gate is opened and the new gate opening is now 0.7 m.

(D.1) Describe and discuss *in words* the 'hydraulic event' occurring (in the 3 km long channel) immediately after the opening of the sluice gate. Use *sketches* to illustrate your discussion. Explain (in words) what type of calculations could be done. *Do not* give any equations. *No* calculation is required.

(D.2) A few hours after the gate opening, the flow characteristics become steady.

(D.2.1) Compute the new free-surface profile.

(D.2.2) On graph paper, plot the channel bottom, the sluice gate and the free-surface profile (obtained in Question D.2.1) along the 3 km long channel. Use an appropriate scale.

(D.3) Compare and discuss the free-surface profiles obtained in Questions C.4/C.5 and D.2. What are the main differences?

P3.3 HYDROLOGY OF WESTERN PERU

P3.3.1 Background

Consult a map of Peru (on the west coast of South America) and focus on the northern coastal city of Trujillo at about 8° South latitude. Being so close to the equator, one might expect that this area would experience equatorial convective rainfall in the southern summer, but not so. Trujillo's dry (desert) climate is like that of Lima (see Table P3.2). It is very pleasant, but what about water supply! Why it does not rain there? Because the ocean's cold Humbolt current flows northward past Peru then eastward across the Pacific driven by the Walker Cell. The proximity of cold water attracts condensation, so atmospheric moisture is scarce. (About eight times per century, a big El Niño event kills the Walker Cell, causing warm Western Pacific water to slash back toward Peru, cutting off (usually about Christmas) the supply of nutrients to Peru's fishing industry, and bringing Ecuador's tropical rain south to Trujillo and Lima.)

Trujillo is on the north side of the Rio Moche estuary. The catchment of this river stretches about 100 km inland to Peru's major Andes plateau at elevation 4.1 km. Its area is about

Table P3.2 Average meteorological conditions in two Peruvian towns

Town	Latitude (deg. S)	Elevation (km)	Coldest temperature (Celsius)	Monsoon rain (wettest) (mm/mo)	Monsoon rain (driest) (mm/mo)	Annual rain (mm/a)
Cajamarca	7.1	2.6	−4 (December)	115 (February)	5 (August)	720
Lima	12.1	0.1	9 (July)	8 (August)	0 (February)	42

$2708 \, km^2$ (about $80 \, km$ by $30 \, km$). The upper half of this catchment is located at elevations above $1500 \, m$ and it would receive rainfall like Cajamarca's. (Because southern Peru's mountains are higher than those in the north, winter snow plays a significant role there, but snow hardly affects the Rio Moche's hydrology.) The average annual discharge of the Rio Moche is about $9.5 \, m^3/s$, but 71% of the annual discharge occurs between February and April and only 1.8% between July and September. The largest recorded flood occurred during the 1925 El Niño, when the width of the river during maximum flood stage would have been about $3.5 \, km$ (about $10 \, km$ upstream of the estuary).

About $35 \, km$ up the coast from the Moche river mouth, the larger Rio Chicama also meets the sea. Its catchment area is about $4400 \, km^2$, and its rainfall characteristics are very similar to those of the adjacent Rio Moche.

Located between the Rio Moche and the Rio Chicama, the Quebrada Rio Seco runs parallel to the Moche river, about $10 \, km$ north-west of the Rio Moche. Its catchment area is about $300 \, km^2$. Because the drainage basin is small, there are short-term flash floods heavily loaded in sediments (from clay to boulders).

P3.3.2 Question A

In tabular form, deduce the seasonal (month-by-month) distribution of streamflows in the Moche river by applying the appropriate monthly rainfalls to the upper two-thirds of the catchment area (none to the remainder). Assume that the monthly (gross) rainfalls are distributed roughly sinusoidally between the wettest and driest values given in Table P3.2. Assume that evaporative processes remove about $1.3 \, mm$ of recent rainfall every day (i.e. $40 \, mm/mo$). (If less than $40 \, mm$ of rain falls in any month, none of this reaches the stream.) For each month's net rainfall, assume that only 40% enters the stream directly as surface run-off in that month, the remainder infiltrates the groundwater system and appears as baseflow in seven subsequent months at rates of 22%, 14%, 9%, 6%, 4%, 3% and 2% respectively (totalling 100%). (*Hint*: delay converting units from mm/mo to kl/s until as late as possible.)

P3.3.3 Question B

If the people of Trujillo wanted to grow maize, they would need a steady flow of irrigation water throughout the 4 month growing season. Because the growing season includes the month of lowest river flow, the maximum area of irrigated crops that could be sustained would be governed by the lowest river flow rate (unless a storage reservoir is provided). For the case of no reservoir, calculate the maximum irrigable area (km^2), given that about $10 \, Ml$ is required for each hectare (spread over 4 months), including transmission losses. *Note*: One square kilometre is 100 hectares. One hectare is 100 ares. One are is a (French) garden plot ($10 \, m \times 10 \, m = 100 \, m^2$).

P3.3.4 Question C

How big a storage volume (in litres) would be required to enable all the Rio Moche water to be used for irrigation at a steady rate, year-round? (*Hint*: find the mean streamflow (mm/mo), subtract this from each monthly flow to deduce which months are deficient, then sum the volumes needed to cover the deficient months (knowing that this deficiency will be replenished in the other months).)

Hydraulics of the Nîmes aqueduct

Preface

The following problem covers lecture material relevant to Basic Principles (Part I), Physical Modelling and Numerical Modelling (Part III) and Design of Hydraulic Structures (Part IV). The assignment was first used in 2002. Based upon real historical facts and archaeological, the geometry of the aqueduct has been slightly simplified to facilitate the calculations.

P4.1 INTRODUCTION

P4.1.1 Presentation

The hydraulic expertise of the Romans contributed significantly to the advance of science and engineering in antiquity. Aqueducts were built primarily for public health and sanitary needs: i.e. public baths, therms and toilets (Hodge, 1992; Fabre *et al.*, 1992, 2000). Many were used for centuries; some are still in use: e.g. at Carthage (Clamagirand *et al.*, 1990). Magnificent aqueduct remains at Rome, in France, Spain and North Africa: e.g. are still standing (e.g. Van deman, 1934; Ashby, 1935; Rakob, 1974; Conseil Général du Rhône, 1987, 1991, 1993, 1996) (Fig. P4.1 and Plates 1 and 2). Aqueduct construction was an enormous task often performed by the army and the design was undertaken by experienced army hydraulicians. The construction cost was gigantic considering the small flow rates (less than $0.5\,\text{m}^3/\text{s}$): it was around 1–3 millions sesterces per km on average (e.g. Fevrier, 1979; Leveau, 1991). (During the Augustan period (BC 33 to AD 14), one sesterce weighted about 1/336 of a pound of silver which would bring the cost of 1 km of aqueduct to about US\$ 23–69 millions, based on US\$ 485.5 per ounce of silver on 25 November 1998! By comparison the pipeline for the Tarong power station (70 km long, $0.9\,\text{m}^3/\text{s}$) in Queensland costed AUD\$ 0.2 millions per km (*Courier Mail* 3 December 1994: p. 13).)

Recent surveys have thrown new light on the longitudinal profiles of Roman aqueducts (Grewe, 1986, 1992; Hodge, 1992; Burdy, 2002). Most aqueducts consisted of very long flat sections with bed slopes around 1–3 m/km, and sometimes short steep portions in between (Chanson, 1998b, 2000c) (Fig. P4.2). Despite arguments suggesting that Roman aqueducts operated with sub-critical flows and that no energy dissipation device was required, hydraulic calculations of aqueduct hydraulics are embryonic. Modern engineering studies suggested that the cur-rent 'misunderstanding' of aqueduct hydraulics derives from the 'ignorance' of historians and archaeologists (Blackman, 1978, 1979; Chanson, 1998b, 2000c). Most hydraulic calculations are feasible by undergraduate engineering students, provided that accurate information on the channel dimensions and flow rate are available (e.g. Henderson, 1966; Chanson, 1999).

Fig. P4.1 Photographs of Roman aqueducts: (a) Gier aqueduct, Lyon, France (86 km long). Arcades de Chaponost, looking upstream from the Beaunant siphon head tank in June 1998. (b) Brévenne aqueduct, Lyon, France (70 km long) – Biternay in September 2000, inside the conduit, looking upstream.

Fig. P4.2 Photograph of a full-size Roman dropshaft (2.1 m drop in invert elevation) (after Chanson, 2002d).

P4.1.2 Hydrology and operation of Roman aqueducts

Hydrology

The hydrology of a catchment is the relationship between rainfall, runoff and stream flows. A hydrological study is required for any water supply system, including an aqueduct. The hydrology of two catchments supplying ancient Roman aqueducts was recently studied: the source de l'Eure at Uzés, supplying the Nîmes aqueduct, and the source de Gorze, supplying the Gorze aqueduct (Metz). Both aqueducts were among the largest Roman aqueducts in Gaul and Germany, with those of Lyon and Cologne. Both were equipped with large-size channel (1.2 m wide at Nîmes, 0.85–1.1 m wide at Gorze). Each aqueduct were supplied by a natural spring, and the catchment area was about 45–60 km^2 (Chanson, 2002c). Furthermore, both aqueducts included a large bridge aqueduct: the Pont du Gard (360 m long and 48.3 m high) and the Pont sur la Moselle (1300 m long and 30 m high).

Today both springs are in use. At Gorze, the average daily flow rate is 93 l/s. Modern data suggest that the aqueduct did not operate at full capacity but for few months per year. During dry periods, the minimum daily flow rate was less than 10% of the maximum flow rate. At the Source de l'Eure (Uzès, Nîmes aqueduct), the average daily flow rate is 343 l/s. Modern data show however large discharge fluctuations. The minimum daily flow rate is about 125 l/s, while the maximum daily flow rate is around 1660 l/s.

For comparison, the average daily flow rate of the Sources de la Siagnole (Mons aqueduct, Fréjus) are 1125 l/s (study period 1981–1993). Again large fluctuations of flow rate are recorded, from no flow in dry periods to a maximum daily flow rate around 17 900 l/s!

Regulation basins

A number of regulation basins were discovered along some aqueducts. At Nîmes and Gorze, two basins were found immediately upstream of the bridge aqueducts. At Nîmes, two further regulation basins were found (Bossy *et al.*, 2000). Most regulation basins consisted of a rectangular pool, a series of valves controlling the downstream aqueduct flow and an overflow system.

Why was the regulation of an aqueduct required? First, the flow rate had to be stopped for maintenance and cleanup. Water quality problems were known in cities (e.g. Frontinus and Vitrivius). One method consisted in the regular cleanup of the channel. Second, regulation systems were possibly used to store water during night times. At Gorze and at Nîmes, the storage capacity in the aqueduct channel was about 20 000 and 55 000 m³, respectively (Chanson, 2002c). This technique would imply a good coordination of gangs of valve operators to open and close the gates twice a day: to open in the morning and to close the flow at night.

P4.1.3 Culvert design

Although the world's oldest culvert is unknown, the Minoans and the Etruscans built ancient culverts in Crete and Northern Italy, respectively (Evans, 1928; O'Connor, 1993). Later the Romans built numerous culverts beneath their roads (Ballance, 1951; O'Connor, 1993). The culvert construction was favoured for small water crossings while a bridge construction was preferred for longer crossings. The common culvert shapes were the arched design and the rectangular (or box) culvert. The Romans also built culverts beneath aqueducts (Chanson, 2002b).

Along the Nîmes aqueduct, a large box culvert was recently excavated at Vallon No. 6, located 17 km downstream of Pont du Gard between the Combe de la Sartanette and Combe Joseph in the Bois de Remoulins (Fabre *et al.*, 1992, 2000; Chanson, 2002b). The culvert was designed to allow stormwater passage beneath the aqueduct in a small valley, locally called *combe*. Note that catchment area was very small: i.e. 0.028 km². At Vallon No. 6, the culvert could pass an intense storm event corresponding to a maximum effective rainfall intensity of nearly 540 mm/h which is consistent with observed maximum rainfall intensity of 800–900 mm/h in the nearby Cévennes range.[1] For comparison, the mean annual rainfall near Nîmes has been about 700–800 mm for the last 50 years. During the same period, recorded intense rainfalls included 430 mm in 7 h (61 mm/h) on 3 October 1988 and 250 mm on the 12 October 1990 (Fabre *et al.*, 2000: pp. 160–161).

The culvert was a multi-cell structure equipped with three rectangular cells with a total cross-sectional area in excess of 1.2 m². The cells were made of large limestone blocks placed on supporting pillars, or dividing walls, founded on worked bedrock. The upstream end of each dividing wall was cut in a chamfer to form cut-waters. Note that the Bornègre bridge on the Nîmes aqueduct, located between Uzès and Pont du Gard, was composed of three arches with two central piers equipped with upstream cut-waters. But the cut-waters of the culvert were better shaped. (The cut-waters of Bornègre bridge were more sturdy and less profiled that those of the multi-cell culvert: i.e. 60° convergence angle at Bornègre, 45° for the culvert.)

---

[1] Such hydrological events are called 'évènements cévenols'. An extreme hydrological event took place between Sunday 8 September and Monday 9 September 2002 in Southern France. More than 37 people died. At Sommières, the water depth of the Virdoule river reached up to 7 m, although the water depth is usually less than 1 m. Interestingly, the old houses in the ancient town of Sommières had no ground floor because of known floods of the Virdoule river.

Historians and archaeologists have no doubt that the multi-cell culvert was built in the early stages of the aqueduct (i.e. 1st century AD). The excavation works showed no sign of refurbishment.

Culverts were seldom used beneath aqueducts and the Vallon No. 6 culvert downstream of Pont du Gard is an unique example. Its unusual features included a box culvert design of large dimensions, a multi-cell structure and a modern, sound design from a hydraulic perspective (Chanson, 2002b).

P4.1.4 The Nîmes aqueduct

The Roman aqueduct supplying the city of Nîmes (*Colonia Augusta Nemausus*) is one of the best documented aqueducts (Fig. P4.3). Classical studies include Esperandieu (1926), Hauck and

(a)

(b)

Fig. P4.3 Photographs of the Nîmes aqueduct: (a) Pont du Gard, Nîmes aqueduct, France in June 1998. View from the right bank. (b) Pont de Bordnègre in September 2000. Inlet view, showing the bridge pier shaped to cut the waters.

Novak (1987), Smith (1992–93) and more importantly the multi-disciplinary works of Fabre *et al.* (1991, 1992, 2000). The notoriety of the aqueduct is connected with its crossing of the Gardon river: i.e. the Pont du Gard which is the most famous three-tier Roman bridge still standing (O'Connor, 1993). Despite some discussion, it is believed that the aqueduct was in use from the 1st century AD up to the 4th or 5th century AD (Fabre *et al.*, 2000).

The Nîmes aqueduct was 49 800 m long, starting at the Source de l'Eure at Uzès which drains a 45–50 km^2 catchment area. The total invert drop was only 14.65 m from the source to the *castellum dividorum* (repartition basin) at Nîmes, which gives the aqueduct one of the flattest gradient among Roman aqueducts (Grewe, 1992; Hodge, 1992; Fabre *et al.*, 2000). The aqueduct channel was typically 1.2 m wide and the maximum flow rate was estimated to be about 0.405 m^3/s (35 000 m^3/day). Fabre *et al.* (1991) showed however an important variability of the spring output at Uzès. During a study period covering July 1967 to May 1968 and January 1976 to December 1978, the average streamflow was 0.343 m^3/s (29 600 m^3/day), while the minimum flow rate was 0.125 m^3/s (10 800 m^3/day) in September 1976 and the maximum discharge was 1.66 m^3/s (143 400 m^3/day) in October 1976.

By its dimensions and capacity, the Nîmes aqueduct was among the largest aqueducts built in Roman Gaul. The list includes the 86 km long Gier aqueduct (Lyon), the Gorze aqueduct (Metz) with its 1300 m long bridge across the Moselle river, and the Mons aqueduct (Fréjus) with a maximum discharge capacity of 0.61 m^3/s (52 500 m^3/day). However the Nîmes aqueduct was smaller than the largest aqueducts at Rome: e.g. the Aqua Marcia and the Aqua Novus (Hodge, 1992; Fabre *et al.*, 1992).

P4.1.5 Bibliography

Ashby, T. (1935) *The Aqueducts of Ancient Rome*. (Clarendon Press: Oxford, UK) I.A. Richmond (ed.), 342 pages.

Ballance, M.H. (1951) The Roman bridges of the Via Flaminia. Papers of the British School at Rome, **19**, 78–117 and plates xiv–xix.

Blackman, D.R. (1978) The volume of water delivered by the four great aqueducts of Rome. *Papers of the British School at Rome*, **46**, 52–72.

Blackman, D.R. (1979) The length of the four great aqueducts of Rome. *Papers of the British School at Rome*, **47**, 12–18.

Bossy, G., Fabre, G., Glard, Y., and Joseph, C. (2000) Sur le Fonctionnement d'un Ouvrage de Grande Hydraulique Antique, l'Aqueduc de Nîmes et le Pont du Gard (Languedoc, France). (On the operation of a large ancient hydraulic structure, the Nîmes aqueduct and the Point du Gard (Languedoc, France).) *Comptes Rendus de l'Académie des Sciences de Paris, Sciences de la Terre et des Planètes*, **330**, 769–775 (in French).

Burdy, J. (1979) Lyon. Lugudunum et ses Quatre Aqueducs. (Lyon. Lugudunum and its Four Aqueducts.) *Dossiers de l'Archéologie*, Séries Les Aqueducs Romains, **38**, October–November, 62–73 (in French).

Burdy, J. (2002) *Les Aqueducs Romains de Lyon*. (Presses Universitaires de Lyon, Lyon, France), 204 pages.

Chanson, H. (1998b) The Hydraulics of Roman Aqueducts: Steep Chutes, Cascades and Dropshafts. *Research Report No. CE156*, Department of Civil Engineering, University of Queensland, Australia, 97 pages.

Chanson, H. (1999) *The Hydraulics of Open Channel Flows: An Introduction*. (Butterworth-Heinemann: Oxford, UK), 512 pages.

Chanson, H. (2000c) Hydraulics of Roman aqueducts: steep chutes, cascades and dropshafts. *American Journal of Archaeology*, **104**(1), January, 47–72.

Chanson, H. (2000d) A hydraulic study of Roman aqueduct and water supply. *Australian Journal of Water Resources*, IEAust, **4**(2), 111–120. Discussion, **5**(2), 217–220.

Chanson, H. (2001) *The Hydraulics of Stepped Chutes and Spillways.* (Balkema: Lisse, The Netherlands), 418 pages.

Chanson, H. (2002a) An experimental study of Roman dropshaft hydraulics. *Journal of Hydraulic Research, IAHR*, **40**(1), 3–12.

Chanson, H. (2002b) Hydraulics of a large culvert beneath the Roman aqueduct of Nîmes. *Journal of Irrigation and Drainage Engineering, ASCE*, **128**(5), October, 326–330.

Chanson, H. (2002c) Certains Aspects de la Conception hydrauliques des Aqueducs Romains. (Some aspect on the hydraulic design of Roman aqueducts.) *Journal La Houille Blanche*, (in print).

Chanson, H. (2002d) An experimental study of Roman dropshaft operation: hydraulics, two-phase flow, acoustics. Report *CH50/02*, Department of Civil Engineering, University of Queensland, Brisbane, Australia, 99 pages.

Chanson, H., and James, D.P. (2002) Historical development of arch dams: from cut-stone arches to modern concrete designs. *Australian Civil Engineering. Transactions, IEAust*, **CE43**, 39–56 and front cover.

Clamagirand, E., Rais, S., Chahed, J., Guefrej, R., and Smaoui, L. (1990) L'Aqueduc de Carthage. ('The carthage aqueduct.') *Journal La Houille Blanche*, **6**, 423–431.

Conseil Général du Rhône (1987) *Préinventaire des Monuments et Richesses Artistiques. I L'Aqueduc Romain du Mont d'Or.* (Pre-inventory of the monuments and art treasures. I The Roman aqueduct of the Mont d'Or.) *Bosc Frères Publication*, Lyon, France, Henri Hours (ed.), 104 pages (in French).

Conseil Général du Rhône (1991) *Préinventaire des Monuments et Richesses Artistiques. II L'Aqueduc Romain de l'Yzeron.* (Pre-inventory of the monuments and art treasures. II The Roman aqueduct of the Yzeron.) *Bosc Frères Publication*, Lyon, France, Henri Hours (ed.), 168 pages and 1 Map (in French).

Conseil Général du Rhône (1993) *Préinventaire des Monuments et Richesses Artistiques. III L'AqueducRomain de la Brévenne.* (Pre-inventory of the monuments and art treasures. III The Roman aqueduct of the Brévenne.) *Bosc Frères Publication*, Lyon, France, Henri Hours (ed.), 230 pages and 1 Map (in French).

Conseil Général du Rhône (1996) *Préinventaire des Monuments et Richesses Artistiques. IV Lyon. L'Aqueduc Romain du Gier.* (Pre-inventory of the monuments and art treasures. IV Lyon. The Roman aqueduct of the Gier.) *Bosc Frères Publication*, Lyon, France, Jean Burdy (ed.), 407 pages and 1 Map (in French).

Evans, A.H. (1928) *The Palace of Minos: A Comparative Account of the Successive Stages of the Early Cretan Civilization as Illustrated by the Discoveries at Knossos.* (Macmillan: London, UK), Vol. II, Part 1, 390 pages and 19 plates.

Fabre, G., Fiches, J.L., and Paillet, J.L. (1991) Interdisciplinary research on the aqueduct of Nîmes and the Pont du Gard. *Journal of Roman Archaeology*, **4**, 63–88.

Fabre, G., Fiches, J.L., Leveau, P., and Paillet, J.L. (1992). The Pont du Gard. Water and the Roman town. *Presses du CNRS*, Caisse Nationale des Monuments Historiques et des Sites, Collection Patrimoine au Présent, Paris, France, p. 127.

Fabre, G., Fiches, J.L., and Paillet, J.L. (2000) *L'Aqueduc de Nîmes et le Pont du Gard. Archéologie, Géosystème, Histoire. (The Nîmes Aqueduct and the Pont du Gard. Archaeology, Geosystem, History.)* (CNRS Editions: CRA Monographies Hors Série, Paris, France), 483 pages and 16 plates (in French).

Fevrier, P.A. (1979) L'Armée Romaine et la Construction des Aqueducs. (The Roman army and the construction of aqueducts.) *Dossiers de l'Archéologie*, Séries Les Aqueducs Romains, **38**, October/November, 88–93 (in French).

Grewe, K. (1986) *Atlas der Römischen Wasserleitungen nach Köln. (Atlas of the Roman Hydraulic Works near Köln.)* (Rheinland Verlag: Köln, Germany), 289 pages (in German).

Grewe, K. (1992) Plannung und Trassierung römisher Wasserleitungen. (*Planning and Surveying of Roman Water Supplies.*) *Verlag Chmielorz GmbH*, Wiesbaden, Germany, Schriftenreihe der Frontinus Gesellschaft, Suppl. II, 108 pages (in German).

Hauck, G.W., and Novak, R.A. (1987). Interaction of flow and incrustation in the Roman aqueduct of Nîmes. *Journal of Hydraulic Engineering, ASCE*, **113**(2), 141–157.

Henderson, F.M. (1966) *Open Channel Flow.* (MacMillan Company: New York, USA).

Hodge, A.T. (1992) *Roman Aqueducts and Water Supply*. (Duckworth: London, UK), 504 pages.

Leveau, P. (1979) La Construction des Aqueducs. (The construction of aqueduct.) *Dossiers de l'Archéologie*, Séries Les Aqueducs Romains, **38**, October–November, 8–19 (in French).

Leveau, P. (1991) Research on Roman aqueducts in the past ten years. *Future Currents in Aqueduct Studies*. Leeds, UK, T. Hodge (ed.), pp. 149–162.

Leveau, P. (1996) The barbegal water mill in its environment: archaeology and the economic and social history of antiquity. *JRA*, **9**, 137–153.

O'Connor, C. (1993) *Roman Bridges*. (Cambridge University Press: Cambridge, UK), 235 pages.

Rakob, F. (1974). Das Quellenheigtum in Zaghouan und die Römische Wasserleitung nach Karthago. *Mitt. des Deutschen Archaeologischen Instituts Roemische Abteilung*, Vol. 81, pp. 41–89, Plates 21–76 & Maps (in German).

Rakob, F. (1979) L'Aqueduc de Carthage. (The aqueduct of carthage.) *Dossiers de l'Archéologie*, Séries Les Aqueducs Romains, **38**, October–November, 34–42 (in French).

Smith, N.A.F. (1992–1993) The Pont du Gard and the Aqueduct of Nîmes. *Transactions of Newcomen Society*, **64**, 53–76. Discussion, **64**, 76–80.

Valenti, V. (1995a) Aqueduc Romain de Mons à Fréjus. 1. Etude Descriptive et Technique. Son Tracé, son Profil, son Assise, sa Source … *Research Report*, Fréjus, France, 97 pages.

Valenti, V. (1995b) Aqueduc Romain de Mons à Fréjus. 2. Etude Hydraulique. Son Débit … de sa Mise en Service à son Déclin. *Research Report*, Fréjus, France, 123 pages.

Van deman, E.B. (1934) *The Building of the Roman Aqueducts*. (Carnegie Institution of Washington, Publication No. 423: Washington DC, USA), 440 pages. (also McGrath Publication Corporation: Washington DC, USA, 440 pages, 1973).

Villanueva, A.V. (1993) *El Abastecimiento de Agua a la Cordoba Romana. I: El Acueducto de Valdepuentes.* (*The Water Supply of the Roman Cordoba. I: Aqueduct of Valdepuentes.*) (Monografias No. 197, Universidad de Cordoba, Servicio de Publicaciones: Cordoba, Spain), 172 pages (in Spanish).

Villanueva, A.V. (1996) *El Abastecimiento de Agua a la Cordoba Romana. II: Acueductos, Ciclo de Distribución y Urbanismo.* (*The Water Supply of the Roman Cordoba. II: Aqueduct, Distribution System and Urbanism.*) (Monografias No. 251, Universidad de Cordoba, Servicio de Publicaciones: Cordoba, Spain), 222 pages (in Spanish).

Internet references

Hydraulics of Roman Aqueducts. Myths, Fables, Realities. A Hydraulician's Perspective	{http://www.uq.edu.au/~e2hchans/rom_aq.html}
Gallery of Photographs in Hydraulic Engineering – Roman Waterworks	{http://www.uq.edu.au/~e2hchans/photo.html #Roman waterworks}
Historical Development of Arch Dams. From Modern Concrete Designs	{http://www.uq.edu.au/~e2hchans/ arch_dam.html} Cut-Stone Arches to
Lacus Curtius	{http://www.ukans.edu/history/index/europe/ ancient_rome/E/Gazetteer/Periods/Roman/Topics/ Engineering/waterworks/home.html}
Traianus	{http://traianus.rediris.es/}
Roman Aqueducts and Water Systems	{http://www.bowdoin.edu/dept/clas/Aqueducts/ index.html}
Gorze Aqueduct, Metz	{http://www.ac-nancy-metz.fr/ia57/jussy/netsco/ English/accueil.htm}

Rome Aqueducts   {http://www.donau.de/privhome/mrechenmacher/
                                        rom/karte1.htm}

Mons Aqueduct, Fréjus   {http://www.chez.com/siagnole/}

P4.2 HYDRAULIC STUDY OF THE NÎMES AQUEDUCT

P4.2.1 Regulation system

The regulation basin upstream of Pont du Gard has roughly a rectangular shape (1.9 m long and 2.1 m wide). The basin invert is 0.1 m below the main canal bed. Upstream and downstream of the basin, the canal is 1.2 m wide with a rectangular cross-section. The inside walls of both canals and basin are lined with mortar.

1.1 The basin outflow is controlled by a sluice gate installed in the outflow canal itself. For a flow rate of 20 000 m³/day, the water depth in the canal, immediately upstream of the gate, is 0.55 m:
   - Calculate the downstream water depth and the force acting on the gate. *Neglect tailwater effects.*
   - Calculate the water depth in the regulation basin.

1.2 The regulation basin is also used as a settling basin to trap sediment matter. The basin can operate successfully as long as the shear velocity is less than 0.005 m/s. Calculate the corresponding maximum flow rate assuming a 1 m flow depth. (*A 1 m flow depth in the basin corresponds to bank full.*)

1.3 The Roman chief engineer decides to undertake a hydraulic model study of the basin to test different gate configurations. Laboratory facilities limit the scale ratio to 4:1. The design flow in prototype is 30 000 m³/day:
   - Determine the maximum model discharge required.
   - Determine the minimum prototype discharge for which negligible scale effects occurs in the model.
   - Discuss your results.

P4.2.2 Free-surface profile calculations

At the upstream end of the aqueduct, the waters from the Eure springs are collected in a large reservoir. The reservoir is controlled by a broad-crested weir (1.0 m wide and 2.1 m long with the crest located 1.4 m above the downstream canal invert) discharging into the main canal. The upstream 5 km of the canal are characterized by large variations in bed slope (Table P4.1).

Considering the real fluid flow in the upstream section of the aqueduct, Table P4.1 provides the geometric characteristics of the 5400 m upstream canal reach. For the design flow of 30 000 m³/day, calculate the critical depth and normal depth in each sub-reach. Report the results in Table P4.2.

Compute the flow depth, flow velocity, Darcy coefficient, Froude number and friction slope at all the positions between the spring and Les Arabades. Plot the backwater curve (i.e. flow depth curve) on graph paper. *Flow resistance calculations must be performed using the Darcy–Weisbach friction factor.*

P4.2.3 Culvert design

Along the Nîmes aqueduct, a large box culvert is located at Vallon No. 6, located 17 km downstream of Pont du Gard. The culvert was designed to allow stormwater passage beneath the

Table P4.1 Nîmes canal geometry between Source de l'Eure and Les Arabades (Commune de St-Maximin)

Location[1]	Bed elevation[1]	x^1 (m)	Roughness[2]	B^2 (m)	Remarks
Source de l'Eure (0)	73.60	0	Mortar	1.0	Immediately downstream of broad-crested weir
2a	71.298	100	Mortar	1.1	
2b	71.220	181.61	Mortar	1.1	
2c	71.128	214.61	Mortar	1.1	
2d	71.054	245.80	Mortar	1.1	
Bassin du Val d'Eure (2A)	70.870	288.41	Mortar	1.1/1.2	Smooth change on canal width
La Montagne (4)	70.810	793.14	Mortar	1.2	
Mas de Préville (7)	70.653	1168.75	Mortar	1.2	
Carrignargues amont (9)	70.110	2029.10	Mortar	1.2	
Les Arabades (19)	68.516	5436.25	Mortar	1.2	

Notes: [1]After Fabre *et al.* (2000); [2]dimension for the new aqueduct.

Table P4.2 Summary table (normal depth and critical depth) for $Q = 30\,000 \, \mathrm{m^3/day}$

Sub-reach[1]	Bed slope[1]	d_c^1 (m)	d_o^2 (m)	B^2 (m)	Remarks
Source de l'Eure (0) to 2a				1.0	
2a–2b				1.1	
2b–2c				1.1	
2c–2d				1.1	
2d to Bassin du Val d'Eure (2A)				1.1	
Bassin du Val d'Eure (2A) to La Montagne (4)				1.2	
La Montagne (4) to Mas de Préville (7)				1.2	
Mas de Préville (7) to Carrignargues amont (9)				1.2	
Carrignargues amont (9) to Les Arabades (19)				1.2	

Notes: [1]After Fabre *et al.* (2000); [2]dimension for a new aqueduct.

aqueduct in a small valley, locally called *combe*. Note that catchment area is very small: i.e. $0.028 \, \mathrm{km^2}$. The culvert is a multi-cell structure equipped with three rectangular cells. The cells are 0.5, 0.8 and 0.6 m wide. Each cell has an internal height of 0.65 m. The barrel length is 3.7 m and the invert slope is about 0.05. (The invert is worked bedrock: $k_s \sim 10 \, \mathrm{mm}$.) The upstream end of each dividing wall is cut in a chamfer to form cut-waters.

Upstream and downstream of the culvert, the natural bed slope is steep (i.e. $S_o \sim 0.16$) and consists of gravels ($k_s = 50 \, \mathrm{mm}$). The valley is narrow and may be approximated as a 3.2 m wide rectangular cross-section.

– For a catchment runoff of $1.5 \, \mathrm{m^3/s}$, calculate the normal depth in the valley.
– For the same runoff, calculate the change in upstream water level caused by the presence of the culvert.
– Did the culvert operate with inlet control or outlet control?

Using the software Hydroculv™, calculate the change in upstream water level caused by the presence of the culvert for a runoff of $1.5 \, \mathrm{m^3/s}$. Compute the flow velocity in the barrel.

Solutions

1.1 The downstream depth equals $d_2 = 0.062 \, \mathrm{m}$. The force acting on the gate is $F = 1.1 \, \mathrm{kN}$. In the regulation basin, the water depth is $d = 0.656 \, \mathrm{m}$.

1.2 The maximum flow rate equals: $Q = 0.22 \, \text{m}^3/\text{s}$.

1.3 Using a Froude similitude, the maximum model discharge equals $(Q_{max})_m = 11 \, \text{l/s}$. The minimum prototype flow rate for which scale effects are negligible is $(Q_{min})_p = 3000 \, \text{m}^3/\text{day}$. Model studies could be conducted from 10% to 100% Q_{max}.

2.1 Normal depth and critical depth calculations

Sub-reach	S_o	d_c (m)	d_o (m)	B	Slope
0–2a	0.023	0.2308	0.1275	1	Steep
0–2a	0.023	0.2166	0.1189	1.1	Steep
2a–2b	0.001	0.2166	0.3544	1.1	Mild
2b–2c	0.0028	0.2166	0.2423	1.1	Mild
2c–2d	0.0024	0.2166	0.2564	1.1	Mild
2d–2A	0.0043	0.2166	0.2083	1.1	Steep
2A–4	0.0001	0.2044	0.7127	1.2	Mild
4–7	0.0004	0.2044	0.4426	1.2	Mild
7–9	0.0006	0.2044	0.3808	1.2	Mild
9–19	0.0005	0.2044	0.4247	1.2	Mild

Notes: Calculations performed assuming $k_s = 1 \, \text{mm}$.

2.2 Backwater computations
   - Critical flow depth occurs at the weir crest, and near 2d.
   - Hydraulic jumps are expected in the sub-reach 2a–2b, in the sub-reach 2A–4.
   - At 0 (weir crest), the downstream flow conditions must be deduced from the Bernoulli principle applied between the crest and downstream flow depth.
   - At 2A, the Bernoulli principle must be used to predict the short, smooth transition in channel width.
   - In the sub-reach 9–19, the sub-critical flow conditions are controlled from downstream, hence from normal flow conditions.

2.3 Culvert calculations
In the flood plain upstream and downstream of the culvert, the normal depth and critical depth equal:
   $d_o = 0.126 \, \text{m}$ (i.e. tailwater depth for the culvert)
   $d_c = 0.281 \, \text{m}$
Note that the valley is a steep slope ($d_o < d_c$).
In the culvert barrel, the normal and critical depths are:
   $d_o = 0.236 \, \text{m}$
   $d_c = 0.40 \, \text{m}$
Note again that the barrel invert slope corresponds to a flat slope.
   In any culvert calculation with near-critical and super-critical inflow conditions, it is essential to check potential 'choking' and hydraulic jump occurrence at the barrel entrance. For the present calculations, application of the Bernoulli principle indicates that the flow into the culvert may operate as a smooth, short contraction without 'choking'.
   Inlet control calculations yield a change in upstream water level of about 0.25 m.
   Outlet control calculations give a change in upstream water level of less than 0.1 m (although the results is not very accurate). In summary, the culvert operate with Inlet control operation.

Remarks: The culvert has little impact onto the runoff flood flow. The afflux is small and the culvert flow is an inlet control operation with super-critical flow in barrel.

Discussion

Hydroculv™ calculations for the culvert predict a change in upstream water level of 0.757 m!!! This is inaccurate. Indeed Hydroculv predicts sub-critical inflow conditions (because the software ignores the upstream steep slope).

Calculations indicate a flow velocity in barrel of about 2.7 m/s.

Remarks

This application is based upon the dimensions of a multi-cell culvert found beneath the Nîmes aqueduct (Fabre *et al.*, 1992). Chanson (2002b) conducted a detailed hydraulic analysis.

Suggestion/correction form

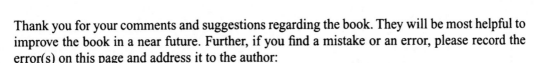

Thank you for your comments and suggestions regarding the book. They will be most helpful to improve the book in a near future. Further, if you find a mistake or an error, please record the error(s) on this page and address it to the author:

Dr H. Chanson
Department of Civil Engineering, The University of Queensland,
Brisbane QLD 4072, Australia
Fax: (61 7) 33 65 45 99
Email: h.chanson@uq.edu.au
Url: http://www.uq.edu.au/~2hchans/
Corrections and updates will be posted at
{http://www.uq.edu.au/~e2hchans/reprints/books3_2.htm}.

SUGGESTION/CORRECTION FORM (2ND EDITION)

Contact

Name	
Address	
Tel.	
Fax	
Email	

Description of the suggestion, correction, error

Part number	
Page number	
Line number	
Figure number	
Equation number	

Proposed correction

Author index

Subject index

(Page number in **bold** refers to figures and tables.)

Lightning Source UK Ltd.
Milton Keynes UK
UKOW03f0005220615

253861UK00009B/128/P